Passive and Active
Network Analysis and Synthesis

ARAM BUDAK

Professor of Electrical Engineering
Colorado State University

Houghton Mifflin Company Boston

Atlanta · Dallas · Geneva, Illinois · Hopewell, New Jersey
Palo Alto · London

Printed in the United States of America

Library of Congress Catalog Card Number: 73–9193

ISBN: 0–395–17203–9

Passive and Active
Network Analysis and Synthesis

To Helga
and
Seta, Lydia, Mark, Sonya, Sylvia, and Sandra

Preface

Twenty years of studying, experimenting, and thinking about circuits resulted in this book. At Colorado State University it is used in a four-quarter sequence on network synthesis at the senior and beginning graduate level, although the introductory chapters may be presented at the junior level. The on-campus lectures are also videotaped and presented off-campus to engineers in various industries in Colorado. In the last five years, over 150 on-campus and 150 off-campus students have taken this course. Several hundred experiments have been conducted by these students and the results were submitted as reports. Through these studies I have come to know which topics need to be strongly emphasized and elaborated upon if the student is to apply theory to practice with understanding and success. I sincerely hope that the approach taken in this book fulfills the wishes of many engineers who want to acquire the much needed know-how to accomplish real things.

Many students, engineers, and colleagues have urged me to write this book. Its principal aim is to provide physical insight and thorough understanding of the complex-frequency domain and its application to circuits. There is hardly an electrical engineer who has not heard about poles and zeros. Yet these frequency-domain concepts are often accepted as mathematical abstractions and not as design tools. Indeed, many an engineer would like to know more about the relationship between a complex pole or a zero and the frequency response of a system. This book presents a clear picture of the correspondence between the time domain and the frequency domain. It does this by examining in detail the changes in the response brought about as a result of changes in pole–zero positions.

Who can deny the immense satisfaction of seeing a circuit perform according to one's expectation? Results that seemingly violate theory usually show insufficient understanding of theory. For example, if theory and practice are to agree, the amplifiers cannot be assumed to be ideal. The frequency dependence of the amplifier gain has marked but predictable effects on the frequency response of active filters. To strengthen the ties between theory and experiment, a number of useful techniques, such as the use of sensitivity functions in studying a response, are developed and applied to specific circuits. These and other matters of importance in the design of circuits are discussed in detail. The reader who studies the book carefully will appreciate the many innovations and the fresh approach that is presented. He will also be motivated to create new circuits. Indeed, the necessary theoretical perspectives and physical concepts for an open-ended approach to practical circuit design are amply provided.

The book covers analysis and synthesis for passive as well as active networks. It also presents and compares a number of useful approximating functions. Particular emphasis is placed on the study of widely used second-order functions and their realizations. A large number of practical circuits is given. The circuit characteristics are discussed in detail. The characteristics of higher-order functions and networks are derived from the properties of the various frequency transformations. The organization of the book is such that much of the information pertaining to a particular subject is contained in a single chapter. The reader can readily refer to a particular chapter to acquire the desired knowledge. I trust that these self-contained chapters will be especially helpful to the practicing engineer. They will give him ideas and show him what has been done. To facilitate learning, many examples are solved throughout the text.

The fundamentals associated with the complex-frequency domain are introduced and discussed in detail in Chapters 1–3. The concepts and analysis techniques presented in these introductory chapters are extensively used later. Therefore, it is essential that they be understood well. In Chapter 4, synthesis is introduced. Foster and Cauer input-impedance realizations for LC and RC networks are developed in this chapter. In Chapter 5, attention is directed to the analysis and synthesis of LC ladder networks with resistive terminations. In Chapter 6, the properties of second-order systems are discussed. In particular, the different kinds of root loci are treated in detail. The concept of sensitivity and its many facets are also introduced in this chapter. Since root loci and sensitivity functions are extensively used throughout the remainder of the text, Chapter 6 lays the foundations and the physical basis for much of the work to come. In Chapters 7 and 8, a detailed study of the operational amplifier is given. Both the ideal and the one-pole rolloff model are used to derive the characteristics of operational amplifier circuits. These chapters are application oriented and show the versatility of the operational amplifier as well as its limitations. In Chapter 9, general techniques are presented for controlling the position of poles and zeros by means of feedforward and feedback amplifiers. In particular, the production of complex critical frequencies with different amplifier configurations is illustrated. The powerful methods presented in these chapters are then applied to obtain realizations of second-order low-pass, high-pass, band-pass, band-stop, and all-pass functions. A detailed discussion of all these functions and their realizations with passive and active networks is given in Chapters 10 through 14. Many practical circuits, their characteristics and limitations are presented in these chapters in a well-organized manner. Chapter 15 discusses *RC* oscillators which are treated as second-order realizations with imaginary-axis poles. Thus, the principles developed in Chapter 9 are again applied to realize a number of oscillator circuits. The study of this chapter is important also because it gives a better understanding of why amplifiers sometimes oscillate. Chapter 16 is an introduction to the properties of approximating functions. In Chapter 17, the low-pass approximations are treated. The characteristics of the maximally flat, the equal-ripple, and the linear-phase approximations are discussed and compared. The realizations of these approximations with passive and active networks are given in Chapter 18. In Chapters 19 through 21, the low-pass to high-pass, the low-pass to band-pass, the low-pass to band-stop transformations, and the resulting networks are presented in detail. A brief discussion of phase equalization is given in Chapter 22.

The chapters are organized to provide a self-contained body of knowledge according to the subject matter. Thus, it is hoped that it is easy to teach oneself fundamentals from the book as well as to use it as a source of information on specific topics and circuits. The highly theoretical reader at times may feel that certain network theoretical facts need precise development. He may find exceptions to the many useful rules in the text. Since the approach taken in this book is how to apply theory to practice and not how to prove it exhaustively, the highly critical reader is encouraged to consult books on synthesis which deal exclusively with the theoretical properties of networks.

I am indebted to many students, engineers, and colleagues who have carefully read the manuscript and spent innumerable hours in the laboratory experimenting with circuits. Special thanks are due to Drs. P. B. Aronhime, D. M. Petrela, and K. E. Waltz whose invaluable contributions in the early stages of writing have made this book possible. The encouragement and the many criticisms provided by my colleagues, Professors E. E. Gray and R. J. Morgan, are gratefully acknowledged. Their enthusiasm has been a continuous source of inspiration. I am also thankful to Professors A. R. Stoudinger, J. W. Steadman, W. S. Wagner, J. L. Allen, W. Lord, and C. C. Britton for their helpful suggestions. Finally, this book would have never gone beyond the stage of class notes without the cheerful assistance provided by my wife. She not only typed and proofread the entire manuscript twice but did all the drawings. I simply had to keep writing to stay ahead of her. It is with great satisfaction and profound relief that I finish the writing of this book.

Aram Budak

Contents

1

The System Function

Networks are used to change the characteristics of signals. In order to determine quantitatively the changes brought about by networks, the networks must be characterized. The system function is used for this purpose. In this chapter, the system function is defined and its usefulness demonstrated by examples. Then, the different methods of exciting a system are presented and examples are given to show how the system function is used to obtain the resulting responses. Finally, the poles and zeros of the system function are defined. Simple but useful analysis techniques are introduced and demonstrated throughout the chapter.

1-1 CHARACTERIZATION OF THE SYSTEM

As an example of a system, consider a network consisting of linear, time-invariant, lumped resistors, inductors, capacitors, and linear dependent sources connected in any arbitrary manner. *The network is dead*; i.e., it does not contain any independent sources. Initial conditions, which may be treated as independent sources, are also zero. At $t = 0$, the network is excited with a *single, independent, time-varying* voltage or current source as shown in Fig. 1-1.

Designate as response in the network any voltage or current. Let $E(s)$ represent the Laplace transform of the excitation, $e(t)$, and $R(s)$ be the Laplace transform of the response, $r(t)$. The system function, $G(s)$, is defined as

$$G(s) = \frac{R(s)}{E(s)}. \tag{1-1}$$

1

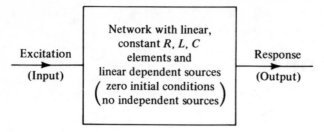

Fig. 1-1   Characterization of a system

As Eq. (1-1) shows, the system function is *the Laplace transform of the response divided by the Laplace transform of the excitation,* provided *all* the constraints shown in Fig. 1-1 are satisfied.

For a network, the system function may take one of the following forms: voltage/voltage, voltage/current, current/current, or current/voltage, where all electrical variables are functions of the complex-frequency variable s. For the same excitation, if a different response in the network is selected, a different system function results. To characterize a particular system function, *the terminals at which the excitation is applied and the terminals at which the response is taken must be specified.* It is meaningless to talk about system functions if these terminals are not designated.

The excitation may be considered as the input signal, $i(t)$, and the response as the output signal, $o(t)$. In terms of these designations, the system function can be alternatively defined as $O(s)/I(s) = G(s)$, where $O(s)$ is the Laplace transform of $o(t)$, and $I(s)$ the Laplace transform of $i(t)$. Henceforth, unless otherwise specified, all signals are considered to be the complex-frequency-domain transforms and are designated by capital letters, for example, $O(s)$, $V(s)$, ... In most instances, the s-dependence, being understood or implied, is not explicitly written, for example, O, V, ... [Time-domain signals are designated by lowercase letters, for example $o(t)$, $v(t)$, ...]

The system function completely characterizes the behavior of the system with regard to the specified input- and output-terminal variables. If $G(s)$ is known (given, calculated, or measured), then the output can be found for a given input,

$$O = GI,$$

or the input can be determined for a given output,

$$I = \frac{O}{G}.$$

Given one variable, the other can be found; this is the analysis aspect of networks. The other aspect is that of synthesis: For a given input, a certain output is desired. What network,

$$G = \frac{O}{I},$$

should be used? More specifically, having found the desired $G(s)$, how is it implemented with R, L, C's, and dependent sources?

EXAMPLE 1-1

The independent source V_1 excites the network shown in Fig. 1-2.

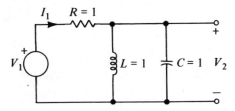

Fig. 1-2

(a) If V_2 is the response, obtain the system function.
(b) If I_1 is the response, obtain the system function.
(c) Obtain V_2/I_1. Is V_2/I_1 a system function?
(d) If $V_1 = 1/s$, obtain responses V_2 and I_1.

SOLUTION

(a) V_1 is divided between the resistor R and the parallel combination of the inductor L with the capacitor C. Using impedances and the voltage-divider rule, obtain V_2:

$$V_2 = V_1 \frac{\dfrac{sL(1/sC)}{sL + (1/sC)}}{R + \dfrac{sL(1/sC)}{sL + (1/sC)}} = V_1 \frac{\dfrac{s(1/s)}{s + (1/s)}}{1 + \dfrac{s(1/s)}{s + (1/s)}} = V_1 \frac{s}{s^2 + s + 1}. \tag{1-2}$$

The system function is

$$\frac{V_2}{V_1} = \frac{s}{s^2 + s + 1}. \quad Ans.$$

(b) Using the rules for parallel and series connection of impedances and the source V_1, find I_1:

$$I_1 = \frac{V_1}{R + \dfrac{sL(1/sC)}{sL + (1/sC)}} = \frac{V_1}{1 + \dfrac{s(1/s)}{s + (1/s)}} = V_1 \frac{s^2 + 1}{s^2 + s + 1}. \tag{1-3}$$

The system function is

$$\frac{I_1}{V_1} = \frac{s^2 + 1}{s^2 + s + 1}. \qquad Ans.$$

(c) To obtain V_2/I_1, divide Eq. (1-2) by Eq. (1-3):

$$\frac{V_2}{I_1} = \frac{V_1 \dfrac{s}{s^2 + s + 1}}{V_1 \dfrac{s^2 + 1}{s^2 + s + 1}} = \frac{s}{s^2 + 1}. \qquad Ans.$$

Since I_1 is not an independent excitation, V_2/I_1 is not a system function.

(d) Substitute $V_1 = (1/s)$ in Eqs. (1-2) and (1-3), and obtain responses V_2 and I_1:

$$V_2 = \frac{1}{s} \frac{s}{s^2 + s + 1} = \frac{1}{s^2 + s + 1}, \qquad Ans.$$

$$I_1 = \frac{1}{s} \frac{s^2 + 1}{s^2 + s + 1} = \frac{s^2 + 1}{s(s^2 + s + 1)}. \qquad Ans.$$

1-2 EXCITATION OF THE SYSTEM

A dead network, one containing no independent sources, is excited by connecting independent voltage and current sources to it. Since dependent sources by themselves do not cause responses, they are considered to be part of the dead network.

A dead network is shown in Fig. 1-3a; a single voltage or current source can be used to excite it. In Fig. 1-3b, a voltage source, v_1, is inserted after cutting a wire in the network. When v_1 is made zero, the original dead network is obtained, as shown in Fig. 1-3c. In Fig. 1-3d, a voltage source, v_2, is connected across an element of the network. In this case, when the voltage source is removed, that is, when v_2 is made zero, a network different from the original dead network results, as shown in Fig. 1-3e.

In Fig. 1-3f, a current source, i_1, is connected across an element of the network. When i_1 is made zero, the original dead network is obtained, as shown in Fig. 1-3g. In Fig. 1-3h, a current source, i_2, is inserted after cutting a wire of the network. In this case, when the current source is removed, that is, when i_2 is made zero, a network different from the original dead network results, as shown in Fig. 1-3i. Thus, *if it is important to preserve the original structure of the network, voltage sources must be inserted by cutting a wire, and current sources must be connected between two nodes of the network.*

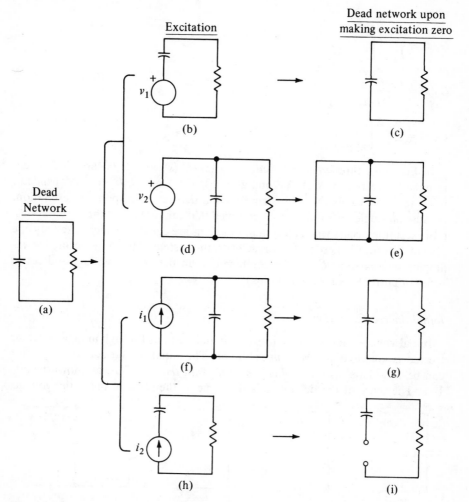

Fig. 1-3   Methods of exciting a dead network

Multiple-Source Excitation

If a linear network is excited by more than one independent source, the principle of superposition is used to determine the response. Each excitation contributes to the response *through its own system function*, and the total response is obtained by summing the individual contributions. Thus, if $E_1(s)$, $E_2(s)$, and $E_3(s)$ represent three independent sources exciting the network, then the resulting response, $R(s)$, can be written in terms of the individual responses:

$$R(s) = R_1(s) + R_2(s) + R_3(s) = E_1(s)G_1(s) + E_2(s)G_2(s) + E_3(s)G_3(s), \qquad (1\text{-}4)$$

where

$$G_1(s) = \frac{R_1(s)}{E_1(s)} \bigg|_{E_2(s)=E_3(s)=0},$$

$$G_2(s) = \frac{R_2(s)}{E_2(s)} \bigg|_{E_1(s)=E_3(s)=0},$$

$$G_3(s) = \frac{R_3(s)}{E_3(s)} \bigg|_{E_1(s)=E_2(s)=0}.$$

In Eq. (1-4), three system functions, $G_1(s)$, $G_2(s)$, and $G_3(s)$ are used to determine the response, $R(s)$. As long as $E_1(s)$, $E_2(s)$, and $E_3(s)$ are independent, it is impossible to obtain *one* system function that relates response to excitation, because no single excitation can be specified that can replace all three excitations. (An exception occurs when all system functions are equal, in which case only one system function is needed to characterize the system. This may occur, for example, when all sources can be combined, as in the case of voltage sources in series or current sources in parallel.)

Initial-Source Excitation

Initial voltages across capacitors and initial currents through inductors can be treated as sources of *independent* excitation. The initial voltage across a capacitor can be considered as a *dc voltage in series with the capacitor*, as shown in Fig. 1-4a. The sense of the dc source is the same as the sense of the initial voltage.

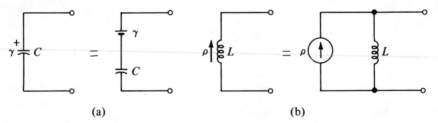

(a)   (b)

Fig. 1-4   Initial conditions represented as independent sources

The initial current through an inductor can be considered as a *dc current source in parallel with the inductor*, as shown in Fig. 1-4b. The sense of the dc source is the same as the sense of the initial current.

EXAMPLE 1-2

In Fig. 1-5, $i(t)$ is an independent current source which is applied to the network at $t = 0$. The initial currents in the inductors are ρ_1 and ρ_2; the initial voltage across the capacitor is γ.

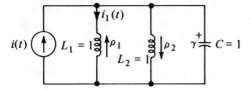

Fig. 1-5

(a) Obtain $I_1(s)$, which is the Laplace transform of the current through the inductor L_1.

(b) What are the individual system functions that are used in obtaining $I_1(s)$?

SOLUTION

(a) Replace all initial conditions by their equivalent sources and redraw Fig. 1-5 as in Fig. 1-6a. Then, keeping the principle of superposition

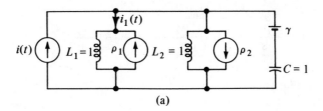

(a)

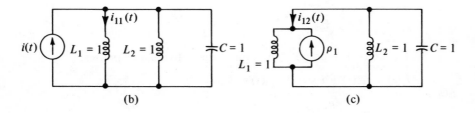

(b)   (c)

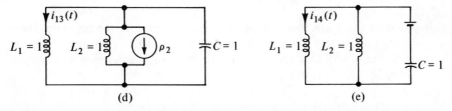

(d)   (e)

Fig. 1-6

in mind, draw Figs. 1-6b through 1-6e to show the contributions to $i_1(t)$ arising from each independent excitation. By using the rules for series and parallel connection of impedances and the rules for voltage and current division, find next $I_{11}(s)$, $I_{12}(s)$, $I_{13}(s)$, and $I_{14}(s)$. Note that the Laplace transforms of the initial conditions are ρ_1/s, ρ_2/s, and γ/s. From Fig. 1-6b,

$$I_{11}(s) = I(s)\,\frac{\dfrac{sL_2(1/sC)}{sL_2 + (1/sC)}}{\dfrac{sL_2(1/sC)}{sL_2 + (1/sC)} + sL_1} = I(s)\,\frac{\dfrac{s(1/s)}{s + (1/s)}}{\dfrac{s(1/s)}{s + (1/s)} + s} = I(s)\,\frac{1}{s^2 + 2}\,.$$

$$(1\text{-}5a)$$

From Fig. 1-6c,

$$I_{12}(s) = -\frac{\rho_1}{s}\,\frac{sL_1}{sL_1 + \dfrac{sL_2(1/sC)}{sL_2 + (1/sC)}} = -\frac{\rho_1}{s}\,\frac{s}{s + \dfrac{s(1/s)}{s + (1/s)}} = -\frac{\rho_1}{s}\,\frac{s^2 + 1}{s^2 + 2}\,.$$

$$(1\text{-}5b)$$

From Fig. 1-6d,

$$I_{13}(s) = -\frac{\rho_2}{s}\,\frac{\dfrac{sL_2(1/sC)}{sL_2 + (1/sC)}}{\dfrac{sL_2(1/sC)}{sL_2 + (1/sC)} + sL_1} = -\frac{\rho_2}{s}\,\frac{1}{s^2 + 2}\,.$$

$$(1\text{-}5c)$$

From Fig. 1-6e,

$$I_{14}(s) = \frac{\gamma}{s}\,\frac{\dfrac{L_2}{L_1 + L_2}}{\dfrac{1}{sC} + s\,\dfrac{L_1 L_2}{L_1 + L_2}} = \frac{\gamma}{s}\,\frac{\dfrac{1}{2}}{\dfrac{1}{s} + s\,\dfrac{1}{2}} = \frac{\gamma}{s}\,\frac{s}{s^2 + 2}\,.$$

$$(1\text{-}5d)$$

The desired response is the sum of the individual responses:

$$I_1(s) = I_{11}(s) + I_{12}(s) + I_{13}(s) + I_{14}(s)$$

$$= \frac{I(s) - \dfrac{\rho_1}{s}\,(s^2 + 1) - \dfrac{\rho_2}{s} + \gamma}{s^2 + 2}\,. \qquad Ans. \qquad (1\text{-}6)$$

(b) Since there are four independent excitations and one response, there are four system functions. To obtain these functions, write the response as the sum of four responses arising from each independent excitation while the others are kept zero:

$$I_1(s) = I(s)G_1(s) + \frac{\rho_1}{s}\,G_2(s) + \frac{\rho_2}{s}\,G_3(s) + \frac{\gamma}{s}\,G_4(s). \qquad (1\text{-}7)$$

Compare Eq. (1-6) with Eq. (1-7) and obtain the four system functions:

$$G_1(s) = \frac{1}{s^2 + 2}, \qquad G_2(s) = -\frac{s^2 + 1}{s^2 + 2}, \qquad G_3(s) = -\frac{1}{s^2 + 2},$$

$$G_4(s) = \frac{s}{s^2 + 2}. \qquad Ans.$$

Dependent-Source Excitation

Dependent sources are current or voltage sources whose values are determined by some other voltage or current in the network. Dependent sources acting alone cannot produce responses in a network. Therefore, in order for them to exist, the network must be excited by independent sources.

When both independent and dependent sources are present in a network, it is best to solve first for the value of the dependent sources in terms of the independent sources. Then the desired response is obtained by superimposing responses arising from all the sources. *The final result should be expressed always in terms of the independent sources only.*

EXAMPLE 1-3

Obtain the system function for the network shown in Fig. 1-7.

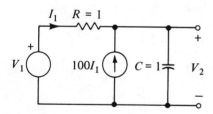

Fig. 1-7

SOLUTION

Two sources act upon the network: the independent voltage source, V_1, and the dependent current source, $100\ I_1$. The response is V_2, and the desired system function is V_2/V_1.

By inspection of the network, obtain

$$V_2 = (I_1 + 100\ I_1)\frac{1}{sC} = 101\frac{I_1}{s}. \qquad (1\text{-}8)$$

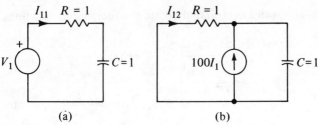

(a)   (b)

Fig. 1-8

Determine I_1, considering the effect of one source at a time. (Turn off the other source, i.e., make its value zero.) Figures 1-8a and 1-8b result. The equation that ties Fig. 1-7 with Fig. 1-8 is

$$I_1 = I_{11} + I_{12}. \tag{1-9}$$

Equation (1-9) is a statement of the superposition principle.

By inspection of Fig. 1-8, obtain I_{11} and I_{12}:

$$I_{11} = \frac{V_1}{R + \dfrac{1}{sC}} = \frac{V_1}{1 + \dfrac{1}{s}} = V_1 \frac{s}{s+1};$$

$$I_{12} = -100 I_1 \frac{\dfrac{1}{sC}}{\dfrac{1}{sC} + R} = -100 I_1 \frac{\dfrac{1}{s}}{\dfrac{1}{s} + 1} = -100 I_1 \frac{1}{s+1}.$$

Substitute I_{11} and I_{12} in Eq. (1-9) and obtain I_1:

$$I_1 = V_1 \frac{s}{s+1} + (-100 I_1) \frac{1}{s+1}. \tag{1-10}$$

Solve Eq. (1-10) explicitly for I_1:

$$I_1 = V_1 \frac{s}{s+101}.$$

Substitute I_1 in Eq. (1-8) and obtain V_2:

$$V_2 = 101 \frac{V_1 \dfrac{s}{s+101}}{s} = V_1 \frac{101}{s+101}.$$

The desired system function is

$$\frac{V_2}{V_1} = \frac{101}{s+101}. \qquad Ans.$$

Common-Mode and Difference-Mode Excitations

Consider a network which is excited by two independent voltage or current sources.

There are instances when it is desirable to calculate the response when the two inputs are identical. The situation is illustrated in Fig. 1-9a, where the two

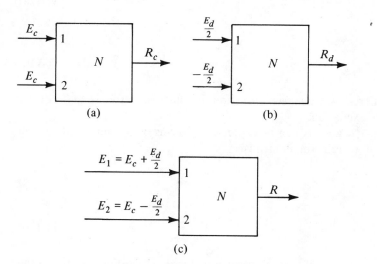

(a)   (b)

(c)

Fig. 1-9   Common-mode and difference-mode excitations

inputs receive a common signal, E_c. The network is said to be excited in the *common mode*. Let G_1 represent the system function when the signal at 2 is zero, and G_2 represent the system function when the signal at 1 is zero. Then, using the principle of superposition, the common-mode response, R_c, is obtained as

$$R_c = E_c G_1 + E_c G_2 = E_c(G_1 + G_2)$$
$$= E_c G_c, \qquad (1\text{-}11a)$$

where

$$G_c = G_1 + G_2 \qquad (1\text{-}11b)$$

is called the common-mode system function. It relates the response to the excitation that is common to both inputs. If the network is designed such that $G_2 = -G_1$, the common-mode system function is zero, and the network does not respond to signals common to both inputs.

There are other instances when it is desirable to calculate the response when the two inputs are equal in value but opposite in polarity. The network is said to be excited in the *difference mode*. The situation is illustrated in Fig. 1-9b where E_d represents the difference-mode signal. (The signals at the inputs are taken as $E_d/2$ and $-E_d/2$, so that the difference is E_d.) By using the principle of superposition, the difference-mode response, R_d, can be obtained:

$$R_d = \left(\frac{E_d}{2}\right) G_1 + \left(-\frac{E_d}{2}\right) G_2 = E_d \left(\frac{G_1 - G_2}{2}\right)$$

$$= E_d G_d, \tag{1-12a}$$

where

$$G_d = \frac{G_1 - G_2}{2} \tag{1-12b}$$

is called the difference-mode system function. It relates the response to the difference of the two excitations.

The functions G_1 and G_2 can be expressed in terms of G_c and G_d by solving Eqs. (1-11b) and (1-12b) simultaneously:

$$G_1 = \frac{G_c}{2} + G_d, \tag{1-13a}$$

$$G_2 = \frac{G_c}{2} - G_d. \tag{1-13b}$$

The most general case of two arbitrary inputs is shown in Fig. 1-9c where the network, N, is excited by E_1 and E_2. The response, R, is given by

$$R = E_1 G_1 + E_2 G_2 = E_1 \left(\frac{G_c}{2} + G_d\right) + E_2 \left(\frac{G_c}{2} - G_d\right)$$

$$= \left(\frac{E_1 + E_2}{2}\right) G_c + (E_1 - E_2) G_d. \tag{1-14}$$

When $E_2 = E_1$, the second term in Eq. (1-14) becomes zero. Since the network is excited in the common mode, Eq. (1-14) must agree with Eq. (1-11a). This requires that

$$E_c = \frac{E_1 + E_2}{2}. \tag{1-15a}$$

When $E_2 = -E_1$, the first term in Eq. (1-14) becomes zero. Since the network is excited in the difference mode, Eq. (1-14) must agree with Eq. (1-12a). This requires that

$$E_d = E_1 - E_2. \tag{1-15b}$$

Hence, Eq. (1-14) can be written as

$$R = E_c G_c + E_d G_d. \tag{1-16}$$

Note also that

$$E_c + \frac{E_d}{2} = E_1, \tag{1-17a}$$

$$E_c - \frac{E_d}{2} = E_2. \tag{1-17b}$$

Thus, any two arbitrary signals, E_1 and E_2, can always be decomposed into common-mode and difference-mode components, according to Eq. (1-17). See also Fig. 1-9c. As Eq. (1-15) shows, *the common-mode component, E_c, is the average of the two independent excitations*, whereas *the difference-mode component, E_d, is the difference*.

EXAMPLE 1-4

To avoid contamination by undesired signals, mostly in the form of 60-Hz pickup, some circuits are designed with two inputs. The desired signal, V_s, is fed in push-pull fashion, while the undesired signal, V_n, appears in equal amounts (push-push) on both input leads, as shown in Fig. 1-10. The network is designed so that its common-mode system function, G_c, is much smaller in magnitude than its difference-mode system function, G_d. Let $G_c = 1$ and $G_d = 10^4$.

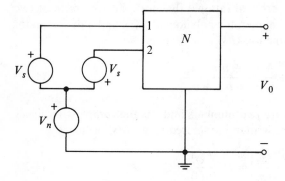

Fig. 1-10

(a) Obtain the output voltage V_o.
(b) Obtain the desired-to-undesired voltage ratio in the output if, at the input,

$$\frac{2V_s}{V_n} = 1.$$

SOLUTION

(a) The common-mode signal is V_n. The difference-mode signal is $2V_s$. Obtain the output by using Eq. (1-16):

$$V_o = V_c G_c + V_d G_d$$
$$= V_n + 2V_s 10^4. \qquad Ans. \qquad (1\text{-}18)$$

(b) From Eq. (1-18), obtain the desired (difference-mode) and the undesired (common-mode) components of the output signal:

$$V_{od} = 2 \times 10^4 V_s,$$
$$V_{oc} = V_n.$$

The desired-to-undesired voltage ratio at the output is

$$\frac{V_{od}}{V_{oc}} = \frac{2 \times 10^4 V_s}{V_n} = 10^4. \qquad Ans.$$

Thus, the signal-to-noise ratio at the output is 10^4 times better than at the input.

1-3 RATIONAL FUNCTIONS, POLES, AND ZEROS

System functions, arising from linear, constant, lumped, and finite networks with linear dependent sources, are real rational functions, i.e., the ratio of two real polynomials. In terms of a numerator polynomial $N(s)$, and a denominator polynomial $D(s)$, the system function $G(s)$ is written as

$$G(s) = \frac{N(s)}{D(s)} = \frac{a_m s^m + a_{m-1} s^{m-1} + \cdots + a_1 s + a_0}{b_n s^n + b_{n-1} s^{n-1} + \cdots + b_1 s + b_0}, \qquad (1\text{-}19)$$

where the a and b coefficients are real numbers and s is the complex-frequency variable. If the polynomials are written in factored form, $G(s)$ becomes

$$G(s) = H \frac{(s - z_1)(s - z_2) \cdots (s - z_m)}{(s - p_1)(s - p_2) \cdots (s - p_n)} = H \frac{N_1(s)}{D_1(s)}, \qquad (1\text{-}20)$$

where $H = a_m/b_n$ is the scale factor. The polynomials $N_1(s)$ and $D_1(s)$ are monic; i.e., the coefficients of the s^m and s^n terms are unity.

The values of s that make the numerator polynomial zero are called the zeros of $N(s)$. If $N(s)$ is of order m, it has m zeros. The factored form, Eq. (1-20), exhibits the zeros explicitly: $z_1, z_2, \ldots, z_m$. The zeros of $N(s)$ are also the finite zeros of $G(s)$.

The values of s that make the denominator polynomial zero are called the zeros of $D(s)$. If $D(s)$ is of order n, it has n zeros. The factored form, Eq. (1-20), displays the zeros of $D(s)$: $p_1, p_2, \ldots, p_n$. The zeros of $D(s)$ are also called the finite poles of $G(s)$. In other words, *the poles of $G(s)$ are those values of s that make $G(s)$ infinite.*

The finite poles and zeros of $G(s)$ are called the *critical frequencies* of $G(s)$. $G(s)$ is completely specified if all the a and b coefficients are given or the scale factor H and all critical frequencies are known. In the s-plane, an x is used to designate the location of a pole, and an o to designate the location of a zero.

The system function, $G(s)$, may also have poles or zeros at infinity; that is, $G(s)$ becomes infinite or zero when s is made infinite. To determine the behavior of $G(s)$ at infinity, the asymptotic form of Eq. (1-19) is obtained:

$$G(s)\big|_{s\to\infty} \to \frac{a_m}{b_n} s^{m-n}.$$

If $m > n$, $G(s) \to \infty$; then, $G(s)$ has $(m - n)$ poles at infinity.

If $m < n$, $G(s) \to 0$; then, $G(s)$ has $(n - m)$ zeros at infinity.

If $m = n$, $G(s) \to (a_m/b_n)$; then, $G(s)$ has neither poles nor zeros at infinity.

For example,

$$G(s) = \frac{5s^4 + 1}{s^2 + s + 1}$$

has a second-order pole at infinity, because

$$G(s)\big|_{s\to\infty} \to 5s^2.$$

On the other hand,

$$G(s) = \frac{2}{s^3 + s^2 + s + 1}$$

has a third-order zero at infinity, because

$$G(s)\big|_{s\to\infty} \to \frac{2}{s^3}.$$

If the zeros or poles at infinity are included, a rational function has as many zeros as it has poles. The number is either m or n, whichever is greater. For example,

$$G(s) = \frac{s^2 + 1}{(s + 1)[(s + 1)^2 + 1]}$$

has altogether three zeros and three poles. Two of the zeros are finite and are at $s = \pm j$; the third zero is at $s = \infty$. All three poles are finite. They are at $s = -1$ and $s = -1 \pm j$.

EXAMPLE 1-5

A rational function has zeros at $s = \pm j2$ and poles at $s = -1, -2$, $(-1 \pm j)$. The scale factor is 2.
(a) Obtain the corresponding rational function.
(b) What is the behavior of this function at infinite frequency?

SOLUTION

(a) Using the given zeros, construct the numerator polynomial:

$$N(s) = (s - j2)(s + j2) = s^2 + 4.$$

Using the given poles, construct the denominator polynomial:

$$D(s) = (s + 1)(s + 2)(s + 1 - j)(s + 1 + j) = s^4 + 5s^3 + 10s^2 + 10s + 4.$$

The desired rational function is

$$G(s) = H \frac{N(s)}{D(s)} \tag{1-21}$$

$$= 2 \frac{s^2 + 4}{(s + 1)(s + 2)[(s + 1)^2 + 1]} = 2 \frac{s^2 + 4}{s^4 + 5s^3 + 10s^2 + 10s + 4}.$$

Ans.

(b) In Eq. (1-21), let s approach infinity:

$$G(s)\big|_{s \to \infty} \to 2 \frac{s^2}{s^4} = \frac{2}{s^2}.$$

$G(s)$ has two zeros at infinity.   *Ans.*

Determination of Poles

The finite poles of the system function $G(s)$ are the values of s which make $G(s)$ infinite. These values of s are also called the *natural or characteristic frequencies* of $G(s)$. In general, these frequencies are determined by the dead network and do not depend upon the position of the excitation or the response. With few exceptions, all system functions of a network have the same finite poles; i.e., the denominator polynomials are identical. The zeros of the denominator polynomial are henceforth called the poles of the network or system, even though it would be more correct to say that they are the poles of a system function. The equation obtained by setting the denominator polynomial equal to zero is called *the characteristic equation*. The roots of the characteristic equation are the poles of the system.

Any system function can be used to determine the poles of the network. First, the dead network is appropriately excited; i.e., an independent voltage source is inserted in any wire of the network or an independent current source is connected between any two nodes of the network. Then an arbitrary response is chosen and the corresponding system function, response/excitation, is calculated. The zeros of the denominator polynomial (the poles of the network) are then obtained. For polynomial orders higher than 2, a digital computer may be used to find the zeros.

EXAMPLE 1-6

Obtain the poles of the network shown in Fig. 1-11. The value of the current source depends upon the voltage across the $\frac{1}{2}$-Ω resistor.

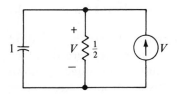

Fig. 1-11

SOLUTION

The network is dead because it contains passive elements and a dependent current source. As it stands, all voltages and currents are zero. Excite the network by inserting an independent voltage source in series with either the resistor or the capacitor. In Fig. 1-12a, the voltage source, V_s,

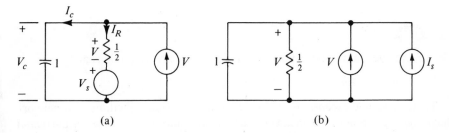

(a)   (b)

Fig. 1-12

is introduced in series with the resistor. Select any current or voltage other than V_s as the response. In Fig. 1-12a, the voltage across the resistor, V, is taken as the response.

The voltage across the capacitor and the current through it are

$$V_c = V + V_s, \qquad I_c = (V + V_s)s.$$

The current through the resistor and the voltage across it are

$$I_R = V - I_c = V - (V + V_s)s,$$

$$V = I_R \tfrac{1}{2} = \tfrac{1}{2}[V - (V + V_s)s],$$

$$V = -\frac{sV_s}{s+1}.$$

The resulting system function is

$$\frac{V}{V_s} = -\frac{s}{s+1}.$$

When $s = -1$, the denominator polynomial is zero. Hence the pole of the network is at $s = -1$.   *Ans.*

The pole of the network can also be determined by exciting the dead network with an independent current source. Since there are only two nodes in the network, the current source, I_s, must be connected as shown in Fig. 1-12b. Select the voltage across the resistor (or any current other than I_s) as the response. By inspection of Fig. 1-12b, obtain V:

$$V = (V + I_s)\frac{\frac{1}{2}\frac{1}{s}}{\frac{1}{2}+\frac{1}{s}} = (V + I_s)\frac{1}{s+2},$$

$$V = \frac{I_s}{s+1}.$$

The system function is

$$\frac{V}{I_s} = \frac{1}{s+1}.$$

From the denominator polynomial, it is seen that the pole is at

$$s = -1.   Ans.$$

There are some exceptions to the rule that all system functions of a network have the same poles. For instance, due to cancellation with zeros, a particular system function of a network may have fewer poles than the other system functions. An example of such a network is given in Fig. 1-13a. Consider the system functions I_1/V_1 and V_2/V_1:

$$\frac{I_1}{V_1} = \frac{1}{2R}\frac{s}{s+\dfrac{1}{RC}}, \qquad \frac{V_2}{V_1} = \frac{1}{2}\frac{s+\dfrac{1}{RC}}{s+\dfrac{1}{RC}} = \frac{1}{2}.$$

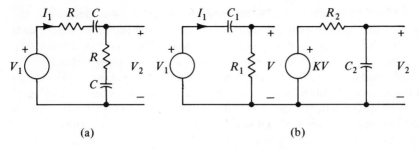

(a)   (b)

Fig. 1-13 (I_1/V_1) and (V_2/V_1) poles differ in both networks

(I_1/V_1) has a pole at $s = -(1/RC)$, whereas (V_2/V_1) has this same pole cancelled by a zero at $s = -(1/RC)$ and therefore has no poles.

Another exception to the rule is shown in Fig. 1-13b, where the dependent source KV transmits the signal forward. Consider the system functions (I_1/V_1) and (V_2/V_1):

$$\frac{I_1}{V_1} = \frac{\dfrac{1}{R}s}{s + \dfrac{1}{R_1 C_1}}, \qquad \frac{V_2}{V_1} = \frac{K}{R_2 C_2} \frac{s}{\left(s + \dfrac{1}{R_1 C_1}\right)\left(s + \dfrac{1}{R_2 C_2}\right)}.$$

(I_1/V_1) has a pole at $s = -(1/R_1 C_1)$ whereas (V_2/V_1) has an additional pole at $s = -(1/R_2 C_2)$.

If in Fig. 1-13b the excitation V_1 is applied in series with R_2, rather than with R_1, the pole at $s = -(1/R_1 C_1)$ does not show in any system function. Since the dependent source does not allow reverse transmission, the $R_1 C_1$ loop does not become excited to reveal the pole at $s = -(1/R_1 C_1)$; the two halves of the network are said to be *decoupled* in this case.

1-4 SUMMARY

The concept of the system function is fundamental to the study of linear networks. The system function is a complex-frequency-domain concept. It is given by response/excitation (output/input), where all variables are functions of s. It is assumed that the system is free of independent sources, including initial-condition sources, and its elements are linear and time-invariant. A system function relates a specific response (output) to a specific excitation (input) in the system.

When more than one independent source is used to excite the system, the principle of superposition can be used to find the desired responses. Dependent sources by themselves do not produce responses. However, in conjunction with independent sources, they produce responses that may be unattainable with passive elements only. The principle of superposition is also applicable to dependent sources. In the final analysis, however, all responses must be expressed as functions of the independent excitations. In certain applications two inputs are used, and the system is designed so that it produces no (or very little) response when the two inputs are alike. Such systems are best analyzed in terms of the common-mode and difference-mode variables and functions.

A rational function, $G(s)$, is a ratio of two polynomials. The values of the complex-frequency variable, s, that make the rational function zero are called the *zeros* of $G(s)$. The values of s that make the rational function infinite are called the *poles* of $G(s)$. System functions arising from linear, time-invariant, finite, and lumped systems are rational functions. The poles of system functions are also called *characteristic* or *natural* frequencies of the system. The poles and zeros together form the *critical* frequencies of the system function.

PROBLEMS

1-1   For each network shown in Fig. 1-14, calculate the system function. The excitations and responses are as shown.

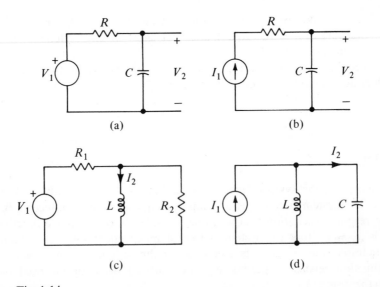

Fig. 1-14

1-2 For the network shown in Fig. 1-15, calculate all system functions. The excitation is as shown.

1-3 In Fig. 1-16, ρ and γ represent initial conditions. Obtain the expression for the output, $V_o(s)$.

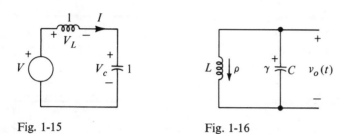

Fig. 1-15   Fig. 1-16

1-4 For each network shown in Fig. 1-17, obtain the desired system function. Only the lefthand source is independent in each network.

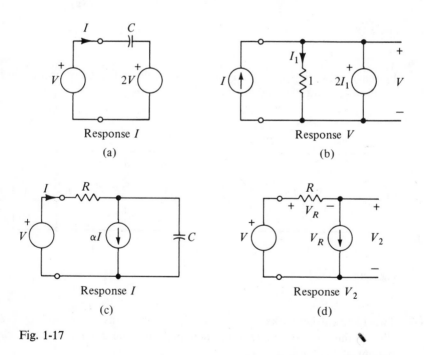

Response *I*

(a)

Response *V*

(b)

Response *I*

(c)

Response V_2

(d)

Fig. 1-17

1-5 For each network shown in Fig. 1-18, obtain the desired response. Except in (d), all sources are independent. γ and ρ represent initial conditions.

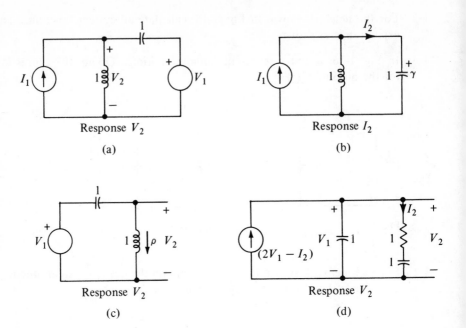

Fig. 1-18

1-6   For each network shown in Fig. 1-19, obtain the indicated response and its common-mode and difference-mode components.

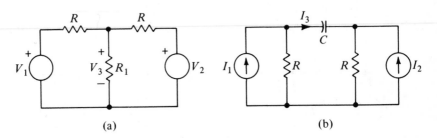

Fig. 1-19

1-7   Two identical networks are connected and driven as shown in Fig. 1-20. What is the common-mode component of I_3 and the difference-mode component of V_3 ?

1-8   An independent force F is applied to the mass M shown in Fig. 1-21. The force F is sensed, converted to a dependent force KF, and fed back as shown. The acceleration of the mass is the response.

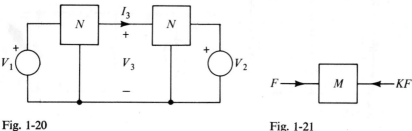

Fig. 1-20   Fig. 1-21

(a) Find the system function.
(b) Explain the significance of the system function if $K = 0, 1, 2, -1$.

1-9 Obtain the poles of the networks shown in Fig. 1-22.

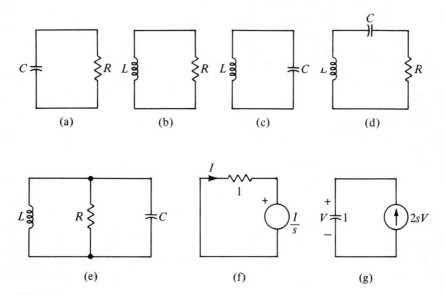

Fig. 1-22

1-10 (a) Find the poles of the network shown in Fig. 1-23 by exciting it with a
        current source.
     (b) Find the poles of the network by exciting it with a voltage source.
        Compare the results with (a).
     (c) Repeat (b) with the voltage source applied to a different branch of the
        network.

1-11 In Fig. 1-24, show that the finite poles of (I_2/I_1) are identical with the
     finite poles of (V_2/I_1).

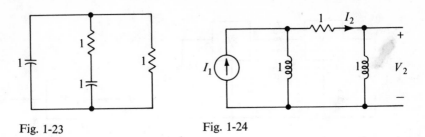

Fig. 1-23   Fig. 1-24

1-12 For the network of Fig. 1-25, compare the poles of (I_1/V), (I_2/V), and (I_3/V). Explain why they are different.

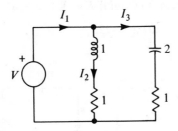

Fig. 1-25

2

Characterization and Discussion of Responses in Networks

Much can be learned about networks by studying the nature of the response. Although observable only as a single waveform on an oscilloscope, its mathematical and physical nature is more conveniently studied by decomposing it into two parts. The characteristics of one part are directly attributable to the poles of the network, whereas the characteristics of the other part are directly related to the excitation. Even though these two parts are not independent of each other, it is possible to exercise control over them by appropriate selection of excitations and networks. Thus, either part of the response or the complete response can be made zero. On the other hand, the waveform of one part can be made to approach the waveform of the other part, and thus exhibit resonance. In most applications, the part arising from network poles vanishes with time. Then, if the input is a sine wave, the output in the steady-state is a sine wave also. The two sine waves have the same frequency but differ in amplitude and phase. The frequency response shows these differences as a function of the frequency of the input sine wave. In this chapter, all these topics are discussed at length and many examples are given to illustrate the principles involved.

2-1 NATURE OF THE RESPONSE

The system function, $G(s)$, relates a single, independent excitation $E(s)$ to a response $R(s)$ by

$$R(s) = E(s)G(s). \tag{2-1}$$

If the system is composed of linear, lumped, constant elements and linear dependent sources forming a finite network, then $G(s)$ is a real rational function—a ratio of 2 real polynomials in s. For a large class of problems, the excitation $E(s)$ is also a rational function. For example,

$$E(s) = \frac{c_1 s + c_0}{d_2 s^2 + d_1 s + d_0}$$

is a very commonly used excitation. By the appropriate choice of the c and d constants, this $E(s)$ can be made to represent the Laplace transform of well-known time functions: impulse, step, ramp, exponential, sinusoidal, damped-sinusoidal.

If both $G(s)$ and $E(s)$ are rational functions, then the response, $R(s)$, is also a rational function and can be written as

$$R(s) = \frac{N(s)}{D(s)} = \frac{N(s)}{(s - p_1)(s - p_2) \cdots (s - p_n)}, \tag{2-2}$$

where $p_1, p_2, \ldots, p_n$ represent the finite poles of the response. Assume that $N(s)$ is of lower order than $D(s)$. [$N(s)$ may, in general, be of higher order than $D(s)$; this case, however, does not arise in this book.]

In order to learn more about the response, $R(s)$ is expanded in partial fractions. If all of the poles are simple, $R(s)$ becomes

$$R(s) = \frac{K_1}{s - p_1} + \frac{K_2}{s - p_2} + \cdots + \frac{K_n}{s - p_n} = \sum_{i=1}^{n} \frac{K_i}{s - p_i}, \tag{2-3}$$

where

$$K_i = R(s)(s - p_i)|_{s = p_i} \qquad (i = 1, 2, \ldots, n). \tag{2-4}$$

K_i is called the residue of the pole p_i.

Since $R(s)$ is a rational function with real coefficients, p_i and K_i, when complex, occur in conjugate pairs. The general form of a pole is $p_i = \alpha_i + j\beta_i$ where the real part of the pole, α_i, and the imaginary part of the pole, β_i, are real numbers. In the s-plane, a pole of $R(s)$ may be at the origin, $\alpha_i = \beta_i = 0$, on the real axis, $\beta_i = 0$, on the imaginary axis, $\alpha_i = 0$, in the left half-plane, $\alpha_i < 0$, or in the right half-plane, $\alpha_i > 0$.

The nature of the response in the time domain can be studied by taking the inverse transform of Eq. (2-3),

$$r(t) = K_1 e^{p_1 t} + K_2 e^{p_2 t} + \cdots + K_n e^{p_n t}$$

$$= \sum_{i=1}^{n} K_i e^{p_i t} \qquad (t > 0).$$ (2-5)

If K_i is complex, it can be written as $|K_i| e^{j\theta_i}$, where $|K_i|$ is the magnitude of K_i, and θ_i its angle. Thus, for k complex poles and $(n - k)$ real poles, $r(t)$ becomes

$$r(t) = \sum_{i=1}^{k} |K_i| e^{\alpha_i t} e^{j(\beta_i t + \theta_i)} + \sum_{i=k+1}^{n} K_i e^{\alpha_i t},$$ (2-6)

where the first summation covers all complex poles and the second summation covers all real poles. When the complex-conjugate pairs are combined, Eq. (2-6) simplifies to

$$r(t) = \sum_{i=1}^{k/2} 2 |K_i| e^{\alpha_i t} \cos(\beta_i t + \theta_i) + \sum_{i=k+1}^{n} K_i e^{\alpha_i t},$$ (2-7)

where

$$2 \cos(\beta_i t + \theta_i) = e^{j(\beta_i t + \theta_i)} + e^{-j(\beta_i t + \theta_i)}.$$

Equation (2-7) is an important result. It shows that if a system is excited by a time-varying source whose transform is a rational function, and *if the response poles are simple, i.e., if there are no multiple-order poles, then the time-domain response is the sum of two kinds of waveforms: sinusoids with exponential amplitudes and exponentials.* The sinusoids may, with time, grow in amplitude ($\alpha_i > 0$), decay in amplitude ($\alpha_i < 0$), or have constant amplitude ($\alpha_i = 0$). The exponentials may also grow, decay, or reduce to a constant. Thus, a simple but a very useful relationship exists between the pole locations in the s-plane and the resulting waveforms in the time domain. This relationship between the two domains is summarized in Fig. 2-1, where the scale factor, K, may be negative.

It is indeed remarkable that these simple waveforms combine to generate an innumerable variety of response waveforms. It is important that the relationship between the two domains be well understood. A response pole at $s = \alpha$ in the complex-frequency domain should immediately call to mind an exponential waveform in the time domain; a pair of poles at $\pm j\beta$ should bring to mind a sinusoid, and so on. Although it is easy to sketch the waveforms due to the individual terms, the complete response may be difficult to visualize if it consists of many terms with different amplitudes, frequencies, and time constants.

If the response $R(s)$ has poles of order k (multiple poles), then the response contains waveforms given by $t^{(j-1)} e^{\alpha_i t} \cos(\beta_i t + \theta_i)$, $j = 1, 2, \ldots, k$. These waveforms can be obtained by multiplying the waveforms given in Fig. 2-1 by t^{j-1}.

The poles of $R(s)$ determine the waveform of each term in the response. What about the zeros of $R(s)$? Since the zeros of $R(s)$ affect the magnitude and angle

Complex frequency domain
or *s*-plane   Time domain

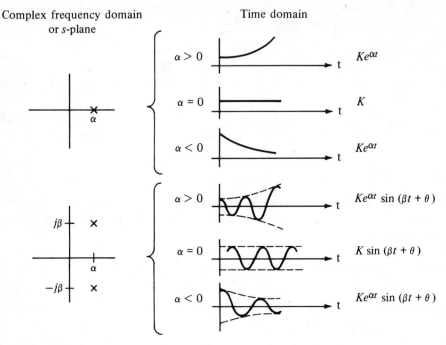

Fig. 2-1   Location of response poles and corresponding time functions

of K_i [see Eq. (2-5)], the zeros determine the amplitude and the starting point of the individual waveforms. This may have a profound effect on the complete response if several terms are present in the response.

EXAMPLE 2-1

Given

$$R(s) = \frac{s + \alpha}{s(s^2 + 4)}.$$

(a) What waveforms are present in $r(t)$?
(b) What effect does the location of the zero have on the response?

SOLUTION

(a) The poles of $R(s)$ are at $s = 0$ and $s = \pm j2$. The pole at $s = 0$ produces a constant term in $r(t)$. Together, the poles at $s = \pm j2$ produce a sinusoidal term in $r(t)$. Therefore $r(t)$ is composed of a constant and a sinusoid.

(b) To determine the effect of the zero quantitatively, expand $R(s)$ in partial fractions and obtain the inverse transform:

$$R(s) = \frac{s + \alpha}{s(s - j2)(s + j2)} = \frac{K_1}{s} + \frac{K_2}{s - j2} + \frac{K_3}{s + j2}.$$

Determine the residues using Eq. (2-4).

$$K_1 = \frac{s + \alpha}{s^2 + 4}\bigg|_{s=0} = \frac{\alpha}{4},$$

$$K_2 = \frac{s + \alpha}{s(s + j2)}\bigg|_{s=j2} = \frac{j2 + \alpha}{j2(j4)} = -\frac{j + \dfrac{\alpha}{2}}{4} = -\frac{1}{4}\sqrt{1 + \left(\frac{\alpha}{2}\right)^2}\, e^{j \tan^{-1}(2/\alpha)},$$

$$K_3 = \text{conjugate of } K_2.$$

The partial-fraction expansion of the response is:

$$R(s) = \frac{\alpha}{4}\frac{1}{s} - \frac{1}{4}\sqrt{1 + \left(\frac{\alpha}{2}\right)^2}\left[\frac{e^{j \tan^{-1}(2/\alpha)}}{s - j2} + \frac{e^{-j \tan^{-1}(2/\alpha)}}{s + j2}\right].$$

The inverse transform of $R(s)$ is

$$r(t) = \frac{\alpha}{4} - \frac{1}{4}\sqrt{1 + \left(\frac{\alpha}{2}\right)^2}\,[e^{j2t}e^{j \tan^{-1}(2/\alpha)} + e^{-j2t}e^{-j \tan^{-1}(2/\alpha)}].$$

Since $e^{j\theta} + e^{-j\theta} = 2 \cos \theta$,

$$r(t) = \frac{\alpha}{4} - \frac{1}{2}\sqrt{1 + \left(\frac{\alpha}{2}\right)^2}\,\cos\left(2t + \tan^{-1}\frac{2}{\alpha}\right).$$

The zero of $R(s)$ is at $s = -\alpha$. Its effect in $r(t)$ is to control the value of the constant term, $(\alpha/4)$, and the amplitude, $\frac{1}{2}\sqrt{1 + (\alpha/2)^2}$, and phase, $\tan^{-1}(2/\alpha)$, of the sinusoid.

2-2 DECOMPOSITION OF THE RESPONSE

The poles of the response $R(s)$ are composed of the poles of the system function and the poles of the excitation. To show this explicitly, $R(s)$ is written as:

$$R(s) = E(s)G(s) = \frac{N(s)}{D(s)} = \frac{N(s)}{D_n(s)D_f(s)}$$

$$= \frac{N(s)}{(s - p_{n1})(s - p_{n2}) \cdots (s - p_{f1})(s - p_{f2}) \cdots}, \tag{2-8}$$

where the subscript n is used for the system-function poles (*natural poles*) and the subscript f for the excitation poles (*forced poles*). Assuming that all poles are simple and $N(s)$ is of lower order than $D(s)$, $R(s)$ can be expanded in partial fractions. Each pole in $R(s)$ contributes to a term in the expansion. Segregating the terms arising from the system poles from those of the excitation poles, $R(s)$ can be written as:

$$R(s) = \left(\frac{K_{n1}}{s - p_{n1}} + \frac{K_{n2}}{s - p_{n2}} + \cdots \right) + \left(\frac{K_{f1}}{s - p_{f1}} + \frac{K_{f2}}{s - p_{f2}} + \cdots \right)$$

$$= R_n(s) + R_f(s). \tag{2-9}$$

The terms arising from the system-function poles are combined to form $R_n(s)$, and the terms arising from the excitation poles are combined to form $R_f(s)$. $R_n(s)$ is called *the natural part of the response*, since the poles associated with $R_n(s)$ arise from the network itself and their values have nothing to do with the excitation; they are *natural* or *characteristic* to the network. $R_f(s)$ is *the forced part of the response*, since its poles depend upon the excitation poles and have nothing to do with the network; the excitation poles force the network to produce the response. In the time domain, Eq. (2-9) becomes

$$r(t) = (K_{n_1} e^{p_{n1}t} + K_{n2} e^{p_{n2}t} + \cdots) + (K_{f1} e^{p_{f1}t} + K_{f2} e^{p_{f2}t} + \cdots) = r_n(t) + r_f(t). \tag{2-10}$$

The individual terms in the natural component, $r_n(t)$, have waveforms solely determined by the network itself, whereas the individual terms in $r_f(t)$ have waveforms identical in shape with the individual parts making up the forcing function (excitation). To have any response at all, the network must be excited. *Thus forced, the network responds in two ways: It reveals its own characteristic behavior in the form of the natural response, and it exhibits the characteristics of the excitation in the form of the forced response.*

A common mistake is to think that a sinusoidally excited linear network always produces a sinusoidal response. The response is sinusoidal *only* when the terms due to the poles of the network (the natural part of the response) are zero. This may never occur, or may occur only after some time has elapsed. For example, when the poles of the network are in the left half-plane, i.e., when the poles have negative real parts, then the natural part of the response eventually vanishes. Only then is the response, since it is the forced part, truly sinusoidal.

EXAMPLE 2-2

Obtain the natural and forced components of the response shown in Fig. 2-2 and discuss the results.

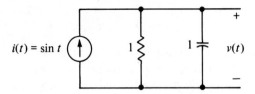

Fig. 2-2

SOLUTION

In the frequency domain the excitation is

$$I(s) = \frac{1}{s^2 + 1}.$$

The system function is

$$G(s) = \frac{V(s)}{I(s)} = \frac{1}{s + 1}.$$

Obtain the response from

$$V = GI = \left(\frac{1}{s + 1}\right)\left(\frac{1}{s^2 + 1}\right).$$

The response has three poles. The pole at $s = -1$ comes from the system function; the poles at $s = \pm j$ belong to the excitation. Segregate the partial-fraction expansion of the response according to the poles to obtain

$$V = V_n + V_f = \frac{1}{2}\frac{1}{s + 1} - \frac{1}{2}\frac{s - 1}{s^2 + 1},$$

$$V_n(s) = \frac{1}{2}\frac{1}{s + 1}, \quad \text{and} \quad V_f(s) = -\frac{1}{2}\frac{s - 1}{s^2 + 1}.$$

The time-domain waveforms are given by

$$v(t) = v_n(t) + v_f(t),$$

$$v_n(t) = \frac{1}{2}e^{-t}, \quad \text{and} \quad v_f(t) = \frac{1}{\sqrt{2}}\sin\left(t - \frac{\pi}{4}\right). \quad Ans.$$

This network is excited by a sinusoidal source. The response contains a sinusoidal component due to the forcing action of the input. It also contains a damped exponential which characterizes the natural behavior of the network. Had the excitation been a step function, the response would still have contained this damped exponential term (with a different amplitude), in addition to a constant term arising from the new forcing function.

2-3 ZERO RESPONSE

Some networks are designed to produce a zero response even though they are excited. This means that the system function for that particular response is identically zero, that is, $G(s) = 0$. Such networks are used to indicate a balanced state. The bridge circuit shown in Fig. 2-3 serves as an example.

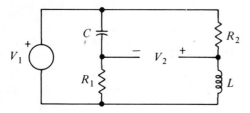

Fig. 2-3 $(V_2/V_1) = 0$, when balanced

The system function for this network is

$$\frac{V_2}{V_1} = G(s) = \frac{s\left(\dfrac{1}{R_1 C} - \dfrac{R_2}{L}\right)}{\left(s + \dfrac{R_2}{L}\right)\left(s + \dfrac{1}{R_1 C}\right)} . \tag{2-11}$$

$G(s) = 0$ when $L = R_1 R_2 C$; this is the condition for balance which results in zero output. However, any departure from the balanced state (e.g., R_1 changes a little from its nominal value) produces a nonzero system function, and, hence, an output results when the system is excited.

2-4 RESPONSE: A SCALED REPLICA OF EXCITATION

Some networks are designed to produce an output which is a scaled-up or scaled-down version of the input. This means that the network has a system function that is constant; that is, $G(s) = K$, where K is a real number. All resistive networks fall in this category. However, it is also possible to have a constant $G(s)$ in nonresistive networks. An example of such a network is shown in Fig. 2-4.

In its basic form, the network consists of $9R$ in series with R. Thus, it acts as a $10 : 1$ attenuator:

$$\frac{V_2}{V_1} = \frac{R}{R + 9R} = \frac{1}{10} .$$

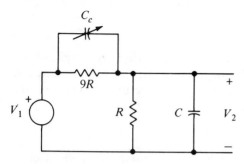

Fig. 2-4   Compensated attenuator

However, it is impossible to construct such networks. The constant presence of the stray capacitor, C, shunting the resistor R, makes the attenuation frequency-dependent:

$$\frac{V_2}{V_1} = \frac{1}{9RC} \frac{1}{s + \dfrac{10/9}{RC}}.$$

To compensate for the undesirable C, the compensating capacitor, C_c, is added across $9R$. From Fig. 2-4, the attenuation is found to be

$$\frac{V_2}{V_1} = \frac{C_c}{C + C_c} \frac{s + \dfrac{1}{9RC_c}}{s + \dfrac{10}{9R(C + C_c)}}. \tag{2-12a}$$

To obtain frequency-independent attenuation, the pole must be cancelled with the zero. This requires that

$$s + \frac{1}{9RC_c} = s + \frac{10}{9R(C + C_c)},$$

which results in

$$C_c = \frac{C}{9}, \tag{2-12b}$$

and

$$\frac{V_2}{V_1} = \frac{C_c}{C + C_c} = \frac{1}{10}.$$

Thus, by adjusting C_c according to Eq. (2-12b), the system function given by Eq. (2-12a) can be made constant. *This pole–zero cancellation scheme is widely used in the design of frequency-independent attenuators.*

2-5 RESPONSE WITH NO FORCED COMPONENT

In order to get a response, a network must be excited (forced). In spite of this forcing action of the input, it is possible to get a response that does not have any forced component. One way to achieve this is to excite the network with the unit-impulse function. Since $E(s) = 1$,

$$R(s) = G(s) = R_n(s),$$

which results in a response containing only the system-function poles. Hence, the response has zero forced component.

Another way of achieving zero forced response is to select a system function with *zeros coincident with all of the poles of excitation*, thereby cancelling them. This is illustrated by the network shown in Fig. 2-5.

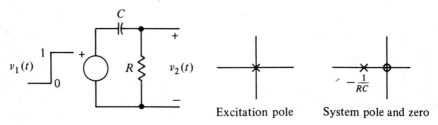

Excitation pole   System pole and zero

Fig. 2-5   Cancelling the excitation pole

The input, $v_1(t)$, is a unit step. Its transform, $(1/s)$, has a pole at the origin. The system function is

$$G(s) = \frac{s}{s + \dfrac{1}{RC}},$$

which has a zero at the origin. The excitation pole is cancelled by the system zero when V_2 is calculated:

$$V_2 = GV_1 = \frac{\not{s}}{s + \dfrac{1}{RC}} \cdot \frac{1}{\not{s}} = \frac{1}{s + \dfrac{1}{RC}},$$

$$v_2(t) = v_n(t) = e^{-t/RC}.$$

The output contains no trace of the input, which is a constant. It contains only the natural response arising from the network pole at $(-1/RC)$.

A practical example of a network which produces a response consisting of the natural component only is the oscillator. An oscillator is a network with poles on the imaginary axis. The excitation pole is at the origin, and arises from the dc sources applied to activate the oscillator. In a properly designed oscillator, the excitation pole does not introduce any dc component in the output.

Another example of a network which produces zero forced response is the frequency-rejection network. The network is tuned so that the system-function zeros coincide with the excitation poles. As a result, the input sinusoidal signal is prevented from reaching the output.

2-6 RESPONSE WITH NO NATURAL COMPONENT

If all the poles of a network are cancelled by the zeros of the excitation, the response will contain only the forced part of the solution. This principle is demonstrated by the network of Fig. 2-6.

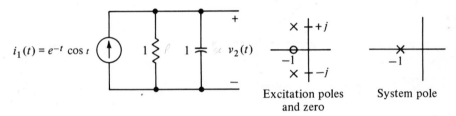

Fig. 2-6   Cancelling the system pole

The input

$$I_1(s) = \frac{s + 1}{(s + 1)^2 + 1}$$

has a zero at -1. The system function

$$G(s) = \frac{1}{s + 1}$$

has a pole at -1. When V_2 is formed, the system pole is cancelled by the excitation zero:

$$V_2 = GI_1 = \frac{1}{(s+1)} \frac{(s+1)}{(s + 1)^2 + 1} = \frac{1}{(s + 1)^2 + 1},$$

$$v_2(t) = e^{-t} \sin t.$$

The output waveform is a damped sinusoid just like the input. The natural part is zero.

2-7 RESPONSE WHEN AN EXCITATION POLE IS NEAR A SYSTEM POLE

System poles describe the natural behavior of the system. Excitation poles impose a forced behavior on the system. Therefore, it is not surprising to see marked changes in the behavior of the natural as well as the forced response as a pole of excitation nears a system pole. To investigate this interaction, an excitation pole at $p + \Delta s$ is made to approach a system pole at p along some path, as shown in Fig. 2-7. Both of these poles are assumed to be simple.

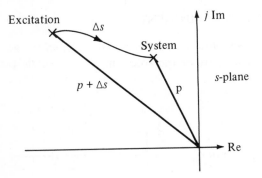

Fig. 2-7   Interaction between poles

The expression for $R(s)$ can be written to show explicitly the two interacting poles. Let $D_n(s)$ and $D_f(s)$ represent the polynomials arising from the system poles and the excitation poles, respectively. $D_n(s)$ contains the pole at $s = p$, and $D_f(s)$ contains the pole at $s = p + \Delta s$. The factors arising from these poles are taken out, to obtain

$$D_n(s) = (s - p)D_{nr}(s), \qquad D_f(s) = (s - p - \Delta s)D_{fr}(s),$$

where $D_{nr}(s)$ and $D_{fr}(s)$ represent the remaining polynomials. $R(s)$ can then be written as:

$$R(s) = \frac{N(s)}{D(s)} = \frac{N(s)}{D_n(s)D_f(s)} = \frac{N(s)}{[(s - p)D_{nr}(s)][(s - p - \Delta s)D_{fr}(s)]}. \qquad (2\text{-}13)$$

$N(s)$ is assumed to have no zero at $s = p$ or at $s = p + \Delta s$, so that cancellation does not occur. The partial-fraction expansion of $R(s)$ is given by

$$R(s) = \frac{K_n}{s - p} + \frac{K_f}{s - p - \Delta s}$$

$$+ \text{[other terms due to the remaining poles of } R(s)], \qquad (2\text{-}14)$$

where

$$K_n = \frac{N(s)}{D_{nr}(s)[(s - p - \Delta s)D_{fr}(s)]}\Bigg|_{s=p} = \frac{-N(p)}{D_{nr}(p)D_{fr}(p)\,\Delta s} = \frac{K(p)}{\Delta s}, \qquad (2\text{-}15)$$

$$K_f = \frac{N(s)}{[(s-p)D_{nr}(s)]D_{fr}(s)}\bigg|_{s=p+\Delta s} = \frac{N(p+\Delta s)}{D_{nr}(p+\Delta s)D_{fr}(p+\Delta s)\,\Delta s} = -\frac{K(p+\Delta s)}{\Delta s}.$$

(2-16)

In the time domain, the response becomes

$$r(t) = \frac{K(p)}{\Delta s}e^{pt} - \frac{K(p+\Delta s)}{\Delta s}e^{(p+\Delta s)t} + \text{(other terms)}.$$

(2-17)

The first term in Eq. (2-17) is due to the system pole; the second term is due to the excitation pole. The other terms are due to the remaining system and excitation poles. As the excitation pole is made to come close to the system pole, by letting $\Delta s \to 0$, two important effects take place. First, *the second term becomes more and more like the negative of the first term.* Second, *the amplitude of each term becomes larger and larger in magnitude* because of the division by Δs. Thus the components of the response get into a highly excited state, exhibiting large natural and forced parts. Whether these large amplitudes in the individual terms produce a significant result when taken together depends upon the location of the two poles.

Both Poles in the Left Half-Plane

If the system and excitation poles are both in the left half-plane and approach each other, the sum of the first two terms of Eq. (2-17) does not show marked behavior. The example given in Fig. 2-8 illustrates this point.

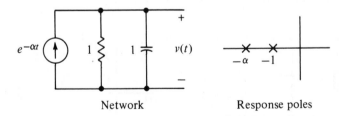

Network   Response poles

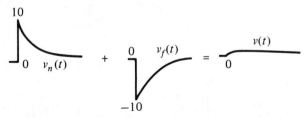

Natural response   Forced response   Response

Fig. 2-8   Decomposition of the response when $\alpha = 1.1$

The system pole is at -1. The excitation pole is at $-\alpha$, and it can be moved by varying the time constant of the input-current waveform. Since there are no other poles involved, the response is entirely due to these two poles. For $\alpha \neq 1$,

$$v_n(t) = \frac{1}{\alpha - 1} e^{-t} \quad \text{and} \quad v_f(t) = -\frac{1}{\alpha - 1} e^{-\alpha t}, \tag{2-18}$$

$$v(t) = v_n(t) + v_f(t).$$

In this example, the amplitude of the natural component is the same as the forced component regardless of α. As $\alpha \to 1$, the amplitude $[1/(\alpha - 1)] \to \infty$. What about the response $v(t)$? In Fig. 2-8, the response and its components are sketched for $\alpha = 1.1$. Note that, in spite of the large $v_n(t)$ and $v_f(t)$ amplitudes, $v(t)$ is small in amplitude. In fact, the response performance is rather disappointing in comparison to the excited behavior of its two components. The disparity becomes more pronounced for $\alpha = 1.001$. Whereas the natural and forced components now have an amplitude of 1000 (100 times larger than for $\alpha = 1.1$), the response differs only slightly from the previous response. *In the response, there is no indication of the tremendous changes occurring in the components.* In fact, even for $\alpha = 1$, when each component has infinite amplitude, indicating coincidence of poles, the response is $r(t) = te^{-t}$, which is not much different from that for $\alpha = 1.1$.

One Pole in the Left Half-Plane, The Other on the Imaginary Axis

The situation is different when one pole is in the left half-plane while the other is on the imaginary axis, as shown in Fig. 2-9. The system poles are at $-0.01 \pm j\sqrt{1 - 10^{-4}}$, and the excitation poles are at $\pm j\omega$.

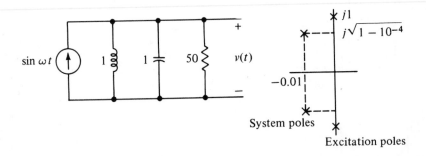

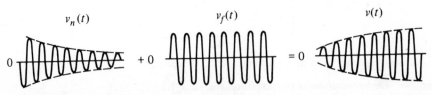

Fig. 2-9   Decomposition of response when $\omega = 1$

The response is given by

$$v_n(t) = \frac{-50\omega}{\sqrt{1 - 10^{-4}}\sqrt{2500(\omega^2 - 1)^2 + \omega^2}} e^{-t/100}$$

$$\times \sin\left\{\sqrt{1 - 10^{-4}}\,t - \tan^{-1}(100\sqrt{1 - 10^{-4}})\right.$$

$$\left. + \tan^{-1}\left[\frac{100\sqrt{1 - 10^{-4}}}{5000(\omega^2 - 1) + 1}\right]\right\}, \tag{2-19}$$

$$v_f(t) = \frac{50\omega}{\sqrt{2500(1 - \omega^2)^2 + \omega^2}} \sin\left\{\omega t - \tan^{-1}\left[\frac{\omega}{50(1 - \omega^2)}\right] + \frac{\pi}{2}\right\}, \tag{2-20}$$

$$v(t) = v_n(t) + v_f(t).$$

The natural response is a damped sinusoid with a time-constant of 100 seconds, whereas the forced response is a pure sinusoid. The amplitudes of both waveforms (disregarding the exponential damping) are very nearly equal, regardless of ω. As the frequency ω of the input signal is brought close to 1, the excitation pole approaches the system pole. At $\omega = 1$, they are very close to each other. At this frequency the amplitude of the forced component is 50, and is at its maximum value. The network is in resonance with the excitation. On either side of resonance, the excitation pole is pulled away from the system pole because the value of ω changes from 1. As a result, the forced-response amplitude falls off sharply.

For $\omega = 1$, Eq. (2-19) and Eq. (2-20) become

$$v_n(t) = -\frac{50}{\sqrt{1 - 10^{-4}}} e^{-t/100} \sin\sqrt{1 - 10^{-4}}\,t, \tag{2-21}$$

$$v_f(t) = 50 \sin t. \tag{2-22}$$

The strong interaction between the poles is reflected in the large signal amplitudes in the natural and forced components of the response. Even though each component starts out with the large amplitude of 50, the response, $v(t)$, for t small, is very small, since the two waveforms are practically equal and opposite. However, as t becomes larger, the natural component gets smaller, and after a few system time constants, it vanishes. Then the response is the same as the forced response, whose large amplitude still reflects the proximity of the system pole to the excitation pole. Thus, in this instance, excited behavior in component parts eventually shows in the response too.

2-8 RESPONSE WHEN EXCITATION AND SYSTEM POLES ARE COINCIDENT

When an excitation pole coincides with a system pole ($\Delta s = 0$ in Fig. 2-7), the corresponding *forced and natural solutions become infinite*. This is demonstrated by letting $\Delta s = 0$ in Eq. (2-17), which is reproduced here for convenience:

$$r(t) = \frac{K(p)}{\Delta s}e^{pt} - \frac{K(p + \Delta s)}{\Delta s}e^{(p + \Delta s)t} + \text{(other terms)}.$$

Although the first and second terms in $r(t)$ become infinite when $\Delta s = 0$, their difference is not infinite. To evaluate the sum of the first two terms, Δs is allowed to *approach* zero, rather than made *equal* to zero:

$$r(t) = -\lim_{\Delta s \to 0}\left[\frac{K(p + \Delta s)e^{(p + \Delta s)t} - K(p)e^{pt}}{\Delta s}\right] + \text{(other terms)}.$$

The limit of the bracketed term is, by definition, the derivative of $[K(s)e^{st}]$ with respect to s, evaluated at $s = p$, that is,

$$\left\{\frac{d}{ds}[K(s)e^{st}]\right\}_{s=p}.$$

Therefore,

$$r(t) = -\frac{d}{ds}[K(s)e^{st}]_{s=p} + \text{(other terms)}$$

$$= -\left[K(s)\frac{d}{ds}(e^{st})\right]_{s=p} - \left\{e^{st}\frac{d}{ds}[K(s)]\right\}_{s=p} + \text{(other terms)}$$

$$= -K(p)te^{pt} - \frac{dK(s)}{ds}\bigg|_{s=p}e^{pt} + \text{(other terms)}. \tag{2-23}$$

The first two terms in Eq. (2-23) represent the response from the two coincident poles. Note that it is no longer possible to separate the response into forced and natural parts, since the poles are now identical and therefore cannot be distinguished individually. However, from a physical viewpoint, it is desirable to recognize that this result is obtained by subtracting an infinite natural response from an infinite forced response.

EXAMPLE 2-3

Study the response and its components for the system shown in Fig. 2-10, and discuss the result.

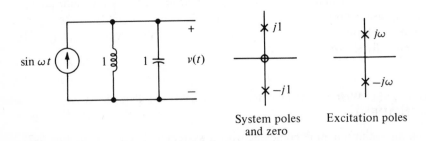

System poles
and zero

Excitation poles

Fig. 2-10

SOLUTION

Since the system function is

$$G(s) = \frac{s}{s^2 + 1},$$

the system poles are at $\pm j1$. Since the excitation is $\omega/(s^2 + \omega^2)$, the excitation poles are at $\pm j\omega$. By varying ω, the excitation poles can be moved along the imaginary axis and brought close to, and made coincident with, the poles of the system. The response is

$$v(t) = v_n(t) + v_f(t) = -\frac{\omega}{1 - \omega^2} \cos t + \frac{\omega}{1 - \omega^2} \cos \omega t \qquad (\omega \neq 1). \qquad Ans.$$

As $\omega \to 1$, the natural- and forced-component amplitudes approach infinity. In other words, the parts of the response show a highly excited state. For $\omega = 1$, $v(t)$ is 0/0. Obtaining the limit of $v(t)$ as $\omega \to 1$ (use L'Hôpital's rule, and differentiate with respect to ω), $v(t)$ becomes

$$v(t) = \frac{t}{2} \sin t, \qquad Ans.$$

which is finite for all $t < \infty$. As long as $\omega \neq 1$, the output is the sum of two sine waves and is therefore bounded. For $\omega = 1$, the output amplitude increases linearly with time. Therefore, in this case, it is possible to determine when coincidence of the poles has occurred, by observing the output waveform as ω is varied.

2-9 THE SINUSOIDAL STEADY-STATE RESPONSE

A large class of networks is designed with left half-plane poles and excited with a sine wave. In these cases, the natural response eventually vanishes (being the sum of exponentially damped terms) while the forced response remains. Attention is then focused on how the input sinusoid is operated upon by the network, to form a desired sinusoidal forced response. Figure 2-11 shows, in block-diagram form, the relationship between the input and the forced component of the output.

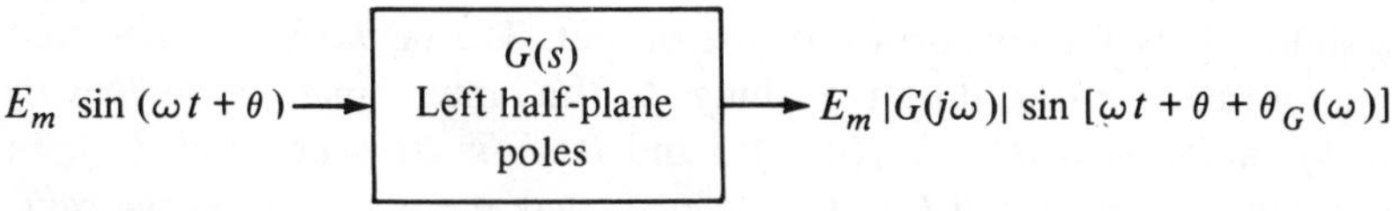

$$E_m \sin(\omega t + \theta) \longrightarrow \boxed{\begin{array}{c} G(s) \\ \text{Left half-plane} \\ \text{poles} \end{array}} \longrightarrow E_m |G(j\omega)| \sin[\omega t + \theta + \theta_G(\omega)]$$

Fig. 2-11   The sinusoidal steady-state response

The excitation has a peak value of E_m and a phase angle of θ. In the complex-frequency domain, it is characterized by poles at $s = \pm j\omega$. The response is given by

$$R(s) = G(s)E(s) = G(s)\left[\frac{E_m(s\sin\theta + \omega\cos\theta)}{s^2 + \omega^2}\right] = G(s)\left[\frac{E_m(s\sin\theta + \omega\cos\theta)}{(s - j\omega)(s + j\omega)}\right].$$

Since the forced response is of interest, only the terms due to poles at $\pm j\omega$ need to be considered:

$$R_f(s) = \frac{K}{s - j\omega} + \text{conjugate},$$

where

$$K = (s - j\omega)R(s)\bigg|_{s=j\omega} = G(j\omega)\left[\frac{E_m(j\omega\sin\theta + \omega\cos\theta)}{2j\omega}\right].$$

Thus,

$$R_f(s) = E_m\frac{G(j\omega)}{2j}(\cos\theta + j\sin\theta)\frac{1}{s - j\omega} + \text{conjugate}$$

$$= \left\{E_m\frac{|G(j\omega)|}{2}e^{j[\theta + \theta_G(\omega) - \pi/2]}\right\}\frac{1}{s - j\omega} + \text{conjugate},$$

where $|G(j\omega)|$ is the magnitude, and $\theta_G(\omega)$ is the angle of $G(j\omega)$. The inverse transform of $R_f(s)$ is $r_f(t)$:

$$r_f(t) = \left\{E_m\frac{|G(j\omega)|}{2}e^{j[\theta + \theta_G(\omega) - \pi/2]}\right\}e^{j\omega t} + \text{conjugate}.$$

The sum of a function and its complex conjugate is twice the real part of the function. Therefore, the forced response becomes

$$r_f(t) = 2\,\text{Re}\left\{E_m\frac{|G(j\omega)|}{2}e^{j[\omega t + \theta + \theta_G(\omega) - \pi/2]}\right\}$$

$$= E_m|G(j\omega)|\cos\left[\omega t + \theta + \theta_G(\omega) - \frac{\pi}{2}\right]$$

$$= E_m|G(j\omega)|\sin[\omega t + \theta + \theta_G(\omega)]. \qquad (2\text{-}24)$$

This result is extremely important, and its significance should be clearly understood. It shows how the network operates on the input sine wave to produce the sinusoidally forced portion of the output. The network characterized by the system function $G(s)$ does two things to the input signal: *it modifies its amplitude by multiplying it with* $|G(j\omega)|$, *and it shifts its phase by* $\theta_G(\omega)$. In order to obtain $|G(j\omega)|$ and $\theta_G(\omega)$, $G(s)$ is evaluated at $s = j\omega$, i.e., *at the value of the upper half-plane excitation pole.*

The results presented in Eq. (2-24) can be generalized (see Problem 2-19) to include, in Fig. 2-11, any input of the form

$$E_m e^{\alpha t} \sin(\omega t + \theta).$$

Note that the poles of excitation are at $\alpha \pm j\omega$. The forced response is given by

$$r_f(t) = E_m e^{\alpha t} |G(\alpha + j\omega)| \sin[\omega t + \theta + \theta_G(\alpha, \omega)], \qquad (2\text{-}25)$$

where $|G(\alpha + j\omega)|$ and $\theta_G(\alpha, \omega)$ are found from evaluating $G(s)$ *at the upper half-plane pole of excitation*, that is,

$$G(s)\Big|_{s=\alpha+j\omega} = G(\alpha + j\omega) = |G(\alpha + j\omega)| e^{j\theta_G(\alpha, \omega)}.$$

By selecting appropriate values of α, ω, and θ, the forced response may be calculated for a variety of inputs. For example, if $\omega = 0$ and $\theta = (\pi/2)$, the input becomes $E_m e^{\alpha t}$. Then the forced response is calculated from:

$$r_f(t) = E_m e^{\alpha t} |G(\alpha)| e^{j\theta_G(\alpha)} = E_m e^{\alpha t} G(\alpha).$$

Since the forced response is identical with the response after all natural response (transient) terms have vanished, *the forced response is also the steady-state response*. The subscript ss is used to denote steady state.

EXAMPLE 2-4

Find the steady-state response for $v_2(t)$ shown in Fig. 2-12.

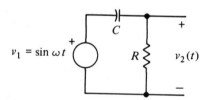

Fig. 2-12

SOLUTION

The system function is

$$G(s) = \frac{V_2(s)}{V_1(s)} = \frac{s}{s + \dfrac{1}{RC}}.$$

Since $G(s)$ has a left half-plane pole at $-(1/RC)$, the natural part vanishes after about three time constants, $3RC$. The response then becomes identical with the forced response, and is given by

$$v_f(t) = |G(j\omega)| \sin[\omega t + \theta_G(\omega)].$$

To find $|G(j\omega)|$ and $\theta_G(\omega)$, $G(s)$ is evaluated at $s = j\omega$:

$$G(j\omega) = \cfrac{1}{1 - j\cfrac{1}{\omega RC}} = \cfrac{1}{\sqrt{1 + \cfrac{1}{\omega^2 R^2 C^2}}} e^{-j[\tan^{-1}(-1/\omega RC)]},$$

$$|G(j\omega)| = \cfrac{1}{\sqrt{1 + \cfrac{1}{\omega^2 R^2 C^2}}}, \qquad \theta_G(\omega) = \tan^{-1}\left(\frac{1}{\omega RC}\right). \qquad (2\text{-}26)$$

Substituting these results in the expression for $v_f(t)$ the steady-state response is obtained:

$$v_{ss}(t) = v_f(t) = \cfrac{1}{\sqrt{1 + \cfrac{1}{\omega^2 R^2 C^2}}} \sin\left[\omega t + \tan^{-1}\left(\frac{1}{\omega RC}\right)\right]. \qquad Ans.$$

EXAMPLE 2-5

Find the inverse transform of

$$R(s) = \frac{1}{s+1}\frac{1}{s^2+1}.$$

SOLUTION

Let $R(s) = G(s)E(s)$. Assume, first,

$$G(s) = \frac{1}{s+1} \qquad \text{and} \qquad E(s) = \frac{1}{s^2+1}.$$

The resulting forced response is, by Eq. (2-24),

$$r_{f1}(t) = |G(j1)|\sin(t + \theta_G)$$

$$= \left|\frac{1}{j+1}\right|\sin(t + \theta_G)$$

$$= \frac{1}{\sqrt{2}}\sin\left(t - \frac{\pi}{4}\right).$$

Assume, next,

$$G(s) = \frac{1}{s^2+1} \qquad \text{and} \qquad E(s) = \frac{1}{s+1}.$$

The resulting forced response is, by Eq. (2-25),

$$r_{f2}(t) = e^{-t}G(-1) = \frac{1}{2}e^{-t}.$$

These two forced responses represent the inverse transforms due to each pole of $R(s)$. Hence $r(t)$ is the sum of the two forced responses, that is,

$$r(t) = \frac{1}{\sqrt{2}} \sin\left(t - \frac{\pi}{4}\right) + \frac{1}{2} e^{-t}. \qquad Ans.$$

The method of inverse transformation presented in this example is applicable also to functions with more than two factors. The only requirement is that the inverse transform of each factor be readily recognizable or obtainable.

Magnitude and Phase of $G(j\omega)$

The sinusoidal steady-state response depends upon the properties of the system function evaluated at the excitation frequency. The magnitude of $G(j\omega)$ determines how the amplitude of the input sinusoid is affected as it goes through the network, while the angle of $G(j\omega)$ determines how the phase is changed. Figure 2-11 shows the input–output relationship. As the frequency of the input sinusoid is varied, $|G(j\omega)|$ and $\theta_G(\omega)$ change, and hence the steady-state output changes. The change brought about by the network can be examined at a glance if $|G(j\omega)|$ and $\theta_G(\omega)$ are plotted against ω. These curves, one for the magnitude and one for the phase, are sufficient to characterize the steady-state sinusoidal response or more briefly, the frequency response of the system under consideration.

The magnitude and phase functions may also be interpreted geometrically from the pole–zero diagram in the s-plane. Let $G(s)$ be a system with m zeros and n poles. Then,

$$G(s) = H \frac{(s - z_1)(s - z_2) \cdots (s - z_m)}{(s - p_1)(s - p_2) \cdots (s - p_n)}.$$

For $s = j\omega$, $G(s)$ becomes:

$$G(j\omega) = H \frac{(j\omega - z_1)(j\omega - z_2) \cdots (j\omega - z_m)}{(j\omega - p_1)(j\omega - p_2) \cdots (j\omega - p_n)}. \qquad (2\text{-}27)$$

Consider the factor $(j\omega - s_i)$ where s_i may be a zero, that is, $s_i = z_i$, or a pole, that is, $s_i = p_i$. Since s_i is, in general, complex, it can be expressed in terms of its real part α_i and imaginary part β_i; that is,

$$s_i = \alpha_i + j\beta_i.$$

The factor $(j\omega - s_i)$ then becomes

$$-\alpha_i + j(\omega - \beta_i) = M_i e^{j\theta_i},$$

where

$$M_i = \sqrt{\alpha_i^2 + (\omega - \beta_i)^2} \qquad \text{and} \qquad \theta_i = \tan^{-1} \frac{\omega - \beta_i}{-\alpha_i}.$$

Hence, Eq. (2-27) may be expressed as

$$G(j\omega) = H \frac{M_{z1}e^{j\theta_{z1}}M_{z2}\,e^{j\theta_{z2}}\cdots M_{zm}\,e^{j\theta_{zm}}}{M_{p1}e^{j\theta_{p1}}M_{p2}\,e^{j\theta_{p2}}\cdots M_{pn}\,e^{j\theta_{pn}}}$$

$$= H \frac{M_{z1}M_{z2}\cdots M_{zm}}{M_{p1}M_{p2}\cdots M_{pn}}\,e^{j(\theta_{z1}+\theta_{z2}+\,\cdots\,+\theta_{zm}-\theta_{p1}-\theta_{p2}-\,\cdots\,-\theta_{pn})} \qquad (2\text{-}28)$$

$$= M(\omega)e^{j\theta(\omega)},$$

where M is the magnitude of $G(j\omega)$ and $\theta(\omega)$ is its phase.

In the s-plane, *the factor $(j\omega - z_1)$ represents the vector from z_1 to $j\omega$;* the factor $(j\omega - p_1)$ represents the vector from p_1 to $j\omega$, and so on. These vectors and the corresponding magnitudes and angles are shown in Fig. 2-13.

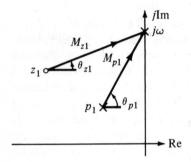

Fig. 2-13   Geometric interpretation

As Eq. (2-28) and Fig. 2-13 indicate, *the magnitude function $M(\omega)$ is H times the product of the zero-to-$j\omega$ distances divided by the product of the pole-to-$j\omega$ distances. The phase function $\theta(\omega)$ is the sum of the zero-to-$j\omega$ angles minus the sum of the pole-to-$j\omega$ angles.* (If H is negative, its magnitude may be associated with M, and π radians added to θ.) Since both the magnitude and angle changes arising from the individual terms can be seen readily by looking at the s-plane diagram, the $M(\omega)$- and $\theta(\omega)$-vs.-ω curves may be sketched by inspection. This is particularly true if a pole or a zero of the network is close to the imaginary axis.

EXAMPLE 2-6

Sketch the frequency response of the network shown in Fig. 2-14.

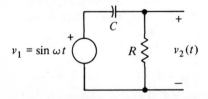

Fig. 2-14

SOLUTION

$G(s)$, $|G(j\omega)|$, and $\theta_G(\omega)$ are found from

$$G(s) = \frac{s}{s + \dfrac{1}{RC}}\bigg|_{s=j\omega} = \frac{j\omega}{j\omega + \dfrac{1}{RC}},$$

$$|G(j\omega)| = \frac{\omega}{\sqrt{\omega^2 + \dfrac{1}{R^2C^2}}} = \frac{M_z}{M_p},$$

$$\theta_G(\omega) = \frac{\pi}{2} - \tan^{-1}(\omega RC) = \frac{\pi}{2} - \theta_p.$$

Either the equations for $|G(j\omega)|$ and $\theta_G(\omega)$, or the geometric interpretation afforded by the pole–zero diagram, is used to sketch the frequency response shown in Fig. 2-15.

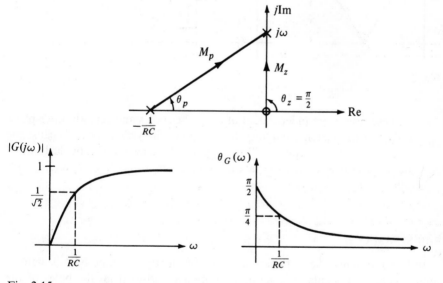

Fig. 2-15

Note that in the steady state, the system discriminates against low frequencies by attenuating them. For frequencies much higher than $(1/RC)$, the input and output are practically the same.

EXAMPLE 2-7

Sketch the frequency response in the vicinity of the system zero (at $j\omega_o$) shown in Fig. 2-16.

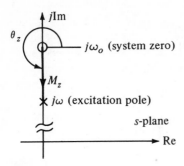

Fig. 2-16

SOLUTION

For ω near ω_0, the magnitude will change as M_z changes. The phase will change as θ_z changes. The frequency response is shown in Fig. 2-17.

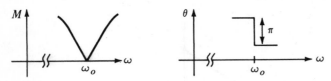

Fig. 2-17

These results are independent of any other poles and zeros in the s-plane as long as the sketch is confined to the immediate vicinity of ω_0 because the other distances and angles will stay practically constant (particularly when the other critical frequencies are quite far from $j\omega_0$), while M_z and θ_z vary rapidly near ω_0.

2-10 SUMMARY

In linear systems, the response can be conveniently decomposed into natural and forced components. The natural component arises from the poles of the system and the forced component from the excitation. It is possible to make the natural component zero by cancelling all system poles with zeros of excitation. This principle is used in the design of attenuators. It is also possible to make the forced component zero by cancelling all the poles of the excitation with system-function zeros. This principle is used in frequency-rejection networks.

When an excitation pole is made to approach a pole of the system, large-amplitude waveforms are generated in both the natural and the forced part of the response. In particular, when the excitation pole is on the imaginary axis (sinusoidal excitation) and the network pole slightly to the left of it, the response reduces, in the steady state, to a large-amplitude sine wave. When the frequency of the input sine wave is either reduced or increased (the upper half-plane excitation pole moved down or up on the imaginary axis), the amplitude of the steady-state sine wave decreases to a marked degree, thus indicating resonant behavior.

When the input is a sine wave, $V_m \sin(\omega t + \phi)$, so is the forced part of the output, $V_{mo} \sin(\omega t + \phi_0)$. V_{mo} and ϕ_0 are related to V_m and ϕ through the system function $G(s)$, evaluated at $s = j\omega$, namely, $V_{mo} = V_m|G(j\omega)|$ and $\phi_0 = \phi +$ angle of $G(j\omega)$. Thus, a frequency-dependent change in the amplitude and phase of the sine wave results.

The frequency response graphically presents the magnitude and phase of $G(j\omega)$ as a function of ω. These curves tell at a glance how much modification will result in the amplitude and phase of a sine wave as it goes through the network.

PROBLEMS

2-1 Construct a network and find an excitation $e(t)$ such that the resulting response is given by

$$r(t) = \frac{1}{2}e^{-t} - \frac{1}{\sqrt{2}}\cos\left(t + \frac{\pi}{4}\right).$$

Give a solution other than $e(t) = r(t)$.

2-2 Given the response

$$R(s) = \frac{s + \alpha}{s(s + 1)}.$$

Discuss how the position of the zero affects the response waveform in the time domain.

2-3 Obtain the response, $i(t)$, in Fig, 2-18. What are its natural and forced components?

2-4 Which system function is zero in Fig. 2-19?

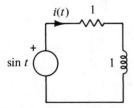

Fig. 2-18

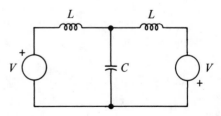

Fig. 2-19

2-5 For the circuit shown in Fig. 2-20, there is a value of K that will make V_2 zero. Find K.

2-6 Derive Eq. (2-12a).

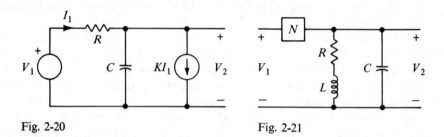

Fig. 2-20   Fig. 2-21

2-7 In Fig. 2-21, the network N is to be designed in such a way that

$$\frac{V_2(s)}{V_1(s)} = \frac{1}{2}.$$

(a) Find N.

(b) Repeat the problem if

$$\frac{V_2(s)}{V_1(s)} = \frac{1}{10}.$$

2-8 Obtain $v_2(t)$ in Fig. 2-22. Discuss the result.

2-9 Construct a network which, when excited by $\sin t$, has a zero forced response. (The natural response is not zero.) Show where the excitation is applied and where the response is observed.

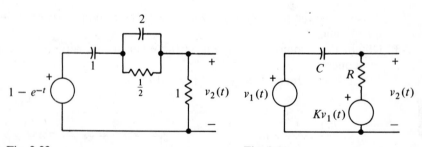

Fig. 2-22   Fig. 2-23

2-10 In Fig. 2-23 the input is

$$v_1(t) = e^{-t/T_1}.$$

Find R, C, and K such that the output $v_2(t)$ is an exponential with a time constant smaller than T_1.

2-11 Construct a network which when excited by $\sin t$ has a zero natural response. (The forced response is not zero.) Show where the excitation is applied and where the response is observed.

2-12 In Fig. 2-24, the response (not shown) is a linear combination of $v_1(t)$ and $i_1(t)$. What nontrivial linear combination gives zero forced response? Zero natural response?

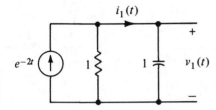

Fig. 2-24

2-13 In Fig. 2-25, for what value of α is the forced response zero?

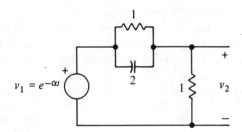

Fig. 2-25

2-14 In Fig. 2-26, $i(t)$ has the waveform shown.
   (a) Sketch the location of the poles and zeros of the excitation.

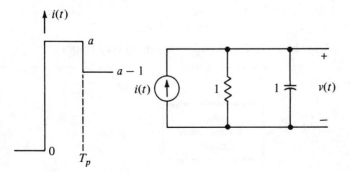

Fig. 2-26

(b) For what value of T_p is the natural component of $v(t)$ zero? Consider $t > T_p$.

(c) Obtain the expression for $v(t)$ and sketch $v(t)$ vs. t for $T_p < \ln(a)$, $T_p = \ln(a)$, and $T_p > \ln(a)$. Explain the results.

2-15   In Fig. 2-27, obtain $v_n(t)$, $v_f(t)$, and $v(t)$. Sketch these waveforms. Explain the behavior of $v_n(t)$, $v_f(t)$, and $v(t)$ as $\omega \to 0$.

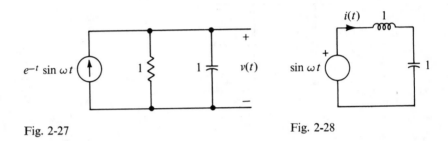

Fig. 2-27   Fig. 2-28

2-16   In Fig. 2-28, determine the current $i(t)$ for $\omega \neq 1$ and $\omega = 1$.

2-17   Obtain the forced response, if $G(s)$ in Fig. 2-11 is allowed to have imaginary-axis and right half-plane poles. Can the forced response be considered the same as the steady-state response when $G(s)$ is thus allowed poles anywhere in the s-plane?

2-18   For the system shown in Fig. 2-29, obtain the steady-state response.

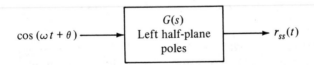

Fig. 2-29

2-19   For the system shown in Fig. 2-30, obtain the forced response.

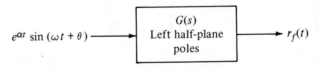

Fig. 2-30

2-20 Apply the result of Problem 2-19 to solve for the forced component of the response for the network shown in Fig. 2-31 if $i(t)$ is
(a) 1,
(b) e^{-t},
(c) $\sin 2t$,
(d) $e^{-t} \cos t$.

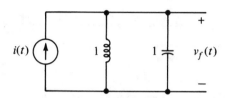

$i(t)$ 1   1 $v_f(t)$

Fig. 2-31

2-21 The differential equation

$$a_n \frac{d^n y}{dt^n} + a_{n-1} \frac{d^{n-1} y}{dt^{n-1}} + \cdots + a_0 y = B + A \sin \omega t$$

has constant coefficients. Initial conditions are zero.
(a) Obtain the system function. Assume that $B + A \sin \omega t$ represents the excitation.
(b) Obtain the forced solution.

2-22 $R(s) = R_1(s) R_2(s)$, where

$$R_1(s) = \frac{s}{s^2 + 1}, \qquad R_2(s) = \frac{1}{s + 1}.$$

(a) Obtain the forced response $r_f(t)$ if $R_2(s)$ is the system function.
(b) Obtain the forced response $r_f(t)$ if $R_1(s)$ is the system function.
(c) Obtain the response $r(t)$.

2-23 Obtain the steady-state response of the system shown in Fig. 2-32. The input is periodic.

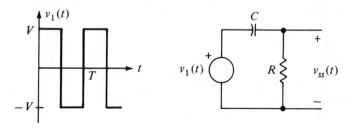

Fig. 2-32

2-24 Obtain the forced response to the periodic input shown in Fig. 2-33.

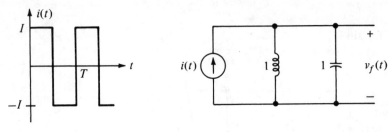

Fig. 2-33

2-25 One of the poles of a system function is located very close to, and slightly to the left of, the imaginary axis. Sketch the frequency response for frequencies in the vicinity of the pole.

2-26 The input is a sine wave. The system poles are equally spaced, as shown in Fig. 2-34. There are a large number of system poles. Sketch the frequency response of this system.

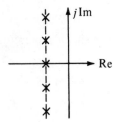

Fig. 2-34

2-27 $G(s) = \dfrac{s^2 + 1}{s^2 + 0.01s + 1}$.

Sketch the frequency response of the system represented by $G(s)$.

3

Properties of
Input Impedance

The input impedance is a unique property of two-terminal networks. It is the ratio of input voltage, $V(s)$, to input current, $I(s)$. For a two-terminal network, it is impossible to specify independently both the input voltage and the input current. When one is independent, the other must be dependent. By means of dependent sources, the input impedance of a passive network can be increased, decreased, or altered. Several methods are presented in this chapter for calculating the input impedance. In particular, the handling of dependent sources is illustrated by solving several example problems. Examples are given also for determining the low-frequency and high-frequency asymptotic behavior of input impedances. The properties of input impedances of LC and RC networks are formulated and discussed. Special emphasis is placed on the pole-zero structures and magnitude and phase characteristics associated with these impedances.

3-1 DEFINITION OF INPUT IMPEDANCE

The input impedance describes the two-terminal properties of linear, two-terminal networks composed of constant R, L, C elements and linear dependent sources. In Fig. 3-1, let the block represent the network. Let the current source, $I(s)$, represent the excitation, and let the input voltage, $V(s)$, be the resulting response.

The input impedance, or simply the impedance, for terminals 1-1' is defined as

$$Z(s) = \frac{V(s)}{I(s)}.$$ (3-1)

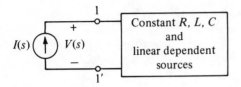

Fig. 3-1   Input impedance

The impedance $Z(s)$ is a rational function if the network is finite and lumped. It completely specifies the complex-frequency-domain terminal behavior of the network.

If a voltage source, $V(s)$, is applied to the network, and the terminal current, $I(s)$, is taken as the response, another system function, $Y(s)$, results:

$$Y(s) = \frac{I(s)}{V(s)}. \qquad\qquad (3\text{-}2)$$

$Y(s)$ is called the input admittance of the network. It is the reciprocal of $Z(s)$, that is,

$$Y(s) = \frac{1}{Z(s)}.$$

Consequently, the zeros of $Y(s)$ are the poles of $Z(s)$, and the poles of $Y(s)$ are the zeros of $Z(s)$. Either function, $Z(s)$ or $Y(s)$, can be used to describe the input characteristics of the network.

The impedance is always associated with two specified terminals of a network. It gives at these terminals the constraint relationship, as imposed by the network, between $V(s)$ and $I(s)$ variables. In a system of interconnected networks, it is necessary to specify which network's impedance is being considered. Thus, the descriptive phrase " the impedance looking into " certain terminals of a network is quite often used. If the terminals at which the impedance is calculated are labeled as the output terminals of a network, then the resulting impedance is called the output impedance of the network.

3-2   CALCULATION OF IMPEDANCE

The impedance looking into a pair of terminals can be calculated by several methods. In some networks, one method is as good as another. In other networks, one particular method of evaluation may require much less effort than another. In still others, it may be impossible to use the simplest method of calculation, namely, the use of series- and parallel-combination rules for impedances. The various methods for calculating impedance are best illustrated by means of an example.

EXAMPLE 3-1

In Fig. 3-2, an amplifier circuit is given. The signal, V_i, is applied at terminals 1-1', and the output, V_o, taken at 2-2'. Calculate the output impedance, Z_0.

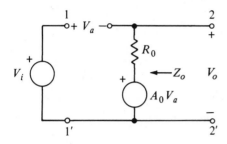

Fig. 3-2

SOLUTION

Note that the network as given contains an independent source, V_i. The output impedance gives the relationship between voltage and current at terminals 2-2' when $V_i = 0$. (When desired, the effect of V_i can be determined separately in the form of a Thévenin-equivalent voltage.)

Since the network contains a dependent source, the rules for series and parallel connection of impedances cannot be used in this case. (See Example 3-2 in Section 3-3 for the application of this method.) Instead, three other methods of calculating Z_0 are given.

Method 1 Remove the independent source, V_i. As shown in Fig. 3-3a, apply at terminals 2-2' a voltage (or current) source, V, and calculate the resulting current, I (or voltage). The output impedance is V/I.

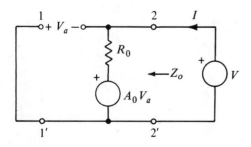

Fig. 3-3a

By inspection of Fig. 3-3a, write

$$I = \frac{V - A_0 V_a}{R_0}.$$

Since V_a is a dependent source, it must be expressed as a function of V. Again, by inspection of the network, obtain

$$V_a = -V.$$

Hence,

$$I = \frac{V + A_0 V}{R_0} = V \frac{(1 + A_0)}{R_0}.$$

From this equation, obtain the output impedance:

$$Z_0 = \frac{V}{I} = \frac{R_0}{1 + A_0}. \qquad Ans.$$

This method is straightforward and is usually the simplest method to use when dependent sources are involved.

Method 2 Refer to Fig. 3-3b. With the switch in position 1, calculate the open-circuit voltage, V_{oc}. Then, using position 2 on the switch, short-circuit terminals 2-2' and calculate the short-circuit current, I_{sc}. The output impedance is V_{oc}/I_{sc}. (Here, the internal independent source, V_i, is used to generate V_{oc} and I_{sc}. Alternatively, V_i can be set to zero, and some other independent voltage or current source can be introduced internally to excite the network and thus obtain V_{oc} and I_{sc}. For example, an independent voltage source can be introduced in series with the resistor R_0 or an independent current source can be connected across it. The resulting V_{oc} and I_{sc} depend upon the position and nature of the excitation but the ratio is always the desired Z_0.)

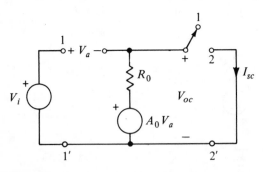

Fig. 3-3b

By inspection of Fig. 3-3b (with switch at 1), obtain

$V_{oc} = A_0 V_a$.

Since V_a is a dependent source, express it in terms of the independent source, V_i. Again, by inspection of the network,

$V_i = V_a + A_0 V_a$.

Hence,

$$V_a = \frac{V_i}{1 + A_0}.$$

Thus, the open-circuit voltage becomes

$$V_{oc} = \frac{A_0}{1 + A_0} V_i.$$

Connect switch to position 2. By inspection of the network, obtain

$$I_{sc} = \frac{A_0 V_a}{R_0}.$$

Obtain the dependency relation between V_a and V_i by noting that

$V_a = V_i$.

Thus, the short-circuit current becomes

$$I_{sc} = \frac{A_0 V_i}{R_0}.$$

The output impedance is

$$Z_0 = \frac{V_{oc}}{I_{sc}} = \left(\frac{A_0}{1 + A_0} V_i\right)\left(\frac{R_0}{A_0 V_i}\right) = \frac{R_0}{1 + A_0}. \qquad Ans.$$

This method requires two calculations. One with terminals open-circuited, the other with terminals short-circuited. Thus, it generally requires more calculation than method 1. However, in some networks, the open-circuiting and short-circuiting of the terminals may reduce the network to a simpler structure which can then be analyzed more readily.

Method 3   Connect a load, R_L, across the output terminals as shown in Fig. 3-3c. Calculate the voltage across R_L. Rearrange the expression for the output voltage so that R_L appears by itself in the denominator as shown below:

$$V_o = V_{oc} \frac{R_L}{R_L + Z_0}.$$

Consequently, Z_0 is determined by inspection of the denominator.

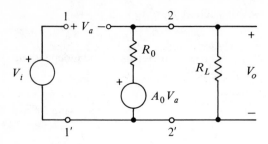

Fig. 3-3c

In addition to Z_0, this method also gives V_{oc} which represents the open-circuit voltage. Thus, while calculating Z_0, the Thévenin-equivalent voltage is also obtained.

By inspection of Fig. 3-3c, obtain V_o.

$$V_o = A_0 V_a \frac{R_L}{R_L + R_0}.$$

Next, express V_a in terms of V_i and V_o:

$$V_a = V_i - V_o.$$

Hence,

$$V_o = A_0(V_i - V_o)\frac{R_L}{R_L + R_0}.$$

Solve for V_o:

$$V_o = \frac{V_i A_0 R_L}{R_L(1 + A_0) + R_0}.$$

In order to isolate R_L in the denominator, divide both numerator and denominator by $(1 + A_0)$:

$$V_o = \left(\frac{V_i A_0}{1 + A_0}\right)\left(\frac{R_L}{R_L + \dfrac{R_0}{1 + A_0}}\right).$$

The output impedance is

$$Z_0 = \frac{R_0}{1 + A_0}. \qquad Ans.$$

Although not requested in this example, the Thévenin-equivalent voltage for terminals 2-2′ is

$$V_{TH} = V_{oc} = V_i \frac{A_0}{1 + A_0}.$$

In this method for calculating the output impedance, the original network is made more complicated by the connection of R_L. Therefore, there is more work involved. However, the calculation also gives the Thévenin-equivalent voltage.

As in Method 2, here too must the network be internally excited to produce the output voltage. Either V_i itself, or some other independent voltage or current source, can be used for this purpose. As long as the excitation is properly introduced, the location of the excitation is unimportant in the calculation of Z_0. However, it should be recognized that the Thévenin-equivalent voltage depends upon the position of the excitation.

3-3 MAGNITUDE AND FREQUENCY SCALING OF IMPEDANCE

The impedance of a resistor is R, of a capacitor $(1/sC)$, of an inductor sL. If the impedance of each resistor, capacitor, and inductor in a network is multiplied by the same constant k (which is equivalent to multiplying R by k, L by k, and dividing C by k), then the input impedance of the network changes from $Z(s)$ to $kZ(s)$. Thus, the magnitude of the impedance can be scaled up $(k > 1)$, or down $(0 < k < 1)$.

If the impedance of each inductor and the admittance of each capacitor in a network is multiplied by the same constant q (which is equivalent to multiplying both L and C by q), then the input impedance changes from $Z(s)$ to $Z(qs)$. Thus, the complex-frequency variable can be scaled to any desired value.

Magnitude and frequency scaling are used to simplify the element values of a network so that analysis becomes less tedious. Scaling is also used to make frequency-response plotting more universally applicable.

EXAMPLE 3-2

Find the impedance, $Z(s)$, of the network shown in Fig. 3-4.

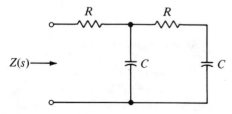

Fig. 3-4

SOLUTION

Note that the two resistors have the same value. To normalize the resistor values to unity, divide all impedances in the network by R. As shown in Fig. 3-5a, this magnitude scaling of impedances changes the values of the two capacitors from C to RC. The input impedance of the resulting network is $Z(s)/R$.

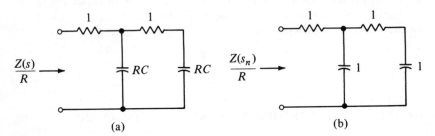

Fig. 3-5

Note that, after the resistor-value normalization the two capacitor values are still equal. To simplify further, normalize the capacitor values to unity by letting $sRC = s_n$. After this frequency scaling, *all* network element values become unity, as shown in Fig. 3-5b. The input impedance of the twice-normalized network is $[Z(s_n)/R]$. Calculate this impedance by using the rules for series and parallel connection of impedances:

$$\frac{Z(s_n)}{R} = 1 + \frac{\dfrac{1}{s_n}\left(1 + \dfrac{1}{s_n}\right)}{\dfrac{1}{s_n} + \left(1 + \dfrac{1}{s_n}\right)} = \frac{s_n^2 + 3s_n + 1}{s_n(s_n + 2)}.$$

To remove the normalizations, substitute sRC for s_n and multiply both sides of the equation by R:

$$Z(s) = R\frac{(sRC)^2 + 3(sRC) + 1}{(sRC)(sRC + 2)}. \qquad Ans.$$

3-4 COMMON-MODE AND DIFFERENCE-MODE IMPEDANCE

Frequently networks are designed so that the output is the difference of two input signals. The input stage of such a network is shown in Fig. 3-6a. To characterize the input properties of this network, the input signals are decomposed into the common- and difference-mode components, as shown in Fig. 3-6b.

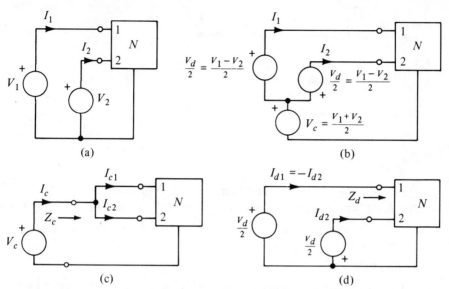

Fig. 3-6   Common- and difference-mode impedances

When $V_1 = V_2$, the network is driven in the common mode, as shown in Fig. 3-6c. The impedance seen by the common source, V_c, is the common-mode impedance:

$$Z_c = \frac{V_c}{I_c}. \qquad (3\text{-}3)$$

The larger the common-mode impedance, the smaller are the input currents due to signals that are common to both inputs.

When $V_1 = -V_2$, the network is driven in the difference mode as shown in Fig. 3-6d. Generally, the input circuitry possesses symmetry so that

$$I_{d1} = -I_{d2} = I_d.$$

Then, the impedance seen by the source V_d is the difference-mode impedance:

$$Z_d = \frac{V_d}{I_d}. \qquad (3\text{-}4)$$

The smaller the difference-mode impedance, the larger are the input currents due to the difference of the two input signals.

EXAMPLE 3-3

The network shown in Fig. 3-7 represents the input equivalent circuit of a differential amplifier. The two inputs are V_1 and V_2.

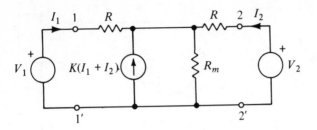

Fig. 3-7

(a) Obtain the common- and difference-mode input impedances.
(b) What is the impedance seen by the source V_1 if $V_2 = V_1$? $V_2 = -V_1$?
$V_2 = 0$?

SOLUTION

Because the dead network possesses symmetry, Fig. 3-7 can be redrawn as in Fig. 3-8a.
(a) To obtain the common-mode input impedance, let $V_1 = V_2$. Then

$$I_1 = I_2 \quad \text{and} \quad I_m = 0.$$

Consequently, the wire in the middle can be cut without altering the currents and voltages in the network. The simplified network of Fig.

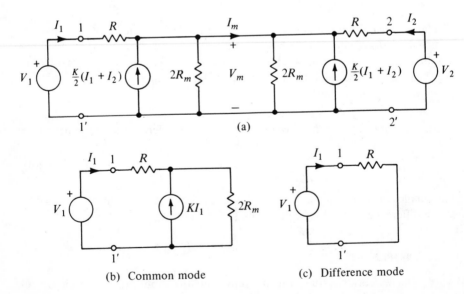

(a)

(b) Common mode

(c) Difference mode

Fig. 3-8

3-8b results. Write Kirchhoff's voltage law for the outer mesh, and solve for I_1:

$$V_1 = I_1 R + I_1(1 + K)2R_m,$$

$$I_1 = \frac{V_1}{R + 2(1 + K)R_m}.$$

The common-mode input impedance is

$$Z_c = \frac{V_1}{2I_1} = \frac{1}{2}[R + 2(1 + K)R_m]. \qquad Ans.$$

[In calculating Z_c, $2I_1$ rather than I_1 is used because the total current supplied by the common source must be considered. See Fig. 3-6c and Eq. (3-3).]

To obtain the difference-mode input impedance, let $V_2 = -V_1$. Then,

$$I_2 = -I_1 \qquad \text{and} \qquad V_m = 0.$$

Consequently, a short circuit can be placed across the middle of Fig. 3-8a without altering the voltage and current variables. Then use the simplified network of Fig. 3-8c to calculate I_1:

$$I_1 = \frac{V_1}{R}.$$

The difference-mode input impedance is

$$Z_d = \frac{2V_1}{I_1} = 2R. \qquad Ans.$$

[In calculating Z_d, $2V_1$ rather than V_1 is used because the voltage between terminals 1-2 must be considered. See Fig. 3-6d and Eq. (3-4).]

(b) When $V_2 = V_1$, the impedance seen by the source V_1 is, from Fig. 3-8b,

$$Z_1 = \frac{V_1}{I_1} = 2Z_c. \qquad Ans.$$

When $V_2 = -V_1$, the impedance seen by the source V_1 is, from Fig. 3-8c,

$$Z_2 = \frac{V_1}{I_1} = \frac{Z_d}{2}. \qquad Ans.$$

(In calculating Z_1 and Z_2, the source V_2 is considered as a dependent source.)

When $V_2 = 0$, the impedance seen by the source V_1 is, from Fig. 3-6a,

$$Z_3 = \frac{V_1}{I_1}\bigg|_{V_2=0}.$$

From Figs. 3-6b, 3-6c, and 3-6d, and using the principle of super-position, I_1 is obtained as

$$I_1 = \frac{I_c}{2} + I_{d\perp}$$

$$= \frac{1}{2}\frac{V_c}{Z_c} + \frac{V_d}{Z_d}$$

$$= \frac{1}{2}\frac{V_1}{2}\frac{1}{Z_c} + \frac{V_1}{Z_d} = V_1\left(\frac{1}{4Z_c} + \frac{1}{Z_d}\right).$$

Hence,

$$Z_3 = \frac{4Z_c Z_d}{4Z_c + Z_d}, \qquad Ans.$$

which is the parallel combination of $4Z_c$ with Z_d. In practice, $|Z_c| \gg |Z_d|$, so that $Z_3 \cong Z_d$.

In this example, five different input impedances are calculated for one network. The impedances are different because the two input voltage sources and the resulting input currents are handled differently in each case.

3-5 THE SIGNIFICANCE OF ZERO AND INFINITE IMPEDANCE

Except for infinite and distributed networks, the input impedance of linear networks is a rational function and can therefore be represented as

$$Z(s) = H\frac{(s - z_1)(s - z_2)\cdots(s - z_m)}{(s - p_1)(s - p_2)\cdots(s - p_n)}. \tag{3-5}$$

$Z(s)$ has zeros at $z_1, z_2, \ldots, z_m$, and poles at $p_1, p_2, \ldots, p_n$. This means, for example, that

$$Z(z_1) = 0 \qquad \text{and} \qquad Z(p_1) = \infty.$$

Since the impedance is a complex-frequency-domain concept, what is the physical meaning of zero or infinite impedance? The answer comes from the time-domain interpretation of the response when the network is excited by an appropriate current source. If the excitation has a simple pole at $s = z_1$, then the forced component of the response due to that excitation is zero, since the pole at z_1 is cancelled by the zero of $Z(s)$, as shown below:

$$V(s) = I(s)Z(s) = \left(\frac{1}{s - z_1}\right) \times H\frac{(s - z_1)(s - z_2)\cdots(s - z_m)}{(s - p_1)(s - p_2)\cdots(s - p_n)}.$$

The time-domain response, $v(t)$, will have no trace of the input current waveform characterized by $e^{z_1 t}$.

On the other hand, if the excitation has a simple pole at $s = p_1$, then the forced component of the response is infinite since excitation and system poles coincide. (Read also the discussion given in Section 2-8.) The resulting $V(s)$ will then have a double pole at $s = p_1$; that is,

$$V(s) = I(s)Z(s) = \left(\frac{1}{s - p_1}\right) \times H \frac{(s - z_1)(s - z_2) \cdots (s - z_m)}{(s - p_1)(s - p_2) \cdots (s - p_n)}.$$

Thus, to speak about the time-domain effect of zero impedance and infinite impedance is meaningless unless the current excitation is chosen such that in the response either pole–zero cancellation results or pole–pole coincidence occurs. In the former case, the forced component in the output (due to the pole in question) is zero. In the latter case, the forced component is infinite. $Z(s)$ is, of course, independent of the excitation unless it is evaluated for a particular value of s, in which case a particular excitation is implied. Consider, for example, the network shown in Fig. 3-9.

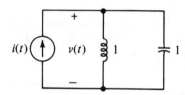

Fig. 3-9 $Z(0) = 0, \; Z(\pm j1) = \infty$

The input impedance is $Z(s) = s/(s^2 + 1)$. Here $Z(s)$ is zero for $s = 0$ and in-finite for $s = \pm j$. That $Z(0) = 0$ does not imply that a short circuit exists across the current source. Rather, it means that, if $i(t)$ is chosen as the unit step, so that $I(s) = (1/s)$ has a pole at $s = 0$, then *the forced response will be zero*. That is,

$$V(s) = I(s)Z(s) = \frac{1}{s}\frac{s}{s^2 + 1} = \frac{1}{s^2 + 1},$$

$v(t) = \sin t.$

Thus, the response, $v(t)$, is identical with the natural response, which arises from the network poles.

Similarly, $Z(j1) = \infty$ does not imply that an open circuit exists across the current source. Rather, it means that if $i(t)$ is chosen as $\sin t$, so that $I(s) = 1/(s^2 + 1)$ has poles at $s = \pm j$, then the forced response will be infinite. That is,

$$V(s) = I(s)Z(s) = \frac{1}{s^2 + 1}\frac{s}{s^2 + 1} = \frac{s}{(s^2 + 1)^2},$$

$v(t) = \dfrac{t}{2}\sin t.$

As the discussion given in Example 2-3 indicates, this response may be interpreted as possessing infinite forced and natural responses.

If $Z(s)$ is not a system function (which is the case when the excitation is a voltage source), then the zero impedance means that an appropriately chosen voltage excitation [one with a pole to match a zero of $Z(s)$] causes infinite forced response. Similarly, infinite impedance means zero forced response if the voltage excitation is appropriately chosen; i.e., it has a pole matching the pole of $Z(s)$. Thus,

$$Z(s) = 0 \quad \text{and} \quad Z(s) = \infty$$

have very definite meanings for the forced response.

If $Z(s)$ is identically equal to zero, regardless of the value of s, then $Z(s)$ is a short circuit. If $Z(s)$ is infinite, regardless of the value of s, then $Z(s)$ is an open circuit.

3-6 ASYMPTOTIC FORMS

Often it is desirable to know how the input impedance behaves for s very small or s very large. These asymptotic forms of $Z(s)$ can be obtained readily from the expression

$$Z(s) = H \frac{(s - z_1)(s - z_2) \cdots (s - z_m)}{(s - p_1)(s - p_2) \cdots (s - p_n)}, \tag{3-6}$$

The complex frequency s is considered small when its *magnitude* is much smaller than the magnitude of all nonzero critical frequencies of $Z(s)$, that is,

$$|s| \ll |z_1|, |z_2|, \ldots, |p_1|, |p_2|, \ldots$$

The complex frequency s is considered large when its *magnitude* is much larger than the magnitude of all critical frequencies, that is,

$$|s| \gg |z_1|, |z_2|, \ldots, |p_1|, |p_2|, \ldots$$

$Z(s)$ for s Small

If $Z(s)$ *has no poles or zeros at the origin*, then for s small, $Z(s)$, given by Eq. (3-6), approaches a real constant R_{eq}:

$$Z(s) \bigg|_{s \text{ small}} \cong H \frac{(-z_1)(-z_2) \cdots (-z_m)}{(-p_1)(-p_2) \cdots (-p_n)} = R_{eq}. \tag{3-7}$$

The network may then be represented for s small by a single resistor of value R_{eq}. For passive networks, R_{eq} is positive.

If $Z(s)$ *has poles of multiplicity q at the origin*, that is, $p_1 = p_2 = \cdots = p_q = 0$, then, for s small, Eq. (3-6) becomes

$$Z(s) \bigg|_{s \text{ small}} \cong H \frac{(-z_1)(-z_2) \cdots (-z_m)}{s^q(-p_{q+1}) \cdots (-p_n)} = \frac{1}{s^q A}, \tag{3-8}$$

where A is a real constant. In particular, in passive networks where there are no dependent sources, $q = 1$, and Eq. (3-8) reduces to

$$Z(s)\Big|_{s\,\text{small}} \cong \frac{1}{sC_{eq}}. \tag{3-9}$$

The network may then be represented for s small by a single capacitor of value C_{eq}.

If $Z(s)$ *has zeros of multiplicity* r *at the origin,* that is, $z_1 = z_2 = \cdots = z_r = 0$, then for s small, Eq. (3-6) becomes

$$Z(s)\Big|_{s\,\text{small}} \cong H \frac{s^r(-z_{r+1})\cdots(-z_m)}{(-p_1)(-p_2)\cdots(-p_n)} = s^r B, \tag{3-10}$$

where B is a real constant. In particular, for passive networks, $r = 1$ and Eq. (3-10) reduces to

$$Z(s) \cong sL_{eq}. \tag{3-11}$$

The network may then be represented for s small by a single inductor of value L_{eq}.

Thus, for s small, the behavior of a two-terminal passive network is indistinguishable from that of a single resistor, a single capacitor, or a single inductor. This equivalent behavior may be obtained directly from the network (by inspection) by using the following rules. Let Z_1 and Z_2 represent two impedances and let $|Z_2| \gg |Z_1|$. Then the series and parallel combinations of these impedances are:

$$Z_1 + Z_2 \cong Z_2, \qquad \frac{Z_1 Z_2}{Z_1 + Z_2} \cong Z_1.$$

Thus, for s small, a resistor in series with a capacitor is replaced by the capacitor, whereas a resistor in parallel with a capacitor is replaced by the resistor, and so on.

In the extreme case when $s = 0$, $Z(s)$ is either constant, infinite, or zero. Its value may be obtained directly from the network by open-circuiting all capacitors and short-circuiting all inductors.

EXAMPLE 3-4

For s small, obtain the expression for the input impedance of the network shown in Fig. 3-10.

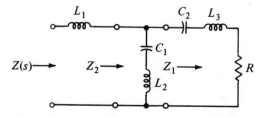

Fig. 3-10

SOLUTION

Starting on the right, combine impedances

$$Z_1 = \left(\frac{1}{sC_2} + sL_3 + R\right)\bigg|_{s \text{ small}} \cong \frac{1}{sC_2},$$

$$Z_2 = \left[\left(\frac{1}{sC_1} + sL_2\right) \text{ in } \| \text{ with } Z_1\right]_{s \text{ small}} \cong \frac{1}{sC_1} \text{ in } \| \text{ with } \frac{1}{sC_2} = \frac{1}{s(C_1 + C_2)},$$

$$Z(s) = (sL_1 + Z_2)\bigg|_{s \text{ small}} \cong Z_2 = \frac{1}{s(C_1 + C_2)}. \qquad Ans.$$

The small-s behavior of $Z(s)$ is capacitive.

Z(s) for s Large

For s large, Eq. (3-6) simplifies to

$$Z(s) \cong Hs^{m-n}. \tag{3-12}$$

Three special cases are of importance:

If $m = n$, $Z(s) \cong H$. The terminal behavior for s large is that of a resistor of value H.

If $m = n - 1$, $Z(s) \cong (H/s)$. The terminal behavior for s large is that of a capacitor of value $(1/H)$.

If $m = n + 1$, $Z(s) \cong Hs$. The terminal behavior for s large is that of an inductor of value H.

When dependent sources are absent in the network, $Z(s)$ for s large behaves like one of the three special cases mentioned above.

In the extreme case when $s = \infty$, $Z(s)$ is either constant, infinite, or zero. Its value may be obtained directly from the network by open-circuiting all inductors and short-circuiting all capacitors.

EXAMPLE 3-5

For s large, obtain the expression for the input impedance of the network given in Fig. 3-10.

SOLUTION

By inspection of the network,

$$Z_1 \cong sL_3,$$

$$Z_2 \cong sL_2 \text{ in } \| \text{ with } sL_3 = s\frac{L_2 L_3}{L_2 + L_3},$$

$$Z_3 \cong sL_1 + s\frac{L_2 L_3}{L_2 + L_3} = s\left(L_1 + \frac{L_2 L_3}{L_2 + L_3}\right). \qquad Ans.$$

The large-s behavior of $Z(s)$ is inductive.

3-7 INPUT IMPEDANCE OF *LC* NETWORKS

The expression for the input impedance and the pole–zero locations of six *LC* networks are given in Fig. 3-11. *Note that the poles and zeros are on the imaginary axis, are simple, and alternate.* Indeed, this pattern for the poles and

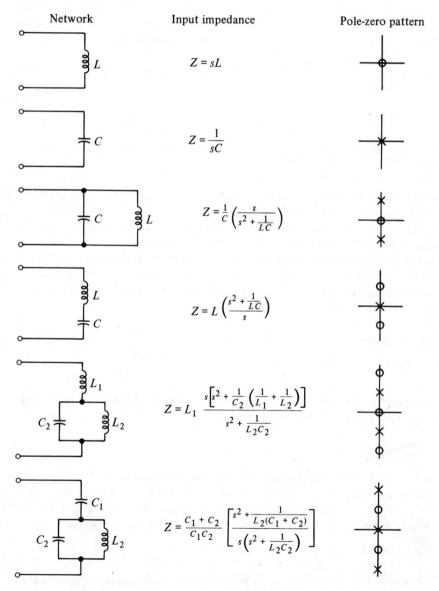

Network	Input impedance	Pole-zero pattern
L	$Z = sL$	
C	$Z = \dfrac{1}{sC}$	
C, L	$Z = \dfrac{1}{C}\left(\dfrac{s}{s^2 + \frac{1}{LC}}\right)$	
L, C	$Z = L\left(\dfrac{s^2 + \frac{1}{LC}}{s}\right)$	
L_1, C_2, L_2	$Z = L_1 \dfrac{s\left[s^2 + \frac{1}{C_2}\left(\frac{1}{L_1} + \frac{1}{L_2}\right)\right]}{s^2 + \frac{1}{L_2 C_2}}$	
C_1, C_2, L_2	$Z = \dfrac{C_1 + C_2}{C_1 C_2}\left[\dfrac{s^2 + \frac{1}{L_2(C_1 + C_2)}}{s\left(s^2 + \frac{1}{L_2 C_2}\right)}\right]$	

Fig. 3-11   The input impedance of *LC* networks

zeros is true for the impedance of any *LC* network.* At the origin, there is either a zero or a pole. Similarly, the highest critical frequency is either a zero or a pole [resulting in a zero or a pole of $Z(s)$ at infinity]. *LC* networks differ in their input impedance only according to the total number of the poles and zeros and their locations on the imaginary axis. In no case can two poles or two zeros be next to each other. *The pattern is always an alternating one: zero, pole, zero, pole, and the critical frequencies are simple.*

A pair of critical frequencies at $\pm j\omega_o$ together form the factor $(s^2 + \omega_o^2)$. Thus, depending upon whether a zero or a pole is at the origin, $Z(s)$, which fits the pole–zero patterns given in Fig. 3-11, is written either as

$$Z(s) = H\frac{(s^2 + \omega_{z1}^2)(s^2 + \omega_{z2}^2)\cdots}{s(s^2 + \omega_{p1}^2)(s^2 + \omega_{p2}^2)\cdots} \qquad (0 < \omega_{z1} < \omega_{p1} < \omega_{z2} < \omega_{p2} < \cdots),$$

$$(3\text{-}13a)$$

or

$$Z(s) = H\frac{s(s^2 + \omega_{z1}^2)(s^2 + \omega_{z2}^2)\cdots}{(s^2 + \omega_{p1}^2)(s^2 + \omega_{p2}^2)\cdots} \qquad (0 < \omega_{p1} < \omega_{z1} < \omega_{p2} < \omega_{z2} < \cdots),$$

$$(3\text{-}13b)$$

where H is a positive, real constant. If the highest critical frequencies are a pair of poles, then the degree of the denominator is one higher than the degree of the numerator. If the highest critical frequencies are a pair of zeros, then the degree of the denominator is one lower than the degree of the numerator. Thus, if the denominator is of degree n, then the numerator polynomial is either of degree $(n + 1)$ or $(n - 1)$. Note that $Z(s)$ is either an even function of s divided by an odd function of s, Eq. (3-13a), or an odd function divided by an even function, Eq. (3-13b). In either case, $Z(s)$ is odd.

The asymptotic forms of $Z(s)$, that is, the expressions for $Z(s)$ for s large or small, are either

$$Z(s) \cong sL \qquad \text{or} \qquad Z(s) \cong \frac{1}{sC}.$$

Consequently, near zero or near infinite frequency, the input impedance is the same as that of an inductor or a capacitor.

Magnitude and Phase of Z_{LC}

Since all the critical frequencies of the input impedance of *LC* networks occur on the imaginary axis, the interesting and the most exciting behavior of $Z(s)$ occurs for $s = j\omega$. In order for $Z(s)$ to exhibit this behavior, it must be driven by a sinusoidal excitation. If the excitation is a current source,

$$i(t) = I_m \sin \omega t,$$

---

* For the proof of these properties, see the detailed and excellent treatment given by David F. Tuttle, Jr., *Network Synthesis* (John Wiley & Sons, Inc., 1958), pp. 231–301.

then the forced component of the input voltage is given by

$$v_f(t) = |Z(j\omega)| I_m \sin[\omega t + \theta_Z(\omega)], \tag{3-14}$$

where $Z(j\omega) = |Z(j\omega)| e^{j\theta_Z(\omega)}$.

If the excitation is a voltage source, $v(t) = V_m \sin \omega t$, then the forced component of the input current is given by

$$i_f(t) = |Y(j\omega)| V_m \sin[\omega t + \theta_Y(\omega)], \tag{3-15a}$$

where $Y(j\omega) = |Y(j\omega)| e^{j\theta_Y(\omega)}$.

Since $|Y(j\omega)| = 1/|Z(j\omega)|$ and $\theta_Y(\omega) = -\theta_Z(\omega)$, then $i_f(t)$ may also be written as

$$i_f(t) = \frac{1}{|Z(j\omega)|} V_m \sin[\omega t - \theta_Z(\omega)]. \tag{3-15b}$$

Equations (3-14) and (3-15b) show that the forced response depends upon $|Z(j\omega)|$ and $\theta_Z(\omega)$. Therefore, it is important to study the $j\omega$-axis properties of $Z(s)$. When $s = j\omega$ is substituted in Eqs. (3-13a) and (3-13b), $Z(s)$ becomes

$$Z(j\omega) = jX(\omega),$$

where

$$X(\omega) = -H \frac{(\omega_{z1}^2 - \omega^2)(\omega_{z2}^2 - \omega^2)\cdots}{\omega(\omega_{p1}^2 - \omega^2)(\omega_{p2}^2 - \omega^2)\cdots} \quad \text{if } Z(s) \text{ has a pole at the origin,} \tag{3-16a}$$

and

$$X(\omega) = H \frac{\omega(\omega_{z1}^2 - \omega^2)(\omega_{z2}^2 - \omega^2)\cdots}{(\omega_{p1}^2 - \omega^2)(\omega_{p2}^2 - \omega^2)\cdots} \quad \text{if } Z(s) \text{ has a zero at the origin.} \tag{3-16b}$$

$X(\omega)$ is called the *reactance function*. Note that $X(\omega)$ changes sign every time the value of ω increases past the magnitude of a zero or a pole. Because of this sign change and the multiplication factor j, the phase of $Z(j\omega)$ alternates between

$$+\frac{\pi}{2} \quad \text{and} \quad -\frac{\pi}{2}.$$

In plotting the magnitude of $Z(j\omega)$, it is customary to use $X(\omega)$ rather than $|X(\omega)|$. Then, a plot of phase becomes unnecessary if it is understood that if $X(\omega) > 0$, the phase is $+(\pi/2)$, and if $X(\omega) < 0$, the phase is $-(\pi/2)$.

The $X(\omega)$-vs.-ω curve is called the *reactance curve*. Its general characteristics are displayed in Fig. 3-12. At low frequencies, Eq. (3-16) simplifies to a reactance that is either capacitive,

$$X(\omega) \cong -\frac{1}{\omega C_L},$$

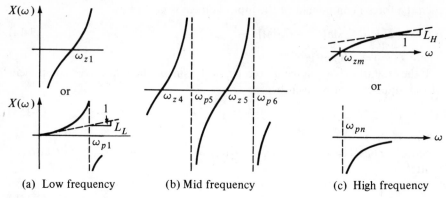

(a) Low frequency   (b) Mid frequency   (c) High frequency

Fig. 3-12   Reactance curve

or inductive,

$$X(\omega) \cong \omega L_L .$$

Figure 3-12a shows the resulting plots. At midfrequencies, the reactance curve goes from $-\infty$ to $+\infty$, as shown in Fig. 3-12b. When the excitation pole, $j\omega$, coincides with a zero of $Z(s)$, for example, $j\omega_{z4}$, $X(\omega)$ is zero. On the other hand, when the excitation pole coincides with a pole of $Z(s)$, for example, $j\omega_{p5}$, $X(\omega)$ is infinite. At high frequencies, Eq. (3-16) simplifies to a reactance that is either inductive,

$$X(\omega) \cong \omega L_H ,$$

or capacitive,

$$X(\omega) = -\frac{1}{\omega C_H} .$$

Figure 3-12c shows the high-frequency behavior. The $X(\omega)$-vs.-ω curve is sufficient to describe $Z(s)$ uniquely.

EXAMPLE 3-6

The input reactance of a network is to vary as shown in Fig. 3-13. Obtain the reactance function, $Z(s)$, and its asymptotic forms.

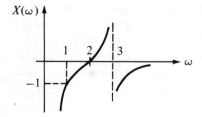

Fig. 3-13

SOLUTION

Figure 3-13 reveals that $Z(s)$ has poles at $s = 0$ and $s = \pm j3$, and zeros at $\pm j2$ and $s = \infty$. Hence, write $Z(s)$ as

$$Z(s) = \frac{H(s + j2)(s - j2)}{s(s + j3)(s - j3)} = H\frac{(s^2 + 4)}{s(s^2 + 9)}.$$

Next, evaluate $Z(s)$ for $s = j\omega$:

$$Z(j\omega) = H\frac{4 - \omega^2}{j\omega(9 - \omega^2)} = j\left[-H\frac{4 - \omega^2}{\omega(9 - \omega^2)}\right].$$

The resulting reactance function and its value at $\omega = 1$ are

$$X(\omega) = -H\frac{4 - \omega^2}{\omega(9 - \omega^2)} \quad \text{and} \quad X(1) = -H\frac{3}{8}.$$

From Fig. 3-13, $X(1) = -1$. Therefore, adjust H to 8/3 to pass the reactance curve through the given point.

The desired reactance and impedance functions are

$$X(\omega) = -\frac{8}{3}\frac{4 - \omega^2}{\omega(9 - \omega^2)}, \quad \textit{Ans.}$$

$$Z(s) = \frac{8}{3}\frac{s^2 + 4}{s(s^2 + 9)}. \quad \textit{Ans.}$$

For s small, $Z(s)$ reduces to

$$Z(s) \cong \frac{32}{27}\frac{1}{s}. \quad \textit{Ans.}$$

Thus, the low-frequency behavior is indistinguishable from the behavior of a (27/32)-farad capacitor.

For s large, $Z(s)$ reduces to

$$Z(s) \cong \frac{8}{3}\frac{1}{s}. \quad \textit{Ans.}$$

Thus, the high-frequency behavior is indistinguishable from the behavior of a (3/8)-farad capacitor.

When an *LC* network is excited by a sinusoidal source, the forced as well as the natural responses contain sinusoidal terms. Since the natural terms never vanish, they are as much a part of the response as the forced terms. Why, then, is no attention paid to the effects arising from the natural terms? The answer is readily available if it is realized that, in practice, inevitable losses associated with capacitors and inductors move the poles and zeros of $Z(s)$ from the imaginary axis to the left half-plane. Consequently, the natural response terms vanish exponentially with time, while the sinusoidally forced response remains for all time. Hence, the interest in the forced response is justified.

The effect of lossy components shows up also in the frequency response. Since the poles and zeros are, in practice, slightly to the left of the imaginary axis, the $|Z(j\omega)|$-vs.-ω curve has neither infinite nor zero magnitude values but rather large and small magnitudes as $j\omega$ gets close to but never quite equals the critical frequencies. Instead of abruptly jumping π radians, the phase undergoes, rather rapidly, nearly π radians of change as $j\omega$ takes on values below and above the critical frequency.

3-8 INPUT IMPEDANCE OF *RC* NETWORKS

The expression for the input impedance and the pole–zero locations of several *RC* networks are given in Fig. 3-14.

A careful study of the pole–zero diagrams reveals the following information about the input impedance of *RC* networks.

1. The zeros are on the negative real axis and are simple.

2. The poles are on the negative real axis. There may be a pole at the origin. All poles are simple.

3. Poles and zeros alternate.

4. The first critical frequency, the one nearest to the origin, is always a pole. The last critical frequency, the one farthest away from the origin, may be either a pole or a zero.

Indeed, *these four properties are sufficient to describe the input impedance of any RC network.*† Analytically, they result in the expression

$$Z(s) = H\frac{(s + \alpha_2)(s + \alpha_4) \cdots}{(s + \alpha_1)(s + \alpha_3) \cdots} \qquad (0 \le \alpha_1 < \alpha_2 < \alpha_3 < \alpha_4 < \cdots), \qquad (3\text{-}17)$$

where H is a positive, real constant. If the highest critical frequency is a pole, then the degree of the denominator polynomial is one higher than the degree of the numerator polynomial. Otherwise the degrees of both polynomials are the same. Thus, if the denominator polynomial is of degree n, then the numerator polynomial is either of degree $(n - 1)$ or of degree n.

For s small,

$$Z(s) \cong R_L \qquad \text{or} \qquad Z(s) \cong \frac{1}{sC_L}.$$

---

† *Ibid.*, pp. 302–367.

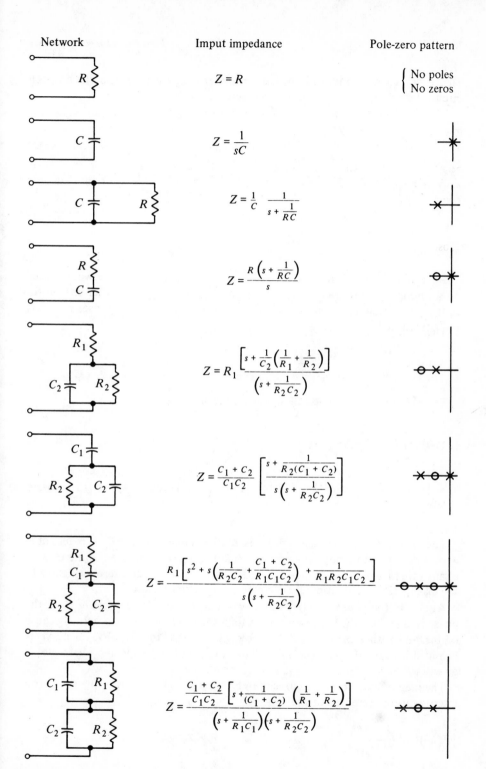

Fig. 3-14 Input impedance of RC networks

Hence, for $s = 0$, the input impedance is either a constant or acts like an open circuit. For s large,

$$Z(s) \cong R_H \qquad \text{or} \qquad Z(s) \cong \frac{1}{sC_H}.$$

Hence, for $s = \infty$, the input impedance is either a constant, R_H, or zero. Near zero or near infinite frequency, the input impedance is the same as that of a resistor or a capacitor.

Magnitude and Phase of Z_{RC}

Since all the critical frequencies of the input impedance of RC networks occur on the nonpositive real axis, the behavior of $Z(s)$, for $s = j\omega$, is rather uneventful compared to that of LC networks. To study the frequency response, $s = j\omega$ is substituted in Eq. (3-17):

$$Z(j\omega) = H \frac{(j\omega + \alpha_2)(j\omega + \alpha_4) \cdots}{(j\omega + \alpha_1)(j\omega + \alpha_3) \cdots}.$$

The magnitude and phase of $Z(j\omega)$ are given by

$$|Z(j\omega)| = H \sqrt{\frac{(\alpha_2^2 + \omega^2)(\alpha_4^2 + \omega^2) \cdots}{(\alpha_1^2 + \omega^2)(\alpha_3^2 + \omega^2) \cdots}}, \tag{3-18}$$

$$\theta_z = \left[\tan^{-1}\left(\frac{\omega}{\alpha_2}\right) + \tan^{-1}\left(\frac{\omega}{\alpha_4}\right) \cdots \right] - \left[\tan^{-1}\left(\frac{\omega}{\alpha_1}\right) + \tan^{-1}\left(\frac{\omega}{\alpha_3}\right) \cdots \right]. \tag{3-19}$$

The magnitude of $Z(j\omega)$, $|Z(j\omega)|$, is infinite at $\omega = 0$ if $\alpha_1 = 0$. It is zero at $\omega = \infty$ if $Z(s)$ has more poles than zeros. $|Z(j\omega)|$ does not become infinite or zero at any other frequency. In fact, because a pole, rather than a zero, is nearer to the origin, $|Z(j\omega)|$ continually decreases as ω varies from 0 to ∞.

At $\omega = 0$, the phase is $-(\pi/2)$ if $\alpha_1 = 0$. Otherwise it is zero. At $\omega = \infty$, the phase is $-(\pi/2)$ if $Z(s)$ has more poles than zeros. Otherwise it is zero. Because of the alternating property of the poles and zeros, the $\theta_z(\omega)$-vs.-ω curve is confined between 0 and $-(\pi/2)$ for all ω. A typical magnitude and phase plot is given in Fig. 3-15.

The magnitude and phase curves can be readily visualized from the geometric interpretation presented in Fig. 3-15a:

$$|Z(j\omega)| = H \frac{M_{z1}}{M_{p1} M_{p2}},$$

$$\theta_z(\omega) = \theta_{z1} - (\theta_{p1} + \theta_{p2}).$$

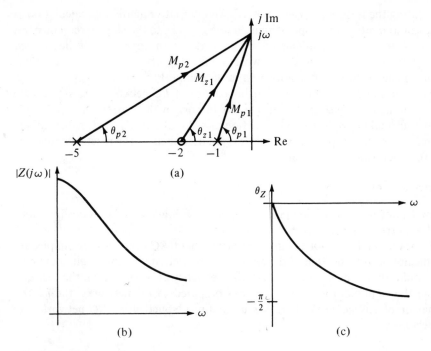

Fig. 3-15   Frequency response

Note that M_{z1}, M_{p1}, M_{p2}, θ_{z1}, θ_{p1}, and θ_{p2} all continuously increase in value as ω is increased from zero to infinity. Because the pole at $s = -1$ is nearer the origin than the zero at $s = -2$,

$$\frac{dM_{p1}}{d\omega} > \frac{dM_{z1}}{d\omega}.$$

Hence, the (M_{z1}/M_{p1}) ratio decreases with ω. Consequently, the $|Z(j\omega)|$-vs.-ω curve is a decreasing function of ω.

3-9 SUMMARY

The properties of a two-terminal network composed of R, L, C elements and linear dependent sources are completely characterized by the input-impedance function. This function uniquely establishes the relationship in the complex-frequency domain between input voltage and input current, provided the excitation is external to the network, i.e., all independent sources within the network are set to zero. (The responses arising from any independent internal sources are treated separately by using the appropriate transfer functions and the principle of superposition.)

The input impedance of a dead network is the ratio of input voltage, $V(s)$, to input current, $I(s)$. It can be obtained by applying a $V(s)$—[or $I(s)$]—and calculating the resulting $I(s)$—[or $V(s)$]. An alternative method of calculation is based on the ratio of the open-circuit voltage, $V_{oc}(s)$, to the short-circuit current, $I_{sc}(s)$, created at the terminals of the network by an appropriately introduced independent source within the network. [When the network, as given, contains an independent source, it can be used to determine $V_{oc}(s)$ and $I_{sc}(s)$.]

The poles and zeros of the input impedance of LC networks are simple, are on the imaginary axis, and alternate. At the origin there is either a pole or a zero, thus making $Z_{LC}(s)$ an odd function of s. At infinity, $Z_{LC}(s)$ is either infinite or zero. The $j\omega$-axis properties of $Z_{LC}(s)$ are described by the reactance function $X(\omega)$, which is obtained from

$$Z_{LC}(j\omega) = jX(\omega).$$

Between any two succeeding poles of $Z_{LC}(s)$, the $X(\omega)$ curve has a positive slope and goes from $-\infty$ to $+\infty$.

The poles and zeros of the input impedances of RC networks are simple, are on the nonpositive portion of the real axis, and alternate. At the origin there may be a pole. In any event, the first critical frequency encountered on the real axis (from the right side) is always a pole. Compared to LC networks, the $j\omega$-axis behavior of RC networks is rather unexciting. Nonetheless, RC networks are widely used.

PROBLEMS

3-1 Let 1-1′ represent two terminals in a dead network. Consider two sets of network poles, one obtained with terminals 1-1′ open-circuited and the other with terminals 1-1′ short-circuited. Show that the impedance looking into terminals 1-1′ can be expressed in terms of these two sets of poles.

3-2 Calculate the input impedance of the networks shown in Fig. 3-16.

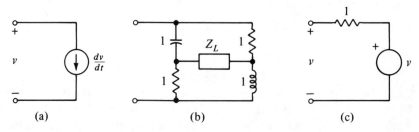

(a)   (b)   (c)

Fig. 3-16

3-3 Find the input impedance of the network shown in Fig. 3-17. The *impedances* of the elements are as shown.

3-4 In Fig. 3-18, the dependent source senses the voltage between *a-a′* and injects it in series with the network as shown.

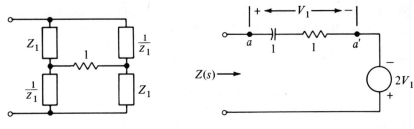

Fig. 3-17   Fig. 3-18

(a) Find the input impedance of the network by exciting it with a voltage source at the input terminals.

(b) Connect a current source across the capacitor to excite the network. Calculate the open-circuit voltage V_{oc}. Short the terminals and calculate the short-circuit current I_{sc}. Obtain the (V_{oc}/I_{sc}) ratio and compare it with a.

(c) Insert a voltage source between the capacitor and the resistor. Obtain (V_{oc}/I_{sc}) for this case and compare it with a.

3-5   In Fig. 3-19, find the input impedance if

(a) $V_2 = V_1$,   (b) $V_2 = -V_1$,   (c) $V_2 = I_1$.

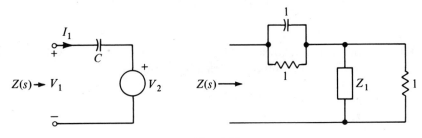

Fig. 3-19   Fig. 3-20

3-6   Show that if Z_1 is appropriately chosen, the input impedance of the network of Fig. 3-20 can be made 1. What is Z_1?

3-7   Calculate the input impedance of the networks shown in Fig. 3-21. In what way do these impedances differ from the ones considered in the text?

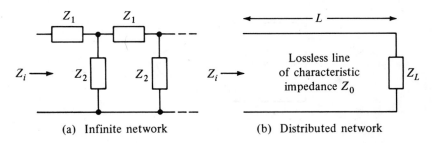

(a) Infinite network   (b) Distributed network

Fig. 3-21

3-8   The network *N* shown in Fig. 3-22 does not contain any independent sources.
The input admittance seen by the source V_1 is

Y_1 if $V_2 = 0$.

It is

Y_2 if $V_2 = V_1$

and

Y_3 if $V_2 = -V_1$.

How is Y_1 related to Y_2 and Y_3?

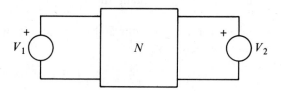

Fig. 3-22

3-9   The network *N* shown in Fig. 3-23 does not contain any independent
sources. The input impedance seen by the source I_1 is

Z_1 if $I_2 = 0$.

It is

Z_2 if $I_2 = I_1$

and

Z_3 if $I_2 = -I_1$.

How is Z_1 related to Z_2 and Z_3?

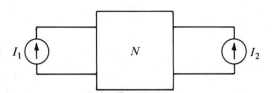

Fig. 3-23

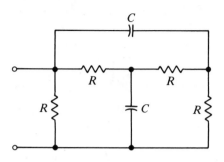

Fig. 3-24

3-10 Calculate the input impedance of the network shown in Fig. 3-24.

3-11 Given a resistor, a capacitor, and an inductor. Connect the three in such a way as to achieve:

(a) inductive behavior at low and resistive behavior at high frequencies;
(b) resistive behavior at low as well as high frequencies;
(c) capacitive behavior at low and inductive behavior at high frequencies.

3-12 The input to the network is a unit-step current, as shown in Fig. 3-25. Obtain $v(t)$.

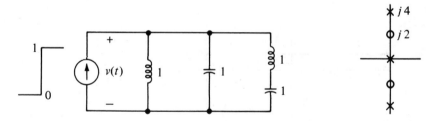

Fig. 3-25   Fig. 3-26

3-13 The impedance of a network has the pole–zero diagram shown in Fig. 3-26. The input is a sinusoidal current excitation of frequency ω. By inspection (do not calculate), show the waveforms that will be present in the input voltage if:

(a) $\omega = 1$,   (b) $\omega = 2$,   (c) $\omega = 3$,   (d) $\omega = 4$.

3-14 Given

$$Z(s) = \frac{(s^2 + 1)(s^2 + 9)(s^2 + 25)}{s(s^2 + 4)(s^2 + 16)},$$

sketch the reactance vs. ω.

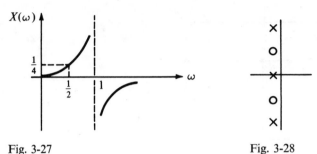

Fig. 3-27   Fig. 3-28

3-15 Obtain a network the input reactance of which behaves as shown in Fig. 3-27.

3-16 For the network shown in Fig. 3-25, sketch the input reactance curve.

3-17 The poles and zeros of $Z(s)$ are as shown in Fig. 3-28. Sketch the magnitude and phase of $Z(j\omega)$ vs. ω.

3-18 Show that the magnitude of the input impedance of all RC networks decreases with ω.

3-19 The impedance of an RC network is sketched for real values of s, as shown in Fig. 3-29. What is $Z(s)$?

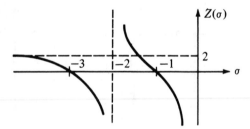

Fig. 3-29

3-20 Let $Z(s)$ be the input impedance of an RC network. Suppose now that the network is modified by shunting each capacitor, C_i, in the network with a resistor, R_i, such that the product $R_i C_i$ is a constant. What is the new $Z(s)$?

3-21 In an RLC network, all capacitors are paralleled with a resistor and a resistor is placed in series with each inductor. The values of all the added resistors (denoted by R_a) are chosen such that the time constant associated with each modification is the same throughout the network, that is,

$$R_{a1} C_1 = R_{a2} C_2 = \cdots = \frac{L_1}{R_{a7}} = \frac{L_2}{R_{a8}} = \cdots = \tau.$$

What effect does this equal-time-constant modification have on the poles and zeros of the input impedance?

3-22 Following the discussion of Section 3-8, list the properties of the input impedance of *RL* networks.

4

Synthesis of
LC and *RC*
Input Impedances

Given a network without any independent sources, the input impedance can be calculated. That is *analysis*. It is equally important to be able to obtain a network when the input impedance is given. That is *synthesis*. Analysis shows that the input impedances of *LC* and *RC* networks must obey certain rules. Conversely, a function obeying these rules can be realized as a two-terminal *LC* or *RC* network. The solution is not unique; several networks that have the same impedance can be found. The procedure is to decompose the rational function into subparts, each of which is readily recognized as the input impedance of a resistor, a capacitor, an inductor, or a simple combination of these. Two such decompositions are: the partial-fraction expansion and the continued-fraction expansion.

The networks resulting from the partial-fraction expansion are called *Foster* networks. Those obtained from the continued-fraction expansion are called *Cauer* networks. In this chapter, Foster and Cauer realizations of *LC* and *RC* input impedance functions are treated in detail.

4-1 *LC* FOSTER NETWORKS

There are two Foster realizations associated with input impedance functions described by simple, imaginary-axis poles and zeros which occur in alternating order. *The first network is obtained by the partial-fraction expansion of Z(s)— (Foster I); the partial-fraction expansion of Y(s) yields the second network— (Foster II).*

To obtain the Foster I network, consider $Z(s)$ for an *LC* network in its most general form:

$$Z(s) = H \frac{(s^2 + \omega_{z1}^2)(s^2 + \omega_{z2}^2) \cdots \begin{cases} [s^2 + \omega_{z(n+1)}^2] \\ \text{or} \\ (s^2 + \omega_{zn}^2) \end{cases}}{s(s^2 + \omega_{p1}^2)(s^2 + \omega_{p2}^2) \cdots (s^2 + \omega_{pn}^2)},$$

$$(0 \le \omega_{z1} < \omega_{p1} < \omega_{z2} < \omega_{p2} < \cdots), \qquad (4\text{-}1)$$

where H is positive.

If ω_{z1} is taken as zero, $Z(s)$ has a zero at the origin instead of a pole. The degree of the numerator polynomial is either one higher than the denominator polynomial (thus causing a *pole* at infinity), or it is one lower (resulting in a *zero* at infinity). Assume that $Z(s)$ has a pole at infinity and $\omega_{z1} \ne 0$. Since the numerator polynomial is of higher degree, division is performed to get

$$Z(s) = Hs + Z_1(s),$$

where $Z_1(s)$ is now in proper form, i.e., its numerator is of lower degree than the denominator. After expanding $Z_1(s)$ in partial fractions and combining the complex-conjugate terms, $Z(s)$ becomes

$$Z(s) = Hs + \frac{K_0}{s} + \frac{K_1 s}{s^2 + \omega_{p1}^2} + \frac{K_2 s}{s^2 + \omega_{p2}^2} + \cdots + \frac{K_n s}{s^2 + \omega_{pn}^2}. \qquad (4\text{-}2)$$

The K's are evaluated as follows:

$$K_0 = sZ(s)\Big|_{s=0},$$

$$K_i = \frac{(s^2 + \omega_{pi}^2)}{s} Z(s)\Big|_{s^2 = -\omega_{pi}^2}, \qquad (i = 1, 2, \ldots, n).$$

Because of the alternating property of the poles and zeros, all the K's are real and positive. As a result, the realization of Eq. (4-2) with positive L's and C's becomes feasible.

The first term in Eq. (4-2) is Hs which is the impedance of an inductor (H)-henries in value.

The second term in Eq. (4-2) is K_0/s which is the impedance of a capacitor $(1/K_0)$- farads in value.

The third term in Eq. (4-2) is

$$Z_3 = \frac{K_1 s}{s^2 + \omega_{p1}^2} = \frac{1}{s\left(\dfrac{1}{K_1}\right) + \dfrac{1}{s\left(\dfrac{K_1}{\omega_{p1}^2}\right)}} = \frac{1}{Y_3},$$

which gives

$$Y_3 = s\left(\frac{1}{K_1}\right) + \frac{1}{s\left(\dfrac{K_1}{\omega_{p1}^2}\right)}. \qquad (4\text{-}3)$$

The admittance Y_3, given by Eq. (4-3), is the sum of two admittances. The first represents the admittance of a $(1/K_1)$-farad capacitor, and the second the admittance of a (K_1/ω_{p1}^2)-henry inductor. The inductor and capacitor are connected in parallel to form Z_3. The remaining terms of Eq. (4-2) are also recognized as the parallel connection of inductors with capacitors. The complete realization of Eq. (4-2) is given in Fig. 4-1.

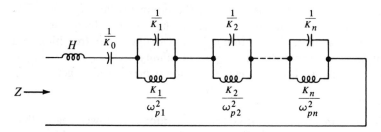

Fig. 4-1   Realization of Foster I

Several important observations can be made by studying the Foster I network given in Fig. 4-1.

1. It represents an input-impedance realization of any rational function with a positive scale factor and with simple poles and zeros (conjugate pairs) arranged in alternating order along the imaginary axis.

2. The first inductor makes $Z(s)$ infinite for $s = \infty$. If it is absent, $Z(\infty) = 0$ because of the all-capacitive path that exists between the input terminals.

3. The first capacitor makes $Z(s)$ infinite for $s = 0$. If it is absent, $Z(0) = 0$ because of the all-inductive path that exists between the input terminals.

4. A minimum number of elements is used to realize $Z(s)$. *There is an element for each pole* (including the one at infinity) *of $Z(s)$.*

5. The number of inductors and the number of capacitors either are equal or differ by one.

6. The network exposes all the poles of $Z(s)$. The first series inductor exposes the pole at infinity. The first series capacitor exposes the pole at the origin. The first parallel LC circuit exposes a pair of poles at $\pm j\omega_{p1}$, and so on. It is, however, not possible to ascertain the location of the zeros of $Z(s)$ by inspection of the Foster I network.

With Fig. 4-1 in mind, the Foster I network can be drawn directly from the given $Z(s)$ by noting the number of poles and the value of $Z(0)$ or $Z(\infty)$. The number of poles, including any at infinity, determines the number of elements.

If the number of elements is odd, either a series inductor or series capacitor is present. To determine which is present, consider either $Z(0)$ or $Z(\infty)$. If $Z(0) = 0$, the series capacitor cannot be present [it makes $Z(0) = \infty$]; hence, the first element is an inductor. On the other hand, if $Z(0) = \infty$, the series capacitor must be the first element. Alternatively, the value of $Z(s)$ at infinite frequency can be used to determine whether a series capacitor or a series inductor is present. If $Z(\infty) = 0$, the series inductor cannot be present [it makes $Z(\infty) = \infty$];

hence, the first element is a capacitor. On the other hand, if $Z(\infty) = \infty$, the series inductor must be present.

If the number of elements is even, the series L and the series C are either both present or both absent. If $Z(0) = 0$, they are both absent [because the series capacitor alone forces $Z(0)$ to be infinite]. If $Z(0) = \infty$, they both are present. Alternatively, $Z(\infty)$ can be used to determine the presence or absence of the two series elements. If $Z(\infty) = 0$, both series elements are absent [because the series inductor alone forces $Z(\infty)$ to be infinite]. If $Z(\infty) = \infty$, they are both present.

The remaining elements are drawn as parallel LC circuits connected in series. Thus, the Foster I network—without element values—can be drawn by inspection of $Z(s)$. The partial-fraction expansion of $Z(s)$ is necessary only to find the element values.

EXAMPLE 4-1

(a) Obtain the Foster I realization for

$$Z(s) = H\frac{s(s^2 + 4)}{(s^2 + 1)(s^2 + 9)}.$$

Assume $H = 1$.
(b) What is the network if $H = 10$?
(c) What is the network if $Z(s)$ is changed by replacing s by $10s$?

SOLUTION

(a) $Z(s)$ has poles at $\pm j1$, $\pm j3$, and zeros at 0, $\pm j2$, ∞. The poles and zeros are simple and alternate on the imaginary axis. The scale factor is positive. Hence, $Z(s)$ is realizable as the input impedance of an LC network. $Z(s)$ has four poles. Hence, the network can be realized with four elements. $Z(0) = 0$; hence, there is no series capacitor. There is also no series inductor because the number of elements is even. [Alternatively, $Z(\infty) = 0$; hence, there is no series inductor.] The network can be drawn, therefore, as two parallel LC circuits connected in series, as shown in Fig. 4-2.

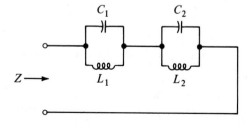

$Z \longrightarrow$

Fig. 4-2

To find the values of the elements, expand $Z(s)$ in partial fractions (and combine complex-conjugate terms):

$$Z(s) = \frac{K_1 s}{s^2 + 1} + \frac{K_2 s}{s^2 + 9};$$

$$K_1 = Z(s)\left(\frac{s^2 + 1}{s}\right)\Big|_{s^2 = -1} = \frac{s^2 + 4}{s^2 + 9}\Big|_{s^2 = -1} = \frac{3}{8},$$

$$K_2 = Z(s)\left(\frac{s^2 + 9}{s}\right)\Big|_{s^2 = -9} = \frac{s^2 + 4}{s^2 + 1}\Big|_{s^2 = -9} = \frac{5}{8}.$$

Then,

$$Z(s) = \frac{\frac{3}{8}s}{s^2 + 1} + \frac{\frac{5}{8}s}{s^2 + 9} = Z_1 + Z_2.$$

Z_1, for s small, is $\dfrac{3}{8} s$; hence, $L_1 = \dfrac{3}{8}$.   *Ans.*

Z_1, for s large, is $\dfrac{3}{8s}$; hence, $C_1 = \dfrac{8}{3}$.   *Ans.*

Z_2, for s small, is $\dfrac{5}{72} s$; hence, $L_2 = \dfrac{5}{72}$.   *Ans.*

Z_2, for s large, is $\dfrac{5}{8s}$; hence, $C_2 = \dfrac{8}{5}$.   *Ans.*

(b) The impedance, $Z(s)$, is scaled up by a factor of 10 if H is changed from 1 to 10. Therefore, the impedance of each element should be multiplied by 10. Thus, L_1 and L_2 become $10L_1$ and $10L_2$; C_1 and C_2 become $(C_1/10)$ and $(C_2/10)$.

(c) Replacement of s by $10s$ makes the impedance of each inductor and the admittance of each capacitor ten times larger. Hence, replace L_1, L_2, C_1, and C_2 with $10L_1$, $10L_2$, $10C_1$, and $10C_2$.

To obtain the Foster II LC network, expand $Y(s)$ in partial fractions. In its most general form, $Y(s)$ is given by

$$Y(s) = H \frac{(s^2 + \omega_{z1}^2)(s^2 + \omega_{z2}^2) \cdots [s^2 + \omega_{z(n+1)}^2]}{s(s^2 + \omega_{p1}^2)(s^2 + \omega_{p2}^2) \cdots (s^2 + \omega_{pn}^2)}, \tag{4-4}$$

which has a partial-fraction expansion (with combined complex-conjugate terms) given by:

$$Y(s) = Hs + \frac{K_0}{s} + \frac{K_1 s}{s^2 + \omega_{p1}^2} + \frac{K_2 s}{s^2 + \omega_{p2}^2} + \cdots + \frac{K_n s}{s^2 + \omega_{pn}^2}. \tag{4-5}$$

Because of the alternating pole–zero structure, the K's are all positive and real, and are found from

$$K_0 = s\,Y(s)\Big|_{s=0},$$

$$K_i = \frac{s^2 + \omega_{pi}^2}{s}\,Y(s)\Big|_{s^2 = -\omega_{pi}^2}, \qquad (i = 1, 2, \ldots, n).$$

As decomposed in Eq. (4-5), $Y(s)$ represents the sum of admittances. Hence the network realization contains structures in parallel.

The first term is Hs, which is the admittance of a capacitor (H)-farads in value.

The second term is (K_0/s), which is the admittance of an inductor $(1/K_0)$-henries in value.

The third term is

$$Y_3 = \frac{K_1 s}{s^2 + \omega_{p1}^2} = \frac{1}{s\left(\dfrac{1}{K_1}\right) + \dfrac{1}{s(K_1/\omega_{p1}^2)}} = \frac{1}{Z_3},$$

$$Z_3 = s\left(\frac{1}{K_1}\right) + \frac{1}{s\left(\dfrac{K_1}{\omega_{p1}^2}\right)}.$$

The impedance Z_3 represents the series combination of a $(1/K_1)$-henry inductor with a (K_1/ω_{p1}^2)-farad capacitor.

The complete realization of Eq. (4-5) is presented in Fig. 4-3.

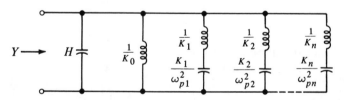

$Y \longrightarrow$

Fig. 4-3   Realization of Foster II

Several important observations can be made from Fig. 4-3. To promote uniformity, $Z(s)$, rather than $Y(s)$, is used in the discussion that follows.

1. The network represents an input-impedance realization of any rational function with a real, positive scale factor and with simple poles and zeros (conjugate pairs) arranged in alternating order along the imaginary axis.

2. The first capacitor makes $Z(s)$ zero for $s = \infty$. [If it is absent, $Z(\infty) = \infty$, because the shunt inductors cause the network to be open-circuited at $s = \infty$.]

3. The first inductor makes $Z(s)$ zero for $s = 0$. [If it is absent, $Z(0) = \infty$, because the shunt capacitors cause the network to become open-circuited at $s = 0$].

4. A minimum number of elements is used to realize $Z(s)$. There is an element for each pole of $Z(s)$, including any at infinity.

5. The number of inductors is the same as the number of capacitors or differs by one.

6. The network exposes all the zeros of $Z(s)$. The first shunt capacitor exposes the zero at infinity; the first shunt inductor exposes the zero at the origin. The first shunt LC combination exposes a pair of zeros at $\pm j\omega_{p1}$, and so on. It is not possible, however, to ascertain the location of the poles of $Z(s)$ by inspection of the Foster II network.

With Fig. 4-3 in mind, the Foster II network can be drawn directly from the given $Z(s)$ by noting the number of poles and the value of $Z(0)$ or $Z(\infty)$. The number of poles, including any at infinity, determines the number of elements.

If the number of elements is odd, either a shunt capacitor or a shunt inductor is present. Since the shunt capacitor is responsible for making $Z(\infty) = 0$, the value of $Z(\infty)$ determines the presence or absence of the shunt capacitor. [Alternatively, the value of $Z(0)$ can be used to determine the presence or absence of the shunt inductor, because it alone is responsible for making $Z(0) = 0$.]

If the number of elements is even, the shunt L and the shunt C are either both present or both absent. If $Z(0) = 0$, they are both present, because only the shunt inductor can cause $Z(0)$ to be zero. If $Z(0) = \infty$, they are both absent. [Alternatively, if $Z(\infty) = 0$, they are both present because only the shunt capacitor can cause $Z(\infty)$ to be zero.]

The remaining elements are drawn as series LC circuits connected in parallel. Thus, the Foster II network—without element values—can be drawn by inspection of $Z(s)$. The partial-fraction expansion of $Y(s)$ is necessary only to find the element values.

EXAMPLE 4-2

Obtain the Foster II realization for

$$Z(s) = \frac{s(s^2 + 4)}{(s^2 + 1)(s^2 + 9)}.$$

SOLUTION

$Z(s)$ has four poles. Hence, the network has four elements. $Z(0) = 0$; hence, the shunt inductor must be present. This requires that the shunt capacitor also be present, since the number of elements is even. (Alternatively, $Z(\infty) = 0$; hence, the shunt capacitor must be present.) The network can be drawn, therefore, as shown in Fig. 4-4.

$Z(s)$, for s small, is $\frac{4}{9}s$. Hence, $L_3 = \frac{4}{9}$.   *Ans.*

$Z(s)$, for s large, is $\frac{1}{s}$. Hence, $C_3 = 1$.   *Ans.*

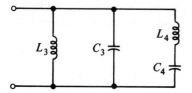

Fig. 4-4

From the partial-fraction expansion of $Y(s)$, the remaining two elements can be found:

$$Y(s) = \frac{(s^2 + 1)(s^2 + 9)}{s(s^2 + 4)} = Y_3 + Y_4 = Y_3 + \frac{K_1 s}{s^2 + 4},$$

where Y_3 represents the admittance of the two shunt elements that have already been evaluated from the asymptotic form of $Z(s)$. (Alternatively, the complete partial-fraction expansion can be used to determine the values of L_3 and C_3):

$$K_1 = Y(s)\left(\frac{s^2 + 4}{s}\right)\Bigg|_{s^2 = -4} = \frac{15}{4},$$

$$Z_4(s) = \frac{1}{Y_4} = \frac{4}{15}s + \frac{16}{15s}.$$

Hence,

$$L_4 = \frac{4}{15}, \qquad C_4 = \frac{15}{16}. \qquad \textit{Ans.}$$

4-2 *LC* CAUER NETWORKS

There are two Cauer networks associated with a given input impedance which is *LC*-realizable. Cauer networks are ladder networks, and the general form is shown in Fig. 4-5.

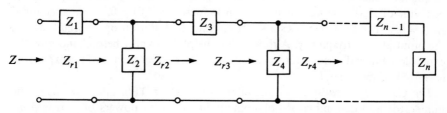

Fig. 4-5   Cauer network

Recognizing the repetitive series-shunt patterns of the ladder network, the impedance can be written by inspection of Fig. 4-5 in the continued-fraction form:

$$Z = Z_1 + \cfrac{1}{Y_2 + \cfrac{1}{Z_3 + \cfrac{1}{Y_4 + \cfrac{\vdots}{Z_{n-1} + \cfrac{1}{Y_n}}}}} \tag{4-6}$$

Referring to Fig. 4-5, it is seen that impedance Z equals Z_1 plus Z_{r1}, where Z_{r1} represents the impedance of the remaining network to the right of Z_1. The admittance Y_{r1} of this network is Y_2 plus Y_{r2}, where Y_{r2} represents the admittance of the network to the right of Z_2. The impedance Z_{r2} is Z_3 plus Z_{r3}, and the pattern is repeated to form Eq. (4-6).

In an *LC* ladder network, Z_1 or Z_n or both may be zero, depending upon the value of the impedance at zero and infinite frequencies. *In the Cauer I network, inductors are in series and capacitors are in shunt. In the Cauer II network, the series elements are capacitors and the shunt elements inductors.* The number of elements is equal to the number of poles of $Z(s)$, counting also the pole at infinity if one is there.

As in the two Foster forms, knowledge of the number of elements and the extreme frequency behavior of $Z(s)$ allows the drawing of the Cauer networks. The continued-fraction expansion merely determines the values of the elements.

The Cauer I expansion is done either on $Z(s)$ or $Y(s)$, after arranging the numerator and denominator polynomials in *descending* powers of s. It is done on $Z(s)$ if the first element is a series inductor, which is the case if an open circuit is to exist at infinite frequencies $[Z(\infty) = \infty]$. It is done on $Y(s)$ if the first element is the shunt capacitor, which is the case if a short circuit is to exist at infinite frequencies $[Z(\infty) = 0]$. Thus, how the network starts depends on the infinite-frequency behavior of $Z(s)$. How it ends is determined by the zero-frequency behavior of $Z(s)$; the last element is an inductor if $Z(0) = 0$, or a capacitor if $Z(0) = \infty$. The former causes a short circuit, the latter an open circuit, across the input at zero frequency. Although it is helpful to know beforehand how the expansion will end, it ends automatically with the correct element if the expansion is started properly.

The Cauer II expansion is done either on $Z(s)$ or $Y(s)$, after arranging the numerator and denominator polynomials in *ascending* powers of s. It is done on $Z(s)$, if $Z(0) = \infty$, which requires that the first element be a series capacitor, to cause an open circuit for $s = 0$. It is done on $Y(s)$, if $Z(0) = 0$, which requires that the first element be a shunt inductor, to cause a short circuit for $s = 0$. How the network starts depends, therefore, on the zero-frequency behavior of $Z(s)$. How it ends is determined by the infinite-frequency behavior of $Z(s)$; the last element is an inductor if $Z(\infty) = \infty$, a capacitor if $Z(\infty) = 0$. The former causes an open circuit, the latter a short circuit at infinite frequencies. Once the expansion is started correctly, it ends correctly, so it is not necessary to determine beforehand how the network will end.

EXAMPLE 4-3

Find the Cauer networks for

$$Z(s) = \frac{s(s^2 + 4)}{(s^2 + 1)(s^2 + 9)}.$$

SOLUTION

Note first that $Z(0) = 0$ and $Z(\infty) = 0$. Note next that $Z(s)$ has four poles; therefore the network realizations contain four elements. Since this number is even, there are two inductors and two capacitors.

Cauer I LC networks start with either an inductor in the series arm of the ladder or a capacitor in shunt. Because $Z(\infty) = 0$ in this case, the first element is a shunt capacitor. Alternatively it can be argued that the first element cannot be a series inductor because $Z(\infty) \neq \infty$. Once the kind and position of the first element is determined correctly, the remaining three elements can be drawn readily, as in Fig. 4-6a. Note that the expansion terminates automatically with an inductor, as it must if $Z(0)$ is to be zero.

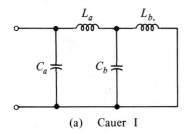 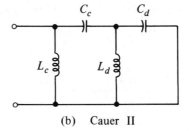

(a) Cauer I (b) Cauer II

Fig. 4-6

Cauer II LC networks start with either a capacitor in the series arm of the ladder or an inductor in shunt. Because $Z(0) = 0$ in this case, the first element is a shunt inductor. Alternatively it can be argued that the first element cannot be a series capacitor because $Z(0) \neq \infty$. Once the kind and position of the first element is determined correctly, the remaining three elements can be drawn readily, as in Fig. 4-6b. Note that the expansion terminates automatically with a capacitor, as it must if $Z(\infty)$ is to be zero.

To find the values of the elements in the Cauer I network, the admittance $Y(s)$—since the first element is in shunt—is arranged in descending powers of s and expanded in continued fractions. Either division, followed by inversion, followed by division, etc., is performed, or the algorithm presented in Appendix C is used.

$$Y(s) = \frac{s^4 + 10s^2 + 9}{s^3 + 4s}$$

$$= s + \cfrac{1}{\cfrac{1}{6}s + \cfrac{1}{\cfrac{36}{15}s + \cfrac{1}{\cfrac{5}{18}s}}}$$

Algorithm

$Y_1 = 1s$

$Z_2 = \dfrac{1}{6}s$

$Y_3 = \dfrac{36}{15}s$

$Z_4 = \dfrac{5}{18}s$

$$
\begin{array}{c|ccc}
 & 1 & 10 & 9 \\
\hline
 & 1 & 4 & \\
\hline
 & 6 & 9 & \\
 & 15 & & \\
 & \overline{6} & & \\
 & 9 & &
\end{array}
$$

From the longhand expansion or from the algorithm, the element values can be read directly:

$$C_a = 1, \qquad L_a = \frac{1}{6}, \qquad C_b = \frac{36}{15}, \qquad L_b = \frac{5}{18}.$$

(The Y's represent the shunt elements, the Z's the series.)

To find the value of the elements in the Cauer II network, the admittance $Y(s)$—since the first element is in shunt—is arranged in ascending powers of s and expanded in continued fractions.

$$Y(s) = \frac{9 + 10s^2 + s^4}{4s + s^3}$$

$$= \frac{9}{4s} + \cfrac{1}{\cfrac{16}{31s} + \cfrac{1}{\cfrac{961}{60s} + \cfrac{1}{\cfrac{15}{31s}}}}$$

Algorithm

$Y_1 = \dfrac{9}{4}\dfrac{1}{s}$

$Z_2 = \dfrac{16}{31}\dfrac{1}{s}$

$Y_3 = \dfrac{961}{60}\dfrac{1}{s}$

$Z_4 = \dfrac{15}{31}\dfrac{1}{s}$

$$
\begin{array}{c|ccc}
 & 9 & 10 & 1 \\
\hline
 & 4 & 1 & \\
\hline
 & 31 & & \\
 & \overline{4} & 1 & \\
 & 15 & & \\
 & \overline{31} & &
\end{array}
$$

From the longhand expansion or from the algorithm, the element values can be written readily:

$$L_c = \frac{4}{9}, \qquad C_c = \frac{31}{16}, \qquad L_d = \frac{60}{961}, \qquad C_d = \frac{31}{15}.$$

(The Y's represent the shunt elements, the Z's the series.)

The networks shown in Fig. 4-2, Fig. 4-4, and Fig. 4-6 all have the same input impedance, namely

$$Z(s) = \frac{s(s^2 + 4)}{(s^2 + 1)(s^2 + 9)}.$$

They all have the same number of inductors and the same number of capacitors. Only the element values and the way they are connected are different.

4-3 MIXED FOSTER AND CAUER NETWORKS

Foster and Cauer forms can be mixed in developing the network which is to realize a given $Z(s)$. One may start with the Foster I expansion and obtain a few elements, and then develop the remaining impedance as Cauer II for a few elements, after which the remainder may be realized as Cauer I and so on. The resulting network uses the same number of inductors and the same number of capacitors as the all-Foster or the all-Cauer networks. Thus, the mixed forms also use a minimum number of elements to realize $Z(s)$. The higher the order of the system, the greater the variety of possible mixed forms.

EXAMPLE 4-4

Obtain a mixed realization for

$$Z(s) = \frac{s(s^2 + 4)(s^2 + 25)}{(s^2 + 1)(s^2 + 9)}.$$

SOLUTION

$Z(s)$ has five poles. Hence, a minimum of five elements is needed to realize $Z(s)$. $Z(\infty) = \infty$; hence, the first element in a Foster I expansion is an inductor. This element is realized and pulled out of the network:

$$Z(s) = s + Z_{r1}.$$

The remaining impedance Z_{r1} is

$$Z_{r1} = Z(s) - s = \frac{19s\left(s^2 + \dfrac{91}{19}\right)}{(s^2 + 1)(s^2 + 9)}.$$

Next $Y_{r1} = (1/Z_{r1})$ is realized in Foster II form:

$$Y_{r1} = \frac{1}{19} \frac{(s^2 + 1)(s^2 + 9)}{s\left(s^2 + \dfrac{91}{19}\right)}$$

$$= \frac{1}{19} s + \frac{\dfrac{9}{91}}{s} + \frac{\dfrac{5760}{32851} s}{s^2 + \dfrac{91}{19}}.$$

The resulting network, utilizing mixed Foster forms, is given in Fig. 4-7.

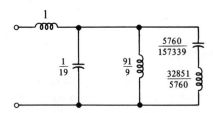

Fig. 4-7

As an independent check, the asymptotic forms obtained from the given $Z(s)$ and the realized network can be compared. From the expression for $Z(s)$,

$$Z(s)\bigg|_{s\ \text{small}} \cong s\,\frac{100}{9},$$

$$Z(s)\bigg|_{s\ \text{large}} \cong s.$$

From the network diagram,

$$Z(s)\bigg|_{s\ \text{small}} \cong s + s\,\frac{91}{9} = s\,\frac{100}{9},$$

$$Z(s)\bigg|_{s\ \text{large}} \cong s.$$

Although the network in Fig. 4-7 was developed as a combination of Foster I and II networks, it is easy to see that it also represents a combination of Cauer I and II expansions, the first two elements coming from the Cauer I development and the remainder from Cauer II.

Another mixed form representing the given $Z(s)$ is shown in Fig. 4-8. Using the Foster II expansion, the values of L_1 and C_1 are determined. The remaining impedance is expanded in Foster I (or Cauer I) form to obtain the values for L_3, L_2, and C_2. This realization also uses three inductors and two capacitors. By comparing Fig. 4-8 with Fig. 4-7, it is seen by inspection that

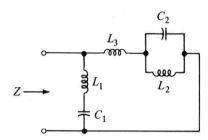

Fig. 4-8

$$\frac{L_1 L_3}{L_1 + L_3} = 1 \qquad \text{(from the asymptotic behavior at high frequencies),}$$

and that

$$L_2 + L_3 = 1 + \frac{91}{9} \qquad \text{(from the asymptotic behavior at very low frequencies).}$$

4-4 *RC* FOSTER AND CAUER NETWORKS

A rational function, that has simple, alternating poles and zeros located on the nonpositive real axis, is realizable as the input impedance of an *RC* network if the critical frequency nearest to or at the origin is a pole.

There are two Foster networks associated with such an impedance. The first network is obtained by expanding $Z(s)$ in partial fractions (Foster I); the second is obtained by expanding $[Y(s)/s]$ in partial fractions (Foster II).

To obtain the Foster I network, consider

$$Z(s) = H \frac{(s + \alpha_2)(s + \alpha_4) \cdots}{(s + \alpha_1)(s + \alpha_3)(s + \alpha_5) \cdots} \qquad (0 \le \alpha_1 < \alpha_2 < \alpha_3 < \alpha_4 \cdots), \qquad (4\text{-}7)$$

where H is positive. Assuming that $\alpha_1 = 0$ and that the degree of the numerator polynomial is equal to the degree of the denominator polynomial, the partial-fraction expansion becomes

$$Z(s) = H + \frac{K_1}{s} + \frac{K_3}{s + \alpha_3} + \frac{K_5}{s + \alpha_5} + \cdots \qquad (4\text{-}8)$$

The K's are given by

$$K_1 = sZ(s)\bigg|_{s=0},$$

$$K_n = (s + \alpha_n)Z(s)\bigg|_{s=-\alpha_n}, \qquad (n = 3, 5, \ldots)$$

All the K's are positive real numbers because of the alternating arrangement of the poles and zeros. Therefore, Eq. (4-8) can be realized with positive R's and positive C's, as shown in Fig. 4-9.

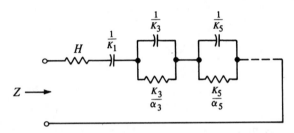

Fig. 4-9   Foster I realization

Figure 4-9 reveals several important results.

1. It gives an RC input-impedance realization for any rational function which obeys Eq. (4-7). The realization uses a minimum number of elements.

2. If $Z(0) = \infty$, the first series capacitor must be present to open-circuit the network at $s = 0$. If $Z(0) \neq \infty$, then $Z(0)$ is a constant, and the first capacitor must be *absent* to allow a resistive path across the input for $s = 0$. Thus, the low- (or zero)-frequency behavior of $Z(s)$ determines the presence or absence of the first capacitor.

3. If $Z(\infty) \neq 0$, the first series resistor must be present to prevent a short circuit at $s = \infty$. If $Z(\infty) = 0$, then the first resistor must be absent, allowing the capacitive path across the input to cause a short circuit for $s = \infty$. Thus the high- (or infinite)-frequency behavior of $Z(s)$ determines the presence or absence of the first resistor.

4. If the asymptotic behavior of $Z(s)$ for s small, as well as for s large, is that of a *resistor*, then the number of resistors in the realization is *one more* than the number of capacitors. If the asymptotic behavior at both extreme frequencies is that of a *capacitor*, then the number of capacitors exceeds by one the number of resistors. If the asymptotic behavior at one extreme frequency is that of a resistor and at the other that of a capacitor, then the number of resistors is *equal* to the number of capacitors. In any event, there must be one capacitor for each pole.

5. The network exposes all the poles of $Z(s)$. The first capacitor, if present,

indicates a pole at the origin. Each RC parallel network causes a pole on the negative real axis located at $(-1/R_i C_i)$—(where $i = 3, 5, \ldots$). It is not possible to say anything about the finite zeros of $Z(s)$ by inspection of the Foster I network.

The Foster I network can be drawn directly from the given $Z(s)$ by noting its asymptotic behavior at the extreme frequencies (s small and s large) and by noting the number of poles. Since each pole represents an energy-storage element, the number of capacitors is thus readily determined. A series capacitor is present if $Z(0) = \infty$. The other capacitors are in parallel with the resistors. A series resistor is present if $Z(\infty) \neq 0$. Thus the network, but not the value of its elements, can be drawn by inspection of $Z(s)$. The partial-fraction expansion can then be used to fix the values of the elements.

To obtain the Foster II network, $[Y(s)/s]$ is expanded in partial fractions [expansion of $Y(s)$ would result in negative K's]. The most general expansion is of the form

$$\frac{Y(s)}{s} = H + \frac{K_0}{s} + \frac{K_2}{s + \alpha_2} + \frac{K_4}{s + \alpha_4} + \cdots, \tag{4-9}$$

where all the K's are positive and real. The K's are found from $K_0 = Y(0)$:

$$K_n = \frac{Y(s)}{s}(s + \alpha_n)\Big|_{s = -\alpha_n}, \qquad (n = 2, 4, \ldots).$$

After the K's are determined, Eq. (4-9) is multiplied through by s, to obtain

$$Y(s) = Hs + K_0 + \frac{K_2 s}{s + \alpha_2} + \frac{K_4 s}{s + \alpha_4} + \cdots, \tag{4-10}$$

which is readily synthesized as the network shown in Fig. 4-10.

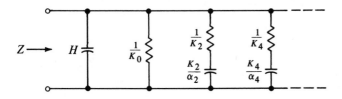

Fig. 4-10   Foster II realization

Note that the Foster II realization exhibits the zeros of $Z(s)$. The first capacitor is responsible for the zero at infinity. The other shunt-connected RC networks are responsible for all the negative real-axis zeros, the locations of which are given by $(-1/R_i C_i)$—(where $i = 2, 4, \ldots$). Just as in the Foster I form, a knowledge of the total number of capacitors and the asymptotic behavior fixes the topology of the network. The element values are then obtained from the expansion.

The two Cauer networks associated with an impedance function which is *RC* realizable are given in Fig. 4-11.

The *n* capacitors represent an *n*-pole system. *The Cauer I network has resistors in series, capacitors in shunt; the Cauer II network has the opposite arrangement.*

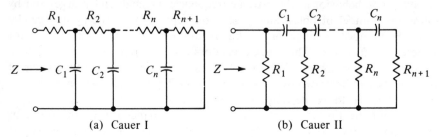

(a) Cauer I   (b) Cauer II

Fig. 4-11   Cauer networks

The low- and high-frequency asymptotic behavior determines how the network starts and how it ends. For example, if $Z(0) = \infty$, R_{n+1} must be infinite in Cauer I, and R_1 must be infinite in Cauer II; if $Z(\infty) = 0$, R_1 must be zero in Cauer I, and R_{n+1} zero in Cauer II (to provide the necessary capacitive path across the input). In Cauer I development, the polynomials are arranged in *descending* powers of *s* whereas, in Cauer II realization, the polynominals are arranged in *ascending* powers of *s*. In either case, the expansion is performed on $Z(s)$ if the first element is a series element (as determined from the asymptotic behavior); if the first element is a shunt element, the expansion is done on $Y(s)$.

It is also possible to get mixed forms by switching from one form of development to another. These mixed Foster and Cauer forms use exactly the same number of resistors and capacitors as either of the two Foster or the two Cauer networks.

EXAMPLE 4-5

Obtain five realizations for

$$Z(s) = \frac{(s + 1)(s + 3)}{s(s + 2)(s + 4)}.$$

SOLUTION

$Z(s)$ is *RC*-realizable, since the poles and zeros are simple and alternate properly on the nonpositive real axis. $Z(s)$ has three poles; hence the realization must contain three capacitors. Since the asymptotic behavior of $Z(s)$ at both extreme frequencies is capacitive,

$$Z(s)\Big|_{s\ small} \cong \frac{3}{8s}, \qquad Z(s)\Big|_{s\ large} \cong \frac{1}{s},$$

the realization contains two resistors. There is but one way of arranging three capacitors and two resistors to achieve the two Foster and the two Cauer forms. The networks shown in Fig. 4-12 result.

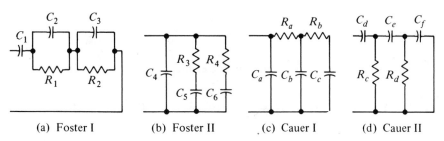

(a) Foster I   (b) Foster II   (c) Cauer I   (d) Cauer II

Fig. 4-12

Comparing the low-frequency asymptotic behavior of the function $Z(s)$ with the impedance of the networks, it is seen that

$$C_1 = \frac{8}{3}, \qquad C_4 + C_5 + C_6 = \frac{8}{3}, \qquad C_a + C_b + C_c = \frac{8}{3}, \qquad C_d = \frac{8}{3}.$$

From the high-frequency asymptotic behavior, it is seen that

$$\frac{1}{C_1} + \frac{1}{C_2} + \frac{1}{C_3} = 1, \qquad C_4 = 1, \qquad C_a = 1, \qquad \frac{1}{C_d} + \frac{1}{C_e} + \frac{1}{C_f} = 1.$$

Because the first element in Fig. 4-12c is in shunt, Cauer I expansion is performed on $Y(s)$. Because the first element in Fig. 4-12d is in series, the Cauer II expansion is performed on $Z(s)$.

The partial-fraction and the continued-fraction expansions can now be performed to find the values of the elements.

Foster I

Expand $Z(s)$ in partial fractions:

$$Z = \frac{(s+1)(s+3)}{s(s+2)(s+4)} = \frac{\frac{3}{8}}{s} + \frac{\frac{1}{4}}{s+2} + \frac{\frac{3}{8}}{s+4}.$$

By comparison with Fig. 4-12a, the element values are

$$C_1 = \frac{8}{3}, \qquad C_2 = 4, \qquad R_1 = \frac{1}{8}, \qquad C_3 = \frac{8}{3}, \qquad R_2 = \frac{3}{32}. \qquad Ans.$$

Foster II

Expand $[Y(s)/s]$ in partial fractions:

$$\frac{Y}{s} = \frac{(s+2)(s+4)}{(s+1)(s+3)} = 1 + \frac{\frac{3}{2}}{s+1} = \frac{\frac{1}{2}}{s+3}.$$

Solve for Y to get

$$Y = s + \frac{\frac{3}{2}s}{s+1} + \frac{\frac{1}{2}s}{s+3}.$$

By comparison with Fig. 4-12b, the element values are

$$C_4 = 1, \qquad C_5 = \frac{3}{2}, \qquad R_3 = \frac{2}{3}, \qquad C_6 = \frac{1}{6}, \qquad R_4 = 2. \qquad Ans.$$

Cauer I

Put the polynomials in descending powers of s and expand $Y(s)$ in continued fractions (the algorithm presented in Appendix C may also be used for this purpose):

$$Y = \frac{s^3 + 6s^2 + 8s}{s^2 + 4s + 3}$$

Algorithm

$$= s + \cfrac{1}{\frac{1}{2} + \cfrac{1}{\frac{4}{3}s + \cfrac{1}{\frac{3}{2} + \cfrac{1}{\frac{1}{3}s}}}}$$

$Y_1 = 1s \longrightarrow$

$Z_2 = \dfrac{1}{2} \longrightarrow$

$Y_3 = \dfrac{4}{3}s \longrightarrow$

$Z_4 = \dfrac{3}{2} \longrightarrow$

$Y_5 = \dfrac{1}{3}s \longrightarrow$

	1	6	8
	1	4	3
	2	5	
	$\frac{3}{2}$	3	
	1		
	3		

By comparison with Fig. 4-12c, the element values are

$$C_a = 1, \qquad R_a = \frac{1}{2}, \qquad C_b = \frac{4}{3}, \qquad R_b = \frac{3}{2}, \qquad C_c = \frac{1}{3}. \qquad Ans.$$

Cauer II

Put the polynomials in ascending powers of s and expand $Z(s)$ in continued fractions:

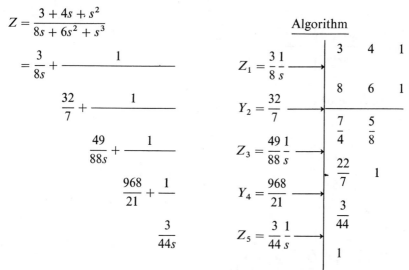

$$Z = \frac{3 + 4s + s^2}{8s + 6s^2 + s^3}$$

$$= \frac{3}{8s} + \cfrac{1}{\cfrac{32}{7} + \cfrac{1}{\cfrac{49}{88s} + \cfrac{1}{\cfrac{968}{21} + \cfrac{1}{\cfrac{3}{44s}}}}}$$

Algorithm

$$Z_1 = \frac{3}{8}\frac{1}{s}$$

$$Y_2 = \frac{32}{7}$$

$$Z_3 = \frac{49}{88}\frac{1}{s}$$

$$Y_4 = \frac{968}{21}$$

$$Z_5 = \frac{3}{44}\frac{1}{s}$$

3	4	1
8	6	1
$\frac{7}{4}$	$\frac{5}{8}$	
$\frac{22}{7}$	1	
3		
44		
1		

By comparison with Fig. 4-12d, the element values are

$$C_d = \frac{8}{3}, \qquad R_c = \frac{7}{32}, \qquad C_e = \frac{88}{49}, \qquad R_d = \frac{21}{968}, \qquad C_f = \frac{44}{3}. \qquad \textit{Ans.}$$

To obtain a fifth network, the mixture of Foster I and Foster II networks shown in Fig. 4-13 is used. First, the pole of Z at $s = -4$ is realized in Foster I form, and the R_7C_7 network pulled out. The remaining impedance Z_r is realized as a Foster II network:

$$Z = \frac{(s + 1)(s + 3)}{s(s + 2)(s + 4)} = \frac{\frac{3}{8}}{s + 4} + Z_r,$$

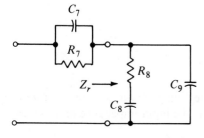

Fig. 4-13

where

$$Z_r = \frac{5s + 6}{8s(s + 2)},$$

$$\frac{Y_r}{s} = \frac{8(s + 2)}{5\left(s + \frac{6}{5}\right)} = \frac{\frac{32}{25}}{s + \frac{6}{5}} + \frac{8}{5},$$

which becomes

$$Y_r = \frac{\frac{32}{25}s}{s + \frac{6}{5}} + \frac{8}{5}s.$$

Hence,

$$Z = \frac{\frac{3}{8}}{s + 4} + \frac{1}{\frac{(32/25)s}{s + (6/5)} + \left(\frac{8}{5}\right)s}.$$

By comparison with Fig. 4-13, the element values are obtained:

$$R_7 = \frac{3}{32}, \qquad C_7 = \frac{8}{3}, \qquad R_8 = \frac{25}{32}, \qquad C_8 = \frac{16}{15}, \qquad C_9 = \frac{8}{5}. \qquad \textit{Ans.}$$

As an additional check, it is seen from the network that

$$Z(s)\Big|_{s\,\text{small}} \cong \frac{1}{s(C_8 + C_9)} = \frac{3}{8s},$$

$$Z(s)\Big|_{s\,\text{large}} \cong \frac{1}{sC_7} + \frac{1}{sC_9} = \frac{1}{s\dfrac{C_7 C_9}{C_7 + C_9}} = \frac{1}{s}.$$

4-5 SUMMARY

A function which is described by simple, imaginary-axis poles and zeros occurring in alternating order, is realizable as the input impedance of a number of different *LC* networks.

To obtain the Foster I network, expand $Z(s)$ in partial fractions. The resulting network is a series connection of an inductor, a capacitor, and several parallel

LC circuits. The series inductor indicates the presence of an impedance pole at infinity. The series capacitor indicates the presence of a pole at the origin. The parallel *LC* circuits indicate the presence of conjugate imaginary-axis poles.

To obtain the Foster II network, expand $Y(s)$ in partial fractions. The resulting network is a parallel connection of an inductor, a capacitor, and several series *LC* circuits. The shunt inductor indicates the presence of an impedance zero at the origin. The shunt capacitor indicates the presence of a zero at infinity. The series *LC* circuits indicate the presence of conjugate imaginary-axis zeros.

To obtain the Cauer I network, arrange the numerator and denominator polynomials of $Z(s)$ or $Y(s)$ in descending powers of s. Expand $Z(s)$ in continued fractions if $Z(s)$ has a pole at infinity. Otherwise, expand $Y(s)$. The resulting ladder network has inductors in the series branches and capacitors in the shunt.

To obtain the Cauer II network, arrange the numerator and denominator polynomials of $Z(s)$ or $Y(s)$ in ascending powers of s. Expand $Z(s)$ in continued fractions if $Z(s)$ has a pole at the origin. Otherwise, expand $Y(s)$. The resulting ladder network has capacitors in the series branches and inductors in the shunt.

A function which is described by simple, alternating poles and zeros located on the nonpositive real axis (with a pole nearest to or at the origin) is realizable as the input impedance of a number of different *RC* networks.

To obtain the Foster I network, expand $Z(s)$ in partial fractions. The resulting network is a series connection of a resistor, a capacitor, and several parallel *RC* circuits. The series capacitor indicates the presence of an impedance pole at the origin. The parallel *RC* circuits indicate the presence of poles on the negative real axis. The series resistor prevents $Z(\infty)$ from being zero.

To obtain the Foster II network, expand $Y(s)/s$ in partial fractions. The resulting network is a parallel connection of a resistor, a capacitor, and several series *RC* circuits. The shunt capacitor indicates the presence of an impedance zero at infinity. The series *RC* circuits indicate the presence of zeros on the negative real axis. The shunt resistor prevents $Z(0)$ from being infinite.

To obtain the Cauer I network, arrange the numerator and denominator polynomials of $Z(s)$ or $Y(s)$ in *descending* powers of s. Expand $Z(s)$ in continued fractions if $Z(\infty) \neq 0$. Otherwise, expand $Y(s)$. The resulting ladder network has resistors in the series branches and capacitors in shunt.

To obtain the Cauer II network, arrange the numerator and denominator polynomials of $Z(s)$ or $Y(s)$ in *ascending* powers of s. Expand $Z(s)$ in continued fractions if $Z(0) = \infty$. Otherwise expand $Y(s)$. The resulting ladder network has capacitors in the series branches and resistors in shunt.

PROBLEMS

4-1 Show that the K's in Eq. (4-2) are all real and positive.

4-2 Without actually calculating $Z(s)$, determine the number of finite poles and zeros associated with the input impedance of each network shown in Fig. 4-14.

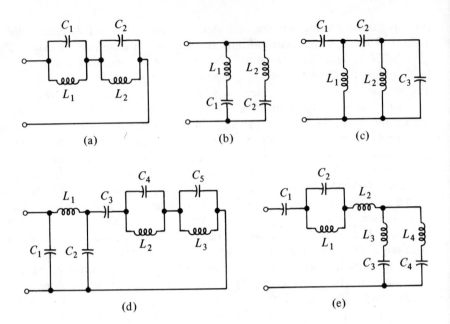

Fig. 4-14

4-3 (a) Obtain the expression for $Z(s)$ of the network shown in Fig. 4-14a
    (b) What happens to $Z(s)$, its poles and zeros, and the network, as
    $L_1 C_1 \rightarrow L_2 C_2$?

4-4 By inspection of the networks shown in Fig. 4-15, sketch the input-
    reactance-vs.-ω curve.

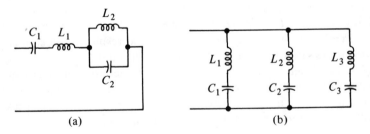

Fig. 4-15

4-5 Obtain the two Cauer networks (without values) associated with each
    network of Fig. 4-15.

4-6 By inspection of Fig. 4-16, obtain (without element values) the two Foster
    and the two Cauer networks which have the same $Z(s)$.

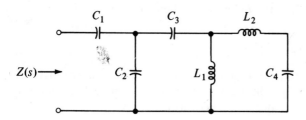

Fig. 4-16

4-7 By inspection of Fig. 4-17, draw (without element values) the two Foster networks.

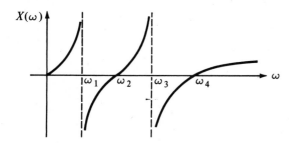

Fig. 4-17

4-8 Obtain four realizations that have an input impedance given by

$$Z(s) = \frac{s(s^2 + 4)(s^2 + 25)}{(s^2 + 1)(s^2 + 9)}.$$

4-9 What network has the input impedance given by

$$Z(s) = \frac{4s}{2s^2 + s + 12}?$$

4-10 The input impedance of a network is given by

$$Z = s + \frac{1}{s + Z}.$$

(a) Find the explicit expression for Z.
(b) Find a network realization for Z.
(c) Sketch the frequency response of Z.
(d) Repeat (a), (b), and (c), if

$$Z = s + \frac{1}{s + Y} \qquad \text{where} \quad Y = \frac{1}{Z}.$$

4-11 The impedance Z of the LC network shown in Fig. 4-18 is

$$Z = H \frac{(s^2 + \omega_1^2)(s^2 + \omega_3^2)}{s(s^2 + \omega_2^2)(s^2 + \omega_4^2)}.$$

It is realized by using a minimum number of elements. The first three elements of the realization are shown. Complete the realization (no values; just draw the remainder of the network).

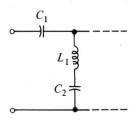

Fig. 4-18

4-12 In Fig. 4-8,

$$Z = \frac{s(s^2 + 4)(s^2 + 25)}{(s^2 + 1)(s^2 + 9)}.$$

Find the values of all the elements.

4-13 Find the Foster and Cauer networks for

$$Z(s) = \frac{(s + 1)(s + 3)}{s(s + 2)}.$$

4-14 Is it possible in Fig. 4-19 to make $Z_1 = Z_2$? Explain.

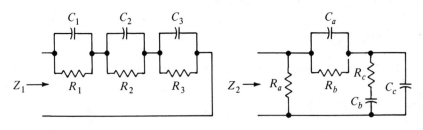

Fig. 4-19

4-15 The input impedance of a network is given by

$$Z(s) = 1 + \frac{1}{s + Y}, \qquad \text{where} \quad Y = \frac{1}{Z}.$$

(a) Find the explicit expression for $Z(s)$.

(b) Find a network realization for $Z(s)$.

4-16 Obtain four realizations for

$$Z(s) = \frac{s(s + 2)(s + 4)}{(s + 1)(s + 3)}.$$

4-17 Consider the input impedance of an LC, RC, or RL network containing three elements. Show that no distinction can be made between the Foster and Cauer realizations, i.e., there are only two realizations.

5

Transfer Functions
of Ladder Networks:
Analysis and Synthesis

Three-terminal ladder networks are used widely for signal processing. Resistively terminated *LC* ladder networks may be used to realize transfer functions with left half-plane poles. Low-pass, high-pass, band-pass, and band-stop filters are examples of such realizations. Imaginary-axis transfer-function zeros may be implemented by the proper choice of *LC* impedances in the series and shunt arms of the ladder. With *RC* ladder networks, negative real-axis poles and zeros can be implemented. In this chapter, the properties of these ladder networks are discussed, and synthesis procedures are given for realizing transfer functions that meet the realizability conditions. Examples are given to demonstrate the various methods of synthesis. The production of complex transfer-function zeros with second-order, *RC*, parallel-ladder networks is discussed and illustrated. A simple procedure is given for constructing *RC* networks with gains greater than unity.

5-1 LADDER NETWORKS

A ladder network is composed of elements connected alternately in series and in parallel. The input is connected at one end of the ladder and the output taken at the other end. Since the input and the output are associated with two different ports of the network, the resulting system function is called the *transfer function*. If the input is a *current* source, the first element is almost always a shunt element (since any series element would change only the input impedance but not the transfer function). If the input is a *voltage* source, the first element is almost always a series element (since any shunt element would affect only the

112

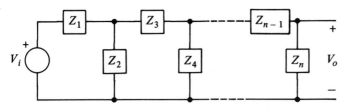

Fig. 5-1   Transfer function

input impedance but not the transfer function). With either type of excitation, the output may be taken as voltage or current. In Fig. 5-1, the input and output are taken as voltages.

Most ladder networks are designed with a common input and output terminal. This desirable feature is shown in the network of Fig. 5-1. All ladder networks considered henceforth are assumed to have this common ground connection.

Another important feature of ladder networks is the ease with which the zeros of the transfer function are recognized or implemented. In general, *a zero of the transfer function occurs for those values of s that make a series impedance infinite or a shunt impedance zero.* Thus, the zeros of V_o/V_i in Fig. 5-1 are identical with the poles of $Z_1, Z_3, \ldots, Z_{n-1}$ and the zeros of $Z_2, Z_4, \ldots, Z_n$.

Quite often, simple networks are used as series or parallel elements in the ladder structure to produce zeros of transfer functions at specific locations in the *s*-plane. Some commonly encountered series networks used for this purpose are given in Fig. 5-2a.

The inductor L_1 causes a transfer-function zero at infinity because its impedance is infinite for $s = \infty$.

The capacitor C_3 causes a transfer-function zero at the origin because its impedance is infinite for $s = 0$.

The parallel $L_5 C_5$ network causes a transfer-function zero at $s = \pm j(1/\sqrt{L_5 C_5})$ because its impedance is infinite there.

The parallel $R_7 C_7$ network causes a transfer-function zero at $s = -(1/R_7 C_7)$ because its impedance is infinite there.

Some commonly encountered shunt networks that produce specific transfer-function zeros are given in Fig. 5-2b.

The inductor L_2 causes a transfer-function zero at the origin because its impedance is zero at $s = 0$.

The capacitor C_4 causes a transfer-function zero at infinity because its impedance is zero at $s = \infty$.

The series $L_6 C_6$ network causes a transfer-function zero at $s = (\pm j /\sqrt{L_6 C_6})$ because its impedance is zero there.

The series $R_8 C_8$ network causes a transfer-function zero at $s = -(1/R_8 C_8)$ because its impedance is zero there.

Thus, *an infinite series impedance or a zero shunt impedance at $s = s_0$ prevents the transmission of an input signal with a pole at $s = s_0$.* The resulting output does not contain the forced response due to the input-signal pole at $s = s_0$. However,

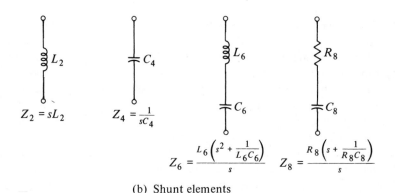

(a) Series elements

(b) Shunt elements

Fig. 5-2 Transfer-function zero-producing sections

the output is not altogether zero because of the natural response terms and terms due to the other excitation poles that may be present.

There are cases when a series impedance is infinite at $s = s_0$ and yet a signal with a pole at $s = s_0$ is transmitted. This occurs when the impedance of the ladder structure to the right of the series impedance is also infinite at $s = s_0$. (The signal voltage is then forced to divide between two infinite impedances.) In still other cases, a shunt impedance that is zero at $s = s_0$ may let the signal through if the impedance of the ladder structure to the right also has a zero at $s = s_0$. (The signal current is then forced to divide between two zero impedances.) These special cases, however, may be regarded as resulting from the cancellation of the transfer-function zero with a pole, thus rendering the zero-producing sections ineffective.

5-2 TRANSFER-FUNCTION ZEROS CAUSED BY *LC* IMPEDANCES

If the network is *LC*, the input-impedance function has poles and zeros that are purely imaginary. Therefore, such networks, when used in the series or in the shunt arms of a ladder, cause transfer-function zeros that are on the imaginary axis only. Consider, for example, the ladder network shown in Fig. 5-3.

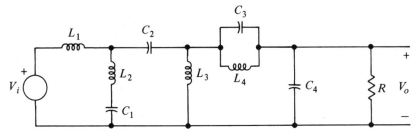

Fig. 5-3 Imaginary-axis zero-producing sections

The impedances in the series and shunt arms are

$$Z_1 = sL_1, \qquad Z_3 = \frac{1}{sC_2}, \qquad Z_5 = \frac{s\dfrac{1}{C_3}}{s^2 + \dfrac{1}{L_4 C_3}},$$

$$Z_2 = \frac{L_2\left(s^2 + \dfrac{1}{L_2 C_1}\right)}{s}, \qquad Z_4 = sL_3, \qquad Z_6 = \frac{1}{sC_4}, \qquad Z_8 = R.$$

The series impedances become infinite for $s = \infty$ (due to sL_1), $s = 0$ [due to $(1/sC_2)$], and $s = \pm j(1/\sqrt{L_4 C_3})$ (due to $L_4 C_3$). The shunt impedances become zero for $s = \pm j(1/\sqrt{L_2 C_1})$ (due to $L_2 C_1$), $s = 0$ (due to sL_3), $s = \infty$ [due to $(1/sC_4)$]. Therefore, assuming that cancellation of terms does not occur, (V_o/V_i) can be written as

$$\frac{V_o}{V_i} = H \frac{s\left(s^2 + \dfrac{1}{L_4 C_3}\right)\left(s^2 + \dfrac{1}{L_2 C_1}\right)s}{s^8 + \cdots + b_0} \qquad (b_0 \neq 0). \tag{5-1}$$

As Eq. (5-1) indicates, quite a bit can be said about the transfer function without elaborate calculations. Note that the denominator polynomial must be of degree 8 so that the two zeros at infinity (caused by L_1 and C_4) are realized. The scale factor H and b_0 are evaluated by comparing the asymptotic behavior of Eq. (5-1) with that of the network given in Fig. 5-3.

For s large, Eq. (5-1) and direct inspection of Fig. 5-3 give, respectively,

$$\frac{V_o}{V_i} \cong \frac{H}{s^2}, \qquad \text{and} \qquad \frac{V_o}{V_i} \cong \frac{1}{sL_1}\frac{1}{sC_4}.$$

These two results must agree. Therefore,

$$H = \frac{1}{L_1 C_4}.$$

For s small, Eq. (5-1) and Fig. 5-3 give, respectively,

$$\frac{V_o}{V_i} \cong \frac{H}{b_0} \frac{1}{L_4 C_3} \frac{1}{L_2 C_1} s^2, \quad \text{and} \quad \frac{V_o}{V_i} \cong sC_2 \cdot sL_3 \,.$$

These two results must also agree. Therefore,

$$b_0 = \frac{1}{L_1 L_2 L_3 L_4 C_1 C_2 C_3 C_4} \,.$$

What about the poles of (V_o/V_i)? If the terminal resistance, R, were infinite or zero, the poles would be on the imaginary axis, since the network would then be purely LC. For any other value of R, the poles are in the left half-plane. Thus, as R is increased from zero to infinity, the poles move from the imaginary axis to the left half-plane and then back to the imaginary axis.

All Zeros at Infinity

A transfer function with all the zeros at infinity can be obtained by a ladder network composed of series inductors and shunt capacitors, as shown in Fig. 5-4.

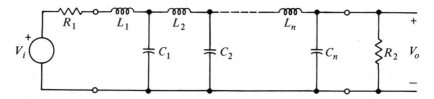

Fig. 5-4   Transfer function with all zeros at infinity

This network has n series inductors and n shunt capacitors. Since each series inductor and each shunt capacitor independently produces a zero at infinity, (V_o/V_i) has $2n$ zeros at infinity; (V_o/V_i) has no other zeros. To force (V_o/V_i) to be zero $2n$ times at infinity, (V_o/V_i) must have $2n$ poles. By inspection of the network given in Fig. 5-4,

$$\left.\frac{V_o}{V_i}\right|_{s\,\text{small}} \cong \frac{R_2}{R_1 + R_2} \quad \text{and} \quad \left.\frac{V_o}{V_i}\right|_{s\,\text{large}} \cong \frac{1}{s^{2n} L_1 C_1 L_2 C_2 \cdots L_n C_n} \,.$$

These results, obtained directly from the network, are satisfied if (V_o/V_i) is written as

$$\frac{V_o}{V_i} = \frac{1}{s^{2n} L_1 C_1 L_2 C_2 \cdots L_n C_n + \cdots + \dfrac{R_1 + R_2}{R_2}} \,. \tag{5-2}$$

As the asymptotic forms indicate, this network passes the low frequencies and attenuates the high frequencies. If the poles of Eq. (5-2) are properly chosen, the network serves as a low-pass filter.

All Zeros at the Origin

A network that attenuates the low frequencies and passes the high frequencies is shown in Fig. 5-5. The n capacitors produce n zeros at $s = 0$; the n inductors cause another n zeros at the origin. Therefore, (V_o/V_i) has $2n$ zeros, all at the origin. It also has $2n$ poles.

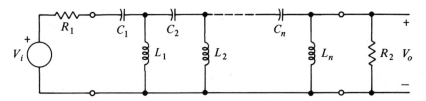

Fig. 5-5   Transfer function with all zeros at the origin

By inspection of the asymptotic behavior of the network, (V_o/V_i) can be written as

$$\frac{V_o}{V_i} = \frac{s^{2n}}{\dfrac{R_1 + R_2}{R_2}s^{2n} + \cdots + \dfrac{1}{C_1 L_1 C_2 L_2 \cdots C_n L_n}}. \tag{5-3}$$

When the poles are properly chosen, the network of Fig. 5-5 may serve as a high-pass filter.

EXAMPLE 5-1

The oscillator shown in Fig. 5-6 contains third-harmonic distortion. Design a filter that eliminates the third-harmonic signal without attenuating the fundamental.

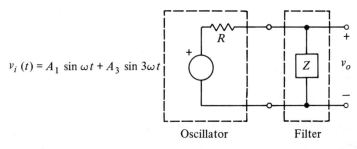

$v_i(t) = A_1 \sin \omega t + A_3 \sin 3\omega t$

Oscillator   Filter

Fig. 5-6

SOLUTION

The filtering is done by the impedance Z. In order to eliminate the third-harmonic signal, Z must have zeros at $s = \pm j3\omega$. In order to pass the fundamental without attenuation, Z must have poles at $s = \pm j\omega$. Thus, the design criteria are met by the following Z:

$$Z(s) = \frac{s^2 + 9\omega^2}{s^2 + \omega^2}.$$

This Z, however, is not realizable with passive components. Its poles and zeros are on the imaginary axis but do not alternate. To make Z realizable, modify it by introducing a zero at the origin:

$$Z(s) = \frac{s(s^2 + 9\omega^2)}{s^2 + \omega^2}. \tag{5-4}$$

The addition of a zero at the origin affects neither the rejection of the third harmonic nor the passing of the fundamental. (If the input had any dc component, the added zero would have taken it out.)

To obtain the elements of the filter, expand $Z(s)$ or $Y(s)$ in partial fractions:

Foster I Network

Expand $Z(s)$ in partial fractions:

$$Z(s) = s + \frac{8\omega^2 s}{s^2 + \omega^2}$$

$$= s + \frac{1}{s\dfrac{1}{8\omega^2} + \dfrac{1}{s8}} = sL_1 + \frac{1}{sC_1 + \dfrac{1}{sL_2}},$$

$$L_1 = 1, \qquad C_1 = \frac{1}{8\omega^2}, \qquad L_2 = 8.$$

The resulting network is shown in Fig. 5-7a.

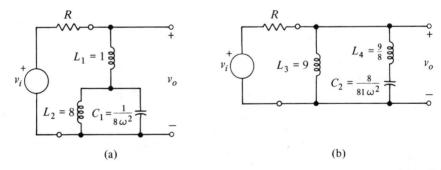

(a)   (b)

Fig. 5-7

Foster II Network

Expand $Y(s)$ in partial fractions:

$$Y(s) = \frac{1}{9}\frac{1}{s} + \frac{8}{9}\frac{s}{s^2 + 9\omega^2}$$

$$= \frac{1}{9s} + \frac{1}{\dfrac{9}{8}s + \dfrac{81\omega^2}{8s}} = \frac{1}{sL_3} + \frac{1}{sL_4 + \dfrac{1}{sC_2}},$$

$$L_3 = 9, \qquad L_4 = \frac{9}{8}, \qquad C_2 = \frac{8}{81\omega^2}.$$

The resulting network is shown in Fig. 5-7b.

Both networks shown in Fig. 5-7 meet the design specifications. However, because of nonideal elements, the actual implementation of these networks will neither eliminate the third harmonic completely, nor pass the fundamental without attenuation.

5-3 SYNTHESIS OF RESISTOR-TERMINATED *LC* LADDER NETWORKS

The poles of *LC* networks are simple and are located on the imaginary axis. These poles can be moved into the left half-plane by inserting a single resistor in the network. The more general case of using resistive terminations at both the source and load end of the lossless *LC* network is shown in Fig. 5-8.

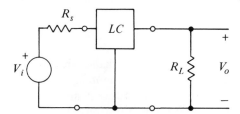

Fig. 5-8   Resistively terminated *LC* network

If the *LC* network is constructed as a ladder network, then the zeros of the transfer function are all on the imaginary axis and at the origin. The voltage ratio is given by

$$\frac{V_o}{V_i} = H \frac{s^k(s^2 + \omega_1^2)(s^2 + \omega_2^2) \cdots}{(s + \gamma_1)(s + \gamma_2) \cdots [(s + \alpha_1)^2 + \beta_1^2][(s + \alpha_2)^2 + \beta_2^2] \cdots}, \tag{5-5}$$

where k is zero or a positive integer.

There may be several zeros at the origin, one for each series capacitor or shunt inductor. The imaginary-axis zeros may occur in any multiplicity, for example, $\omega_1 = \omega_2$. All poles have negative real parts. There can be no pole at the origin or at infinity (try to construct a network). The number of finite zeros is always equal to or less than the number of poles. The scale factor, H, is positive.

In the special case when

$$R_S = 0 \quad \text{and} \quad R_L = \infty,$$

the network of Fig. 5-8 becomes purely LC. Then, the transfer function is given by

$$\frac{V_o}{V_i} = H \frac{s^k(s^2 + \omega_1^2)(s^2 + \omega_2^2)\cdots}{(s^2 + \beta_1^2)(s^2 + \beta_2^2)\cdots}, \tag{5-6}$$

where k is zero or a positive, even integer.

In this lossless case, the number of zeros at the origin must be even because zeros cannot be produced by a shunt inductor across the input or a series capacitor at the output. As a result, the voltage ratio is always an *even function* of s. The imaginary-axis poles are simple but the imaginary-axis zeros may be of any multiplicity. Unlike the input impedance of LC networks, the poles and zeros *do not have to alternate*. The number of finite zeros is always equal to or less than the number of poles. The scale factor, H, is positive. It should be realized that because of the inevitable losses associated with L's and C's, it is impossible to realize Eq. (5-6) in practice.

Decomposition of the Denominator Polynomial

Consider a polynomial, $P(s)$, *with left-half-plane zeros*. Such polynomials form the denominator polynomial of RLC networks. Partition the polynomial into its even and odd parts:

$$P(s) = \text{Ev}(s) + \text{Od}(s). \tag{5-7}$$

The even and odd parts of this polynomial possess remarkable properties.* The zeros of the even part are simple and are located on the imaginary axis; so are the zeros of the odd part. Furthermore, the zeros of the even part alternate with the zeros of the odd part. The resulting implication is obvious. The $\text{Ev}(s)/\text{Od}(s)$ or $\text{Od}(s)/\text{Ev}(s)$ ratio has precisely the same properties as the input impedance of LC networks. Thus, *ratios constructed from the even and odd parts of a polynomial, which has left-half-plane zeros, are realizable as the input impedance of LC networks.*

It should also be clear from this discussion that a polynomial constructed from the sum of the numerator and denominator polynomials of the input impedance of an LC network is a polynomial with left-half-plane zeros.

---

* For detailed treatment of the properties, see Louis Weinberg, *Network Analysis and Synthesis* (McGraw-Hill Book Company, 1962), p. 237.

EXAMPLE 5-2

Given the polynomial

$$P(s) = s^5 + 7s^4 + 20s^3 + 30s^2 + 24s + 8.$$

Show that the ratio formed from the even and odd parts of this polynomial is realizable as the input impedance of an *LC* network.

SOLUTION

The polynomial, in factored form, is

$$P(s) = (s + 1)(s + 2)^2[(s + 1)^2 + 1].$$

Note that all the zeros of $P(s)$ are in the left half-plane. Therefore, the ratio formed from its even and odd parts is *LC*-realizable. To show this, decompose $P(s)$ into its even and odd parts and factor the resulting polynomials:

$$P(s) = (7s^4 + 30s^2 + 8) + (s^5 + 20s^3 + 24s);$$
$$\mathrm{Ev}(s) = 7s^4 + 30s^2 + 8 = 7(s^2 + \tfrac{2}{7})(s^2 + 4) = 7(s^2 + 0.535^2)(s^2 + 2^2),$$
$$\mathrm{Od}(s) = s^5 + 20s^3 + 24s = s(s^2 + 10 - 2\sqrt{19})(s^2 + 10 + 2\sqrt{19})$$
$$= s(s^2 + 1.132^2)(s^2 + 4.326^2).$$

The zeros of $\mathrm{Ev}(s)$ are at:

$$s = \pm j0.535, \ \pm j2.$$

The zeros of $\mathrm{Od}(s)$ are at

$$s = \pm j1.132, \ \pm j4.326, \ 0.$$

The zeros of both parts are simple and purely imaginary. Furthermore, they can be arranged in alternating order. Therefore, the rational function formed by the ratio of the even-to-odd part or its inverse is realizable as the input impedance of an *LC* network; for example,

$$Z_{LC}(s) = \frac{\mathrm{Ev}(s)}{\mathrm{Od}(s)} = \frac{7(s^2 + 0.535^2)(s^2 + 2^2)}{s(s^2 + 1.132^2)(s^2 + 4.326^2)}. \qquad Ans.$$

Resistive Termination on the Source Side

Since a single resistor is sufficient to move the poles of an *LC* network into the left half-plane, this resistor can be placed at a convenient location. One such location is at the input, where the resistor may absorb any resistance present in the source. Without loss of generality, let this source resistance be 1Ω. (Later, the impedance of each element in the network can be multiplied by R_S to comply with the actual value of source resistance; such magnitude scaling does not affect the voltage ratio.) The problem is the following:

Given a transfer function that meets the conditions set forth in Eq. (5-5), develop a synthesis technique to realize it in the form shown in Fig. 5-9.

To learn more about the network of Fig. 5-9, analyze it first. Let $T_{LC}(s)$ represent the transfer function from port 1 to port 2. This transfer function, since

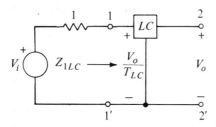

Fig. 5-9   Resistive-source termination

it is the voltage ratio of an LC ladder network, is an even function of s [see Eq. (5-6)]. Let the voltage at port 2 be V_o; then, the voltage at port 1 is V_o/T_{LC} and is given by

$$\frac{V_o}{T_{LC}} = V_i \frac{Z_{1LC}}{1 + Z_{1LC}},$$

where Z_{1LC} represents the input impedance of the LC network. The transfer function is

$$\frac{V_o}{V_i} = \frac{Z_{1LC} T_{LC}}{1 + Z_{1LC}}, \qquad (Z_{1LC} T_{LC} \text{ is odd}). \tag{5-8}$$

Because T_{LC} is even and Z_{1LC} is odd, $Z_{1LC} T_{LC}$ is odd. Equation (5-8) can now be used as a guide in developing a synthesis procedure. First, the denominator polynomial of the given (V_o/V_i) is decomposed into even and odd parts:

$$\frac{V_o}{V_i} = \frac{N}{D} = \frac{N}{\text{Ev} + \text{Od}}. \tag{5-9}$$

Then, depending on whether N is odd or even, both the numerator and the denominator of Eq. (5-9) are divided by Ev or Od:

$$\frac{V_o}{V_i} = \frac{\dfrac{N}{\text{Ev}}}{1 + \dfrac{\text{Od}}{\text{Ev}}} \qquad \left(\frac{N}{\text{Ev}} \text{ is odd}\right), \tag{5-10a}$$

$$\frac{V_o}{V_i} = \frac{\dfrac{N}{\text{Od}}}{1 + \dfrac{\text{Ev}}{\text{Od}}} \qquad \left(\frac{N}{\text{Od}} \text{ is odd}\right). \tag{5-10b}$$

Comparison of Eq. (5-10) with Eq. (5-8) shows that

$$T_{LC} = \frac{N}{\text{Od}}, \qquad Z_{1LC} = \frac{\text{Od}}{\text{Ev}} \qquad \text{when } N \text{ is odd,} \qquad (5\text{-}11a)$$

$$T_{LC} = \frac{N}{\text{Ev}}, \qquad Z_{1LC} = \frac{\text{Ev}}{\text{Od}} \qquad \text{when } N \text{ is even.} \qquad (5\text{-}11b)$$

Thus, in terms of the N, Ev, and Od parts of the voltage ratio [see Eq. (5-9)] to be realized, T_{LC} and Z_{1LC} can be constructed using Eq. (5-11). Note that both T_{LC} and Z_{1LC} deal with only the LC part of the network shown in Fig. 5-9. Then, the constructed input impedance, Z_{1LC} is developed as a ladder whose branches realize all the zeros of (V_o/V_i). [The finite zeros of T_{LC} may differ from the finite zeros of (V_o/V_i) because of cancellation.]

EXAMPLE 5-3

Obtain a network that will realize

$$\frac{V_o}{V_i} = H \frac{s}{(s + 1)[(s + \frac{1}{2})^2 + \frac{3}{4}]}.$$

The source resistance is 50Ω. What is the H of the resulting realization?

SOLUTION

The poles of (V_o/V_i) are in the left half-plane at $s = -1, (-1 \pm j\sqrt{3})/2$. One zero is at the origin and two zeros are at infinity. The function meets the realizability conditions expressed by Eq. (5-5). To obtain the network, determine first the N, Ev, and Od of the given function:

$$\frac{V_o}{V_i} = H \frac{s}{(s + 1)[(s + \frac{1}{2})^2 + \frac{3}{4}]} = \frac{Hs}{s^3 + 2s^2 + 2s + 1}$$

$$= \frac{Hs}{(2s^2 + 1) + (s^3 + 2s)} = \frac{N}{\text{Ev} + \text{Od}};$$

$$N = Hs, \qquad \text{Ev} = 2s^2 + 1, \qquad \text{Od} = s^3 + 2s.$$

Since N is odd, use Eq. (5-11a) to construct the input impedance of the LC network:

$$Z_{1LC} = \frac{\text{Od}}{\text{Ev}} = \frac{s(s^2 + 2)}{2s^2 + 1} = \frac{1}{2} \frac{s(s^2 + 2)}{s^2 + \frac{1}{2}}.$$

Expand Z_{1LC} in partial fractions to obtain the Foster I network (in this case, there being only three elements, the Foster I network is identical with the Cauer I):

$$Z_{1LC} = \frac{1}{2} s + \frac{\frac{3}{4}s}{s^2 + \frac{1}{2}} = \frac{1}{2} s + \frac{1}{\frac{4}{3} s + \frac{2}{3s}} = sL_1 + \frac{1}{sC_1 + \frac{1}{sL_2}} \; ; \qquad (5\text{-}12a)$$

$$L_1 = \tfrac{1}{2}, \qquad C_1 = \tfrac{4}{3}, \qquad L_2 = \tfrac{3}{2}. \qquad (5\text{-}12b)$$

Draw the Z_{1LC} network and arrange the output port in such a way that two zeros at infinity and one zero at the origin are realized. The result is the network shown in Fig. 5-10a. The zeros at infinity are caused by L_1 and C_1; the zero at the origin is caused by L_2.

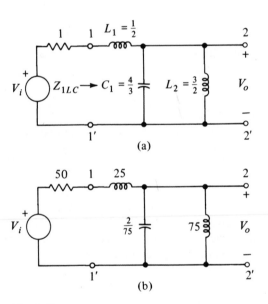

(a)

(b)

Fig. 5-10

By inspection of the network and the function, obtain (V_o/V_i) for s large:

$$\left. \frac{V_o}{V_i} \right|_{s\,\text{large}} \cong \frac{1}{sL_1} \frac{1}{sC_1} = \frac{3}{2} \frac{1}{s^2}.$$

Hence, H is $\tfrac{3}{2}$.

Multiply all impedances by 50 to obtain a realization with 50-Ω source resistance. The resulting network is given in Fig. 5-10b. It realizes the desired transfer function using a 50-Ω source resistance.

To obtain an alternative solution to the problem, expand Y_{1LC} in partial fractions and get the Foster II network (in this case, there being only three elements, the Foster II network is identical with Cauer II):

$$Y_{1LC} = \frac{2s^2 + 1}{s(s^2 + 2)} = \frac{\frac{1}{2}}{s} + \frac{\frac{3}{2}s}{s^2 + 2}$$

$$= \frac{1}{2s} + \frac{1}{\frac{2}{3}s + \frac{4}{3s}} = \frac{1}{sL_a} + \frac{1}{sL_b + \frac{1}{sC_a}};$$ (5-13a)

$$L_a = 2, \qquad L_b = \tfrac{2}{3}, \qquad C_a = \tfrac{3}{4}.$$ (5-13b)

Draw the Y_{1LC} network and arrange the output port such that two zeros at infinity and one zero at the origin result, as shown in Fig. 5-11a. The

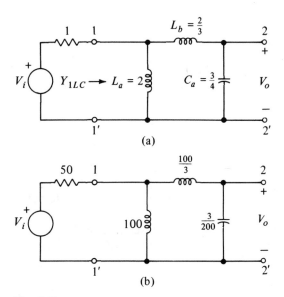

Fig. 5-11

zeros at infinity are caused by L_b and C_a; the zero at the origin is caused by L_a.

By inspection of the network and the function, obtain (V_o/V_i) for s large:

$$\left.\frac{V_o}{V_i}\right|_{s\,\text{large}} \cong \frac{1}{sL_b}\frac{1}{sC_a} = \frac{2}{s^2}.$$

Hence H is 2.

Finally, multiply all impedances by 50 to obtain a realization with 50-Ω source resistance. The resulting network is given in Fig. 5-11b.

Resistive Termination on the Load Side

Another convenient place to introduce a resistor in an LC network is at the output port, where it may serve as a load. Let this load resistance be 1Ω. A given transfer function, (V_o/V_i), that meets the realizability conditions set forth by Eq. (5-5), may then be realized in the form shown in Fig. 5-12a.

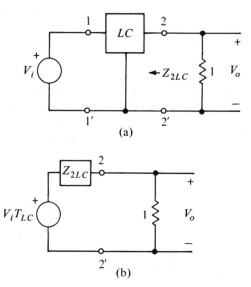

Fig. 5-12   Resistive-load termination

Let $T_{LC}(s)$ represent the transfer function (voltage ratio) from port 1 to port 2 *when port 2 is open-circuited*. Let Z_{2LC} represent the output impedance of the LC network $(V_i = 0)$. Using T_{LC} and Z_{2LC}, the Thévenin-equivalent circuit, looking in from port 2, is drawn as in Fig. 5-12b, and the transfer function is calculated as

$$\frac{V_o}{V_i} = \frac{T_{LC}}{1 + Z_{2LC}}. \tag{5-14}$$

Because T_{LC} represents a voltage ratio in an LC ladder network, it is, by Eq. (5-6), an even function of s. In order to obtain T_{LC} and Z_{2LC} from the given transfer function, it is decomposed as follows:

$$\frac{V_o}{V_i} = \frac{N}{D} = \frac{N}{\text{Ev} + \text{Od}}; \tag{5-15}$$

$$\frac{V_o}{V_i} = \frac{\dfrac{N}{\text{Ev}}}{1 + \dfrac{\text{Od}}{\text{Ev}}} \quad (N \text{ is even}); \tag{5-16a}$$

$$\frac{V_o}{V_i} = \frac{\dfrac{N}{Od}}{1 + \dfrac{Ev}{Od}} \qquad (N \text{ is odd}).$$ (5-16b)

Comparison of Eq. (5-14) with Eq. (5-16) gives

$$T_{LC} = \frac{N}{Ev}, \qquad Z_{2LC} = \frac{Od}{Ev}, \qquad \text{when } N \text{ is even};$$ (5-17a)

$$T_{LC} = \frac{N}{Od}, \qquad Z_{2LC} = \frac{Ev}{Od}, \qquad \text{when } N \text{ is odd.}$$ (5-17b)

Thus, in terms of the N, Ev, and Od parts of the given voltage ratio, T_{LC} and Z_{2LC} can be constructed by using Eq. (5-17). Then the constructed output impedance, Z_{2LC}, is developed as a ladder, the branches of which realize all the zeros of (V_o/V_i). [The finite zeros of T_{LC} may differ from the finite zeros of (V_o/V_i) because of cancellation.]

EXAMPLE 5-4

Working between a 50-Ω load and a zero source resistance, design a network that realizes:

$$\frac{V_o}{V_i} = H \frac{s}{(s + 1)[(s + \frac{1}{2})^2 + \frac{3}{4}]}.$$

What is the H of the resulting realization?

SOLUTION

The given transfer function is identical with the one used in Example 5-3. As before, decompose (V_o/V_i) and obtain N, Ev, and Od:

$$\frac{V_o}{V_i} = \frac{Hs}{(2s^2 + 1) + (s^3 + .2s)} = \frac{N}{Ev + Od};$$

$$N = Hs, \qquad Ev = 2s^2 + 1, \qquad Od = s^3 + 2s.$$

Because N is odd, use Eq. (5-17b) to construct the output impedance of the LC network:

$$Z_{2LC} = \frac{Ev}{Od} = \frac{2s^2 + 1}{s(s^2 + 2)}.$$

Expand Z_{2LC} in partial fractions to obtain the Foster I (or Cauer II) network representation:

$$Z_{2LC} = \frac{\frac{1}{2}}{s} + \frac{\frac{3}{2}s}{s^2 + 2}$$

$$= \frac{1}{2s} + \frac{1}{\frac{2}{3}s + \frac{4}{3s}} = \frac{1}{sC_1} + \frac{1}{sC_2 + \frac{1}{sL_1}};$$

$$C_1 = 2, \qquad C_2 = \tfrac{2}{3}, \qquad L_1 = \tfrac{3}{4}.$$

Draw the Z_{2LC} network and arrange the input port such that two zeros at infinity and one zero at the origin are realized. The result is the network shown in Fig. 5-13a. (Note that the component values could have been

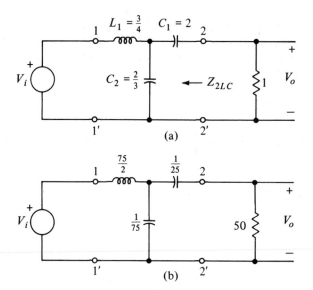

Fig. 5-13

taken directly from the values of Fig. 5-11a and assigned in the reverse direction.) The zeros at infinity are caused by L_1 and C_2; the zero at the origin is caused by C_1.

By inspection of the network, obtain (V_o/V_i) for s large:

$$\left.\frac{V_o}{V_i}\right|_{s\text{ large}} \cong \frac{1}{sL_1}\frac{1}{sC_2} = \frac{2}{s^2}.$$

Hence, $H = 2$. To obtain the final network with 50-Ω load, multiply all impedances by 50. The resulting network is shown in Fig. 5-13b.

To obtain an alternative solution to the problem, expand Y_{2LC} in partial fractions, and get the Foster II (or Cauer I) network:

$$Y_{2LC} = \frac{s(s^2 + 2)}{2s^2 + 1} = \frac{1}{2}\frac{s(s^2 + 2)}{s^2 + \frac{1}{2}}$$

$$= \frac{1}{2}s + \frac{\frac{3}{4}s}{s^2 + \frac{1}{2}} = \frac{1}{2}s + \frac{1}{\frac{4}{3}s + \frac{2}{3s}} = sC_a + \frac{1}{sL_a + \frac{1}{sC_b}};$$

$$C_a = \tfrac{1}{2}, \qquad L_a = \tfrac{4}{3}, \qquad C_b = \tfrac{3}{2}.$$

Draw the Y_{2LC} network and arrange the input port so that two zeros at infinity and one zero at the origin are realized. The result is the network shown in Fig. 5-14a. (Note that the component values could have been taken

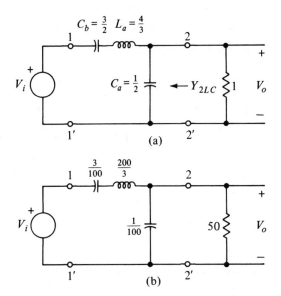

Fig. 5-14

directly from the values of Fig. 5-10a and assigned in the reverse direction.) The zeros at infinity are caused by L_a and C_a; the zero at the origin is caused by C_b.

By inspection of the network, obtain (V_o/V_i) for s large:

$$\left.\frac{V_o}{V_i}\right|_{s\text{ large}} \cong \frac{1}{sL_a}\frac{1}{sC_a} = \frac{3}{2}\frac{1}{s^2}.$$

Hence, $H = (3/2)$. To obtain the final network with 50-Ω load, multiply all impedances by 50. The resulting network is shown in Fig. 5-14b.

5-4 TRANSFER-FUNCTION ZEROS CAUSED BY *RC* IMPEDANCES

The poles of *RC* networks are on the nonpositive real axis; the zeros of *RC* impedances are on the negative real axis. Therefore, when *RC* impedances are used as series or shunt elements in a ladder, the resulting transfer-function zeros are on the nonpositive real axis. Consider, for example, the *RC* ladder shown in Fig. 5-15.

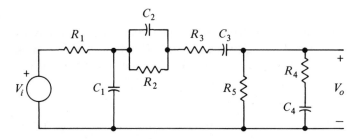

Fig. 5-15 Nonpositive-real-axis zero-producing sections

The shunt capacitor C_1 causes a zero of (V_o/V_i) at infinity because

$$\frac{1}{sC_1}\bigg|_{s\to\infty} \to 0.$$

The $R_2 C_2$ network in the series branch of the ladder causes a zero of (V_o/V_i) at $s = -(1/R_2 C_2)$ because its impedance $\{(1/C_2)/[s + (1/R_2 C_2)]\}$ becomes infinite for $s = -(1/R_2 C_2)$.

The series capacitor C_3 causes a zero of (V_o/V_i) at $s = 0$ because

$$\frac{1}{sC_3}\bigg|_{s\to 0} \to \infty.$$

The $R_4 C_4$ network in the shunt branch of the ladder causes a zero of (V_o/V_i) at $s = -(1/R_4 C_4)$ because its impedance $\{R_4[s + (1/R_4 C_4)]/s\}$ becomes zero for $s = -(1/R_4 C_4)$.

Altogether (V_o/V_i) has three finite zeros and one zero at infinity. To cause the zero at infinity, V_o/V_i must have four poles. Therefore, (V_o/V_i) is written as

$$\frac{V_o}{V_i} = H \frac{\left(s + \dfrac{1}{R_2 C_2}\right)s\left(s + \dfrac{1}{R_4 C_4}\right)}{s^4 + \cdots + b_0}. \qquad (5\text{-}18)$$

By inspection of the network,

$$\frac{V_o}{V_i}\bigg|_{s\text{ large}} \cong \frac{1}{R_1}\frac{1}{sC_1}\frac{\dfrac{R_4 R_5}{R_4 + R_5}}{R_3 + \dfrac{R_4 R_5}{R_4 + R_5}},$$

which fixes H in Eq. (5-18) as

$$\frac{R_4 R_5}{R_1 C_1 [R_3 (R_4 + R_5) + R_4 R_5]}.$$

Similarly, by inspection of the network,

$$\left.\frac{V_o}{V_i}\right|_{s \text{ small}} \cong sC_3 R_5,$$

which requires that, in Eq. (5-18),

$$b_0 = \frac{1}{R_1 C_1 R_2 C_2 R_3 C_3 R_4 C_4 \left(1 + \dfrac{R_5}{R_4} + \dfrac{R_5}{R_3}\right)}.$$

The poles of (V_o/V_i) are all on the negative real axis. It is interesting to note that if R_5 were infinite, then C_3 no longer would cause a zero at the origin, since the network to the right of C_3 has also a pole at $s = 0$ (caused by C_4). In this case, b_0 would be zero in Eq. (5-18), and the factorable s term in the denominator would cancel the zero in the numerator, thus reducing V_o/V_i to a third-order system.

If all the capacitors in an *RC* ladder network are directly in shunt, the zeros of the transfer function are all at infinity. On the other hand, if all capacitors appear directly in the series branches of the ladder, the transfer-function zeros are all at the origin.

5-5 SYNTHESIS OF *RC* LADDER NETWORKS

Any transfer function of the form

$$\frac{V_o}{V_i} = H \frac{s^k \left(\dfrac{s}{\alpha_{z1}} + 1\right)\left(\dfrac{s}{\alpha_{z2}} + 1\right) \cdots \left(\dfrac{s}{\alpha_{zm}} + 1\right)}{(s + \alpha_{p1})(s + \alpha_{p2}) \cdots (s + \alpha_{pn})} \qquad (0 < \alpha_{p1} < \alpha_{p2} < \cdots < \alpha_{pn})$$

$$(5\text{-}19)$$

is realizable to within a scale factor as an *RC* ladder network if $k + m \le n$, and H is a positive number. Some or all of the zeros may be put at infinity by selecting the k and α_z's appropriately.

The synthesis procedure starts with the construction of an input *RC* admittance function, Y_i. Since Y_i, not Z_i, is a system function, its poles are selected to agree with the poles of Eq. (5-19). The zeros of Y_i are selected so that Y_i is *RC*-realizable.

$$Z_i = \frac{1}{Y_i}$$

is then expanded in the form of a ladder network, so that all the zeros of V_o/V_i are realized.

For an example, consider

$$\frac{V_o}{V_i} = H\frac{s^n}{(s + \alpha_1)(s + \alpha_2)\cdots(s + \alpha_n)}, \tag{5-20}$$

which has n zeros at the origin. The n poles call for n capacitors, which must be put in the series branches of the ladder to cause the n zeros for $s = 0$. To find the element values, the system function Y_i is constructed with poles identical with Eq. (5-20). Its zeros are arbitrary as long as they alternate with the poles. Hence, the input impedance Z_i is written as

$$Z_i = \frac{(s + \alpha_1)(s + \alpha_2)\cdots(s + \alpha_n)}{(s + \alpha_{p1})(s + \alpha_{p2})\cdots(s + \alpha_{pn})}, \tag{5-21}$$

where the α_p numbers are chosen such that

$$0 \leq \alpha_{p1} < \alpha_1 < \alpha_{p2} < \alpha_2 \cdots < \alpha_{pn} < \alpha_n.$$

The Cauer II expansion of Eq. (5-21) not only realizes the constructed input impedance but also satisfies Eq. (5-20), since the capacitors automatically appear in the series branches, as shown in Fig. 5-16. In the realization shown in Fig. 5-16, α_{p1} is taken as zero; otherwise an additional resistor appears across the

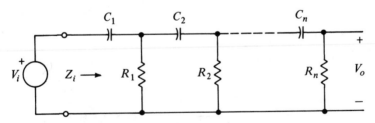

Fig. 5-16   Transfer function with all the zeros at the origin

input. It is also possible to use an additional pole for Z_i, in which case an extra capacitor appears across the output. Neither the resistor across the input nor the capacitor across the output would affect the poles and zeros of (V_o/V_i). However, the element values in the network would be different.

Inspection of the high-frequency asymptotic behavior of Fig. 5-16 reveals that $H = 1$.

If all the poles of (V_o/V_i) are on the nonpositive real axis and all its zeros are at infinity, then, in the ladder realization, all capacitors appear in the shunt arms and all resistors in the series arms.

EXAMPLE 5-5

Obtain a realization of

$$\frac{V_o}{V_i} = H\frac{s^2}{(s + 1)(s + 3)}.$$

What is the resulting H?

SOLUTION

(V_o/V_i) is of second order with poles on the negative real axis. Hence, an RC realization with two capacitors can be used. (V_o/V_i) has two zeros at $s = 0$. Hence, the capacitors must appear in the series arms of the ladder network. Two sections of the network shown in Fig. (5-16) will realize the desired transfer function. To obtain the element values, the input impedance must be constructed and developed. The input impedance must have zeros at $s = -1$ and $s = -3$. [The input admittance is a system function, and therefore must have the same poles as (V_o/V_i).] Let

$$Z_i = H\frac{(s + 1)(s + 3)}{(s + \alpha_1)(s + \alpha_2)} \qquad (0 \le \alpha_1 < 1 < \alpha_2 < 3).$$

To simplify results, let $H = 1$, $\alpha_1 = 0$, and $\alpha_2 = 2$. Then

$$Z_i = \frac{(s + 1)(s + 3)}{s(s + 2)}.$$

Develop this impedance in Cauer II form to assure that capacitors end up in the series arms. Since the first element is in the series arm, perform the continued-fraction expansion on Z_i:

$$Z_i = \frac{3 + 4s + s^2}{2s + s^2} = \frac{3}{2s} + \cfrac{1}{\frac{4}{5} + \cfrac{1}{(25/2s) + 5}} = \frac{1}{sC_1} + \cfrac{1}{\frac{1}{R_1} + \cfrac{1}{(1/sC_2) + R_2}};$$

hence

$$C_1 = \tfrac{2}{3}, \qquad R_1 = \tfrac{5}{4}, \qquad C_2 = \tfrac{2}{25}, \qquad R_2 = 5.$$

The resulting realization is shown in Fig. 5-17.

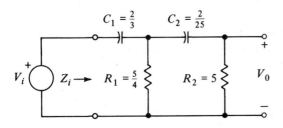

Fig. 5-17

To evaluate H, obtain (V_o/V_i) from Fig. 5-17 for s large, and compare it with the large-s value of the given function:

$$\left.\frac{V_o}{V_i}\right|_{s \text{ large}} \cong 1 = H.$$

Thus $H = 1$.

5-6 SOME *RC* NETWORKS WITH GAIN GREATER THAN UNITY

It is possible to construct *RC* networks so that the transfer-function magnitude,

$$\left|\frac{V_o(j\omega)}{V_i(j\omega)}\right|,$$

is greater than 1 over a band of frequencies. Such networks are said to possess *gain*. This greater-than-unity gain characteristic can be used with a voltage amplifier, having a gain less than unity, to construct oscillators. (See Section 15-5.)

To see how a gain greater than unity may be achieved, consider the frequency response of the low-pass network shown in Fig. 5-18.

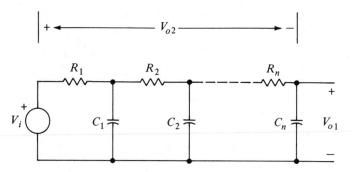

Fig. 5-18   Low-pass network

All the zeros are at infinity because of the shunt capacitors; therefore, the transfer function (V_{o1}/V_i) is written as

$$\frac{V_{o1}}{V_i} = \frac{H}{(s + \alpha_1)(s + \alpha_2) \cdots (s + \alpha_n)} \qquad (0 < \alpha_1 < \alpha_2 < \cdots < \alpha_n). \qquad (5\text{-}22)$$

Comparison of the high-frequency and low-frequency asymptotic behavior of Eq. (5-22) with that of the network of Fig. 5-18 reveals that

$$H = \frac{1}{R_1 C_1 R_2 C_2 \cdots R_n C_n},$$

$$\alpha_1 \alpha_2 \cdots \alpha_n = \frac{1}{R_1 C_1 R_2 C_2 \cdots R_n C_n}.$$

For $s = j\omega$, (V_{o1}/V_i) becomes

$$\frac{V_{o1}}{V_i} = M(\omega)e^{j\theta(\omega)},$$

where

$$M(\omega) = \frac{1}{\sqrt{\left(\dfrac{\omega^2}{\alpha_1^2} + 1\right)\left(\dfrac{\omega^2}{\alpha_2^2} + 1\right) \cdots \left(\dfrac{\omega^2}{\alpha_n^2} + 1\right)}} \qquad (5\text{-}23)$$

and

$$\theta(\omega) = -\left[\tan^{-1}\frac{\omega}{\alpha_1} + \tan^{-1}\frac{\omega}{\alpha_2} + \cdots + \tan^{-1}\frac{\omega}{\alpha_n}\right]. \qquad (5\text{-}24)$$

Both the magnitude characteristic, $M(\omega)$, and the phase characteristic, $\theta(\omega)$, are decreasing functions of ω and are sketched in Fig. 5-19a and b.

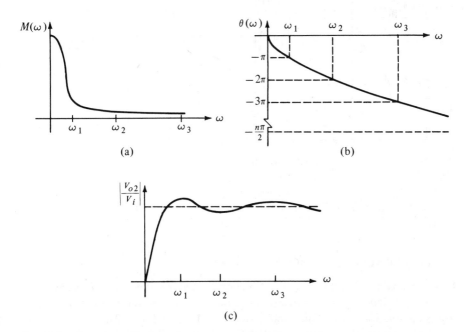

Fig. 5-19   Steps in achieving greater-than-unity gain

Note that the network is capable of shifting the phase by as much as $(n\pi/2)$, $(\pi/2)$ radians for each pole. If $n > 2$, there is always one or more frequencies $(\omega_1, \omega_3, \text{etc.})$, at which the steady-state output is exactly 180° out of phase with the input. At these frequencies, if the output V_{o1} is subtracted from the input, V_i, the two sinusoidal signals add to produce a steady-state signal greater than the input. But $(V_1 - V_{o1})$ is V_{o2}, the voltage across the series branches of the ladder, as shown in Fig. 5-18. Hence, the transfer function (V_{o2}/V_i) has a magnitude characteristic that exceeds 1 at ω_1, ω_3, etc. In fact, the magnitude is greater than 1 not only at these frequencies but also at frequencies adjacent to them. (Two sine waves of the same frequency, when combined, produce a sine wave of amplitude greater than the amplitude of either wave for a wide range of phase differences.) The magnitude of (V_{o2}/V_i), which is sketched in Fig. 5-19c, shows alternate bands of frequencies over which the magnitude is less or greater than 1. It is possible to achieve a gain greater than unity also with a two-section ladder (see Problem 5-19).

Figure 5-18 can be redrawn so that the input V_i and the output V_{o2} have common ground. This is shown in Fig. 5-20 for a three-section ladder, using equal resistors and equal capacitors.

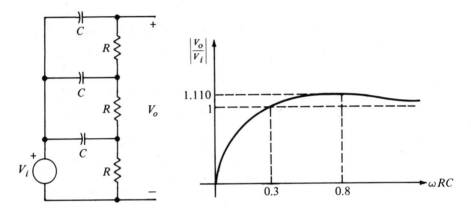

Fig. 5-20   An RC network with gain greater than unity

The magnitude characteristic becomes greater than 1 for frequencies slightly greater than $(0.3/RC)$. The maximum gain occurs at $\omega = (0.8/RC)$, and it is equal to 1.110.

5-7 *RC* PARALLEL-LADDER NETWORKS

The zeros of RC ladder transfer functions are all on the nonpositive real axis. When such ladders are connected in parallel, it is possible to obtain complex transfer-function zeros. One such network is given in Fig. 5-21a.

The ladder network composed of $R_1 C_1 R_2$ is in parallel with the ladder network composed of C_2. Alternatively, the network may be considered as a single ladder excited at both ends, and the output may be taken between the two ends, as shown in Fig. 5-21b. Using superposition, V_o is written as

$$V_o = V_i T_l + V_i T_r = V_i \frac{N_l}{D} + V_i \frac{N_r}{D}, \tag{5-25}$$

where $T_l = (N_l/D)$ and $T_r = (N_r/D)$ are the transfer functions associated with the lefthand and the righthand sources, respectively. By inspection of Fig. 5-21b,

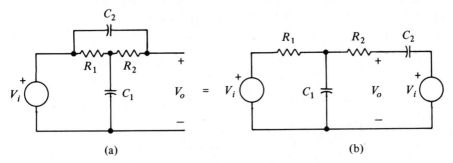

Fig. 5-21   Network with complex transfer-function zeros

the two zeros of T_l are seen to be at infinity (caused by C_1 and C_2), whereas the two zeros of T_r are at

$$s = 0 \qquad \text{(caused by } C_2\text{)}$$

and at

$$s = -\frac{R_1 + R_2}{R_1 R_2 C_1} \qquad \text{(caused by the } R_2 R_1 C_1 \text{ network)}.$$

Using these zeros and examining the high- and low-frequency asymptotic behavior, V_o is readily obtained as

$$V_o = V_i \frac{\dfrac{1}{R_1 R_2 C_1 C_2}}{s^2 + \cdots + \dfrac{1}{R_1 R_2 C_1 C_2}} + V_i \frac{s\left(s + \dfrac{R_1 + R_2}{R_1 R_2 C_1}\right)}{s^2 + \cdots + \dfrac{1}{R_1 R_2 C_1 C_2}}. \tag{5-26}$$

The zeros of (V_o/V_i) are given by the roots of

$$s^2 + s\left[\frac{R_1 + R_2}{R_1 R_2 C_1}\right] + \frac{1}{R_1 R_2 C_1 C_2} = 0. \tag{5-27}$$

The zeros are complex if

$$\left(\frac{R_1 + R_2}{R_1 R_2 C_1}\right)^2 - \frac{4}{R_1 R_2 C_1 C_2} < 0, \tag{5-28}$$

which simplifies to:

$$\frac{C_1}{C_2} > \frac{1}{2} + \frac{1}{4}\left(\frac{R_1}{R_2} + \frac{R_2}{R_1}\right). \tag{5-29}$$

Thus, for $R_1 = R_2$, complex zeros in the left half-plane result if $C_1 > C_2$. The resistors may be used to adjust the real part of the zeros, and C_2 may be used to control the imaginary part.

Another left-half-plane complex-zero-producing parallel-ladder network is given in Fig. 5-22. For this network, the zeros are complex if

$$\frac{R_2}{R_1} > \frac{1}{2} + \frac{1}{4}\left(\frac{C_1}{C_2} + \frac{C_2}{C_1}\right). \tag{5-30}$$

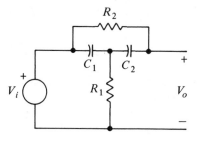

Fig. 5-22   Transfer function with complex zeros

5-8 SUMMARY

A voltage ratio with left-half-plane poles and imaginary-axis zeros is realized as a resistively-terminated LC network, provided the number of finite zeros does not exceed the number of poles.

When a resistive-source (load) termination is desired and the number of imaginary-axis zeros is odd (even), develop the Od/Ev ratio as the input (output) impedance of the LC ladder network. Arrange the impedances in the series and shunt arms of the ladder so that the desired transfer-function zeros are produced. If all the zeros are at infinity, develop the impedance in Cauer I form so that all inductors are in the series arms and all capacitors in the shunt arms. On the other hand, if all the zeros are at the origin, develop the impedance in Cauer II form so that all capacitors are in the series arms and all inductors in the shunt arms. By mixing Cauer I and II forms, it is possible to place some of the zeros at the origin and the rest at infinity.

When a resistive-source (load) termination is desired and the number of imaginary-axis zeros is even (odd), develop the Ev/Od ratio as the input (output) impedance of the LC ladder network.

To realize complex zeros with RC networks, use parallel-ladder structures.

To obtain gains greater than unity with RC networks, construct RC low-pass (all capacitors in shunt) or high-pass (all capacitors in series) ladder networks employing at least two capacitors, and take as output the voltage across all the series branches.

PROBLEMS

5-1   For the network shown in Fig. 5-23, obtain by inspection all the zeros of (V_o/V_i). How many poles does (V_o/V_i) have?

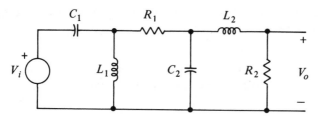

Fig. 5-23

5-2 For the network shown in Fig. 5-24, obtain by inspection all the zeros of (V_o/V_i) and the number of its poles.

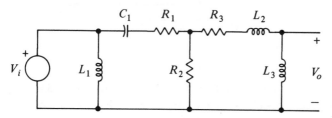

Fig. 5-24

5-3 Obtain by inspection the zeros and the number of poles of (V_o/V_i) for all the networks shown in Fig. 5-25. Then, calculate (V_o/V_i) for all the net-

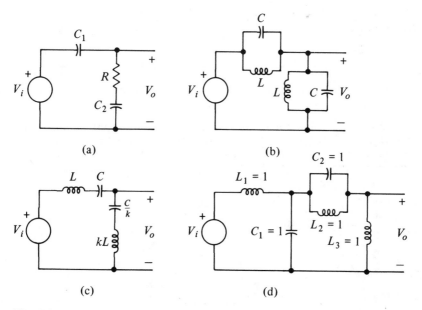

(a) (b)

(c) (d)

Fig. 5-25

works. Compare the calculated zeros with those found by inspection. Explain any differences.

5-4   In the steady state, the third harmonic of the oscillator shown in Fig. 5-26 is to be suppressed by the filter, without attenuating the fundamental. Using as few elements as possible, obtain two different designs for Z.

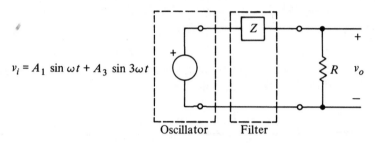

Fig. 5-26

5-5   In Fig. 5-27, obtain in factored form the numerator polynomial for (V_o/V_i). How many poles does (V_o/V_i) have? What is (V_o/V_i) for s small and for s large?

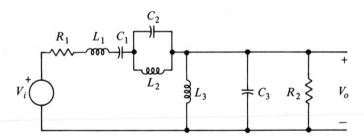

Fig. 5-27

5-6   In Fig. 5-28, obtain in factored form the numerator polynomial for (V_o/V_i). How many poles does (V_o/V_i) have? What is (V_o/V_i) for s small and for s large?

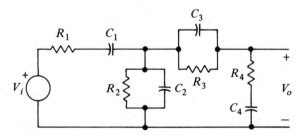

Fig. 5-28

5-7 Determine the transfer-function zeros caused by the two *RLC* networks shown in Fig. 5-29.

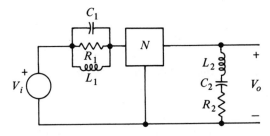

Fig. 5-29

5-8 An *LC* network is excited, starting at $t = 0$, by a sine wave, as shown in Fig. 5-30. Design the network in such a way that its output is a nonzero dc voltage.

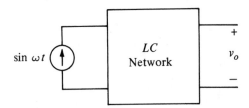

Fig. 5-30

5-9 If losses are neglected, the equivalent circuit of the quartz crystal is as shown in Fig. 5-31a.

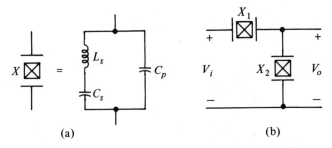

Fig. 5-31

(a) Sketch the reactance of the crystal vs. ω.
(b) Two quartz crystals, X_1 and X_2, are connected as shown in Fig. 5-31b.

X_2 is cut so that its complex *poles* (for the impedance) match exactly with the complex *zeros* of X_1. Sketch

$$\left|\frac{V_o}{V_i}\right| \quad \text{vs.} \quad \omega \quad \text{(no values).}$$

5-10 How do the transfer functions $(I_o/V_i,\ V_o/I_i,\ I_o/I_i)$ of the ladder network shown in Fig. 5-32 differ from the transfer functions of voltage ratios of *LC* ladder networks?

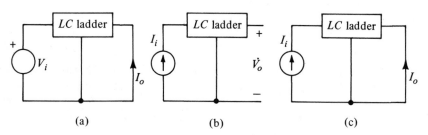

(a)   (b)   (c)

Fig. 5-32

5-11 Sketch the $|T(j\omega)|$-vs.-ω curve:

$$T(s) = \frac{H}{(s^2 + \omega_1^2)(s^2 + \omega_2^2)(s^2 + \omega_3^2)}.$$

5-12 Obtain network realizations for each transfer function given below, by developing the input impedance function, Z_i. The zeros of Z_i must agree with the poles of (V_o/V_i) (why?), and its poles must be chosen such that the function is *LC*-realizable. Determine also the H of the resulting realizations.

(a) $\dfrac{V_o}{V_i} = \dfrac{H}{(s^2 + 1)(s^2 + 3)(s^2 + 5)};$

(b) $\dfrac{V_o}{V_i} = H\dfrac{s^6}{(s^2 + 1)(s^2 + 3)(s^2 + 5)};$

(c) $\dfrac{V_o}{V_i} = H\dfrac{s^2}{(s^2 + 1)(s^2 + 3)(s^2 + 5)}.$

5-13 Obtain two realizations for $(V_o/V_i) = 1/[(s + 1)(s^2 + s + 1)]$.

5-14 Without elaborate calculations, find all the transfer-function zeros of the network shown in Fig. 5-33.

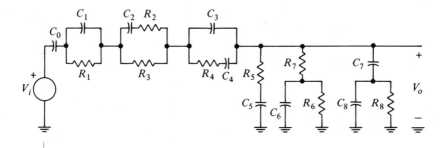

Fig. 5-33

5-15 Obtain an *RC* realization, and the value of *H* for:

(a) $\dfrac{V_o}{V_i} = \dfrac{H}{(s+1)(s+3)}$;

(b) $\dfrac{V_o}{V_i} = H\dfrac{s}{(s+1)(s+3)}$;

(c) $\dfrac{I_o}{I_i} = H\dfrac{s^2}{(s+1)(s+3)}$;

(d) $\dfrac{I_o}{V_i} = H\dfrac{s^3}{(s+1)(s+3)}$;

(e) $\dfrac{V_o}{I_i} = \dfrac{H}{s(s+2)(s+4)}$;

(f) $\dfrac{V_o}{V_i} = H\dfrac{s+1}{s+2}$.

5-16 Without calculating the element values, obtain networks that have the indicated transfer functions:

(a) $\dfrac{V_o}{V_i} = \dfrac{s^2}{(s+\alpha_1)(s+\alpha_2)}$;

(b) $\dfrac{V_o}{V_i} = \dfrac{1}{(s^2+\alpha_1)(s^2+\alpha_2)}$;

(c) $\dfrac{V_o}{V_i} = \dfrac{s^2}{(s^2+\alpha_1)(s^2+\alpha_2)}$.

5-17 Show that complex zeros are produced by the network of Fig. 5-22 if Eq. (5-30) is satisfied.

5-18 For the network shown in Fig. 5-18, the values of the resistors and capacitors are adjusted to produce

$$\dfrac{V_{o1}}{V_i} = \dfrac{6}{(s+1)(s+2)(s+3)}.$$

Obtain the poles and zeros of (V_{o2}/V_i).

5-19 Show that in Fig. 5-34 $|V_o(j\omega)/V_i(j\omega)|$ can become greater than 1.

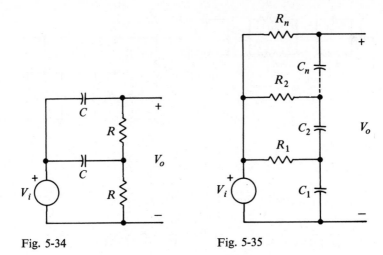

Fig. 5-34   Fig. 5-35

5-20 For the network shown in Fig. 5-35, sketch $|V_o(j\omega)/V_i(j\omega)|$ vs. ω.

5-21 A bridged-T network is shown in Fig. 5-36. Under what condition does (V_o/V_i) have a pair of imaginary-axis zeros? At what frequency do they occur?

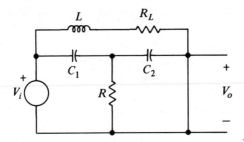

Fig. 5-36

6

Properties of
Second-Order Systems

The study of second-order systems is important because of their simplicity and usefulness. Not only are their mathematical properties readily obtainable, but their physical characteristics are also easily measurable and adjustable. Furthermore, they serve as building blocks for higher-order systems.

In this chapter, second-order systems are described in terms of the α and β, or the ω_o and Q of the poles and zeros. The physical meaning of these parameters is demonstrated in terms of the step response and the frequency response. It is shown that the loci of complex poles or zeros, as a function of a single element in the network, are either circles or constant-α lines. Therefore, the changes caused in the α and β, or the ω_o and Q, of the characteristic frequencies are readily predictable when an element of the network changes. Various sensitivity functions are used to describe these changes. These sensitivity functions indicate the direction and magnitude of incremental changes in the poles and zeros. They also show the incremental changes occurring in the magnitude and phase characteristics. Consequently, they may be used as an aid in tuning the network, or they may serve as a figure of merit in comparing the element sensitivities of different networks that realize a given system function. Throughout the chapter, the physical interpretation of the mathematical properties of second-order systems is emphasized and demonstrated with examples.

6-1 DESCRIPTION OF SECOND-ORDER SYSTEMS

The most general second-order function is the *biquadratic function*, which is the ratio of two second-order polynomials:

$$T(s) = \frac{N(s)}{D(s)} = \frac{a_2 s^2 + a_1 s + a_0}{s^2 + b_1 s + b_0}. \tag{6-1}$$

When the poles are properly placed, the numerator $N(s)$ of Eq. (6-1) can be adjusted to make $T(s)$ low-pass ($a_2 = a_1 = 0$), band-pass ($a_2 = a_0 = 0$), high-pass ($a_1 = a_0 = 0$), all-pass ($a_2 = 1$, $a_1 = -b_1$, $a_0 = b_0$), and band-stop ($a_1 = 0$). These designations refer to the magnitude of the frequency response, $|T(j\omega)|$, and indicate how the amplitudes of sine waves with different frequencies are affected in the steady state as they are processed by the network. For example, a low-pass function has a magnitude characteristic that passes the low frequencies and attenuates the high frequencies.

The interesting and most useful properties of $T(s)$ occur when its critical frequencies are complex. A pair of variables is necessary to describe these critical frequencies. Either the pair (α, β), or the pair (ω_o, Q), is generally used for this purpose. In terms of these variables, the denominator polynomial $D(s)$ can be written as:

$$D(s) = (s + \alpha)^2 + \beta^2, \tag{6-2}$$

$$D(s) = s^2 + s\frac{\omega_o}{Q} + \omega_o^2. \tag{6-3}$$

In Eq. (6-2), $-\alpha$ represents the real part of the poles and β the imaginary part. In Eq. (6-3), ω_o represents the *magnitude* of the poles (the distance from the origin to the poles), and Q is a measure of the slope of the radial lines that connect the poles to the origin. The higher the Q, the steeper the radial lines. When $Q \leq \frac{1}{2}$, the roots of Eq. (6-3) are real. When $Q > \frac{1}{2}$, the roots are complex. Figure 6-1 illustrates the relationship between the (α, β) and (ω_o, Q) designations.

If α and β are given, the ω_o and Q are found from

$$\omega_o = \sqrt{\alpha^2 + \beta^2}, \qquad Q = \frac{\sqrt{\alpha^2 + \beta^2}}{2\alpha}. \tag{6-4a}$$

If ω_o and Q are given, α and β are found from

$$\alpha = \frac{\omega_o}{2Q}, \qquad \beta = \omega_o\sqrt{1 - \frac{1}{4Q^2}}. \tag{6-4b}$$

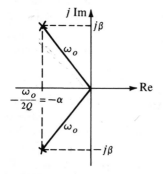

Fig. 6-1   Pole designations

Both representations contain the same information. However, the (α, β) designation lends itself more readily to physical interpretation and measurement when the step response of the system is considered. On the other hand, the (ω_o, Q) representation yields more directly measurable results when the sinusoidal steady-state response is considered. To illustrate the differences in interpretation, consider first the step response shown in Fig. 6-2.

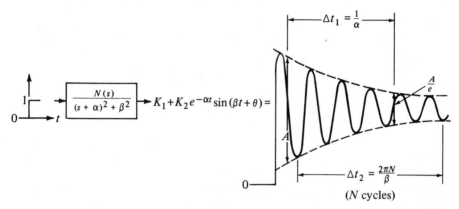

Fig. 6-2   Step response when poles are complex

The response contains a constant term, K_1, arising from the pole of excitation. It also contains a damped sine wave, $K_2 e^{-\alpha t} \sin (\beta t + \theta)$, arising from the poles of the system. The numerator polynomial, $N(s)$, determines the values of the scale factors, K_1 and K_2 ; it has no other influence. The real part of the complex poles, $-\alpha$, determines the rate at which the amplitude of the sine wave decays. The imaginary part of the complex poles, β, determines the frequency of the sine wave. If the β/α ratio is high, greater than 10, the value of α and β can be determined fairly accurately from measurements taken on an oscilloscope display of the step response. As shown in Fig. 6-2, α is found by measuring the time interval, Δt_1, between any two points on the envelope which have amplitudes of A and A/e, respectively; β is found by measuring the time interval, Δt_2, between N successive peaks of the sine wave. Thus,

$$\alpha = \frac{1}{\Delta t_1} \quad \text{and} \quad \beta = N \frac{2\pi}{\Delta t_2}. \tag{6-5}$$

The longer it takes for the envelope to decay, the smaller is α, and therefore the closer the poles are to the imaginary axis. The larger the number of cycles of the damped sine wave in a given time interval, the larger is β, and the *farther* away are the poles from the real axis.

Now consider the sinusoidal steady-state response shown in Fig. 6-3.

The steady-state amplitude of the output sine wave is proportional to $M(\omega)$, which is the magnitude of the system function evaluated at $s = j\omega$, that is,

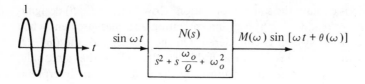

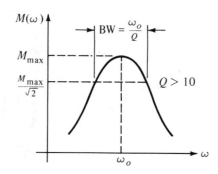

Fig. 6-3   Sinusoidal steady-state response

$$M(\omega) = \left| \left[\frac{N(s)}{s^2 + s\dfrac{\omega_o}{Q} + \omega_o^2} \right]_{s=j\omega} \right|$$

$$= \frac{|N(j\omega)|}{\sqrt{(\omega_o^2 - \omega^2)^2 + \left(\dfrac{\omega_o}{Q}\omega\right)^2}}.$$

If the Q of the poles is high, greater than 10, and the zeros of $N(s)$ are far enough from the poles, which would be the case if

$$N(s) = H, \; Hs, \; Hs^2, \; \text{or} \; H(s + \omega_o)^2,$$

then, for all practical purposes, the $M(\omega)$-vs.-ω characteristic peaks at $\omega = \omega_o$ (see Problem 6-1). As shown in Fig. 6-3, there are two frequencies, one on each side of ω_o, at which the magnitude is $1/\sqrt{2}$ times the peak value. These frequencies are called the 3-dB-[20 log $(\sqrt{2}) = 3.010 \cong 3$] frequencies. Their difference, BW, is called the 3-dB bandwidth which is related to Q (see Problem 6-1) by:

$$BW = \frac{\omega_o}{Q} \qquad (Q > 10). \tag{6-6}$$

Thus, *only two measurements are necessary to fix the position of the complex poles.* The measurement of the frequency of peaking determines the magnitude of the poles, ω_o, and the measurement of the 3-dB bandwidth determines ω_o/Q.

The higher the Q, the narrower is the bandwidth and the more selective is the system for sinusoidal signals. High Q also implies that the poles are very close to the imaginary axis. For example, a pole with a Q of 50 will be practically next to the imaginary axis, since $\alpha \cong 0.01\beta$. Consequently, high-Q poles result in a natural response which is a slowly decaying sine wave.

Sometimes, the pair (α, ω_o) is also used to describe the pole locations. In this case, $-\alpha$ is the real part of the poles, and ω_o the magnitude, and $D(s)$ is written as

$$D(s) = s^2 + 2\alpha s + \omega_o^2.$$ (6-7)

6-2 POLE–ZERO LOCI OF SECOND-ORDER SYSTEMS

A system function is a rational function which relates the output variable to the input variable. Its implementation with electrical components results in a physical system that, in general, departs from the desired system characteristics for many reasons, some of which are:

1. Actual elements are approximations to the ideal elements. For example, the inductor, defined by the mathematical relationship $v = L(di/dt)$, does not exist in practice; the terminal behavior of the actual element, at best, *approximates* the desired model for the element.
2. Actual elements are not constant. They depend on environmental conditions such as temperature, humidity, etc. Furthermore, their values may change with age.
3. Due to economic and practical considerations, the actual element values cannot be made equal to the exact values specified by the design. Rather, tolerance limits are imposed on the element values. Thus, all the resistors and capacitors in a network may differ by as much as $\pm 5\%$ from the exact values, or the open-loop gain of an amplifier may differ from one unit to another by a factor of 10.

For these reasons, a system which is designed with a certain pole–zero distribution to achieve a specified frequency response may, in practice, exhibit characteristics that are different. It is, therefore, important that the relationship between changes in element values and changes in frequency response be established clearly. How do the changes in each element of the network affect the pole–zero positions? What element should be adjusted to cause a particular correction in the frequency-response characteristics?

It is possible to acquire very close control over the system if the order is not too high, or if a pair of dominant critical frequencies shapes the characteristics over a given band of frequencies. For example, consider the system shown in Fig. 6-4a.

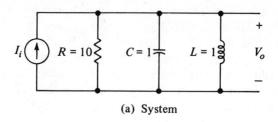

(a) System

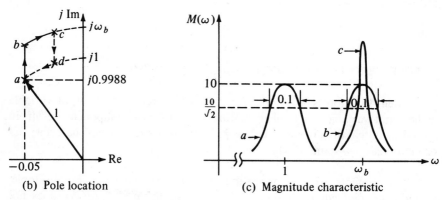

(b) Pole location   (c) Magnitude characteristic

Fig. 6-4 Effect of element changes

The system is described by

$$\frac{V_o}{I_i} = \frac{1}{C} \frac{s}{s^2 + s \dfrac{1}{RC} + \dfrac{1}{LC}}.$$ (6-8)

The pole positions are given by

$$\omega_o = \frac{1}{\sqrt{LC}} \text{ and } Q = R \sqrt{\frac{C}{L}},$$

or by

$$\alpha = \frac{1}{2RC} \text{ and } \beta = \sqrt{\frac{1}{LC} - \left(\frac{1}{2RC}\right)^2}.$$

Note that the magnitude of the poles, ω_o, is independent of R, and the real part of the poles, $-\alpha$, is independent of L. On the other hand, Q and β are dependent upon all three elements in the system. Therefore, it is possible to vary the Q of the pole and keep its ω_o constant by changing R alone; or the β of the pole can be varied and its α kept constant by changing L alone. For the given set of nominal values, that is, $R = 10$, $C = 1$, and $L = 1$, the upper half-plane pole is sketched as point a in Fig. 6-4b. The pole's ω_o is 1 and its Q is 10. The corresponding magnitude response is drawn as curve a in Fig. 6-4c. The frequency of peaking is $\omega_o = 1$, and the 3-dB bandwidth is

$$\omega_o/Q = 0.1.$$

Now, if L is decreased slightly from its nominal value of 1, the pole will move in a constant-α manner from point a to point b in Fig. 6-4b, resulting in a new pole with a larger ω_o (further out from the origin) and a higher Q (steeper radial line). Therefore, the magnitude characteristic peaks at a higher frequency, $\omega = \omega_b$. The 3-dB bandwidth stays the same, since

$$BW = \omega_o/Q = 1/RC$$

is independent of L. The result is curve b in Fig. 6-4c. If R is now increased slightly from its nominal value of 10, the pole will move in a constant-ω_o manner from point b to point c in Fig. 6-4b, thereby increasing the Q. The result is a sharpening of the magnitude characteristic about $\omega = \omega_b$ (curve c).

Once the relationship between the pole position and the resulting magnitude characteristic is understood, and the effect of each element on the pole position is known, a network can be tuned accurately to achieve the desired response. For example, suppose the desired response is curve a of Fig. 6-4c but actual measurements give curve c. The following two-step adjustment procedure can be used to obtain the desired curve. First, R may be decreased in order to decrease the Q without affecting the frequency of peaking (pole moves from c to b); then L may be increased to move the center frequency from ω_b to 1 (pole moves from b to a). On the other hand, L may be increased first, causing the pole to move from c to d, thereby resulting in a shift of the frequency of peaking from ω_b to 1; then R may be decreased, causing the pole to move from d to a, resulting in curve a. In practice, it may be necessary to repeat these adjustments until the desired response is obtained. However, at every stage of tuning, the experimenter has full control of the system. By watching the magnitude characteristic on an oscilloscope while the input frequency is swept, he knows what element to vary and in what direction to vary it to bring about a desired change in the magnitude curve.

This example shows that the pole movement brought about by changes in R is different from that caused by changes in L. If C had been changed, the pole would have moved in still a different direction since C affects the denominator polynomial in a manner different from R or L. This can be demonstrated by multiplying the numerator and the denominator of Eq. (6-8) by RLC and rearranging the resulting denominator polynomial so that the effect of each element is exposed explicitly, as follows:

$$D(s) = (sL) + R(LCs^2 + 1), \tag{6-9a}$$

$$D(s) = (R) + L(RCs^2 + s), \tag{6-9b}$$

$$D(s) = (sL + R) + C(RLs^2). \tag{6-9c}$$

If $D(s)$ is written as $c_2 s^2 + c_1 s + c_0$, then the set given by Eq. (6-9) shows that R affects the c_2 and c_0 coefficients in the $D(s)$ polynomial, whereas L affects the c_2 and c_1 coefficients. These results can be put in a more formal form, by writing:

$$D(s) = P_1(s) + qP_2(s), \tag{6-10}$$

where q represents the element that is varying and $P_1(s)$ and $P_2(s)$ are polynomials that do not contain q. Thus, if q is taken as R, then

$$P_1(s) = sL \quad \text{and} \quad P_2(s) = LCs^2 + 1,$$

neither of which depend on R. Equation (6-10) shows that each element in the network appears in $D(s)$ in a linear manner, that is, q is to the first power. Indeed, this is true of denominator polynomials of any order. *A single element never enters more than once in any term of the polynomial.* This is not to be confused with the case where more than one element in the network is designated by the same symbol, in which case the polynomial may contain terms with the element value raised to powers greater than one.

General Formulation of Root Loci

In general, an element in the network appears in a linear manner in both the numerator and the denominator polynomials of a system function. Therefore, Eq. (6-1) can be written as

$$T(s) = \frac{P_a(s) + qP_b(s)}{P_c(s) + qP_d(s)}, \tag{6-11}$$

where q is an element in the network, and $P_a(s)$, $P_b(s)$, $P_c(s)$, and $P_d(s)$ are polynomials that do not contain q. A change in q causes a change in the poles and zeros of $T(s)$. If $T(s)$ is biquadratic, then the general form of the numerator or the denominator polynomial is

$$P(s) = P_1(s) + qP_2(s)$$
$$= (a_2 s^2 + a_1 s + a_0) + q(b_2 s^2 + b_1 s + b_0), \tag{6-12}$$

where the a and b coefficients are independent of q. As long as the second-order nature of $P(s)$ is preserved, some of the a and b coefficients in Eq. (6-12) may be zero. For example, when R is the variable in the network of Fig. 6-4 and $P(s)$ is the denominator polynomial, then $a_2 = a_0 = b_1 = 0$, because

$$P(s) = sL + R(LCs^2 + 1).$$

But for $q = L$, the zero coefficients are a_2, a_1, and b_0 because

$$P(s) = R + L(RCs^2 + s).$$

The roots of Eq. (6-12) either are both real or are complex conjugates. The root positions in the s-plane change as q varies. The resulting curve is called the *root locus*. The variable q may be positive, as in the case of a resistor, or negative, as in the case of the gain of an inverting amplifier. The two roots of Eq. (6-12), s_1 and s_2, are given by:

$$s_{1,2} = \frac{-(a_1 + b_1 q) \pm \sqrt{(a_1 + b_1 q)^2 - 4(a_2 + b_2 q)(a_0 + b_0 q)}}{2(a_2 + b_2 q)} \qquad (6\text{-}13)$$

When the roots are complex conjugates, the arithmetic average of the two roots, for example $(s_1 + s_2)/2$, is equal to

$$-\frac{a_1 + b_1 q}{2(a_2 + b_2 q)},$$

which is the real part of the roots. On the other hand, their geometric average, $\sqrt{s_1 s_2}$, is equal to

$$\sqrt{\frac{a_0 + b_0 q}{a_2 + b_2 q}},$$

which represents the magnitude of the roots, since it is the same as ω_o.

Keeping in mind that these roots may be either the poles or the zeros of a second-order system function, two important questions can be raised:

1. How do the roots, when complex, move if q changes by an incremental amount from its nominal value?
2. What are the loci of the roots as q varies over a wide range of values?

If a network is to be tuned finely, so that a prescribed frequency response is obtained accurately, then the answer to question 1 is essential. To describe properties of second-order systems in general requires that question 2 be answered. Since the answer to question 2 also answers question 1 qualitatively, it is considered first.

In general, q may assume any value between $-\infty$ and $+\infty$. To cover all possible cases, let q vary from $-\infty$ to $+\infty$. The resulting loci of the roots of Eq. (6-13) are shown in Fig. 6-5a. If the value of q is selected appropriately,

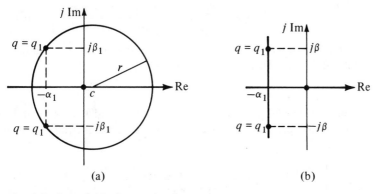

(a)   (b)

Fig. 6-5   Root loci of second-order systems

a root can be placed anywhere on the real axis. Similarly, *complex-conjugate roots can be positioned anywhere on a circle of radius r and center c.* (When c and r are infinite, the circle degenerates to the vertical line shown in Fig. 6-5b.) The a and b coefficients of Eq. (6-12) determine r and c.

To show that all of the real axis is part of the root loci, pick any point, for example, $s = \alpha$, on the real axis. If α is a root of Eq. (6-12), it must satisfy

$$P(\alpha) = (a_2 \alpha^2 + a_1 \alpha + a_0) + q(b_2 \alpha^2 + b_1 \alpha + b_0) = P_1(\alpha) + qP_2(\alpha) = 0. \qquad (6\text{-}14)$$

Since $P_1(\alpha)$ and $P_2(\alpha)$ are real numbers, a positive or a negative value of q can always be found such that Eq. (6-14) is satisfied. [If $P_1(\alpha)$ and $P_2(\alpha)$ are numbers of the same sign, then a real root results for q negative; if $P_1(\alpha)$ and $P_2(\alpha)$ are numbers of opposite sign, then a real root results for q positive.] Thus, any real-axis point α, can be made a root of $P(s)$ by appropriately choosing q. To determine which sections of the real axis are part of the root loci for q positive or for q negative, the P_1 and P_2 polynomials of Eq. (6-14) are written in factored form:

$$P(s) = a_2 \left[(\alpha - s_a)(\alpha - s_b) + q \frac{b_2}{a_2} (\alpha - s_c)(a - s_d) \right] = 0 \qquad (a_2 \neq 0), \qquad (6\text{-}15)$$

where s_a, s_b, s_c, and s_d are the zeros of the $P_1(s)$ and $P_2(s)$ polynomials. Suppose s_a is real. Then, the factor $(\alpha - s_a)$ changes sign when α becomes greater than s_a. On the other hand, if s_a is complex, then s_b is its conjugate, and the product of the factors, $(\alpha - s_a)(\alpha - s_b)$, is always positive and no change in sign occurs. Consequently, *if $q(b_2/a_2)$ is positive, those sections of the real axis to the left of an odd number of $P_1(s)$ and $P_2(s)$ zeros are part of the root loci.* Conversely, *if $q(b_2/a_2)$ is negative, those sections of the real axis to the left of an even number of $P_1(s)$ and $P_2(s)$ zeros are part of the root loci.* If both positive and negative values are allowed for $q(b_2/a_2)$, the entire real axis, from $-\infty$ to $+\infty$, becomes part of the root loci provided that $q(b_2/a_2)$ is allowed to span the necessary range of values.

To show that, when the roots are complex, they fall on a circle, consider first the real and imaginary parts of the roots.

$$-\alpha = -\frac{(a_1 + b_1 q)}{2(a_2 + b_2 q)}, \qquad \beta = \frac{\sqrt{4(a_2 + b_2 q)(a_0 + b_0 q) - (a_1 + b_1 q)^2}}{2(a_2 + b_2 q)}. \qquad (6\text{-}16)$$

Both α and β are functions of q, and therefore they change as q is changed. If the locus is indeed a circle, then α and β must satisfy

$$(-\alpha - c)^2 + \beta^2 = r^2,$$

where c is the center of the circle and r its radius. Both c and r must be independent of q. When the expressions for $-\alpha$ and β are substituted, the equation of the circle becomes

$$\left[-\frac{(a_1 + b_1 q)}{2(a_2 + b_2 q)} - c \right]^2 + \left[\frac{4(a_2 + b_2 q)(a_0 + b_0 q) - (a_1 + b_1 q)^2}{4(a_2 + b_2 q)^2} \right] = r^2,$$

which simplifies to:

$$[a_1 c + a_0 + (c^2 - r^2)a_2] + q[b_1 c + b_0 + (c^2 - r^2)b_2] = 0.$$

To satisfy this equation for all values of q which result in complex-conjugate roots requires that

$$a_1 c + a_0 + (c^2 - r^2)a_2 = 0,$$

$$b_1 c + b_0 + (c^2 - r^2)b_2 = 0.$$

The center c and the radius r of the circle are obtained by simultaneously solving these two equations:

$$c = -\frac{a_0 b_2 - a_2 b_0}{a_1 b_2 - a_2 b_1}, \tag{6-17a}$$

$$r = \sqrt{c^2 + c\frac{a_1}{a_2} + \frac{a_0}{a_2}} = \sqrt{c^2 + c\frac{b_1}{b_2} + \frac{b_0}{b_2}}, \tag{6-17b}$$

where the first form of the expression for r must be used if $a_2 \neq 0$, and the second form if $b_2 \neq 0$.

The second question, posed previously, has therefore a very interesting and significant answer:

As a single element is varied in a second-order system, the poles and zeros move on loci that are either part of the real axis or part of a circle.

(When $c = \infty$, then $r = \infty$, and the circle degenerates into a vertical line, and the roots travel in a constant-α manner as a function of q.) There are no other forms of loci for the roots. It is important that this result be understood. For example, as q is changed from 0 to ∞, the poles may travel partly on the real axis and partly on an arc of a circle of radius r_1 and center c_1, while the zeros may travel only on part of a circle with radius r_2 and center c_2, or they may stay put on the real axis. The nature of the $P_1(s)$ and $P_2(s)$ polynomials and the range of q-values determine whether the locus is part of a circle, part of the real axis, or a point.

The first question can now be answered qualitatively. When an element in a network is changed infinitesimally, *the complex poles or zeros of the system function move in a direction that is tangent to a circle of radius r and center c,* where c and r are given by Eq. (6-17). The direction of the change can be established by determining how the real part of the root, $-\alpha$, and its magnitude, ω_o, vary with q. From Eq. (6-13) $-\alpha$ and from Eq. (6-12) ω_o^2 are obtained and differentiated with respect to q:

$$-\alpha = -\frac{(a_1 + qb_1)}{2(a_2 + qb_2)}, \qquad \omega_o^2 = \frac{a_0 + qb_0}{a_2 + qb_2} > 0,$$

$$\frac{d(-\alpha)}{dq} = \frac{a_1 b_2 - a_2 b_1}{2(a_2 + qb_2)^2}, \qquad \frac{d\omega_o}{dq} = \frac{a_2 b_0 - a_0 b_2}{2\omega_o(a_2 + qb_2)^2}. \tag{6-18}$$

Thus, a positive change in q, that is, $\Delta q > 0$, causes the roots to move to the right if

$$a_1 b_2 - a_2 b_1 > 0.$$

It also causes the magnitude of the roots to increase if $a_2 b_0 - a_0 b_2 > 0$.

EXAMPLE 6-1

Sketch, as a function of the capacitor value, C, the loci of the poles and zeros of V_o/I_i given in Fig. 6-4a. For what value of C is the imaginary part of the poles insensitive to a very small change in C?

SOLUTION

From Eq. (6-8), obtain (V_o/I_i):

$$\frac{V_o}{I_i} = \frac{RLs}{Ls + R + C(RLs^2)}.$$

The zeros of (V_o/I_i) are at the origin and at infinity and are independent of C. So they stay put.

The loci of the poles are determined by the roots of

$$P(s) = (Ls + R) + C(RLs^2) = 0.$$

The corresponding P_1 and P_2 polynomials are

$$P_1(s) = L\left(s + \frac{R}{L}\right), \qquad P_2(s) = RLs^2.$$

Mark in the s-plane the zeros of $P_1(s)$ and $P_2(s)$. Use the letter B (for beginning) to designate the zeros of $P_1(s)$ and the letter E (for ending) to designate the zeros of $P_2(s)$. Thus, place a B at $s = -R/L$ and two E's at $s = 0$. When $C = 0$, $P(s) = P_1(s)$. The two roots are at $s = -R/L$ and $s = -\infty$. (The latter root can be found by observing the asymptotic behavior of the roots as C approaches zero.) When $C \to \infty$, $P(s) \to P_2(s)$. The two roots approach $s = 0$. So the roots begin at $s = -R/L$ and $s = -\infty$ and end at $s = 0$.

Since the variable C is always positive, and the coefficients of the highest power in the P_1 and P_2 polynomials are also positive, those sections of the real axis to the left of an odd number of P_1 and P_2 zeros are part of the root loci. There is one such section in this case, namely the real axis to the left of $s = -R/L$. When $C = L/(4R^2)$, the roots are both at $s = -2R/L$. When $C > L/(4R^2)$, the roots are complex and fall on a circle. Because the roots leave the real axis at $s = -2R/L$ [corresponding to $C = L/(4R^2)$] and return to the origin for $C = \infty$, the center of the circle must be at $s = -R/L$ and its radius equal to R/L. The complete loci of the poles is shown in Fig. 6-6. The poles move in the direction of the arrows as C varies from 0 to ∞.

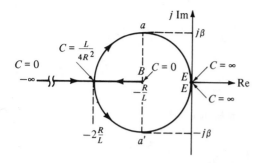

Fig. 6-6   The loci of the poles as a function of C

As an independent check, Eq. (6-17) can be used to determine the radius of the circle and the location of its center.

When the complex poles are at a and a', a slight change in C moves the poles horizontally; hence the imaginary part of the poles, β, remains constant. At these points the real part of the poles is

$$-\frac{1}{2RC} = -\frac{R}{L}.$$

Hence, C must be chosen equal to $L/(2R^2)$ in order to make the imaginary part of the poles insensitive to incremental changes in C.

EXAMPLE 6-2

(a) For $L = 0$, find the dc gain, the 3-dB bandwidth, and the gain-bandwidth product of the transimpedance shown in Fig. 6-7.

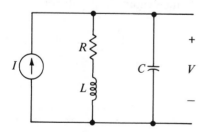

Fig. 6-7

(b) Draw the loci of the poles as a function of L. Show that the introduction of L increases the gain-bandwidth product. This method of extending the gain-bandwidth product is called *shunt peaking*.

SOLUTION

(a) The transimpedance for $L = 0$ is

$$\frac{V}{I} = \frac{1}{C} \frac{1}{s + \dfrac{1}{RC}}.$$

The resulting dc gain and 3-dB bandwidth are

dc gain $= G = R$,

3-dB bandwidth $= BW = \dfrac{1}{RC}$ rad/s.

The gain-bandwidth product, GBW, is

$$\mathrm{GBW} = R\left(\frac{1}{RC}\right) = \frac{1}{C}.$$

If the GBW is used as a figure of merit for this circuit (the larger the GBW, the more the bandwidth for a given dc gain), its value will be limited by C which cannot be made smaller than the stray shunt capacitance associated with the circuit.

(b) The insertion of an inductor in series with the resistor introduces another pole which makes the system second-order. The dc gain is still R and therefore is independent of L, but the 3-dB bandwidth becomes wider, thereby improving the GBW. The modified system function is:

$$\frac{V}{I} = \frac{1}{C} \frac{s + \dfrac{R}{L}}{s^2 + s\dfrac{R}{L} + \dfrac{1}{LC}} = R \frac{(sRC) + \dfrac{R^2C}{L}}{(sRC)^2 + (sRC)\dfrac{R^2C}{L} + \dfrac{R^2C}{L}}.$$

To sketch the poles as a function of L, the denominator polynomial of the transimpedance is partitioned as

$$R\left(s + \frac{1}{RC}\right) + L(s^2).$$

For $L = 0$, the poles are at

$$s = -\infty \qquad \text{and} \qquad s = -\frac{1}{RC}.$$

For $L = \infty$, both poles are at the origin. The pole loci are sketched in Fig. 6-8. The poles become complex for $L/R^2C > \tfrac{1}{4}$.

As long as the two poles are widely separated on the real axis, the 3-dB bandwidth is determined essentially by the pole nearest to the origin. The closer this pole is to the origin, the narrower is the bandwidth. Because the pole at $s = -1/RC$ is moved farther away from the origin as

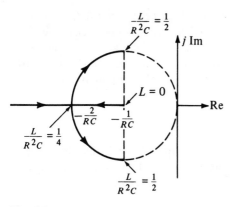

Fig. 6-8

L is increased from 0, an increase in the bandwidth results. However, as L becomes larger, the other pole at $s = -\infty$ is brought more to the right, and it also starts influencing the 3-dB bandwidth. Furthermore, the zero of V/I at $s = -R/L$, since it is dependent on L, moves closer to the origin also, thereby producing an additional effect on the 3-dB bandwidth. The overall effect is best studied by plotting the magnitude characteristics with $[L/(R^2C)]$ as a parameter as shown in Fig. 6-9. Note that for

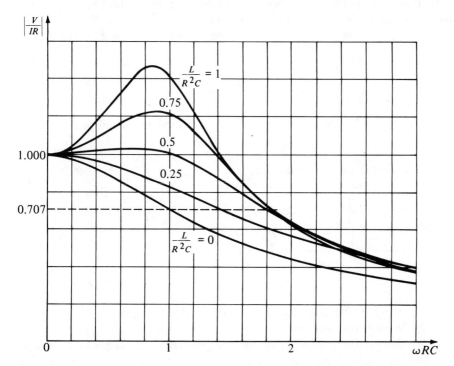

Fig. 6-9

$[L/(R^2 C)] = 0.5$, the bandwidth is almost twice (1.8 times) as wide as for $L = 0$. As $[L/(R^2 C)]$ is increased further, the magnitude characteristics start peaking noticeably (indicating that the poles are getting closer to the imaginary axis) without significantly affecting the bandwidth. Therefore, the gain-bandwidth product of the system is improved if $[L/(R^2 C)] \leq 0.5$.

Constant-α Root Loci

It is possible to construct second-order systems in such a way that the complex poles or zeros move on a constant-α line, a vertical line, while an element, q, in the network is varied. If the numerator or denominator polynomial is arranged as

$$P(s) = (a_2 s^2 + a_1 s + a_0) + q(b_2 s^2 + b_1 s + b_0) = P_1(s) + qP_2(s),$$

then the real part, $-\alpha$, of the complex critical frequencies is given by:

$$-\alpha = -\frac{a_1 + b_1 q}{2(a_2 + b_2 q)}. \tag{6-19}$$

To make $-\alpha$ independent of q requires that

$$a_1 b_2 = a_2 b_1. \tag{6-20}$$

Note that this condition can also be obtained from Eq. (6-17a) by making the radius of the circle infinite. The easiest way to satisfy Eq. (6-20) is to make $b_2 = b_1 = 0$, in which case q *varies only the constant coefficient in* $P(s)$, that is,

$$P(s) = a_2 s^2 + a_1 s + (a_0 + qb_0). \tag{6-21}$$

EXAMPLE 6-3

The network shown in Fig. 6-10 is often used as a band-pass network.

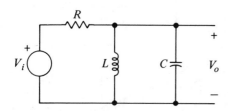

Fig. 6-10

(a) At what frequency does the magnitude function peak?
(b) What is the 3-dB bandwidth?
(c) Discuss the possibility of moving the frequency of peaking without altering the bandwidth.

SOLUTION

(a) Obtain the transfer function first, and put it in standard form:

$$\frac{V_o}{V_i} = \frac{\dfrac{1}{RC}s}{s^2 + s\dfrac{1}{RC} + \dfrac{1}{LC}} \tag{6-22a}$$

$$= H \frac{s}{s^2 + s\dfrac{\omega_o}{Q} + \omega_o^2}, \tag{6-22b}$$

where

$$H = \frac{1}{RC}, \qquad \omega_o = \frac{1}{\sqrt{LC}}, \qquad Q = R\sqrt{\frac{C}{L}}. \tag{6-22c}$$

Substitute next $s = j\omega$ and obtain the magnitude function

$$\frac{V_o}{V_i} = H \frac{j\omega}{-\omega^2 + j\omega\dfrac{\omega_o}{Q} + \omega_o^2},$$

$$\left|\frac{V_o}{V_i}\right| = \frac{H\omega}{\sqrt{(\omega_o^2 - \omega^2)^2 + \left(\dfrac{\omega\omega_o}{Q}\right)^2}}. \tag{6-23}$$

The magnitude is zero at $\omega = 0$ and $\omega = \infty$. Therefore, it must peak at some frequency between the extreme frequencies. To obtain the peaking frequency, differentiate $|V_o/V_i|^2$ with respect to ω^2 and set the result to zero:

$$\frac{d}{d\omega^2}\left|\frac{V_o}{V_i}\right|^2 = \frac{d}{d\omega^2}\left[\frac{H^2\omega^2}{(\omega_o^2 - \omega^2)^2 + \left(\dfrac{\omega\omega_o}{Q}\right)^2}\right]$$

$$= \frac{H^2\left\{\left[(\omega_o^2 - \omega^2)^2 + \left(\dfrac{\omega\omega_o}{Q}\right)^2\right] - \omega^2\left[-2(\omega_o^2 - \omega^2) + \dfrac{\omega_o^2}{Q^2}\right]\right\}}{\left[(\omega_o^2 - \omega^2)^2 + \left(\dfrac{\omega\omega_o}{Q}\right)^2\right]^2}$$

$$= \frac{H^2(\omega_o^2 + \omega^2)(\omega_o^2 - \omega^2)}{\left[(\omega_o^2 - \omega^2)^2 + \left(\dfrac{\omega\omega_o}{Q}\right)^2\right]^2} = 0,$$

which gives the peaking frequency as

$$\omega = \omega_o = \frac{1}{\sqrt{LC}}. \qquad Ans. \qquad (6\text{-}24)$$

(b) In order to find the 3-dB bandwidth, first obtain the peak magnitude by substituting $\omega = \omega_o$ in Eq. (6-23):

$$\left|\frac{V_o}{V_i}\right|_{peak} = H\frac{Q}{\omega_o}.$$

Next, obtain the two frequencies, ω_1 and ω_2, at which the magnitude, given by Eq. (6-23), is down by 3-dB from the peak value, that is,

$$\left|\frac{V_o}{V_i}\right| = \frac{1}{\sqrt{2}}\frac{HQ}{\omega_o},$$

$$\left|\frac{V_o}{V_i}\right|^2 = \frac{1}{2}\left(\frac{HQ}{\omega_o}\right)^2 = \frac{H^2\omega_{1,2}^2}{(\omega_o^2 - \omega_{1,2}^2)^2 + \left(\frac{\omega_{1,2}\omega_o}{Q}\right)^2},$$

which simplifies to

$$(\omega_o^2 - \omega_{1,2}^2)^2 = \left(\frac{\omega_{1,2}\omega_o}{Q}\right)^2,$$

$$\omega_o^2 - \omega_{1,2}^2 = \pm\left(\frac{\omega_{1,2}\omega_o}{Q}\right),$$

$$\omega_{1,2}^2 \pm \left(\frac{\omega_o}{Q}\right)\omega_{1,2} - \omega_o^2 = 0,$$

$$\omega_{1,2} = \pm\frac{1}{2}\left(\frac{\omega_o}{Q}\right) \pm \sqrt{\frac{1}{4}\left(\frac{\omega_o}{Q}\right)^2 + \omega_o^2}. \qquad (6\text{-}25)$$

Equation (6-25) gives four frequencies, two of which are negative and therefore of no practical significance. The two positive frequencies are

$$\omega_1 = -\frac{1}{2}\left(\frac{\omega_o}{Q}\right) + \sqrt{\frac{1}{4}\left(\frac{\omega_o}{Q}\right)^2 + \omega_o^2},$$

$$\omega_2 = +\frac{1}{2}\left(\frac{\omega_o}{Q}\right) + \sqrt{\frac{1}{4}\left(\frac{\omega_o}{Q}\right)^2 + \omega_o^2}.$$

The 3-dB bandwidth is the difference of these two frequencies:

$$BW = \omega_2 - \omega_1 = \frac{\omega_o}{Q} = \frac{1}{RC}. \qquad Ans.$$

(c) From (a) and (b), obtain the frequency of peaking and the 3-dB bandwidth:

$$\omega_o = \frac{1}{\sqrt{LC}}, \qquad BW = \frac{1}{RC}.$$

Since L appears in the expression of ω_o and not in the expression for BW, the frequency of peaking can be moved while maintaining constant bandwidth by changing L.

To determine the effect of L on the pole locations, obtain the denominator polynomial from Eq. (6-22a):

$$s^2 + s\frac{1}{RC} + \frac{1}{LC}.$$

Note that L affects only the constant term in the quadratic polynomial. The coefficient of the s term, $(1/RC)$, is equal to the 3-dB bandwidth [see part (b)] and is independent of L. Hence, in order to achieve constant bandwidth but variable center frequency, the poles of the second-order band-pass function must be moved in a constant-α manner, where $\alpha = \frac{1}{2}$BW.

Constant-ω_o Root Loci

By changing a single element in the network, the complex critical frequencies of second-order systems can be varied along a circle centered at the origin of the s-plane. Thus the Q of the critical frequencies varies while ω_o is held constant. Let $P(s)$ represent the numerator or the denominator polynomial. Arrange it as

$$P(s) = (a_2 s^2 + a_1 s + a_0) + q(b_2 s^2 + b_1 s + b_0) = P_1(s) + qP_2(s). \tag{6-26}$$

Then, ω_o is identified as the square root of the constant term divided by the coefficient of the s^2 term:

$$\omega_o = \sqrt{\frac{a_0 + qb_0}{a_2 + qb_2}}. \tag{6-27}$$

It is assumed that the quantity under the radical is a positive number. Otherwise the roots are not complex. To make ω_o independent of q requires that

$$a_0 b_2 = a_2 b_0. \tag{6-28}$$

Note that Eq. (6-28) can also be obtained from Eq. (6-17b) by requiring that the center of the circle be at the origin. The easiest way to satisfy Eq. (6-27) is to make $b_2 = b_0 = 0$, and $(a_0/a_2) > 0$, in which case q *varies only the coefficient of the s term in $P(s)$*, i.e., Eq. (6-26) becomes

$$P(s) = a_2 s^2 + (a_1 + qb_1)s + a_0 \qquad (a_2/a_0 > 0).$$

EXAMPLE 6-4

(a) Discuss the possibility of moving the poles of the network shown in Fig. 6-10 in a constant-ω_o manner.
(b) Draw the pole loci as a function of R.

SOLUTION

(a) Obtain the characteristic (denominator) polynomial from Eq. (6-22):

$$s^2 + s\frac{1}{RC} + \frac{1}{LC} = s^2 + s\frac{\omega_o}{Q} + \omega_o^2,$$

where

$$\omega_o = \frac{1}{\sqrt{LC}}, \qquad Q = R\sqrt{\frac{C}{L}}.$$

Note that ω_o is independent of R. Hence, the poles, when complex, can be moved in a constant-ω_o manner if R is varied.

(b) To draw the pole-loci as a function of R, rearrange the characteristic polynomial so that the effect of R is explicitly noted:

$$s^2 + s\frac{1}{RC} + \frac{1}{LC} = \frac{1}{RLC}[sL + R(1 + s^2LC)] = \frac{1}{RLC}[P_1(s) + RP_2(s)],$$

where

$$P_1(s) = sL, \qquad P_2(s) = 1 + s^2LC.$$

When $R = 0$, the poles are at $s = 0$ [see $P_1(s)$] and $s = -\infty$. Mark these points with the letter B. When $R = \infty$, the poles are at $\pm j(1/\sqrt{LC})$ [see $P_2(s)$]. Mark these points with the letter E. Since R can take on only positive values, the real-axis sections of the loci are to the left of an odd number of $P_1(s)$ and $P_2(s)$ zeros. Hence, the entire negative real axis is part of the pole loci. The poles become complex when $R > \frac{1}{2}\sqrt{L/C}$ (which makes $Q > \frac{1}{2}$). When complex, the poles are on a circle, centered at the origin of the s-plane. The radius of the circle is $\omega_o = 1/\sqrt{LC}$ [see part (a)]. The resulting pole loci are shown in Fig. 6-11.

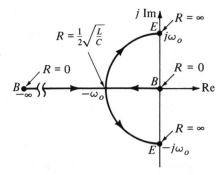

Fig. 6-11

Constant-β and Constant-Q Root Loci

As long as one element is varied, the s-plane root loci of complex poles or zeros are either circles with the center at some point on the real axis or vertical lines. Therefore, it is impossible to obtain constant-β root loci, since such loci are horizontal lines. Similarly, it is impossible to obtain constant-Q root loci, since such loci are radial lines. However, when the element change is infinitesimal, the roots may be displaced horizontally or radially by an infinitesimal amount. To see how this can be achieved, refer to Fig. 6-12.

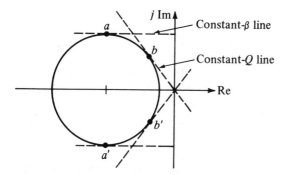

Fig. 6-12   Points of constant-β and constant-Q

Assume that the loci of the roots as a function of the element q form a circle in the left (or right) half-plane and q is adjusted to put the roots at a and a'. Then, for infinitesimal changes in q, the roots are displaced along the constant-β lines shown. At the point a, $(d\beta/dq) = 0$. On the other hand, if q is adjusted to put the roots at b and b', where the radial lines are tangent to the circle, then for infinitesimal changes in q, the roots are displaced along the constant-Q lines shown. At the point b, $(dQ/dq) = 0$. If part of the circle is in the right half-plane and the rest in the left half-plane, then it is impossible to have constant-Q at any point.

EXAMPLE 6-5

Discuss the possibility of moving the poles of the network shown in Fig. 6-10 in a constant-β and constant-Q manner.

SOLUTION

Constant-β and constant-Q operation may result only at certain pole locations in the s-plane. Also, the constraint for infinitesimal changes in element values must be observed.

Obtain first the characteristic polynomial, from Eq. (6-22a),

$$s^2 + \frac{1}{RC}s + \frac{1}{LC}.$$

To find β, equate this polynomial to the standard form given by

$$(s + \alpha)^2 + \beta^2 = s^2 + s2\alpha + \alpha^2 + \beta^2,$$

and solve for β:

$$\beta = \sqrt{\frac{1}{LC} - \left(\frac{1}{2RC}\right)^2}.$$

If constant-β operation is to result at a point, the derivative of β with respect to an element must be zero at that point. So, check the three possible derivatives:

$$\frac{d\beta}{dR} = \frac{1}{4\beta R^3 C^2} \neq 0, \qquad \frac{d\beta}{dL} = -\frac{1}{2}\frac{1}{\beta L^2 C} \neq 0.$$

Neither $(d\beta/dR)$ nor $(d\beta/dL)$ are zero for any finite value of the elements R or L. Hence, constant-β operation is not possible when R or L is the variable:

$$\frac{d\beta}{dC} = \left.\frac{\frac{L}{2R^2} - C}{2\beta LC^3}\right|_{C = L/(2R^2)} = 0.$$

Since $(d\beta/dC) = 0$ when C is adjusted to $[L/(2R^2)]$, constant-β operation is possible. The poles are then at $s = -(R/L)(1 \pm j)$.

To find Q, equate the characteristic polynomial to the standard form given by

$$s^2 + s\frac{\omega_o}{Q} + \omega_o^2,$$

and solve for Q:

$$Q = R\sqrt{\frac{C}{L}}.$$

If constant-Q operation is to result at a point, the appropriate derivatives must be zero at that point. So, check the three possible derivatives:

$$\frac{dQ}{dR} = \sqrt{\frac{C}{L}} \neq 0 \qquad \text{for any finite value of } R,$$

$$\frac{dQ}{dL} = -\frac{1}{2}\frac{Q}{L} \neq 0 \qquad \text{for any finite value of } L,$$

$$\frac{dQ}{dC} = \frac{1}{2}\frac{Q}{C} \neq 0 \qquad \text{for any finite value of } C.$$

Thus, for this network it is impossible to achieve constant-Q operation for its complex poles (if one element is varied at a time). Refer also to the loci of the poles given in Fig. 6-6 to see that for $C \neq \infty$ it is impossible to draw a line (from the origin) that is tangent to the circle.

6-3 THE ROOT-SENSITIVITY FUNCTION

Networks are constructed to produce a specified input–output relationship. For example, a network may be designed to pass a band of frequencies and reject the rest (sinusoidal steady-state, of course). Since the solution to the design problem is not unique, different networks can be constructed to produce the same input impedance or transfer function. As long as ideal elements are used under ideal conditions, one network works just as well as the other. In practice, however, one network may outperform another because it is less sensitive to element variations and to environmental changes. This network may be no more expensive to construct than the other. Therefore, a quantitative measure is needed in order to compare networks with regard to element variations from the ideal. Sensitivity functions are used for this purpose. These functions provide a numerical measure of how much an important aspect of the system or response varies as an element or a combination of elements varies from the nominal (design) values. One such function is the root-sensitivity function.

The root-sensitivity function, S_q^r, is associated with the roots of an equation. S_q^r gives the change, $\Delta r/|r|$, occurring in a root per $\Delta q/q$ (the per-unit change in the element q). It is defined by

$$S_q^r = \lim_{\Delta q \to 0} \frac{\dfrac{\Delta r}{|r|}}{\dfrac{\Delta q}{q}} = \frac{q}{|r|} \frac{\delta r}{\delta q} \qquad (q \neq 0, q \neq \infty, r \neq 0), \qquad (6\text{-}29)$$

where r represents a simple root, which may be either a pole or a zero of a system function. S_q^r is read as the sensitivity of the root with respect to the element q. Its physical significance is demonstrated in Fig. 6-13, where the root is shown as a pole.

The pole at a moves to b when q changes by Δq. (For the sake of clarity, the changes are exaggerated in Fig. 6-13.) The per-unit change in the pole

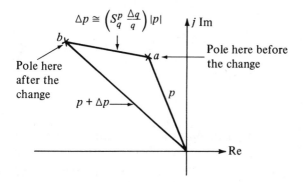

Fig. 6-13   Pole sensitivity

position, $\Delta p/|p|$, is directly proportional to the sensitivity of the pole and the per-unit change in q, that is,

$$\frac{\Delta p}{|p|} \cong S_q^p \frac{\Delta q}{q}.$$

Therefore, if the pole is to stay put as the element q changes, S_q^p evaluated at a must be zero. It is impossible to make $S_q^p = 0$ for every element of the network; the pole position must depend upon some of the elements. Note that S_q^p is a complex number. The smaller its magnitude, the smaller the magnitude of the pole displacement. It may be possible to minimize pole displacement if the change due to one element is offset by an opposite change due to another element. For example, if changes are due to temperature, some elements with positive temperature coefficient and others with negative temperature coefficient may be used.

Since the angle of S_q^r gives the direction of the root displacement, it can be used in tuning the network. For example, if $S_{R1}^r = -j2$, then increasing R_1 slightly causes the root to move straight down.

If a polynomial of any order n is decomposed into its $P_1(s)$ and $P_2(s)$ parts,

$$P(s) = P_1(s) + qP_2(s),$$

then the sensitivity of any root of $P(s) = 0$ with respect to q can be expressed in terms of the $P_1(s)$ and $P_2(s)$ polynomials, thereby allowing interpretation of S_q^r in terms of $P_1(s)$ and $P_2(s)$. To obtain the desired relationship, let $s = s_i$ be a root of $P(s) = 0$, that is,

$$P(s_i) = P_1(s_i) + qP_2(s_i) = 0.$$

If q changes by Δq, then the root, s_i, changes by Δs_i such that

$$P(s_i + \Delta s_i) = P_1(s_i + \Delta s_i) + (q + \Delta q)P_2(s_i + \Delta s_i) = 0$$

is satisfied. $P_1(s)$ and $P_2(s)$ may be expanded in a Taylor's series about s_i giving

$$\left[P_1(s_i) + \left.\frac{\delta P_1(s)}{\delta s}\right|_{s=s_i} \times \Delta s_i + \frac{1}{2} \left.\frac{\delta^2 P_1(s)}{\delta s^2}\right|_{s=s_i} \times (\Delta s_i)^2 + \cdots \right]$$

$$+ (q + \Delta q)\left[P_2(s_i) + \left.\frac{\delta P_2(s)}{\delta s}\right|_{s=s_i} \times \Delta s_i + \frac{1}{2} \left.\frac{\delta^2 P_2(s)}{\delta s^2}\right|_{s=s_i} \times (\Delta s_i)^2 + \cdots \right] = 0,$$

$$[P_1(s_i) + qP_2(s_i)]$$

$$+ \left[\frac{\delta P_1(s)}{\delta s} + q\frac{\delta P_2(s)}{\delta s} \right]_{s=s_i} \times \Delta s_i + \Delta qP_2(s_i) + (\text{higher-order terms}) = 0.$$

The first bracketed term is the same as $P(s_i)$; hence, it is zero. The second bracketed term is $\delta P(s)/\delta s|_{s=s_i}$. As $\Delta q \to 0$, $\Delta s_i \to 0$, and the higher-order terms may be neglected. Using these simplifications, the sensitivity of the root at s_i with respect to q, $S_q^{s_i}$, is written as

$$S_q^{s_i} = \frac{q}{|s_i|} \frac{\delta s_i}{\delta q} = \frac{q}{|s_i|} \lim_{\Delta q \to 0} \frac{\Delta s_i}{\Delta q} = -\frac{q}{|s_i|} \frac{P_2(s)}{\dfrac{\delta P(s)}{\delta s}} \Bigg|_{s=s_i} = \frac{1}{|s_i|} \frac{P_1(s)}{\dfrac{\delta P(s)}{\delta s}} \Bigg|_{s=s_i}, \qquad (6\text{-}30)$$

where $q \neq 0$, $q \neq \infty$, $s_i \neq 0$.

Equation (6-30) is an important result. It gives the sensitivity of a polynomial's zero to changes in the element q in terms of the $P_1(s)$- and $P_2(s)$-parts of the polynomial. Note that the root sensitivity is zero if s_i is a root of $P_1(s) = 0$ [in which case s_i is also a root of $P_2(s) = 0$]; in this case; this root can be factored out and thereby be made independent of q.

If $P(s)$ is of second order, that is, if

$$P(s) = a_2 s^2 + a_1 s + a_0 + q(b_2 s^2 + b_1 s + b_0) = P_1(s) + qP_2(s)$$

and s_i is a root of $P(s) = 0$, then

$$S_q^{s_i} = -\frac{q}{|s_i|} \left[\frac{b_2 s_i^2 + b_1 s_i + b_0}{2a_2 s_i + a_1 + q(2b_2 s_i + b_1)} \right] = \frac{1}{|s_i|} \frac{a_2 s_i^2 + a_1 s_i + a_0}{2a_2 s_i + a_1 + q(2b_2 s_i + b_1)}. \qquad (6\text{-}31)$$

Equations (6-29), (6-30), or (6-31) may be used to calculate the root-sensitivity function of second-order systems. If the general expression for the root is known, it may be easier to use the direct differentiation indicated by Eq. (6-29). For polynomials of order higher than two, the general expression for the root may not be available and therefore differentiation cannot be performed. In that case, Eq. (6-30) must be used.

If a network contains ten elements, then there are ten sensitivity functions associated with each pole or zero of the system function. If the system has four poles and three zeros, there are altogether 70 root-sensitivity functions to deal with. Even though some of these may form complex-conjugate pairs, the total number of root sensitivities becomes formidable if the order of the system is high. However, it is usually not necessary to calculate all of the root-sensitivity functions since a pair of complex-conjugate poles or zeros may dominate the response characteristics in a given band of frequencies. In this case, the significant sensitivity functions are associated with these dominant critical frequencies.

When the nominal value of q is zero, the per-unit change of q, $(\Delta q/q)$, since it is infinite, cannot be interpreted meaningfully. (This case may arise when a small element is introduced in the network.) Therefore, the sensitivity function, as defined by Eq. (6-29) is not useful. In this case, the normalization with respect to q is dropped, and the function

$$\mathscr{S}_q^r = \frac{1}{|r|} \frac{\delta r}{\delta q} \qquad (6\text{-}32)$$

is used as a measure of the roots' sensitivity with respect to the element q. Note that script $\mathscr{S}$ is used to differentiate this function from the regular sensitivity function which is designated by the capital S.

If $P(s)$ is split into the $P_1(s)$ and $P_2(s)$ polynomials,

$$P(s) = P_1(s) + qP_2(s),$$

then, $\mathscr{S}_q^r$ can be expressed in terms of these polynomials as

$$\mathscr{S}_q^r = \frac{1}{|r|}\frac{\delta r}{\delta q} = -\frac{1}{|r|}\frac{P_2(s)}{\dfrac{\delta P(s)}{\delta s}}\Bigg|_{s=r} \qquad (q \neq \infty, r \neq 0). \qquad (6\text{-}33)$$

Equation (6-33) is obtained from Eq. (6-30) by dividing it by q. The sensitivity function given by Eq. (6-33) may be used for polynomials of any order and for any finite value of q. In particular, if $q = 0$, Eq. (6-33) simplifies to

$$\mathscr{S}_q^r = \frac{1}{|r|}\frac{\delta r}{\delta q} = -\frac{1}{|r|}\frac{P_2(s)}{\dfrac{\delta P_1(s)}{\delta s}}\Bigg|_{s=r} \qquad (q \neq \infty, r \neq 0). \qquad (6\text{-}34)$$

Note that in all sensitivity expressions, the case $q = \infty$ is excluded because the derivative of a function is undefined at infinity.

EXAMPLE 6-6

The zeros of the cubic polynomial

$$s^3 + k(s^2 + s) + 1 = 0$$

are at $s = -1$ and $s = -\frac{1}{2} \pm j(\frac{1}{2}\sqrt{3})$, when $k = 2$.

(a) Obtain the sensitivity of the three zeros with respect to k when $k = 2$.
(b) Obtain the approximate position of the zeros when $k = 2.02$.
(c) Compare the approximate zero locations with the exact values.

SOLUTION

(a) Segregate the polynomial into the $P_1(s)$ and $P_2(s)$ parts:

$$P(s) = s^3 + k(s^2 + s) + 1$$
$$= (s^3 + 1) + k(s^2 + s) = P_1(s) + kP_2(s). \qquad (6\text{-}35)$$

Use Eq. (6-30) to obtain $S_k^{z_i}$, where z_i represents any one of the three zeros:

$$S_k^{z_i} = -\frac{k}{|z_i|}\frac{P_2(s)}{\dfrac{\delta P(s)}{\delta s}}\Bigg|_{s=z_i}$$

$$= -\frac{k}{|z_i|}\frac{(s^2 + s)}{3s^2 + k(2s + 1)}\Bigg|_{k=2,\,s=z_i} = -\frac{2}{|z_i|}\frac{z_i^2 + z_i}{3z_i^2 + 2(2z_i + 1)}.$$

For $z_1 = -1$, $z_{2,3} = -\frac{1}{2} \pm j(\frac{1}{2}\sqrt{3})$, the sensitivity functions become

$$S_k^{z_1} = -\frac{2}{|-1|} \frac{(-1)^2 + (-1)}{3(-1)^2 + 2[2(-1) + 1]} = 0, \qquad Ans.$$

$$S_k^{z_{2,3}} = -\frac{2}{\left|-\frac{1}{2} \pm j\frac{\sqrt{3}}{2}\right|} \frac{\left(-\frac{1}{2} \pm j\frac{\sqrt{3}}{2}\right)^2 + \left(-\frac{1}{2} \pm j\frac{\sqrt{3}}{2}\right)}{3\left(-\frac{1}{2} \pm j\frac{\sqrt{3}}{2}\right)^2 + 2\left[2\left(-\frac{1}{2} \pm j\frac{\sqrt{3}}{2}\right) + 1\right]}$$

$$= -1 \mp j\frac{\sqrt{3}}{3}, \qquad Ans.$$

where the minus sign is associated with the upper half-plane zero.

(b) Use the definition of $S_k^{z_i}$ to obtain the incremental displacements of the zeros:

$$S_k^{z_i} = \frac{k}{|z_i|} \frac{\delta z_i}{\delta k},$$

$$S_k^{z_i} \cong \frac{k}{|z_i|} \frac{\Delta z_i}{\Delta k},$$

$$\Delta z_i \cong |z_i| \, S_k^{z_i} \frac{\Delta k}{k}.$$

For $z_1 = -1$, $z_{2,3} = -\frac{1}{2} \pm j(\frac{1}{2}\sqrt{3})$, $k = 2$, and $\Delta k = 0.02$, the zero displacements become

$$\Delta z_1 = |-1|0\frac{0.02}{2} = 0,$$

$$\Delta z_{2,3} = \left|-\frac{1}{2} \pm j\frac{\sqrt{3}}{2}\right|\left(-1 \mp j\frac{\sqrt{3}}{3}\right)\frac{0.02}{2} = -0.01 \mp j\frac{0.01\sqrt{3}}{3}.$$

The approximate positions of the new zeros are

$$(z_1)_n \cong z_1 + \Delta z_1 = -1 + 0 = -1, \qquad Ans.$$

$$(z_{2,3})_n \cong z_{2,3} + \Delta z_{2,3} = -\frac{1}{2} \pm j\frac{\sqrt{3}}{2} + \left(-0.01 \mp j\frac{0.01\sqrt{3}}{3}\right)$$

$$= -\frac{1}{2}(1 + 0.02) \pm j\frac{\sqrt{3}}{2}\left(1 - \frac{0.02}{3}\right). \qquad Ans.$$

Thus, the real-axis zero does not move, whereas the upper half-plane zero moves to the left and down. These positions are approximate because the zeros are assumed to move along tangential lines drawn to the root loci at the nominal zero positions. Because k changes by a small amount, 1%, the approximate zero locations are very near the actual zero locations.

(c) In this example, the exact zero locations can be determined because $s = -1$ is always a zero of the polynomial regardless of the value of k. Factor this zero of Eq. (6-35):

$$P(s) = (s^3 + 1) + k(s^2 + s) = (s + 1)(s^2 - s + 1) + ks(s + 1)$$
$$= (s + 1)[s^2 + (k - 1)s + 1].$$

The exact zero locations, expressed as a function of k, are

$$z_1 = -1, \qquad z_{2,3} = -\frac{k - 1}{2} \pm j\sqrt{1 - \frac{(k - 1)^2}{4}}.$$

For $k = 2.02$, the zeros become

$$z_1 = -1, \qquad z_{2,3} = -\frac{1}{2}(1 + 0.02) \pm j\frac{\sqrt{3}}{2}\sqrt{1 - \frac{0.0404}{3}}. \qquad Ans.$$

When these values are compared with the values given in (b), it is seen that the approximation is very good. The agreement would not be as close had k changed 10% rather than 1%.

6-4 ω_o SENSITIVITY, Q SENSITIVITY, AND OTHER SENSITIVITIES

Instead of the root-sensitivity function, the ω_o-and Q-sensitivity functions can be used to determine the effect of element variations on the systems performance. These are defined as

$$S_q^{\omega_o} = \lim_{\Delta q \to 0} \left\{ \frac{\dfrac{\Delta\omega_o}{\omega_o}}{\dfrac{\Delta q}{q}} \right\} = \frac{q}{\omega_o}\frac{\delta\omega_o}{\delta q} \qquad (q \neq 0, q \neq \infty, \omega_o \neq 0), \qquad (6\text{-}36a)$$

$$S_q^{Q} = \lim_{\Delta q \to 0} \left\{ \frac{\dfrac{\Delta Q}{Q}}{\dfrac{\Delta q}{q}} \right\} = \frac{q}{Q}\frac{\delta Q}{\delta q} \qquad (q \neq 0, q \neq \infty, Q \neq 0), \qquad (6\text{-}36b)$$

where ω_o and Q are associated with a pole or a zero. Since ω_o and Q are real numbers, the defining equations are set up on a per-unit basis. (This is not done with the root sensitivity because the root is a complex number.) Note that $s_q^{\omega_o}$ and S_q^{Q} are real numbers. The physical significance of $S_q^{\omega_o}$, the sensitivity of ω_o with respect to q, and of S_q^{Q}, the sensitivity of Q with respect to q, becomes clear by referring to Fig. 6-14, where the root under consideration is a pole.

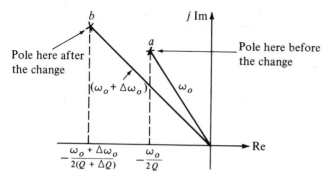

Fig. 6-14 ω_o and Q sensitivity

The pole at point a described by (ω_o, Q) will move to point b when q is changed by Δq. At the new location, the pole is described by $(\omega_o + \Delta\omega_o, Q + \Delta Q)$. The amount of change in ω_o and Q is directly proportional to the respective sensitivity function. Using incremental rather than differential changes, the approximate expressions for the changes are given by

$$\frac{\Delta\omega_o}{\omega_o} \cong S_q^{\omega_o}\frac{\Delta q}{q}, \qquad \frac{\Delta Q}{Q} \cong S_q^{Q}\frac{\Delta q}{q}. \tag{6-37}$$

The smaller the value of the ω_o-sensitivity and the Q-sensitivity functions at a point, the smaller the changes in the ω_o and Q of the pole for a given change in q. $S_q^{\omega_o}$ is zero at the point a if the pole moves at right angles (as q is changed infinitesimally) to the radial line connecting point a to the origin. S_q^{Q} is zero if the pole moves radially outward or inward from the point a (as q is changed infinitesimally).

Since, for high-Q circuits, ω_o is the frequency at which the magnitude of the frequency response peaks and ω_o/Q is a measure of its bandwidth, the $S_q^{\omega_o}$ and S_q^{Q} figures become highly significant because they indicate the important changes occurring in the frequency response.

The root, ω_o, and Q sensitivities are peculiar to the movement of a specific root. To obtain an overall measure of the system performance, the system-function sensitivity may be used. It is defined as

$$S_q^{G(s)} = \frac{q}{G(s)}\frac{\delta G(s)}{\delta q}, \tag{6-38}$$

where $G(s)$ is the system function and q is the element whose effect on $G(s)$ is under investigation. Since $G(j\omega)$ takes on special significance when sinusoidal excitation is used [its magnitude, $|G(j\omega)|$, multiplies the amplitude of the input sine wave while its angle $\theta(\omega)$ adds to the angle of the sine wave], the $s = j\omega$ evaluation of Eq. (6-38) gives important information:

$$S_q^{G(j\omega)} = \frac{q}{G(j\omega)} \frac{\delta G(j\omega)}{\delta q} = \frac{q}{|G(j\omega)|e^{j\theta(\omega)}} \frac{\delta}{\delta q}[|G(j\omega)|e^{j\theta(\omega)}]$$

$$= \frac{q}{|G(j\omega)|e^{j\theta(\omega)}} \left[j|G(j\omega)|e^{j\theta(\omega)} \frac{\delta\theta(\omega)}{\delta q} + e^{j\theta(\omega)} \frac{\delta}{\delta q}|G(j\omega)| \right]$$

$$= \frac{q}{|G(j\omega)|} \frac{\delta}{\delta q}|G(j\omega)| + jq\frac{\delta\theta(\omega)}{\delta q}$$

$$= S_q^{|G(j\omega)|} + j\theta(\omega)S_q^{\theta(\omega)}. \tag{6-39}$$

Equation (6-39) relates the sensitivity of $G(j\omega)$ to the sensitivity of its magnitude, $S_q^{|G(j\omega)|}$, and to the sensitivity of its phase, $S_q^{\theta(\omega)}$. Note that

$$S_q^{|G(j\omega)|} = \mathrm{Re}\{S_q^{G(j\omega)}\}, \tag{6-40}$$

that is, *the sensitivity of the magnitude of the system function may be obtained by first calculating the sensitivity of the system function for $s = j\omega$ and then taking its real part.* It may be easier to do it this way rather than to calculate first $|G(j\omega)|$ and then its sensitivity.

$S_q^{|G(j\omega)|}$ is a function of ω. So different values of magnitude sensitivities are obtained for different values of ω. The sensitivity may be low over some band of frequencies and high over another band. Thus, a plot of $S_q^{|G(j\omega)|}$ vs. ω gives at a glance an overall picture of the sensitivity characteristic of the transfer function's magnitude with respect to a particular parameter in the system.

EXAMPLE 6-7

For the network shown in Fig. 6-15, calculate the pole, ω_o, Q, and transfer-function sensitivities with respect to each element and discuss the results. Assume that the poles are complex.

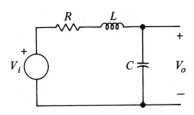

Fig. 6-15

SOLUTION

Calculate first the transfer function and then the pole locations:

$$\frac{V_o}{V_i} = \frac{\dfrac{1}{LC}}{s^2 + s\dfrac{R}{L} + \dfrac{1}{LC}} = \frac{\omega_o^2}{s^2 + s\dfrac{\omega_o}{Q} + \omega_o^2} = T(s);$$

$$p_{1,2} = -\frac{R}{2L} \pm j \sqrt{\frac{1}{LC} - \left(\frac{R}{2L}\right)^2};$$

$$|p_{1,2}| = \omega_o, \qquad \omega_o = \frac{1}{\sqrt{LC}}, \qquad Q = \frac{1}{R}\sqrt{\frac{L}{C}}.$$

Pole Sensitivities Using Eq. (6-29), calculate the three pole-sensitivity functions:

$$S_R^{p_{1,2}} = \frac{R}{|p_{1,2}|}\frac{\delta p_{1,2}}{\delta R} = \frac{1}{|p_{1,2}|}\left[-\frac{R}{2L} \mp j\frac{\left(\dfrac{R}{2L}\right)^2}{\sqrt{\dfrac{1}{LC} - \left(\dfrac{R}{2L}\right)^2}}\right]$$

$$= -\frac{1}{2Q}\left(1 \pm j\frac{1}{\sqrt{4Q^2 - 1}}\right),$$

$$S_L^{p_{1,2}} = \frac{L}{|p_{1,2}|}\frac{\delta p_{1,2}}{\delta L} = \frac{1}{|p_{1,2}|}\left[\frac{R}{2L} \mp j\frac{\left(\dfrac{R}{2L}\right)^2\left(\dfrac{2L}{R^2C} - 1\right)}{\sqrt{\dfrac{1}{LC} - \left(\dfrac{R}{2L}\right)^2}}\right]$$

$$= \frac{1}{2Q}\left(1 \mp j\frac{2Q^2 - 1}{\sqrt{4Q^2 - 1}}\right),$$

$$S_C^{p_{1,2}} = \frac{C}{|p_{1,2}|}\frac{\delta p_{1,2}}{\delta C} = \frac{1}{|p_{1,2}|}\left[\mp j\frac{\dfrac{1}{2LC}}{\sqrt{\dfrac{1}{LC} - \left(\dfrac{R}{2L}\right)^2}}\right] = \mp j\frac{Q}{\sqrt{4Q^2 - 1}}.$$

Note that all pole-sensitivity functions are dependent on the Q of the poles. For $Q \gg 1$, these functions simplify to

$$S_R^{p_{1,2}} = -\frac{1}{2Q} \mp j\frac{1}{4Q^2},$$

$$S_L^{p_{1,2}} = \frac{1}{2Q} \mp j\frac{1}{2},$$

$$S_C^{p_{1,2}} = \mp j\frac{1}{2}.$$

Thus, the upper half-plane pole moves to the left and down if R is increased slightly. It moves to the right and down if L is increased slightly. It moves only down if C is increased slightly. The three sensitivity vectors are shown in Fig. 6-16a, which may be used as an aid in tuning high-Q

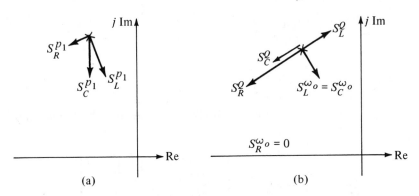

(a)   (b)

Fig. 6-16

circuits. It should be emphasized that this diagram is exact only when infinitesimal changes (about the pole's nominal position) are considered. (When large changes are considered, pole-loci diagrams should be used.)

ω_o and Q Sensitivities   Using Eq. (6-36), calculate the ω_o- and Q-sensitivity functions:

$$S_R^{\omega_o} = \frac{R}{\omega_o}\frac{\delta\omega_o}{\delta R} = 0,$$

$$S_L^{\omega_o} = \frac{L}{\omega_o}\frac{\delta\omega_o}{\delta L} = -\frac{1}{2},$$

$$S_C^{\omega_o} = \frac{C}{\omega_o}\frac{\delta\omega_o}{\delta C} = -\frac{1}{2},$$

$$S_R^{Q} = \frac{R}{Q}\frac{\delta Q}{\delta R} = -1,$$

$$S_L^{Q} = \frac{L}{Q}\frac{\delta Q}{\delta L} = \frac{1}{2},$$

$$S_C^{Q} = \frac{C}{Q}\frac{\delta Q}{\delta C} = -\frac{1}{2}.$$

Note that all these sensitivity functions are constant. Recalling that ω_o is constant when the pole moves on a circle centered at the origin of the s-plane and that Q is constant when the pole moves along a radial line, the sensitivity vectors can be drawn as shown in Fig. 6-16b. Once again, it should be emphasized that these vectors provide a measure for infinitesimal changes in the element values. The two vector diagrams of Fig. 6-16 present somewhat different but equally valid views about the movement of the upper half-plane pole as a function of per-unit change in an element value.

Transfer-Function Sensitivities   Using Eq. (6-38), calculate the three transfer-function sensitivities:

$$S_R^{T(s)} = \frac{R}{T(s)} \frac{\delta T(s)}{\delta R} = \frac{R}{T(s)} \left[\frac{-\dfrac{s}{L^2 C}}{\left(s^2 + s\dfrac{R}{L} + \dfrac{1}{LC}\right)^2} \right] = -\frac{sT(s)}{\omega_o Q},$$

$$S_L^{T(s)} = \frac{L}{T(s)} \frac{\delta T(s)}{\delta L} = \frac{L}{T(s)} \left[\frac{-\dfrac{s^2}{L^2 C}}{\left(s^2 + s\dfrac{R}{L} + \dfrac{1}{LC}\right)^2} \right] = -\left(\frac{s}{\omega_o}\right)^2 T(s),$$

$$S_C^{T(s)} = \frac{C}{T(s)} \frac{\delta T(s)}{\delta C} = \frac{C}{T(s)} \left[\frac{-\dfrac{1}{LC^2}\left(s^2 + s\dfrac{R}{L}\right)}{\left(s^2 + s\dfrac{R}{L} + \dfrac{1}{LC}\right)^2} \right]$$

$$= -\frac{s\left(s + \dfrac{\omega_o}{Q}\right)}{\omega_o^2} T(s) = T(s) - 1.$$

Note that

$$S_C^{T(s)} = S_R^{T(s)} + S_L^{T(s)}.$$

In order to obtain a physical interpretation of the transfer-function sensitivity, consider the sensitivity of the magnitude of $T(j\omega)$ with respect to L. Using Eq. (6-40), obtain

$$S_L^{|T(j\omega)|} = \text{Re}\{S_L^{T(j\omega)}\}$$

$$= \text{Re}\left\{ -\left(\frac{s}{\omega_o}\right)^2 T(s)\Big|_{s=j\omega} \right\}$$

$$= \text{Re}\left[\frac{\omega^2}{(\omega_o^2 - \omega^2) + j\dfrac{\omega\omega_o}{Q}} \right]$$

$$= \frac{\left(\dfrac{\omega}{\omega_o}\right)^2 \left[1 - \left(\dfrac{\omega}{\omega_o}\right)^2\right]}{\left[1 - \left(\dfrac{\omega}{\omega_o}\right)^2\right]^2 + \left(\dfrac{\omega}{\omega_o}\dfrac{1}{Q}\right)^2}.$$

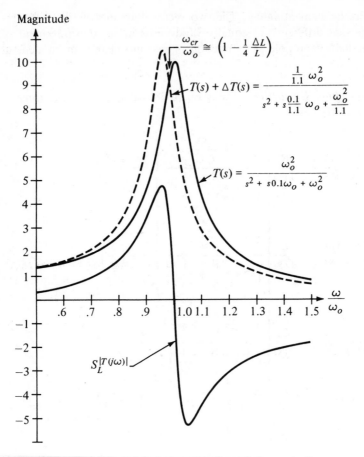

Fig. 6-17   Two ways of depicting the effect of change in L

This sensitivity function is plotted in Fig. 6-17. To obtain the per-unit change in the magnitude function at a given frequency, the value of the sensitivity function obtained at that frequency must be multiplied by the per-unit change in L, that is,

$$\frac{\Delta|T(j\omega)|}{|T(j\omega)|} \cong \frac{\Delta L}{L} S_L^{|T(j\omega)|}.$$

The smaller $\Delta L/L$, the more accurate is $\Delta|T(j\omega)|/|T(j\omega)|$. As the sensitivity-function curve indicates, the most pronounced changes in the magnitude characteristic occur near $\omega = \omega_o$. Because of the shape of the $S_L^{|T(j\omega)|}$-vs.-(ω/ω_o) curve, the magnitude function will peak at a lower frequency if the value of L is increased slightly. This is shown by the two magnitude curves, also plotted in Fig. 6-17. The solid curve represents the magnitude of $T(j\omega)$ for $Q = 10$. The dotted curve represents the magnitude characteristic after L is increased 10%. Note the correspondence between the difference of the two magnitude curves and the sensitivity function. Note also that the sensitivity is zero at $\omega = \omega_o$, a frequency that is somewhat higher than the frequency where the two magnitude curves cross each other. As shown in Fig. 6-17 (and in Problem 6-9), this frequency of crossing is

$$\omega_{cr} \cong \omega_o \left(1 - \frac{1}{4} \frac{\Delta L}{L} \right), \tag{6-41}$$

which reduces to ω_o as $\Delta L \to 0$. Once again, it should be emphasized that $S_L^{|T(j\omega)|}$ reflects infinitesimal changes whereas the two magnitude curves differ because of incremental changes.

EXAMPLE 6-8

Except for the scale factors, the two networks shown in Fig. 6-18 realize the same transfer function as the passive network given in Fig. 6-15. Compare the three networks with regard to ω_o and Q sensitivities.

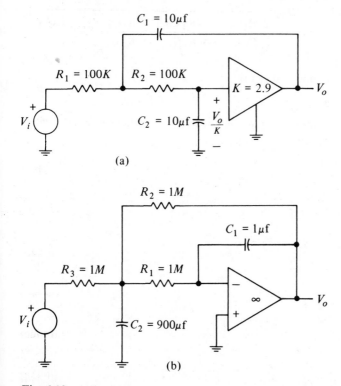

(a)

(b)

Fig. 6-18

SOLUTION

The transfer function of the network in Fig. 6-18a is given by (see Section 10-3)

$$T(s) = \cfrac{\cfrac{K}{R_1 R_2 C_1 C_2}}{s^2 + s \cfrac{1}{\sqrt{R_1 R_2 C_1 C_2}} \left(\sqrt{\dfrac{R_1 C_1}{R_2 C_2}} + \sqrt{\dfrac{R_2 C_2}{R_1 C_1}} + \sqrt{\dfrac{R_1 C_2}{R_2 C_1}} - K \sqrt{\dfrac{R_1 C_1}{R_2 C_2}} \right) + \cfrac{1}{R_1 R_2 C_1 C_2}}, \quad (6\text{-}42)$$

where K is the gain of the amplifier. For the indicated element values, Eq. (6-42) reduces to the nominal transfer function

$$T(s) = \frac{2.9}{s^2 + 0.1s + 1}.$$

In terms of the elements R_1, R_2, C_1, C_2, and K, the ω_o and Q of the poles are given by

$$\omega_o = \frac{1}{\sqrt{R_1 R_2 C_1 C_2}}, \qquad Q = \cfrac{1}{\sqrt{\dfrac{R_1 C_1}{R_2 C_2}} + \sqrt{\dfrac{R_2 C_2}{R_1 C_1}} + \sqrt{\dfrac{R_1 C_2}{R_2 C_1}} - K \sqrt{\dfrac{R_1 C_1}{R_2 C_2}}}.$$

$$(6\text{-}43)$$

The resulting nominal values are

$$\omega_o = 1, \qquad Q = 10.$$

For $R_1 = R_2 = R$ and $C_1 = C_2 = C$, the five ω_o and the five Q sensitivities are found by appropriate differentiations:

$$S_{R_1}^{\omega_o} = \frac{R_1}{\omega_o} \frac{\delta \omega_o}{\delta R_1} = -\frac{1}{2}, \tag{6-44a}$$

$$S_{R_2}^{\omega_o} = \frac{R_2}{\omega_o} \frac{\delta \omega_o}{\delta R_2} = -\frac{1}{2}; \tag{6-44b}$$

$$S_{C_1}^{\omega_o} = \frac{C_1}{\omega_o} \frac{\delta \omega_o}{\delta C_1} = -\frac{1}{2}, \tag{6-44c}$$

$$S_{C_2}^{\omega_o} = \frac{C_2}{\omega_o} \frac{\delta \omega_o}{\delta C_2} = -\frac{1}{2}; \tag{6-44d}$$

$$S_K^{\omega_o} = \frac{K}{\omega_o} \frac{\delta \omega_o}{\delta K} = 0; \tag{6-44e}$$

$$S_{R_1}^{Q} = \frac{R_1}{Q} \frac{\delta Q}{\delta R_1} = Q - \frac{1}{2} = 9.5, \tag{6-45a}$$

$$S_{R_2}^Q = \frac{R_2}{Q}\frac{\delta Q}{\delta R_2} = -Q + \frac{1}{2} = -9.5; \tag{6-45b}$$

$$S_{C_1}^Q = \frac{C_1}{Q}\frac{\delta Q}{\delta C_1} = 2Q - \frac{1}{2} = 19.5, \tag{6-45c}$$

$$S_{C_2}^Q = \frac{C_2}{Q}\frac{\delta Q}{\delta C_2} = -2Q + \frac{1}{2} = -19.5; \tag{6-45d}$$

$$S_K^Q = \frac{K}{Q}\frac{\delta Q}{\delta K} = 3Q - 1 = 29. \tag{6-45e}$$

The transfer function of the network in Fig. 6-18b is given by (see Section 10-7)

$$T(s) = -\frac{\dfrac{1}{R_1 R_3 C_1 C_2}}{s^2 + \sqrt{\dfrac{C_1}{C_2}}\left(\sqrt{\dfrac{R_1}{R_2}} + \sqrt{\dfrac{R_2}{R_1}} + \dfrac{\sqrt{R_1 R_2}}{R_3}\right)\dfrac{s}{\sqrt{R_1 C_1 R_2 C_2}} + \dfrac{1}{R_1 C_1 R_2 C_2}}, \tag{6-46}$$

which, for the indicated element values, reduces to

$$T(s) = -\frac{1}{s^2 + 0.1s + 1}.$$

In terms of the elements R_1, R_2, R_3, C_1, and C_2, the ω_o and Q of the poles are given by

$$\omega_0 = \frac{1}{\sqrt{R_1 C_1 R_2 C_2}}, \qquad Q = \frac{1}{\sqrt{\dfrac{C_1}{C_2}}\left(\sqrt{\dfrac{R_1}{R_2}} + \sqrt{\dfrac{R_2}{R_1}} + \dfrac{\sqrt{R_1 R_2}}{R_3}\right)}. \tag{6-47}$$

The five ω_o and the five Q sensitivities are found by appropriate differentiations:

$$S_{R_1}^{\omega_o} = S_{R_2}^{\omega_o} = S_{C_1}^{\omega_o} = S_{C_2}^{\omega_o} = -\tfrac{1}{2}, \qquad S_{R_3}^{\omega_o} = 0, \tag{6-48}$$

$$S_{R_1}^Q = -\frac{1}{2} + \frac{1}{1 + \dfrac{R_1}{R_2} + \dfrac{R_1}{R_3}} = -\frac{1}{6},$$

$$S_{R_2}^Q = -\frac{1}{2} + \frac{1}{1 + \dfrac{R_2}{R_1} + \dfrac{R_2}{R_3}} = -\frac{1}{6},$$

$$S_{R_3}^Q = \frac{1}{1 + \dfrac{R_3}{R_1} + \dfrac{R_3}{R_2}} = \frac{1}{3}; \qquad S_{C_1}^Q = -S_{C_2}^Q = -\frac{1}{2}. \tag{6-49}$$

For comparison purposes, the ω_o- and Q-sensitivity figures of the three networks realizing the same low-pass function (for any given ω_o and Q) are listed below.

Passive Network	*Active Network*	
	Amplifier K	*Amplifier A*
(Fig. 6-15)	(Fig. 6-18a)	(Fig. 6-18b)
$\left\|S^{\omega_o}_{\text{any element}}\right\| \leq \frac{1}{2}$	$\left\|S^{\omega_o}_{\text{any element}}\right\| \leq \frac{1}{2}$	$\left\|S^{\omega_o}_{\text{any element}}\right\| \leq \frac{1}{2}$
	$\left\|S^Q_{R_1}\right\| = \left\|S^Q_{R_2}\right\| = Q - \frac{1}{2},$	
$\left\|S^Q_{\text{any element}}\right\| \leq 1.$	$\left\|S^Q_{C_1}\right\| = \left\|S^Q_{C_2}\right\| = 2Q - \frac{1}{2},$	$\left\|S^Q_{\text{any element}}\right\| \leq \frac{1}{2}.$
	$S^Q_K = 3Q - 1.$	

As far as ω_o sensitivities are concerned, all three networks are good; no sensitivity figure is greater than $\frac{1}{2}$ in magnitude. However, when Q sensitivities are compared, large differences are observed between the first active network (with $K = 2.9$) and the other two networks. The first active network's Q sensitivities are dependent upon Q itself, resulting in large sensitivity figures even for moderately large Q's. On the other hand, the other two networks possess excellent Q sensitivities, none being greater than 1 in magnitude. This example shows that when a choice, based only on sensitivity figures, is to be made between the two active realizations, the second active network is much better.

It should be realized, however, that in this example, the amplifiers were considered ideal. Additional sensitivity figures, based on the nonideal characteristics of amplifiers, should also be considered. Furthermore, a high-sensitivity figure alone may not always imply a touchy circuit because the actual change also depends upon how much the element changes. For example, the change in Q as a result of a change in K may not be large (even though S^Q_K is large) if K is not allowed to change much. This can be seen when the per-unit changes are considered, that is,

$$\frac{\Delta Q}{Q} \cong S^Q_K \frac{\Delta K}{K} = (3Q - 1)\frac{\Delta K}{K}.$$

The K-amplifier is usually designed such that the value of K depends upon the ratio of two resistors which track very well in their characteristics (with temperature, aging, etc.), thus making $(\Delta K/K)$ a very small number. Consequently, even though Q may be large, $\Delta Q/Q$ may be kept small.

6-5 WORST-CASE ANALYSIS

Each element in a network affects in its own way the characteristic frequencies of the system functions. If there are k elements in the network, then there are k sensitivity functions (root, ω_o, or Q) associated with each pole or zero. When all the elements depart from their nominal (design) values, as is the case when the temperature varies, the resulting change is given by the sum of the individual changes. For example, if the pole p_1 is under consideration, then $(\Delta p_1/|p_1|)$ is given by

$$\frac{\Delta p_1}{|p_1|} \cong \sum_{i=1}^{k} S_{q_i}^{p_1} \frac{\Delta q_i}{q_i}, \tag{6-50}$$

where q_i is an element in the network and $S_{q_i}^{p_1}$ is the sensitivity function of the pole p_1 with respect to q_i. The expression is approximate because incremental rather than infinitesimal changes are considered.

For another example, consider the ω_o of the poles of the network given in Fig. 6-18a. It depends on the elements R_1, R_2, C_1, and C_2. If they all change randomly, the per-unit change in ω_o is given by

$$\frac{\Delta \omega_o}{\omega_o} \cong \sum_{i=1}^{4} S_{q_i}^{\omega_o} \frac{\Delta q_i}{q_i} \tag{6-51}$$

$$= S_{R_1}^{\omega_o} \frac{\Delta R_1}{R_1} + S_{R_2}^{\omega_o} \frac{\Delta R_2}{R_2} + S_{C_1}^{\omega_o} \frac{\Delta C_1}{C_1} + S_{C_2}^{\omega_o} \frac{\Delta C_2}{C_2}.$$

For this network all sensitivity figures are equal to $-\frac{1}{2}$. Therefore,

$$\frac{\Delta \omega_o}{\omega_o} \cong -\frac{1}{2} \left(\frac{\Delta R_1}{R_1} + \frac{\Delta R_2}{R_2} + \frac{\Delta C_1}{C_1} + \frac{\Delta C_2}{C_2} \right).$$

The worst case occurs when all changes are either positive or negative, giving

$$\left| \frac{\Delta \omega_o}{\omega_o} \right|_{\text{worst}} \cong \frac{1}{2} \left(\frac{|\Delta R_1|}{R_1} + \frac{|\Delta R_2|}{R_2} + \frac{|\Delta C_1|}{C_1} + \frac{|\Delta C_2|}{C_2} \right).$$

If all the resistance changes are known to be of one sign and capacitance changes of opposite sign, as could be the case when changes due to temperature are considered, then a more realistic and less pessimistic per-unit change in ω_o is:

$$\left| \frac{\Delta \omega_o}{\omega_o} \right| \cong \frac{1}{2} \left[\left| \frac{\Delta R_1}{R_1} + \frac{\Delta R_2}{R_2} \right| - \left| \frac{\Delta C_1}{C_1} + \frac{\Delta C_2}{C_2} \right| \right].$$

In this case, by proper matching of temperature coefficients, ω_o may be made independent of temperature.

6-6 SOME USEFUL RULES FOR CALCULATING SENSITIVITY FUNCTIONS

In sensitivity calculations, considerable time and effort can be saved by applying certain "sensitivity-algebra" rules. Let f be a function of q and define the sensitivity of f with respect to q as

$$S_q^f = \frac{q}{f} \frac{\delta f}{\delta q}.$$

The calculation of S_q^f is simplified when the following rules are used. In these rules, c and n are constants and f, f_1, and f_2 are differentiable with respect to q.

1. $S_q^c = 0$;   2. $S_q^{cq} = 1$;

3. $S_q^{(cf)^n} = n S_q^{cf} = n S_q^f$ $\left(S_q^{\sqrt{cf}} = \tfrac{1}{2} S_q^f ; S_q^{1/f} = -S_q^f\right)$;

4. $S_q^{f_1 + f_2 + \cdots} = \dfrac{f_1 S_q^{f_1} + f_2 S_q^{f_2} + \cdots}{f_1 + f_2 + \cdots}$;

5. $S_q^{f_1 f_2 \cdots} = S_q^{f_1} + S_q^{f_2} + \cdots$;   6. $S_q^{f_1 / f_2} = S_q^{f_1} - S_q^{f_2}$.

Note that these rules do not apply directly to the root-sensitivity function because its definition uses normalization with respect to the magnitude of the root and not the root itself.

EXAMPLE 6-9

Show that

$$S_q^{f_1 + f_2 + \cdots} = \frac{f_1 S_q^{f_1} + f_2 S_q^{f_2} + \cdots}{f_1 + f_2 + \cdots} .$$

SOLUTION

Apply the definition for sensitivity, and get

$$S_q^{f_1 + f_2 + \cdots} = \left(\frac{q}{f_1 + f_2 + \cdots}\right) \frac{\delta}{\delta q} (f_1 + f_2 + \cdots)$$

$$= \frac{f_1 \left(\dfrac{q}{f_1} \dfrac{\delta f_1}{\delta q}\right) + f_2 \left(\dfrac{q}{f_2} \dfrac{\delta f_2}{\delta q}\right) + \cdots}{f_1 + f_2 + \cdots}$$

$$= \frac{f_1 S_q^{f_1} + f_2 S_q^{f_2} + \cdots}{f_1 + f_2 + \cdots} . \qquad \textit{Ans.}$$

EXAMPLE 6-10

Obtain the sensitivity of

$$f = \sqrt{a} + \frac{1}{\sqrt{a}}$$

with respect to a.

SOLUTION

Use rule 4 (sensitivity of the sum of two functions) to obtain

$$S_a^f = \frac{\sqrt{a}\, S_a^{\sqrt{a}} + \dfrac{1}{\sqrt{a}}\, S_a^{1/\sqrt{a}}}{\sqrt{a} + \dfrac{1}{\sqrt{a}}}. \tag{6-52}$$

Use rule 6 (sensitivity of the ratio of two functions) to obtain

$$S_a^{1/\sqrt{a}} = S_a^1 - S_a^{\sqrt{a}}.$$

But, according to rule 1 (sensitivity of a constant),

$$S_a^1 = 0.$$

Hence,

$$S_a^{1/\sqrt{a}} = -S_a^{\sqrt{a}}.$$

Equation (6-52) then simplifies to

$$S_a^f = \frac{\sqrt{a} - \dfrac{1}{\sqrt{a}}}{\sqrt{a} + \dfrac{1}{\sqrt{a}}}\, S_a^{\sqrt{a}} = \frac{a-1}{a+1}\, S_a^{\sqrt{a}}. \tag{6-53}$$

Use rule 3 (sensitivity of a function raised to a power, $\frac{1}{2}$ in this case) to obtain

$$S_a^{\sqrt{a}} = \tfrac{1}{2} S_a^a.$$

But according to rule 2 (sensitivity of a function with respect to itself),

$$S_a^a = 1.$$

Hence,

$$S_a^{\sqrt{a}} = \tfrac{1}{2}.$$

When this value is substituted in Eq. (6-53), it becomes

$$S_a^f = \frac{1}{2}\frac{a-1}{a+1}. \qquad Ans.$$

Note that not even once was differentiation used to get this answer. With a little practice, many of the intermediate steps can be left out and the answer can be found in one or two steps.

6-7 SUMMARY

The (α, β) or (ω_o, Q) designations may be used to describe the position of the poles or zeros of a rational function. In these representations, $-\alpha$ is the real part, and β is the imaginary part of the critical frequencies; ω_o represents the magnitude of the critical frequencies when they are complex.

When a pair of poles of a system are close to the imaginary axis (high Q), the magnitude characteristic presents a definite peaking very near or at ω_o and the resulting bell-shaped curve about ω_o has a 3-dB bandwidth of (ω_o/Q).

The complex portion of the loci of the poles and zeros of second-order systems (as one system element is varied) is either a circle centered somewhere on the real axis or a vertical line. When only the coefficient of the s term in the numerator or denominator polynomial is affected by changes in an element value, the resulting change in the zeros of the polynomial is along a circle centered at the origin (if the zeros are complex). This represents a constant-ω_o type of root locus. On the other hand, when only the constant term is affected, the resulting displacements of the zeros are along a constant-α line. It is impossible to achieve constant-β or constant-Q root loci as a function of a single element in the system.

The root sensitivity, the ω_o- and Q-sensitivity functions are used to obtain a measure of the rate of change of the roots with respect to a particular element in the system. The larger the value of a sensitivity function (for a given set of element values), the more the root under investigation departs from its nominal value when a small change in the element is made. The root-sensitivity function gives information about the rate of change of the real and imaginary parts of complex roots. The ω_o- and Q-sensitivity functions give information about the rate of change of the magnitude and Q of the complex roots. The directional information provided by these functions, e.g., increase in the imaginary part or increase in Q, can be used to determine which element to vary in order to bring about a desired change in the frequency-response characteristics. These functions are also used to make a comparative study of different network realizations that produce the same system function.

PROBLEMS

6-1 The poles and zeros of

$$T(s) = \frac{a_2 s^2 + a_1 s + a_0}{s^2 + s \dfrac{\omega_o}{Q} + \omega_o^2}$$

are not close to each other. The Q is high.
(a) Show that $|T(j\omega)|$ peaks very nearly at $\omega = \omega_o$.
(b) Show that the 3-dB bandwidth is very nearly (ω_o/Q).

6-2 The number of cycles in Δt_1, shown in the step response of Fig. 6-2, is N. Show that Q is very nearly equal to πN.

6-3 $P(s) = [(s + 2)^2 + 1] + q[(s + 1)^2 + 1]$.

    (a) Plot the loci of the roots of $P(s)$ as q varies from 0 to ∞.

    (b) Repeat (a) if q varies from 0 to $-\infty$.

6-4 (a) $P(s) = [s(s + 3)] + q[(s + 1)(s + 2)]$.

    Plot of the loci of the roots of $P(s)$ as q takes on values from $-\infty$ to $+\infty$.

    (b) Repeat (a) if $P(s) = (s^2 + 1) + qs^2$.

6-5 $P(s) = P_1(s) + qP_2(s)$. The polynomials $P_1(s)$ and $P_2(s)$ are independent of q. For the $P_1(s)$ and $P_2(s)$ root locations shown in Fig. 6-19, sketch the roots of $P(s)$ as a function of q.

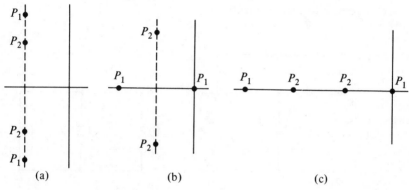

(a)   (b)   (c)

Fig. 6-19

6-6 $P(s) = P_1(s) + qP_2(s)$. Under what conditions are the $P(s)$ zeros independent of q?

6-7 A complex-conjugate pair of zeros form

$$P(s) = (s + \alpha)^2 + \beta^2 = s^2 + s\frac{\omega_o}{Q} + \omega_o^2,$$

where α, β, ω_o, and Q are functions of q. Sketch the zeros of $P(s)$ as a function of q $(q > 0)$ if the q dependence is given by

    (a) $\alpha = q\alpha_o$, $\beta = \beta_o$ (α_o and β_o are constants);

    (b) $\alpha = \alpha_o$, $\beta = q\beta_o$;

    (c) $\omega_o = q\omega_c$, $Q = Q_o$ (ω_c and Q_o are constants);

    (d) $\omega_o = \omega_c$, $Q = qQ_o$.

Note that two of the root loci are not circles or vertical lines. Why?

6-8 Show that the root sensitivity of a pair of complex-conjugate roots, $r_{1,2}$, can be expressed in terms of the ω_o and Q sensitivities as follows:

$$S_q^{r_{1,2}} = \frac{r_{1,2}}{|r_{1,2}|}\left(S_q^{\omega_o} \mp j\frac{S_q^Q}{\sqrt{4Q^2 - 1}}\right).$$

6-9 Derive Eq. (6-41).

6-10

$$T(s) = H \frac{(s - z_1)(s - z_2)\cdots(s - z_m)}{(s - p_1)(s - p_2)\cdots(s - p_n)}.$$

Show that

$$S_q^{T(s)} = S_q^H - \sum_{i=1}^{m} \frac{|z_i|}{s - z_i} S_q^{z_i} + \sum_{i=1}^{n} \frac{|p_i|}{s - p_i} S_q^{p_i}.$$

6-11 (a) Obtain (V_o/V_i) for the network shown in Fig. 6-20.

    (b) Draw the loci of the zeros of (V_o/V_i) as a function of each element in the network as it varies from 0 to ∞.

    (c) Obtain the root, ω_o, and Q sensitivities of the zeros of (V_o/V_i) with respect to each element.

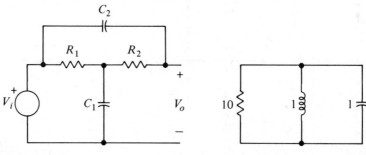

Fig. 6-20   Fig. 6-21

6-12 Obtain the pole sensitivities with respect to each element for the circuit shown in Fig. 6-21.

6-13 For f a function of q and for c and n constant, show that

    (a) $S_q^{(cf)^n} = n S_q^f$;

    (b) $S_q^{1/f} = -S_q^f$.

6-14 (a) Derive Eq. (6-42).

    (b) From the result of (a), obtain ω_o and Q.

    (c) Show that for $R_1 = R_2 = R$ and $C_1 = C_2 = C$, the Q sensitivities are as given by Eq. (6-45).

6-15 (a) Derive Eq. (6-46).

    (b) From the result of (a), obtain ω_o and Q.

    (c) Show that the Q sensitivities are as given by Eq. (6-49).

6-16 (a) Obtain the RLC ladder realization with resistive-load termination, R, for the transfer function:

$$T(s) = \frac{H}{s^2 + \dfrac{\omega_o}{Q} s + \omega_o^2}.$$

    (b) Calculate the pole, ω_o, Q, and transfer-function sensitivities for each element, and compare them with the results obtained for the resistive-source realization given in Example 6-7.

7

The Operational Amplifier: Modelling and Applications

Poles and zeros of system functions obtained from RLC networks cannot be placed arbitrarily anywhere in the s-plane. They must obey certain constraints. For example, poles cannot be placed in the right half-plane. If the network is LC, the poles must be simple and on the imaginary axis. If the network is RC, the poles must be simple and on the nonpositive real axis.

The introduction of dependent sources, however, removes all these restrictions and allows the placement of poles and zeros anywhere in the s-plane. (Complex critical frequencies must, of course, occur in conjugate pairs.) For example, by means of dependent sources the poles of RC networks can be lifted off the nonpositive real axis and moved around at will; in particular, they can be placed anywhere in the left half-plane and near the imaginary axis.

In so doing, RC-dependent source networks simulate system functions obtained from RLC networks, thereby eliminating the need for inductors. This feature is particularly important for low-frequency filter work, where the use of inductors becomes impractical because of their weight, bulk, and considerable departure from ideal behavior. Moreover, RC-dependent source networks can be adjusted to produce a pair of imaginary-axis poles, a feat that cannot be achieved in practice with passive networks, since actual LC networks look more like RLC networks because of losses. The resulting system, with imaginary-axis poles, is used in the construction of sinusoidal oscillators. In this chapter, operational amplifiers are used to implement dependent sources. The model for the ideal operational amplifier is presented first and used in the development of many practical circuits. Then the one-pole rolloff model is discussed, and its effect on amplifier characteristics treated in detail. The aim of this chapter is to provide a good understanding of the operational amplifier and to provide proficiency in the use of this very versatile and important device. Many examples are given to demonstrate the principles involved and to discuss practical applications as well as limitations.

7-1 THE IDEAL OPERATIONAL AMPLIFIER

The ideal operational amplifier, shown in Fig. 7-1a, is an important building block for generating dependent sources. It has two input ports. One is between the *negative terminal (terminal 1) and ground*, the other is between the *positive terminal (terminal 2) and ground*. It has one output port, the voltage across which is designated as V_o. The ∞ sign is put on the amplifier to indicate that it has infinite gain. The dc sources, V_c and V_d, are necessary to establish proper operating conditions. As far as signals are considered, terminals 4 and 5 are at ground voltage.

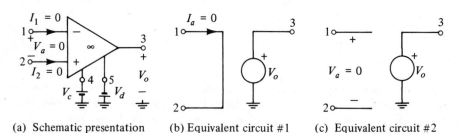

(a) Schematic presentation   (b) Equivalent circuit #1   (c) Equivalent circuit #2

Fig. 7-1   The ideal operational amplifier

When operating in the linear region, the ideal operational amplifier has the following characteristics:

1. It has infinite impedance at both input ports, i.e., neither the negative nor the positive terminal draws any current ($I_1 = I_2 = 0$).
2. It has infinite gain. As a consequence, the *voltage between the negative and the positive terminals*, V_a, is zero.
3. It has zero output impedance; i.e., the output voltage, V_o, is independent of the current drawn from the output.

These three features are expressed by either of the equivalent circuits presented in Fig. 7-1b and 7-1c. The circuit in (b) is drawn with a short circuit between the negative and positive input terminals, *thus automatically satisfying the $V_a = 0$ condition*; the $I_a = 0$ constraint is indicated on the diagram to show that *neither input draws any current in spite of the short circuit existing between them*. The circuit in (c) is drawn with an open circuit between the negative and positive input terminals, *thus automatically satisfying the condition that neither input draws any current*; the $V_a = 0$ constraint is indicated on the diagram to show that *the voltage between the negative and positive input terminals is zero*.

Henceforth, the ideal operational amplifier is drawn as in Fig. 7-1a, without the zero-current and the zero-voltage markings, these being understood. To obtain a system function of a network, each operational amplifier schematic presentation is removed, and either the equivalent circuit of Fig. 7-1b or the equivalent circuit of Fig. 7-1c is substituted; then, the desired input–output relationship is calculated by ordinary circuit-theory techniques. With a little practice, the network equations may be written directly without making any substitutions of equivalent circuits.

The amplifier is called an operational amplifier because operations like inversion, summation, integration, etc., are readily performed when these amplifiers are used with *RC* networks.

7-2 OPERATIONAL AMPLIFIER APPLICATIONS

The operational amplifier is quite often used with *resistive feedback from the output terminal to the negative input terminal.* Depending upon where the input signals are applied, the resulting network is used to amplify a signal, to amplify and invert a signal, or to take the difference between two signals and amplify it. The amplification may be greater than unity, or it may be unity, in which case the significant feature of the circuit is its low output impedance (ideally zero). It is also possible to obtain gain magnitudes which are less than unity.

The Noninverting Amplifier

The noninverting amplifier is shown in Fig. 7-2a. The output, V_o, is fed back to the (inverting) negative terminal of the operational amplifier through the R_2R_1 network. The input is applied to the (noninverting) positive terminal.

To calculate the transfer function, the operational amplifier is replaced by the model given in Fig. 7-1b or c. The resulting equivalent circuits are shown in Fig. 7-2b and c. Either circuit can be used to obtain V_o/V_i.

If the equivalent circuit of Fig. 7-2b is used, *the expression for the current* I_a *must be obtained and set equal to zero.* The resulting equation can then be solved for the transfer function. Thus,

$$I_a = \frac{V_i}{R_1} + \frac{V_i - V_o}{R_2} \qquad \text{(Kirchhoff's Current Law applied to node 1).}$$

But

$$I_a = 0 \qquad \text{(constraint imposed by operational amplifier).}$$

Hence,

$$\frac{V_i}{R_1} + \frac{V_i - V_o}{R_2} = 0,$$

$$\frac{V_o}{V_i} = 1 + \frac{R_2}{R_1}.$$

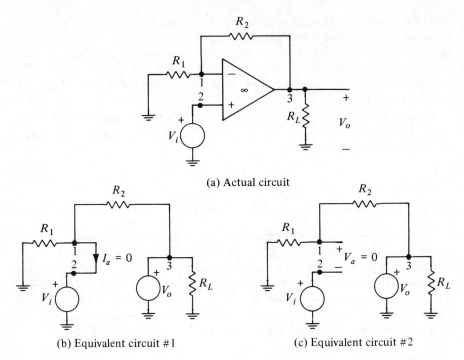

(a) Actual circuit

(b) Equivalent circuit #1   (c) Equivalent circuit #2

Fig. 7-2   Noninverting amplifier and equivalent representations

On the other hand, if the equivalent circuit of Fig. 7-2c is used, *the expression for the voltage V_a must be obtained and set equal to zero.* Thus

$$V_a = V_o \frac{R_1}{R_1 + R_2} - V_i \qquad \text{(by inspection of network).}$$

But

$$V_a = 0 \qquad \text{(constraint imposed by operational amplifier).}$$

Hence,

$$V_o \frac{R_1}{R_1 + R_2} - V_i = 0,$$

$$\frac{V_o}{V_i} = 1 + \frac{R_2}{R_1}.$$

Both methods of solution yield the same answer; namely, *the gain of the noninverting amplifier is*

$$G = \frac{V_o}{V_i} = 1 + \frac{R_2}{R_1}. \qquad (7\text{-}1)$$

By reference to Eq. (7-1) and Fig. 7-2a, several important observations can be made about the noninverting amplifier.

1. The gain is positive (hence, the name, *noninverting amplifier*).
2. The gain is equal to or greater than one.
3. Unity gain is achieved by making $R_2 = 0$ or $R_1 = \infty$. Both of these conditions are met by short-circuiting the output directly to the negative terminal.
4. The gain depends upon the resistance ratio of two resistors. Hence, it can be accurately set and maintained.
5. The gain does not depend upon any impedance the source may have, because the positive input terminal does not draw any current. Or (what amounts to the same result) the source, V_i, sees infinite impedance.
6. The output impedance is zero. Hence, the gain is independent of any load that is connected to the output. However, the operational amplifier must be able to supply the output current demanded by the feedback network $[V_o/(R_1 + R_2)]$ and the load (V_o/R_L).

The gain of the noninverting amplifier becomes frequency-dependent if impedances $Z_2(s)$ and $Z_1(s)$ are used instead of R_2 and R_1. In that case, (V_o/V_i) becomes

$$\frac{V_o}{V_i} = 1 + \frac{Z_2(s)}{Z_1(s)}. \tag{7-2}$$

The Inverting Amplifier

The inverting amplifier is shown in Fig. 7-3a. The output, V_o, is fed back to the negative terminal of the operational amplifier through the $R_2 R_1$ network. The input is connected to R_1 while the positive terminal of the operational amplifier is grounded.

To calculate the transfer function, the operational amplifier is replaced by the model given in Fig. 7-1b or c. The resulting equivalent circuits are shown in Fig. 7-3b and c. Either circuit can be used to obtain (V_o/V_i). Or, by keeping the $I_a = 0$ and $V_a = 0$ constraints in mind, V_o can be calculated directly from the actual circuit of Fig. 7-3a, as follows: Because $V_a = 0$, the current through R_1 is

$$I_1 = \frac{V_1}{R_1}.$$

Because $I_a = 0$, the current through R_2 is $I_2 = I_1$. The output voltage is

$$V_o = -I_2 R_2 + V_a = -\frac{V_i}{R_1} R_2.$$

Hence, *the gain of the inverting amplifier is*

$$G = \frac{V_o}{V_i} = -\frac{R_2}{R_1}. \tag{7-3}$$

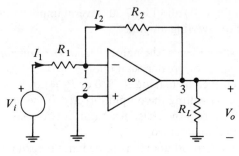

(a) Actual circuit

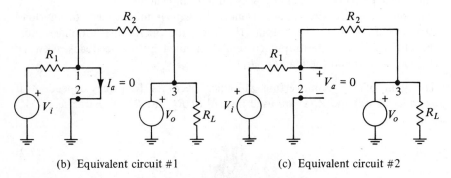

(b) Equivalent circuit #1   (c) Equivalent circuit #2

Fig. 7-3   Inverting amplifier and equivalent representations

By reference to Eq. (7-3) and Fig. 7-3a, several important observations can be made about the inverting amplifier.

1. The gain is negative (hence, the name *inverting amplifier*).
2. The magnitude of the gain can be made less than unity ($R_2 < R_1$), equal to unity ($R_2 = R_1$), or greater than unity ($R_2 > R_1$).
3. The gain depends upon the resistance ratio of two resistors. Hence, it can be accurately set and maintained.
4. The source, V_i, sees an impedance of R_1. Hence it must be able to deliver a current of (V_i/R_1). (When the source contains an internal resistance R_s, then V_i can be taken as the open-circuit source voltage, in which case R_s should be treated as being part of R_1.)
5. The output impedance is zero. Therefore, the load, R_L, does not enter the expression for the gain. However, the operational amplifier must be able to deliver the output current demanded by the feedback network (V_o/R_2) and the load (V_o/R_L).

The gain of the inverting amplifier becomes frequency-dependent if impedances $Z_2(s)$ and $Z_1(s)$ are used instead of R_2 and R_1. In that case, (V_o/V_i) becomes:

$$\frac{V_o}{V_i} = -\frac{Z_2(s)}{Z_1(s)}. \tag{7-4}$$

The Difference Amplifier

To obtain the difference between two signals, V_1 and V_2, the circuit shown in Fig. 7-4 can be used.

As in the case of the other circuits, the feedback is applied from V_o to the negative terminal of the amplifier. Since the noninverting gain, $1 + (R_2/R_1)$, is higher than the magnitude of the inverting gain, R_2/R_1, the signal at the positive

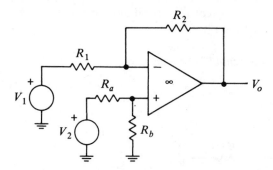

Fig. 7-4   Difference amplifier

terminal must be attenuated if exact subtraction is to take place. The $R_a R_b$ network is used for this purpose. Using superposition, and the gain expressions given by Eq. (7-1) and Eq. (7-3), V_o can be written, by inspection:

$$V_o = V_1\left(-\frac{R_2}{R_1}\right) + V_2\left(\frac{R_b}{R_a + R_b}\right)\left(1 + \frac{R_2}{R_1}\right)$$

$$= \frac{R_2}{R_1}\left[-V_1 + \left(\frac{1 + \dfrac{R_1}{R_2}}{1 + \dfrac{R_a}{R_b}}\right)V_2\right].$$

If

$$\frac{R_a}{R_b} = \frac{R_1}{R_2},$$

then

$$V_o = \frac{R_2}{R_1}(V_2 - V_1). \tag{7-5}$$

Thus, not only is the difference of the two signals obtained, but also the difference is amplified by (R_2/R_1). The source V_1 must be able to deliver a current of

$$I_1 = \frac{V_1 - V_2\left(\dfrac{R_b}{R_a + R_b}\right)}{R_1}, \tag{7-6a}$$

and the source V_2 must be able to supply

$$I_2 = \frac{V_2}{R_a + R_b}.$$ (7-6b)

The Summing Amplifier

To add two signals, V_1 and V_2, the circuit shown in Fig. 7-5 can be used.

To obtain the expression for V_o, the current I_a is found and set equal to zero. The resulting equation is then solved for V_o:

$$I_a = \frac{V_1}{R_1} + \frac{V_2}{R_2} + \frac{V_o}{R_3} = 0 \quad \text{(from Kirchhoff's Current Law)},$$

$$V_o = \left(-\frac{R_3}{R_1}\right) V_1 + \left(-\frac{R_3}{R_2}\right) V_2.$$ (7-7)

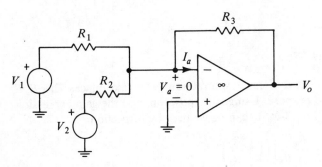

Fig. 7-5   The inverting summing amplifier

The signal sources do not interact, V_1 delivers a current dependent on V_1 only, namely (V_1/R_1), and V_2 delivers a current of (V_2/R_2). Equation (7-7) could have been written directly by applying the inverting gain to each source independently. Note that, when V_1 is on and V_2 is zero, no current is taken by R_2 since the voltage across it is zero (being the voltage across the amplifier inputs); therefore, R_2 has no effect on the gain associated with the source V_1. Similarly, R_1 has no effect on the gain associated with the source V_2.

If R_2 is made to equal R_1, Eq. (7-7) reduces to

$$V_o = -\frac{R_3}{R_1} (V_1 + V_2).$$ (7-8)

Thus, the output becomes proportional to the inverted sum of the input signals.

Transimpedance and Transadmittance Amplifiers

Operational amplifiers are also used to convert voltage to current and current to voltage. A voltage-to-current converter (transadmittance amplifier) is shown in Fig. 7-6a. Because $V_a = 0$, the input current is V_i/R. Because $I_a = 0$, the current through the feedback resistor, R_L, is the same as the input current. Therefore, *the current through R_L, I_o, is independent of R_L*. Thus, the input voltage, V_i, is converted to current, V_i/R, which is then used to drive the load R_L. Stated differently, R_L sees a current source the value of which is determined by the input signal and input resistance. Note that neither end of the load is grounded.

A current-to-voltage converter (transimpedance amplifier) is shown in Fig. 7-6b. Because $I_a = 0$, the current in the feedback resistor R is the input current I_i. Because $V_a = 0$, the output voltage is $-I_i R$. Hence, *the current source I_i is converted to the voltage source $V_o = -I_i R$*. This source is then used to drive the load R_L.

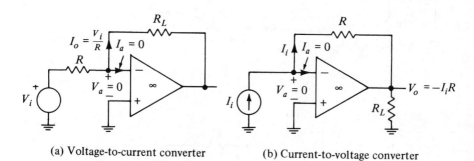

(a) Voltage-to-current converter   (b) Current-to-voltage converter

Fig. 7-6   Voltage-to-current and current-to-voltage converters

EXAMPLE 7-1

The amplifier shown in Fig. 7-7 is to be designed such that any value of gain between -10 and $+10$ can be achieved by adjusting the potentiometer.
(a) Obtain the values for r_1 and r_2.
(b) For what value of k is the gain zero?
(c) Convert element values to practical values.

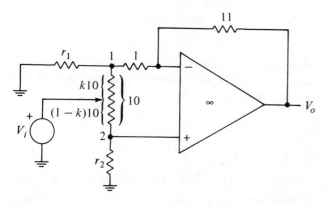

Fig. 7-7

SOLUTION

(a) Because the source resistance is zero, Fig. 7-7 can be redrawn as shown in Fig. 7-8a. To simplify further, obtain the Thévenin-equivalent representations of the circuits to the left of terminals 1 and 2. Draw the resulting network as in Fig. 7-8b, and obtain the inverting and non-

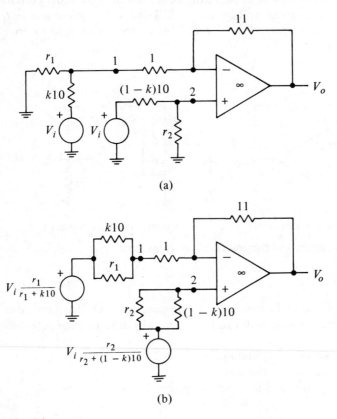

(a)

(b)

Fig. 7-8

inverting gains (note that the impedance in series with the positive lead of the amplifier does not affect the expressions for the gains because this lead does not draw any current):

$$G^- = -\frac{11}{1 + \dfrac{k10r_1}{k10 + r_1}} \; ; \qquad G^+ = 1 - G^-. \tag{7-9}$$

By considering one input at a time, obtain the expression of the output voltage and the resulting gain:

$$V_o = V_i \frac{r_1}{r_1 + k10} G^- + V_i \frac{r_2}{r_2 + (1-k)10} G^+,$$

$$G(k) = \frac{V_o}{V_i} = \frac{r_1}{r_1 + k10} \left(-\frac{11}{1 + \dfrac{k10r_1}{k10 + r_1}} \right)$$

$$+ \frac{r_2}{r_2 + (1-k)10} \left[1 + \frac{11}{1 + \dfrac{k10r_1}{k10 + r_1}} \right]. \quad (7\text{-}10)$$

At one extreme position of the potentiometer, $k = 0$. In this position, all of V_i appears at terminal 1 while terminal 2 receives only part of V_i (after being attenuated by the potentiometer and r_2). Thus, the output is mainly determined by the signal at 1, and it is at its most negative value. Therefore, set Eq. (7-10) to -10, and solve for r_2:

$$G(0) = -11 + 12\frac{r_2}{r_2 + 10} = -10,$$

$$r_2 = \frac{10}{11}. \qquad Ans.$$

At the other extreme position of the potentiometer, $k = 1$. In this position all of V_i appears at terminal 2 while terminal 1 receives only part of V_i (after being attenuated by the potentiometer and r_1). Thus, the output is mainly determined by the signal at 2, and it is at its most positive value. Therefore, set Eq. (7-10) to $+10$, and solve for r_1:

$$G(1) = \frac{r_1}{r_1 + 10} \left(-\frac{11}{1 + \dfrac{10r_1}{10 + r_1}} \right) + \left(1 + \frac{11}{1 + \dfrac{10r_1}{10 + r_1}} \right) = 10,$$

$$r_1 = \frac{20}{99}. \qquad Ans.$$

(b) Using the values of r_1 and r_2 obtained in (a), simplify Eq. (7-10):

$$G(k) = \frac{10(145k - 24)}{(2 + 119k)(12 - 11k)}.$$

The gain reduces to zero when k is adjusted to

$$k = \frac{24}{145}. \qquad Ans.$$

(c) If V_i is higher than a few tenths of a millivolt, the currents in Fig. 7-7 may become excessive, putting an unnecessary burden on the input source and the output of the operational amplifier. To remedy this problem, multiply all resistances by 1000. Since impedance scaling does not affect voltage ratios, the expression for gain remains the same.

EXAMPLE 7-2

Obtain (V_o/V_i) for the network shown in Fig. 7-9.

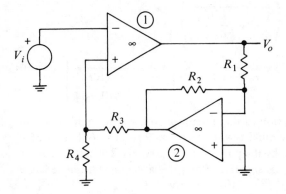

Fig. 7-9

SOLUTION

Operational amplifier #1 is driven by two signals. One is the input signal V_i; the other is derived from the output V_o. Operational amplifier #2 receives V_o as its input and converts it to $-(R_2/R_1)V_o$ at its output. Consequently, Fig. 7-9 can be simplified to Fig. 7-10. Note that feedback is applied to the positive terminal of amplifier #1 after the output signal is inverted by amplifier #2.

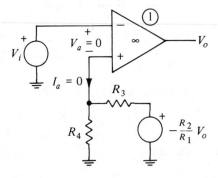

Fig. 7-10

Since the current taken by the positive terminal of the amplifier, I_a, is zero, the voltage across R_4 is

$$V_{R_4} = - \left(\frac{R_2}{R_1}\right) V_o \left(\frac{R_4}{R_3 + R_4}\right).$$

Since the voltage across the amplifier input terminals, V_a, is zero, the voltage across R_4 must equal the input voltage V_i:

$$-\left(\frac{R_2}{R_1}\right)V_o\left(\frac{R_4}{R_3 + R_4}\right) = V_i.$$

Solving for (V_o/V_i), the desired transfer function is obtained:

$$\frac{V_o}{V_i} = -\frac{R_1}{R_2}\left(1 + \frac{R_3}{R_4}\right). \qquad Ans.$$

EXAMPLE 7-3

Consider the difference amplifier shown in Fig. 7-11.
(a) What impedance does the source V_1 see if $V_2 = 0$? $V_2 = V_1$? $V_2 = -V_1$?
(b) What impedance does the source V_2 see if $V_1 = 0$? $V_1 = V_2$? $V_1 = -V_2$?

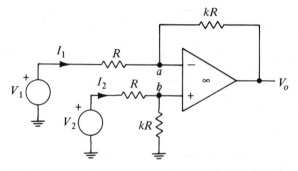

Fig. 7-11

SOLUTION

(a) To obtain the impedance seen by V_1, I_1 must be calculated. Note that the value of I_1 depends not only on V_1 but also on V_2, which determines the voltage at a. Because $V_{ab} = 0$, the voltage at a equals the voltage at b, which is readily calculated:

$$V_b = V_2 \frac{kR}{R + kR} = V_2 \frac{k}{1 + k}.$$

Thus,

$$V_a = V_b = V_2 \frac{k}{1 + k}.$$

The current delivered by the source V_1 is

$$I_1 = \frac{V_1 - V_a}{R} = \frac{V_1 - V_2 \dfrac{k}{1+k}}{R} = \frac{V_1 \left(1 - \dfrac{k}{1+k}\dfrac{V_2}{V_1}\right)}{R}.$$

The impedance seen by the source V_1 can be calculated only when V_2 is zero or dependent on V_1, in which case

$$Z_1 = \frac{V_1}{I_1} = \frac{R}{1 - \dfrac{k}{1+k}\dfrac{V_2}{V_1}}.$$

The three impedances are

$$Z_1 \bigg|_{V_2=0} = R, \qquad Z_1 \bigg|_{V_2=V_1} = R(1+k), \qquad Z_1 \bigg|_{V_2=-V_1} = R\frac{1+k}{1+2k}.$$
$$Ans.$$

(b) To obtain the impedance seen by the source V_2, calculate I_2. Since the value of I_2 does not depend upon V_1, it is readily obtained:

$$I_2 = \frac{V_2}{R + kR} = \frac{V_2}{(1+k)R}.$$

Thus, regardless of the value of V_1, the impedance seen by the source V_2 is

$$Z_2 = \frac{V_2}{I_2} = R(1+k). \qquad Ans.$$

All the applications presented so far are based on the assumption that the operational amplifier is operating linearly, in which case the ideal equivalent circuits given in Fig. 7-1b and 7-1c are valid and can be used to solve network problems. However, operational amplifiers do not always operate in the linear, active mode. Sometimes they oscillate. At other times, they may become saturated, i.e., the output may stay at some fixed, positive or negative voltage. In the latter case, the output equivalent circuit of the operational amplifier is not a dependent voltage source, V_o, but rather a fixed dc source.

Now the question can be raised: How is it known that the operational amplifier is active or saturated? A sure and obvious answer to this question is obtained by building the circuit and testing it. If the circuit works as expected, then the amplifier is operating as assumed. For example, if the amplifier is connected in the inverting mode (see Fig. 7-3a) and assumed to be operating linearly, then the input–output relationship must obey

$$\frac{V_o}{V_i} = -\frac{R_2}{R_1}.$$

On the other hand, in another feedback arrangement, if the expected steady-state output (based on linear behavior) is a sine wave but instead a dc voltage is observed, then the amplifier is saturated, and the assumption that it is active is not valid.

If the ideal, linear model is used for operational amplifiers in a network and subsequent calculation shows that the poles of the system function are all in the left half-plane, it does not necessarily follow that the amplifiers will operate in the linear mode in the actual implementation of the circuit. For example, if the feedback is applied to the positive input terminal of the amplifier in Fig. 7-2 and the negative input terminal is grounded, then even though the resulting network (being purely resistive) has no poles, the output will indicate a saturated state in an actual experimental setup. This apparent discrepancy between theory and practice is resolved if a more realistic model, rather than the ideal model, is used to represent the operational amplifier. Even then, it should be kept in mind that the ultimate test is whether the actual circuit works as expected.

In this text, the ideal operational amplifier model is quite often used because it affords simplicity of analysis and interpretation. However, to provide better agreement between theory and practice, a more accurate, and consequently more complicated, model is necessary. This is particularly true if the amplifier is used in applications other than dc.

7-3 THE ACTUAL OPERATIONAL AMPLIFIER

The actual operational amplifier differs in many ways from the ideal. Foremost among these is the frequency-dependent nature of the gain. As a result, the gain of the operational amplifier is no longer infinite but $A(s)$, and the input voltage is no longer zero but $V_a(s)$. These changes are incorporated in the presentation of the operational amplifier and its equivalent circuit as shown in Fig. 7-12.

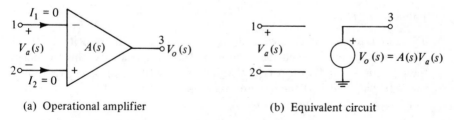

(a) Operational amplifier   (b) Equivalent circuit

Fig. 7-12 Improved modelling of the operational amplifier

As the gain, $A(s)$, approaches infinity, the input voltage, $V_a(s)$, approaches zero, and the output, $V_o(s)$, remains finite and dependent upon the external circuit parameters. In other words, the improved model reduces to the ideal model.

One-Pole Rolloff Model

So far nothing has been said about the nature of $A(s)$. Most operational amplifiers are designed so that, for small signal operation, $A(s)$ can be represented by a dominant pole, that is,

$$A(s) = -\frac{A_0 \omega_a}{s + \omega_a} = -\frac{GB}{s + \omega_a}. \tag{7-11}$$

The magnitude and phase characteristics of Eq. (7-11) are sketched in Fig. 7-13.

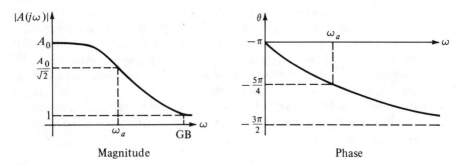

| Magnitude   Phase

Fig. 7-13   Open-loop gain characteristics

From Eq. (7-11) and Fig. 7-13, the following important observations can be made.

1. For $\omega \ll \omega_a$, the magnitude of the gain is A_0. Note that $-A_0$ is the open-loop dc gain of the amplifier. (The loop is open because no external feedback is applied.) A typical value of A_0 is 10^5.
2. When $s = j\omega_a$,

$$|A(j\omega_a)| = A_0/\sqrt{2}.$$

Thus, at ω_a, the magnitude of the gain is down 3 dB from its dc value. The frequency ω_a is called the open-loop 3-dB frequency. At this frequency the phase is $-(5\pi/4)$ rad. A typical value of ω_a is 10 rad/s. ω_a also represents the 3-dB bandwidth of the operational amplifier.
3. For $\omega \gg \omega_a$, the gain is given by:

$$A(j\omega) = -\frac{A_0 \omega_a}{j\omega} = -\frac{GB}{j\omega}.$$

The magnitude of the gain falls off inversely with ω. Its phase is constant, $-(\pi/2)$. Because GB is the product of the magnitude of the dc gain, A_0, with the 3-dB bandwidth, ω_a, it is called the gain-bandwidth product of the operational amplifier. A typical value of GB is 10^6.

4. The gain characteristic is low-pass in nature. Because the magnitude characteristic rolls off at -6 dB/octave at high frequencies, the model is called the one-pole rolloff model.
5. At $s = jGB$, the magnitude of the gain is practically unity. Therefore, $\mathbf{GB}$ can be interpreted as the frequency at which the open-loop gain of the operational amplifier becomes unity. Considering that the dc gain may be 10^5, a gain of 1 at $\omega = \mathbf{GB}$ represents a reduction of 100 dB in gain.

In linear applications, the operational amplifier is not used in the open-loop mode. One reason is that its large gain, particularly for low-frequency signals, causes signals, even with low amplitudes, to overdrive the amplifier, thus exceeding its dynamic range. For example, with $A_0 = 10^5$, a 150-μV dc signal produces at the output -15 V, which may saturate the amplifier. Furthermore, with large signal levels, the operation of the amplifier is more likely to be nonlinear, and, consequently, the small-signal model presented by Eq. (7-11) may no longer be valid. Another reason for not using the operational amplifier in the open-loop mode is that A_0 may vary considerably from one unit to another; furthermore, in a given unit, A_0 may change with temperature. These and other undesirable features (discussed later) make it necessary to operate the amplifier under closed-loop conditions, which render results much less susceptible to the variable characteristics of the open-loop amplifier.

Gain and Bandwidth of the Noninverting Amplifier

The gain of the noninverting amplifier becomes frequency-dependent when a one-pole rolloff model is used for the operational amplifier. To obtain this frequency dependence, consider the noninverting amplifier circuit shown in Fig. 7-14a. Its small-signal equivalent presentation is given in Fig. 7-14b.

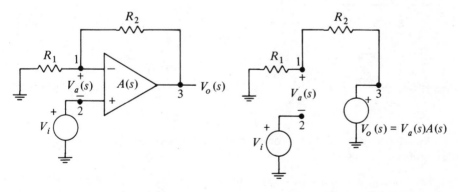

(a) Schematic presentation   (b) Equivalent presentation

Fig. 7-14   Analysis of noninverting amplifier

Before the expression of the gain, V_o/V_i, can be obtained, the dependent source, $V_a A(s)$, must be expressed in terms of the independent excitation, V_i. By inspection of Fig. 7-14b,

$$V_a = V_a A \frac{R_1}{R_1 + R_2} - V_i,$$

which is solved explicitly for V_a:

$$V_a = K_0 \frac{-V_i}{K_0 - A}, \tag{7-12}$$

where

$$K_0 = 1 + \frac{R_2}{R_1}.$$

Since $V_o = A V_a$, the gain is

$$\frac{V_o}{V_i} = K_0 \frac{-A}{K_0 - A}. \tag{7-13}$$

Note that as $|A| \to \infty$,

$$V_a \to 0 \quad \text{and} \quad \frac{V_o}{V_i} \to K_0.$$

Thus, the expressions for the input voltage and the gain revert to the expressions expected from an ideal operational amplifier. See Fig. 7-2 and Eq. (7-1).

When the one-pole rolloff model, $A = -\text{GB}/(s + \omega_a)$ is used, Eqs. (7-12) and (7-13) become

$$\frac{V_a}{V_i} = -\frac{s + \omega_a}{s + \omega_a + \dfrac{\text{GB}}{K_0}}, \tag{7-14}$$

$$\frac{V_o}{V_i} = K_0 \frac{\dfrac{\text{GB}}{K_0}}{s + \omega_a + \dfrac{\text{GB}}{K_0}}. \tag{7-15}$$

As long as $K_0 \ll \text{GB}/\omega_a$ (or in reduced form, $K_0 \ll A_0$), these equations simplify to

$$\frac{V_a}{V_i} = -\frac{s + \omega_a}{s + \dfrac{\text{GB}}{K_0}}, \text{ and} \tag{7-16}$$

$$\frac{V_o}{V_i} = K_0 \frac{\dfrac{GB}{K_0}}{s + \dfrac{GB}{K_0}}. \tag{7-17}$$

To simplify further, normalize the complex-frequency variable with respect to the gain-bandwidth product, i.e., let $s_n = s/GB$.

$$\frac{V_a}{V_i} = -\frac{s_n + \dfrac{1}{A_0}}{s_n + \dfrac{1}{K_0}} \cong -\frac{s_n}{s_n + \dfrac{1}{K_0}}, \qquad \text{where } |s_n| \gg \frac{1}{A_0}, \tag{7-18}$$

and

$$\frac{V_o}{V_i} = K_0 \frac{\dfrac{1}{K_0}}{s_n + \dfrac{1}{K_0}}. \tag{7-19}$$

Only one approximation is used in arriving at Eq. (7-19), namely $K_0 \ll A_0$. Alternatively, Eq. (7-19) can be obtained directly from Eq. (7-13) by using a simplified model, namely,

$$A(s) = -\frac{GB}{s + \omega_a} \cong -\frac{GB}{s} = -\frac{1}{s_n},$$

in which case the operational amplifier acts merely as an integrator.

The magnitude and phase responses resulting from Eqs. (7-18) and (7-19) are plotted as a function of K_0 in Fig. 7-15. Because of the frequency normalization with respect to GB, these curves are universally applicable. A study of these equations and curves provides a list of important and significant characteristics that are associated with the noninverting amplifier.

1. As Eq. (7-15) shows, the dc gain is

$$\left.\frac{V_o}{V_i}\right|_{s=0} = K_0 \frac{\dfrac{GB}{K_0}}{\omega_a + \dfrac{GB}{K_0}} = K_0 \frac{1}{1 + \dfrac{K_0}{A_0}} \cong K_0\left(1 - \frac{K_0}{A_0}\right). \tag{7-20}$$

The smaller the (K_0/A_0) ratio, the more the closed-loop dc gain approaches K_0, and it is therefore independent of A_0. Thus, for $K_0 \le 100$ and $A_0 \ge 10^5$, the dc gain is K_0 with an accuracy of better than 1 part in 1000. Note that for an ideal operational amplifier, the closed-loop dc gain is precisely K_0.

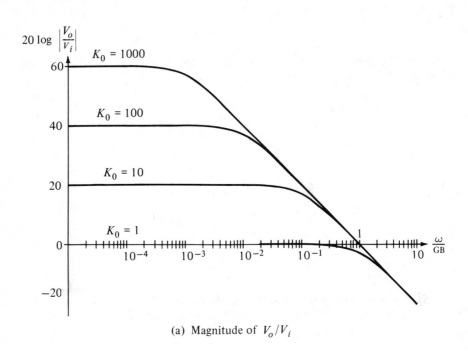

(a) Magnitude of V_o/V_i

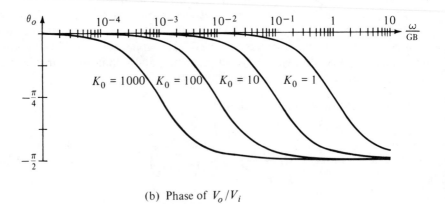

(b) Phase of V_o/V_i

Fig. 7-15 (a) and (b)   Characteristics of noninverting amplifier

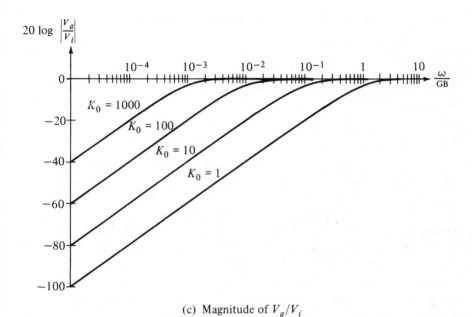

(c) Magnitude of V_a/V_i

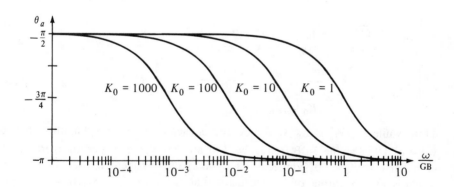

(d) Phase of V_a/V_i

FIG. 7-15 (c) and (d)

2. From Eq. (7-15), the 3-dB frequency is

$$\omega_{3\,dB} = \omega_a + \frac{GB}{K_0} = \frac{GB}{K_0}\left(1 + \frac{K_0}{A_0}\right). \tag{7-21}$$

For the usual case of $K_0/A_0 \ll 1$, Eq. (7-21) simplifies to

$$\omega_{3\,dB} = \frac{GB}{K_0}. \tag{7-22}$$

3. The smaller the closed-loop dc gain, the wider the closed-loop bandwidth. In equation form

$$K_0\,\omega_{3\,dB} \qquad = \qquad GB$$

Gain-bandwidth product$|_{\text{closed loop}}$ = gain-bandwidth product$|_{\text{open loop}}$. $\qquad$ (7-23)

For example, if the operational amplifier has a gain-bandwidth product of $2\pi \times 10^6$ rad/s, then, a noninverting amplifier designed for a dc gain of 10 has a bandwidth of $2\pi \times 10^5$ rad/s.

4. For frequencies much higher than the 3-dB frequency, all magnitude responses fall off at the rate of 6 dB/octave. Study the curves in Fig. 7-15a.

5. The phase of the gain changes from 0 to $-(\pi/2)$, being $-(\pi/4)$ at the 3-dB frequency. Study the curves in Fig. 7-15b.

All these observations apply to the gain function. It is equally important to realize what is happening at the input of the amplifier. *Whereas the voltage between the + and − terminals of the ideal amplifier is zero, it is a function of frequency in the nonideal case.* As Fig. 7-15c shows, the magnitude of (V_a/V_i)—from which V_a is calculated for a given V_i—increases with frequency. In particular, Eq. (7-16) gives

$$\left.\frac{V_a}{V_i}\right|_{s=0} = -\frac{\omega_a}{\dfrac{GB}{K_0}} = -\frac{K_0}{A_0}, \tag{7-24}$$

$$\left.\frac{V_a}{V_i}\right|_{s=j\omega_{3dB}} \cong \left.-\frac{j\omega}{j\omega + \dfrac{GB}{K_0}}\right|_{\omega=GB/K_0} = -\frac{j}{1+j} = -\frac{1}{\sqrt{2}}e^{j\pi/4}. \tag{7-25}$$

Thus, while $|V_a/V_i|$ is K_0/A_0 for dc and is therefore negligibly small, it has a value of $1/\sqrt{2}$ at the 3-dB frequency. For an example, consider an operational amplifier with $A_0 = 10^5$ connected as a unity-gain amplifier, $K_0 = 1$. If the input signal is 1V, an output of 1V is generated at dc with only $-K_0/A_0 = -10\ \mu V$ appearing at the input of the amplifier. On the other hand, at the 3-dB frequency, the magnitude of the output is $(1/\sqrt{2})V$ and it takes $(1/\sqrt{2})V$ at the amplifier input terminals to produce it. At this frequency, the voltage at the + and − input terminals of the operational amplifier is identical in magnitude with the voltage at its output and to have used the ideal model would have resulted in gross error.

Comparison of Fig. 7-15a and c shows the remarkable action of feedback. *It provides an ever-increasing level of signal at the input terminals of the amplifier to maintain a constant output over the pass-band of the amplifier.*

Gain and Bandwidth of the Inverting Amplifier

To obtain the characteristics of the inverting amplifier, consider its schematic and small-signal equivalent presentations given in Fig. 7-16.

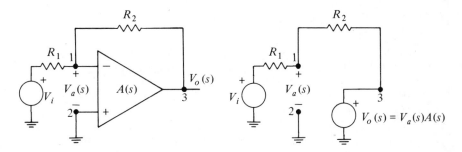

(a) Schematic presentation   (b) Equivalent presentation

Fig. 7-16   Analysis of inverting amplifier

To obtain the transfer function, $V_a(s)$ must be calculated first. By inspection of Fig. 7-16b, and using the principle of superposition, V_a is obtained as

$$V_a = \frac{V_i R_2 + V_a A R_1}{R_1 + R_2},$$

which results in

$$V_a = -K_0 \frac{-V_i}{1 + K_0 - A}, \tag{7-26}$$

where

$$K_0 = \frac{R_2}{R_1}.$$

Since $V_o = A V_a$, the gain is

$$\frac{V_o}{V_i} = -K_0 \frac{-A}{1 + K_0 - A}. \tag{7-27}$$

Note that as $|A| \to \infty$,

$$V_a \to 0 \quad \text{and} \quad V_o/V_i \to -K_0,$$

which agree with the values obtained from an ideal operational amplifier.

When the one-pole rolloff model is used for the operational amplifier, Eqs. (7-26) and (7-27) become

$$\frac{V_a}{V_i} = \frac{K_0}{1 + K_0} \frac{s + \omega_a}{s + \omega_a + \dfrac{GB}{1 + K_0}} \tag{7-28}$$

and

$$\frac{V_o}{V_i} = -K_0 \frac{\dfrac{GB}{1 + K_0}}{s + \omega_a + \dfrac{GB}{1 + K_0}}. \tag{7-29}$$

For $(1 + K_0) \ll A_0$ (the usual case), these equations simplify to

$$\frac{V_a}{V_i} = \frac{K_0}{1 + K_0} \frac{s + \omega_a}{s + \dfrac{GB}{1 + K_0}}, \tag{7-30}$$

and

$$\frac{V_o}{V_i} = -K_0 \frac{\dfrac{GB}{1 + K_0}}{s + \dfrac{GB}{1 + K_0}}. \tag{7-31}$$

To simplify further, let $s_n = s/GB$:

$$\frac{V_a}{V_i} = \frac{K_0}{1 + K_0} \frac{s_n + \dfrac{1}{A_0}}{s_n + \dfrac{1}{1 + K_0}} \cong \frac{K_0}{1 + K_0} \frac{s_n}{s_n + \dfrac{1}{1 + K_0}}, \qquad \text{where } |s_n| \gg \frac{1}{A_0}, \tag{7-32}$$

$$\frac{V_o}{V_i} = -K_0 \frac{\dfrac{1}{1 + K_0}}{s_n + \dfrac{1}{1 + K_0}}. \tag{7-33}$$

Equation (7-33) can be obtained directly from Eq. (7-27) by modelling the operational amplifier as an integrator, that is,

$$A(s) = -\frac{GB}{s + \omega_a} \cong -\frac{GB}{s} = -\frac{1}{s_n}.$$

The poles of the noninverting amplifier should be the same as the poles of the inverting amplifier, since the networks are identical when the sources are reduced to zero. Indeed, the denominator polynomial of Eq. (7-33) is identical with the denominator polynomial of Eq. (7-19) if it is realized that in the former equation

$$K_0 = 1 + \frac{R_2}{R_1},$$

and in the latter

$$K_0 = \frac{R_2}{R_1}.$$

The magnitude and phase responses resulting from Eqs. (7-32) and (7-33) are plotted as a function of K_0 in Fig. 7-17. Because of the frequency normalization with respect to GB, these curves are universally applicable. A study of these equations and curves provides a list of important and significant characteristics that are associated with the inverting amplifier.

1. As Eq. (7-29) shows, the dc gain is

$$\left.\frac{V_o}{V_i}\right|_{s=0} = -K_0 \frac{\dfrac{GB}{1+K_0}}{\omega_a + \dfrac{GB}{1+K_0}} = -K_0 \frac{1}{1 + \dfrac{1+K_0}{A_0}} \cong -K_0\left(1 - \frac{1+K_0}{A_0}\right). \quad (7\text{-}34)$$

The smaller the $(1 + K_0)/A_0$ ratio, the more the dc gain approaches the ideal value of the gain. Suppose, for example, that two different operational amplifiers, with

$$A_{01} = 10^5 \quad \text{and} \quad A_{02} = 10^6,$$

respectively, are used in the inverting configuration to provide a closed-loop dc gain of -10. The actual gains will be

$$-10(1 - 11 \times 10^{-5}) \quad \text{and} \quad -10(1 - 11 \times 10^{-6}),$$

respectively. Thus, a 10-to-1 variation in the open-loop dc gain affects the closed-loop dc gain only by 99 parts per million.

2. From Eq. (7-29), the 3-dB frequency is:

$$\omega_{3\,dB} = \omega_a + \frac{GB}{1+K_0} = \frac{GB}{1+K_0}\left(1 + \frac{1+K_0}{A_0}\right), \quad (7\text{-}35)$$

which, for $(1 + K_0)/A_0 \ll 1$, simplifies to:

$$\omega_{3\,dB} = \frac{GB}{1+K_0}. \quad (7\text{-}36)$$

3. The smaller the closed-loop dc gain, the wider the closed-loop bandwidth. In equation form,

$$(1 + K_0)\omega_{3\,dB} = GB$$

$$(1 + |\text{dc gain}|) \times \text{bandwidth}|_{\text{closed loop}} = \text{gain-bandwidth product}|_{\text{open loop}}.$$

$$(7\text{-}37)$$

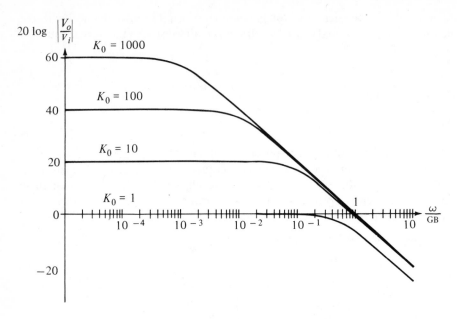

(a) Magnitude of V_o/V_i

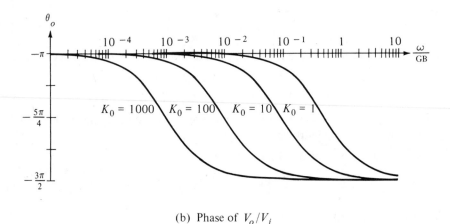

(b) Phase of V_o/V_i

Fig. 7-17 (a) and (b)   Characteristics of inverting amplifier

For example, if the operational amplifier has a gain-bandwidth product of $2\pi \times 10^6$ rad/s, then the inverting amplifier designed for a dc gain of -1 has a bandwidth of $2\pi \times 5 \times 10^5$ rad/s.

4. For frequencies much higher than the 3-dB frequency, all magnitude responses fall off at the rate of 6 dB/octave. Study the curves of Fig. 7-17a.

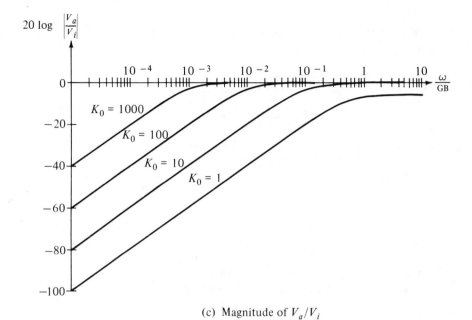

(c) Magnitude of V_a/V_i

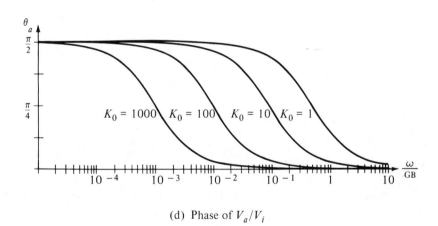

(d) Phase of V_a/V_i

FIG. 7-17 (c) and (d)

5. The phase of the gain changes from $-\pi$ to $-(3\pi/2)$, being $-(5\pi/4)$ at the 3-dB frequency. Study the curves of Fig. 7-17b.

A study of the $|V_a/V_i|$-vs.-ω curves shown in Fig. 7-17c indicates that to maintain a constant output, the voltage at the $+$ and $-$ terminals of the operational amplifier must increase with frequency. This behavior is similar to the one observed in the noninverting amplifier.

EXAMPLE 7-4

The inverting summing amplifier shown in Fig. 7-18 employs a one-pole rolloff operational amplifier.

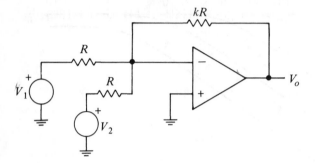

Fig. 7-18

(a) Obtain the expression for the output.
(b) What is the 3-dB bandwidth of the amplifier?

SOLUTION

(a) Obtain the Thévenin equivalent of the two sources and redraw Fig. 7-18 as in Fig. 7-19.

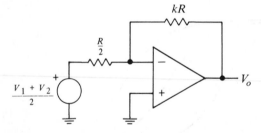

Fig. 7-19

Make use of Eq. (7-31) to obtain the output:

$$V_o = \frac{V_1 + V_2}{2}\left[-K_0 \frac{\dfrac{GB}{1 + K_0}}{s + \dfrac{GB}{1 + K_0}} \right]. \qquad (7-38)$$

Since K_0 is the ratio of the feedback resistor to the input resistor, that is,

$$K_0 = \frac{kR}{\dfrac{R}{2}} = 2k,$$

Eq. (7-38) simplifies to

$$V_o = -k(V_1 + V_2) \frac{\dfrac{GB}{1 + 2k}}{s + \dfrac{GB}{1 + 2k}}. \qquad Ans. \qquad (7\text{-}39)$$

Note that for an ideal operational amplifier, $GB = \infty$, and Eq. (7-39) reduces to

$$V_o = -k(V_1 + V_2).$$

(b) Any low-pass function of the form

$$\frac{K}{s + \alpha}$$

has a 3-dB bandwidth given by $\omega = \alpha$. Hence, obtain the bandwidth directly from Eq. (7-39):

$$\omega_{3\,dB} = \frac{GB}{1 + 2k}. \qquad Ans.$$

The more the negative terminal of the operational amplifier is loaded, the narrower becomes the bandwidth. Thus, if n signals with identical source resistances are connected to the negative terminal for summing, the bandwidth of the amplifier becomes

$$\omega_{3\,dB} = \frac{GB}{1 + nk}.$$

EXAMPLE 7-5

Obtain the step response of the amplifier shown in Fig. 7-20. Sketch $v_o(t)$ and $v_a(t)$ vs. t. The operational amplifier is characterized by a gain-bandwidth product of 10^6 rad/s.

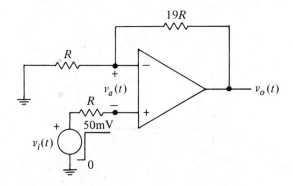

Fig. 7-20

SOLUTION

Obtain the two transfer functions from Eqs. (7-18) and (7-19):

$$\frac{V_a}{V_i} = -\frac{\dfrac{s}{GB}}{\dfrac{s}{GB} + \dfrac{1}{K_0}}, \qquad (7\text{-}40)$$

$$\frac{V_o}{V_i} = K_0 \frac{\dfrac{1}{K_0}}{\dfrac{s}{GB} + \dfrac{1}{K_0}}, \qquad (7\text{-}41)$$

where

$$K_0 = 1 + \frac{19R}{R} = 20.$$

With $v_i(t) = 50 \times 10^{-3}$, obtain the expressions for $V_a(s)$ and $V_o(s)$ by using Eqs. (7-40) and (7-41):

$$V_a(s) = \frac{50 \times 10^{-3}}{s}\left(-\frac{s}{s + \dfrac{10^6}{20}}\right) = -\frac{50 \times 10^{-3}}{s + \dfrac{10^6}{20}}, \qquad (7\text{-}42)$$

$$V_o(s) = \frac{50 \times 10^{-3}}{s}\left(20\,\frac{\dfrac{10^6}{20}}{s + \dfrac{10^6}{20}}\right) = \frac{1}{s} - \frac{1}{s + \dfrac{10^6}{20}}. \qquad (7\text{-}43)$$

Inverse transform Eqs. (7-42) and (7-43) to get

$$v_a(t) = -0.05e^{-t\mu s/20},$$

$$v_o(t) = 1 - e^{-t\mu s/20}. \qquad Ans.$$

$v_a(t)$ and $v_o(t)$ are sketched in Fig. 7-21.

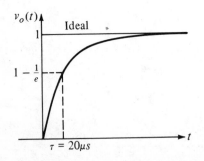

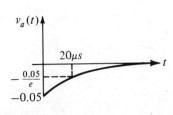

Fig. 7-21

Because of the finite gain-bandwidth product, the output rises exponentially to 1 with a time constant of $K_0/\mathrm{GB} = 20\ \mu s$. (Ideally it should have jumped to 1V as shown.) The voltage between the + and − terminals of the amplifier jumps down to 50 mV and then reduces to zero exponentially with a time constant of 20 μs. (Ideally this voltage should have been zero.) Note that the time constant is the inverse of the 3-dB bandwidth expressed in rad/s, that is, $\tau = K_0/\mathrm{GB} = 1/\omega_{3\,dB}$.

EXAMPLE 7-6

When the operational amplifier is ideal, the network shown in Fig. 7-22 acts as an ideal integrator, i.e., it has a pole at the origin. Assuming that the operational amplifier is characterized by a dominant pole, find the poles of the nonideal integrator.

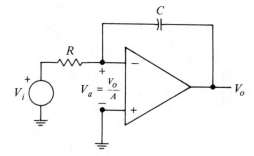

Fig. 7-22

SOLUTION

Using the principle of superposition, obtain V_a and equate it to V_o/A:

$$V_a = \frac{V_i \dfrac{1}{sC} + V_o R}{\dfrac{1}{sC} + R} = \frac{V_o}{A}.$$

Solve for the transfer function:

$$\frac{V_o}{V_i} = -\frac{1}{RC}\frac{1}{s - \dfrac{1}{A}\left(s + \dfrac{1}{RC}\right)}.$$

Substitute the one-pole representation for A, that is, $A = -\mathrm{GB}/(s + \omega_a)$:

$$\frac{V_o}{V_i} = -\frac{1}{RC}\frac{1}{s + \dfrac{(s + \omega_a)}{\mathrm{GB}}\left(s + \dfrac{1}{RC}\right)}$$

$$= -\frac{1}{RC}\frac{\mathrm{GB}}{s^2 + s\left(\mathrm{GB} + \omega_a + \dfrac{1}{RC}\right) + \dfrac{\omega_a}{RC}}.$$

In the usual case, $GB \gg \left(\omega_a + \dfrac{1}{RC}\right)$; therefore simplify Eq. (7-44) to

$$\frac{V_o}{V_i} = -\frac{1}{RC}\frac{GB}{s^2 + sGB + \dfrac{\omega_a}{RC}}.$$

Solve for the zeros of the denominator polynomial:

$$s_{1,2} = -\frac{GB}{2} \pm \sqrt{\left(\frac{GB}{2}\right)^2 - \frac{\omega_a}{RC}} = -\frac{GB}{2}\left(1 \pm \sqrt{1 - \frac{\omega_a}{RC}\frac{4}{(GB)^2}}\right),$$

which, for

$$\frac{\omega_a}{RC}\frac{4}{(GB)^2} \ll 1,$$

simplifies to

$$s_{1,2} \cong -\frac{GB}{2}\left[1 \pm \left(1 - \frac{\omega_a}{RC}\frac{2}{(GB)^2}\right)\right]. \tag{7-45}$$

Equation (7-45) is obtained by recognizing that

$$\sqrt{1-x} \cong 1 - \frac{x}{2} \qquad \text{for } |x| \ll 1.$$

Obtain the poles from Eq. (7-45):

$$s_1 \cong -\frac{1}{RC}\frac{\omega_a}{GB}, \qquad s_2 \cong -GB. \qquad \textit{Ans.}$$

Both poles are on the negative real axis. The first pole, since it is at $-(1/RC)(\omega_a/GB)$, is very close to the origin. The second pole, being at $-GB$, is far out. As the $GB \to \infty$, the first pole approaches the origin and the second infinity.

Using the approximate pole locations, Eq. (7-44) can be written as

$$\frac{V_o}{V_i} \cong -\frac{1}{RC}\frac{GB}{\left(s + \dfrac{1}{RC}\dfrac{\omega_a}{GB}\right)(s + GB)}. \tag{7-46}$$

Note that as $GB \to \infty$, Eq. (7-46) reduces to:

$$\left.\frac{V_o}{V_i}\right|_{GB \to \infty} = -\frac{1}{sRC},$$

which is the transfer function of the ideal integrator. (See also Problem 7-20.)

7-4 RESISTIVE FEEDBACK TO THE POSITIVE TERMINAL

In all the applications discussed so far, feedback is employed from the output of the operational amplifier to the negative input terminal. (Although in Example 7-2, feedback is from the output to the positive input terminal of amplifier #1, the signal is first inverted by operational amplifier #2.) *If the signal is fed back to the positive input terminal through a resistive network, an unstable situation arises which drives the output to a constant value.* To show this, the amplifier of Fig. 7-23 is assumed to be operating linearly, and (V_o/V_i) is calculated:

$$V_o = A(s)V_a = A(s)\left[V_i - V_o \frac{R_1}{R_1 + R_2}\right],$$

$$\frac{V_o}{V_1} = \frac{A(s)}{1 + \frac{R_1}{R_1 + R_2}A(s)}.$$

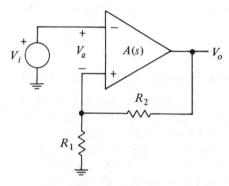

Fig. 7-23   An unstable circuit

If the operational amplifier is assumed to be ideal, that is, if $|A(s)| = \infty$, then

$$\frac{V_o}{V_i} = 1 + \frac{R_2}{R_1}.$$

This would be the correct answer if the operational amplifier were operating linearly. That this is not the case is demonstrated when a more realistic model is used for the operational amplifier, namely

$$A(s) = -\frac{GB}{s + \omega_a}.$$

Then, (V_o/V_i) becomes

$$\frac{V_o}{V_i} = -\frac{GB}{s + \omega_a\left(1 - \frac{A_0}{1 + (R_2/R_1)}\right)}.$$

The pole of the network is located at

$$s = \alpha = \omega_a \left(\frac{A_0}{1 + \dfrac{R_2}{R_1}} - 1 \right).$$

Since $A_0 \gg 1 + (R_2/R_1)$, the network has a pole on the positive real axis. Therefore, when the network is excited, the response will contain the term $Ke^{\alpha t}$ (the natural component), which eventually does drive the output to the limit of its dynamic range. Note that the pole is in the left half-plane if $A_0 < 1 + (R_2/R_1)$. But this case is impractical since it makes (V_o/V_i) highly dependent on A_0, a quantity that varies with time and from unit to unit.

Even when V_i is made zero, the amplifier's output stays at a constant value. Turning on the power supply is sufficient to drive the output to this state. Experience with actual circuits agrees with the result obtained by assuming a one-pole rolloff model for the operational amplifier. This is why resistive feedback from the output to the positive terminal alone is avoided if linear operation is desired.

7-5 THE NEED FOR A DOMINANT POLE IN OPERATIONAL AMPLIFIERS

The uncompensated operational amplifier is a complex electrical device. An accurate model is generally not available to describe its characteristics. The location of the poles and zeros may not be known, may vary from one unit to another, and may change with time. To remove, at least partly, these ambiguities and to make results less dependent on the operational amplifier, resistive feedback is employed from the output to the negative terminal, as shown in Fig. 7-24. However, if the operational amplifier is not properly compensated, feedback may turn the amplifier into an oscillator instead of desensitizing it against parameter variations. This is now demonstrated, and methods of avoiding this problem are given.

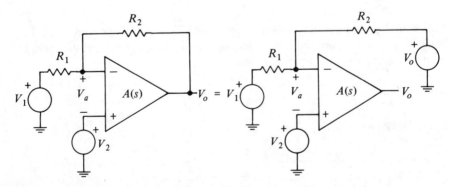

Fig. 7-24   Resistive feedback: general analysis

General Analysis of Feedback

In Fig. 7-24, the voltage that is fed back from the output terminal of the amplifier to the negative terminal of the input is

$$V_o \frac{R_1}{R_1 + R_2} = V_o \beta.$$

Since β is equal to $R_1/(R_1 + R_2)$, it can take on values only from 0 to 1. Note that β represents the fraction of the output voltage that is fed back. To obtain V_o, the voltage V_a is calculated and equated to $V_o/A(s)$:

$$V_a = \frac{V_1 R_2 + V_o R_1}{R_1 + R_2} - V_2 = \frac{V_o}{A(s)},$$

$$V_o = \underbrace{V_1 \frac{R_2}{R_1} \beta A(s)}_{V_{o1}} + \underbrace{V_2[-A(s)]}_{V_{o2}} + \underbrace{V_o \beta A(s)}_{V_{o3}}. \tag{7-47}$$

The component of the output produced by the output itself is

$$V_{o3} = V_o \beta A(s).$$

The transfer function $V_{o3}/V_o = \beta A(s)$ is called the *loop gain of the amplifier* since it represents the gain around the loop, i.e., from output to output. When explicitly solved for V_o, Eq. (7-47) becomes

$$V_o = \frac{V_1 \frac{R_2}{R_1} \beta A(s) - V_2 A(s)}{1 - \beta A(s)},$$

$$V_o = \left[V_1 \left(-\frac{R_2}{R_1} \right) + V_2 \left(1 + \frac{R_2}{R_1} \right) \right] \left[\frac{\beta A(s)}{\beta A(s) - 1} \right]. \tag{7-48}$$

To achieve desensitivity, i.e., to make $V_o(j\omega)$ independent of the operational amplifier, requires that

$$|\beta A(j\omega)| \gg 1, \tag{7-49}$$

in which case the second bracketed term in Eq. (7-48) reduces to unity, and the expression for the output becomes dependent only on the ratio of the external resistors.

While it may be easy to satisfy Eq. (7-49) for low frequencies, there is always a frequency beyond which it is impossible to meet this condition, because operational amplifier gains fall off at higher frequencies. Consequently, at higher frequencies, V_o is no longer independent of $A(s)$; *the degree of desensitivity drops off with frequency.*

Whereas it is important to satisfy Eq. (7-49) to achieve desensitivity at low frequencies, it is equally important that all the zeros of $[\beta A(s) - 1]$ be in the left half-plane so that the natural-response component of V_o, caused by the application of the V_1 and V_2 signals, does vanish with time. Note that $\beta A(s) - 1 = 0$ is the characteristic equation. If $[\beta A(s) - 1]$ has any zeros in the right half-plane, the natural response component of V_o will grow until the amplifier is driven into its nonlinear regions, thereby producing either sustained oscillations or a constant output (in the case of a positive real axis pole). What good is it, then, to have a desensitized amplifier (for low-frequency work) if the amplifier oscillates or latches to a constant voltage.

In a properly designed amplifier, the loop gain, $\beta A(s)$, must meet two conditions:

1. $|\beta A(j\omega)| \gg 1$ over as wide a band of frequencies as possible.
2. The zeros of $\beta A(s) - 1 = 0$ must be all in the left half-plane.

These two requirements usually conflict with each other, i.e., the more condition 1 is satisfied, the more likely the amplifier is to become an oscillator. Generally, the requirement set forth by condition 1 is relaxed sufficiently to allow for a stable system.

Amplifier Characterization with Three Poles

To provide a quantitative discussion about stability, consider an uncompensated operational amplifier characterized by three poles, that is,

$$A(s) = - \frac{A_0 \alpha_1 \alpha_2 \alpha_3}{(s + \alpha_1)(s + \alpha_2)(s + \alpha_3)}, \qquad 0 < \alpha_1 \leq \alpha_2 \leq \alpha_3. \qquad (7\text{-}50)$$

Actual operational amplifiers have many more poles and zeros than indicated by Eq. (7-50). However, here they are assumed to be far enough from α_3 so that the factors involving these critical frequencies can be treated as constants for frequencies up to and including α_3, that is, for $|s| \leq \alpha_3$. Note that the operational amplifier itself is stable, having only left half-plane poles. The dc gain, $A(0)$, is negative because A_0 is positive. For sinusoidal steady-state signals, the magnitude of the gain of the operational amplifier (open-loop gain) is practically flat up to $\omega = \alpha_1$; then it starts falling, first at the rate of 6 dB/octave, then at 12 dB/octave, and, for $\omega > \alpha_3$, at 18 dB/octave as shown in Fig. 7-25. (In this

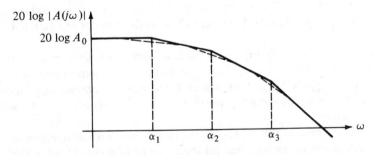

Fig. 7-25   Characteristics of a three-pole operational amplifier

figure, it is assumed that $-\alpha_1$, $-\alpha_2$, and $-\alpha_3$ are separated enough to make the three distinct ranges stand out.)

The closed loop poles, i.e., the poles of (V_o/V_1), with $V_2 = 0$, or of (V_o/V_2), with $V_1 = 0$, are the open loop poles of the operational amplifier $(-\alpha_1, -\alpha_2, -\alpha_3)$ repositioned by the action of feedback, and are given by the roots of the characteristic equation:

$$\beta A(s) - 1 = 0. \tag{7-51}$$

The important factor in Eq. (7-51) is not the amplifier gain by itself but rather the loop gain, $\beta A(s)$. The closed-loop poles of the system are those values of s that make the loop gain exactly 1. If the amplifier is to be stable, these poles must be in the left half-plane. To determine their location, Eq. (7-50) is substituted in Eq. (7-51), and the resulting equation is ordered:

$$\frac{-\beta A_0 \alpha_1 \alpha_2 \alpha_3}{(s + \alpha_1)(s + \alpha_2)(s + \alpha_3)} - 1 = 0,$$

$$(s + \alpha_1)(s + \alpha_2)(s + \alpha_3) + \beta A_0 \alpha_1 \alpha_2 \alpha_3 = 0,$$

$$s^3 + s^2(\alpha_1 + \alpha_2 + \alpha_3) + s(\alpha_1\alpha_2 + \alpha_1\alpha_3 + \alpha_2\alpha_3) + \alpha_1\alpha_2\alpha_3(1 + \beta A_0) = 0. \tag{7-52}$$

The roots of this cubic equation depend upon β, which characterizes the external (feedback) network. Since β appears in conjunction with A_0, *the magnitude of the dc loop gain, βA_0, is taken as the crucial parameter that determines the pole locations* for a given α_1, α_2, and α_3. βA_0 can take on any value between zero, corresponding to no feedback ($\beta = 0$), and A_0, corresponding to maximum feedback ($\beta = 1$). When $\beta A_0 = 0$, the roots of Eq. (7-52) are at $-\alpha_1$, $-\alpha_2$, and $-\alpha_3$; consequently, for βA_0 small, the roots should still be in the left half-plane. A large βA_0 results in a large constant term in the polynomial. If this constant term is made sufficiently large, two of the roots of Eq. (7-52) move into the right half-plane. This can be seen by applying Routh's stability criterion to the cubic equation, expressed as

$$a_3 s^3 + a_2 s^2 + a_1 s + a_0 = 0. \tag{7-53}$$

If Eq. (7-53) is to have left half-plane roots only, then all coefficients must be positive, and

$$a_2 a_1 - a_3 a_0 > 0. \tag{7-54}$$

To obtain the condition for imaginary-axis roots, let $s = j\omega$ in Eq. (7-53):

$$a_3(j\omega)^3 + a_2(j\omega)^2 + a_1(j\omega) + a_0 = 0,$$

$$(a_0 - a_2 \omega^2) + j\omega(a_1 - a_3 \omega^2) = 0. \tag{7-55}$$

If $s = j\omega$ is to be a root, then the real and imaginary parts of Eq. (7-55) must equal zero:

$$a_0 - a_2 \omega^2 = 0, \qquad a_1 - a_3 \omega^2 = 0.$$

Elimination of ω from these two equations yields the condition for imaginary-axis roots:

$$a_3 a_0 = a_2 a_1. \tag{7-56}$$

The two roots are then at

$$s = \pm j\sqrt{\frac{a_0}{a_2}} = \pm j\sqrt{\frac{a_1}{a_3}}. \tag{7-57}$$

Assume that all the a coefficients are positive. If all the coefficients except a_0 are kept constant, and a_0 is gradually increased from 0, the condition expressed by Eq. (7-56) is eventually satisfied, and the cubic has a pair of purely imaginary roots. Let a_{0c} denote this critical value of a_0. If a_0 is made larger than a_{0c}, then the condition expressed by Eq. (7-54) is not satisfied and the cubic has a pair of right half-plane roots.

Critical value of βA_0   Since the magnitude of the dc value of the loop gain, βA_0, controls the constant coefficient of the characteristic equation, Eq. (7-52), βA_0 determines the stability of the closed-loop system. There is a critical value of βA_0, designated by $(\beta A_0)_c$, which makes the system oscillatory. To obtain this value, apply Eq. (7-56) to Eq. (7-52):

$$a_0 = \frac{a_2 a_1}{a_3},$$

$$\alpha_1 \alpha_2 \alpha_3 [1 + (\beta A_0)_c] = (\alpha_1 + \alpha_2 + \alpha_3)(\alpha_1 \alpha_2 + \alpha_1 \alpha_3 + \alpha_2 \alpha_3),$$

$$(\beta A_0)_c = 2 + \frac{\alpha_1}{\alpha_2} + \frac{\alpha_1}{\alpha_3} + \frac{\alpha_2}{\alpha_1} + \frac{\alpha_2}{\alpha_3} + \frac{\alpha_3}{\alpha_1} + \frac{\alpha_3}{\alpha_2}. \tag{7-58}$$

When βA_0 equals $(\beta A_0)_c$, the amplifier oscillates at the frequency determined by Eq. (7-57):

$$\omega_{os} = \sqrt{\frac{a_1}{a_3}} = \sqrt{\alpha_1 \alpha_2 + \alpha_1 \alpha_3 + \alpha_2 \alpha_3}. \tag{7-59}$$

When $\beta A_0 > (\beta A_0)_c$, the amplifier has right half-plane poles and therefore is unstable.

Equation (7-58) is a very important result. It shows that solely ratios of amplifier pole locations alone determine how high a loop gain can be used before the system becomes unstable. The minimum value of Eq. (7-58) occurs when all poles are located at the same place, resulting in $(\beta A_0)_c = 8$ (see Problem 7-13). For example, when $A_0 = 10^5$ and $\alpha_1 = \alpha_2 = \alpha_3 = 10^7$ rad/s, then the network oscillates with the frequency of

$$\omega_{os} = \alpha_1 \sqrt{3} = 10^7 \sqrt{3} \text{ rad/s},$$

with as little as 80 parts per million feedback ($\beta_c = 8/A_0 = 8 \times 10^{-5}$). On the other hand, if $10000\alpha_1 = \alpha_2 = \alpha_3$, the critical loop gain becomes

$$(\beta A_0)_c \cong 2\frac{\alpha_2}{\alpha_1} = 20000,$$

and the network is stable as long as

$$\beta_c < \frac{20000}{A_0} = 0.2.$$

Since $\beta = R_1/(R_1 + R_2)$, $\beta < 0.2$ implies that $(R_2/R_1) > 4$, which means that the network of Fig. 7-24 must be operated with noninverting gains greater than 5 (or inverting gain magnitudes greater than 4) if the amplifier is not to oscillate.

If the amplifier is to be stable under any resistive feedback condition, that is, $0 \leq \beta \leq 1$, then the most stringent condition on A_0 must be used by letting $\beta = 1$ in Eq. (7-58):

$$A_0 < 2 + \frac{\alpha_1}{\alpha_2} + \frac{\alpha_1}{\alpha_3} + \frac{\alpha_2}{\alpha_1} + \frac{\alpha_2}{\alpha_3} + \frac{\alpha_3}{\alpha_1} + \frac{\alpha_3}{\alpha_2}. \tag{7-60}$$

It is clear from Eq. (7-58) that the poles should be widely separated so that a large A_0 could be used to desensitize the amplifier without turning it into an oscillator. To achieve this, α_1 is deliberately made much smaller than α_2 and α_3. This is usually achieved by compensating internally the operational amplifier with an RC network. Equation (7-60) can then be approximated to

$$A_0 < \frac{\alpha_2}{\alpha_1} + \frac{\alpha_3}{\alpha_1}.$$

To obtain as conservative a figure for A_0 as possible, let $\alpha_3 = \alpha_2$, which results in

$$A_0 < 2\alpha_2/\alpha_1 \tag{7-61}$$

Equation (7-61) sets the limit on the dc gain of the operational amplifier if the closed-loop system is to remain stable. Conversely, Eq. (7-61) can be used to determine where the lowest pole should be placed for a given A_0 and α_2. Suppose

$$A_0 = 2 \times 10^5 \quad \text{and} \quad \alpha_2 = 10^7 \text{ rad/s}.$$

Then, by Eq. (7-61), $\alpha_1 < 100$ rad/s. Since $\alpha_1 \ll \alpha_2$, $\omega = \alpha_1$ represents the open-loop 3-dB frequency. This numerical example indicates why most operational amplifiers are designed with such a low value for the 3-dB frequency.

Equation (7-61) is too restrictive if it is known that the amplifier is not going to be used with the unity feedback condition, which calls for the most precaution. More generally, the condition for stability is obtained from Eq. (7-58) by retaining only those terms which have α_1 in the denominator:

$$\beta A_0 \alpha_1 < (\alpha_2 + \alpha_3). \tag{7-62}$$

When expressed as Eq. (7-62), the variable that determines stability is $\beta A_0 \alpha_1$, the negative of the dc loop gain-bandwidth product. Consequently, oscillations in the amplifier may be stopped by decreasing the value of β, A_0, or α_1 until left half-plane poles are obtained.

Pole Loci   The results obtained thus far are summarized in the pole-loci diagram shown in Fig. 7-26.

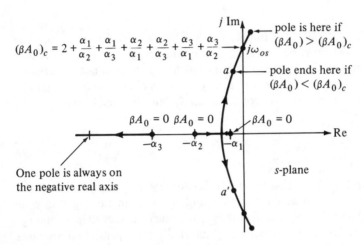

$$(\beta A_0)_c = 2 + \frac{\alpha_1}{\alpha_2} + \frac{\alpha_1}{\alpha_3} + \frac{\alpha_2}{\alpha_1} + \frac{\alpha_2}{\alpha_3} + \frac{\alpha_3}{\alpha_1} + \frac{\alpha_3}{\alpha_2}$$

pole is here if $(\beta A_0) > (\beta A_0)_c$

pole ends here if $(\beta A_0) < (\beta A_0)_c$

$\beta A_0 = 0$ $\beta A_0 = 0$ $\beta A_0 = 0$

One pole is always on the negative real axis

s-plane

Fig. 7-26   System pole loci as a function of βA_0

If the value of βA_0 is slightly less than the critical value of $(\beta A_0)_c$, which is fixed by the open-loop pole positions, then two of the poles are located very close to the imaginary axis (slightly to the left of it). See points a, a' in Fig. 7-26. This causes a peaking in the magnitude of the steady-state response near $\omega = \omega_{os}$ when the system is driven by a sine wave, or it causes ringing (damped sine wave of frequency $\cong \omega_{os}$) when the system is driven by a step function.

Gain and Phase Margins   The pole loci present a clear picture of what happens to the system poles, i.e., the closed-loop poles, when the amount of feedback is varied by changing βA_0. With the understanding obtained from the pole-loci diagram, *the system stability can be discussed in terms of open-loop conditions.* To do this, the feedback loop is broken and the network is driven with an external *sinusoidal* source as shown in Fig. 7-27a. It is assumed here that the output impedance of the operational amplifier is zero. If not, a load of $(R_1 + R_2)$ can be placed across the output to simulate the current drain on the output. The system function for this configuration, that is, V_o/V, is the loop gain, $\beta A(s)$. Typical magnitude and phase characteristics of $\beta A(j\omega)$ are sketched in Figs. 7-27b and 7-27c for three values of β. Since β is a real number (the feedback network is resistive), it affects only the magnitude characteristic and not the

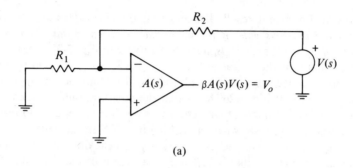

(a)

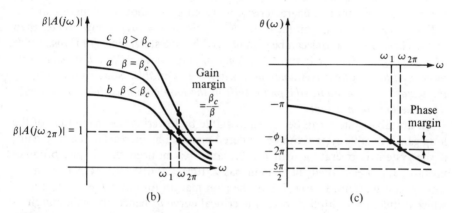

(b)   (c)

Fig. 7-27   Loop gain and its characteristics

phase; therefore, only one phase curve is shown. The effect of β on the magnitude characteristic is to scale it up or down, since β appears as a multiplying factor in the loop-gain expression.

The phase of the loop gain varies between $-\pi$ and $-(5\pi/2)$, a change of $-(3\pi/2)$, $-(\pi/2)$ for each pole of $A(s)$. Therefore, at some frequency labelled as $\omega_{2\pi}$ in Fig. 7-27c, the phase is -2π, which means that input and output are in phase. At this frequency, the magnitude of the loop gain is unity, less than unity, or greater than unity, depending upon the value of β. If β is adjusted to the critical value, β_c, the closed-loop system has a pair of imaginary-axis poles as shown on the pole-loci diagram of Fig. 7-26. Therefore, $s = j\omega_{2\pi}$ (and its conjugate) must be a root of Eq. (7-51), that is,

$$\beta_c A(s)\bigg|_{s=j\omega_{2\pi}} - 1 = 0. \qquad (7\text{-}63)$$

Since the angle of $A(j\omega_{2\pi})$ is -2π, Eq. (7-63) can be simplified to

$$\beta_c |A(j\omega_{2\pi})| = 1.$$

This is an important result. It states that *the closed-loop system will have a pair of imaginary-axis poles if the magnitude of the loop gain, evaluated at the frequency of 360° phase shift, is unity.* See curve *a* in Fig. 7-27b. In this case, the system oscillates at the frequency $\omega_{2\pi} = \omega_{os}$, where ω_{os} is given by Eq. (7-59). If the amount of feedback is reduced, that is, β is made less than β_c, then the pole-loci diagram of Fig. 7-26 shows that the poles are pulled back into the left half-plane. Since β is smaller than before, this results in a reduced loop-gain magnitude characteristic, as shown by curve *b* in Fig. 7-27b. In particular, the magnitude of the loop gain at $\omega = \omega_{2\pi}$ is less than unity. On the other hand, if the amount of feedback is increased, that is, if $\beta > \beta_c$, then the pole-loci diagram of Fig. 7-26 shows that the complex poles move into the right half-plane. Correspondingly, the loop-gain magnitude characteristic is scaled up as shown by curve *c* in Fig. 7-27b. Note that the magnitude of the loop gain at $\omega = \omega_{2\pi}$ is now greater than unity. These observations can be summarized by one statement: *If the magnitude of the loop gain, evaluated at the frequency of 360° phase shift, is less than unity, the closed-loop system is stable; otherwise, it is unstable.* Stated differently, *if the system is to be stable, the phase shift in the loop gain must be less than 360° when the loop gain is unity.*

An unstable system can be rendered stable by lowering the magnitude of the loop gain. A stable system can be rendered unstable by increasing the loop gain; if the system is operating with $\beta = 0.1$ and it is noticed that it just becomes unstable for $\beta = 1$, then the system has a 10:1 (20-dB) loop-gain margin (or simply gain margin). In Fig. 7-27b, the gain margin for curve *b* is the factor by which it must be multiplied to get the critical curve *a*; hence, the gain margin is (β_c/β), or [20 log(β_c/β) dB].

Curve *b* passes through unity gain at $\omega = \omega_1$; at this frequency, the phase is $-\phi_1$, as shown in Fig. 7-27c. Suppose that the phase of the loop gain is changed without changing the magnitude of the loop gain. Then it would require an additional phase lag of $(2\pi - \phi_1)$ to move the phase down enough so that at $\omega = \omega_1$ it becomes -2π radians. Curve *b* is said to have a phase margin of $(2\pi - \phi_1)$ radians. If it were possible to produce this additional phase shift without changing the magnitude characteristic, the system would oscillate at $\omega = \omega_1$. The gain and phase margins give a measure of how much the gain (at constant phase) or the phase (at constant gain) can be changed before left half-plane poles are moved to the imaginary axis (on different pole-loci curves, of course).

Dominant-Pole Representation   As Eq. (7-58) shows, large dc loop gains (for achieving desensitivity) can be used only if the open-loop poles are widely separated. If this is done by deliberately making $\alpha_1 \ll \alpha_2$ and α_3, then the three-pole $A(s)$ can be approximated with a single pole, that is,

$$A(s) = -\frac{A_0 \alpha_1 \alpha_2 \alpha_3}{(s + \alpha_1)(s + \alpha_2)(s + \alpha_3)}$$

$$\cong -\frac{A_0 \alpha_1}{s + \alpha_1} = -\frac{GB}{s + \omega_a}. \tag{7-64}$$

The pole at $-\alpha_1$ plays a dominant role in the behavior of the system; therefore, in most instances, the one-pole representation given by Eq. (7-64) is adequate to describe the operational amplifier if it is not operated at too high frequencies. It should be kept in mind that although a model like Eq. (7-64) by itself predicts absolute stability under resistive feedback regardless of A_0, it is based on the assumption that the three-pole system is stable.

The condition for stability, $\alpha_1 < 2\alpha_2/A_0$, shows that the farther away the second pole $-\alpha_2$ is from the first pole $-\alpha_1$, the larger α_1 can be made for a given A_0. The larger α_1, the larger becomes the gain-bandwidth product (GB) of the operational amplifier. This in turn allows for larger bandwidths of operation under closed-loop conditions, as is shown in Section 7-3.

EXAMPLE 7-7

An operational amplifier has an open-loop gain given by

$$A(s) = -\frac{A_0 \alpha_1 \alpha_2 \alpha_3}{(s + \alpha_1)(s + \alpha_2)(s + \alpha_3)}.$$

By means of internal compensation techniques using *RC* networks, it is possible to modify the expression for the gain by the factor

$$\frac{s + \alpha_1}{s + \alpha_0} \quad \text{or} \quad \frac{s + \alpha_2}{s + \alpha_4},$$

where $0 < \alpha_0 < \alpha_1 < \alpha_2 < \alpha_3 < \alpha_4$. What is gained by these compensation schemes?

SOLUTION

If $A(s)$ is modified by $(s + \alpha_1)/(s + \alpha_0)$, it becomes

$$A(s) = \frac{(s + \alpha_1)}{(s + \alpha_0)}\left[\frac{-A_0 \alpha_1 \alpha_2 \alpha_3}{(s + \alpha_1)(s + \alpha_2)(s + \alpha_3)}\right] = -\frac{A_0 \alpha_1 \alpha_2 \alpha_3}{(s + \alpha_0)(s + \alpha_2)(s + \alpha_3)}.$$

Thus the pole at $-\alpha_1$ is taken out, and instead a pole at $-\alpha_0$ substituted. Since $\alpha_0 < \alpha_1$, the poles of $A(s)$ are now more widely separated. Therefore higher gain and phase margins result for a given loop gain. Stated differently, it now takes a larger loop gain to make the amplifier unstable under resistive feedback [see Eq. (7-58)].

If $A(s)$ is modified by $(s + \alpha_2)/(s + \alpha_4)$, it becomes

$$A(s) = \frac{(s + \alpha_2)}{(s + \alpha_4)}\left[\frac{-A_0 \alpha_1 \alpha_2 \alpha_3}{(s + \alpha_1)(s + \alpha_2)(s + \alpha_3)}\right] = -\frac{A_0 \alpha_1 \alpha_2 \alpha_3}{(s + \alpha_1)(s + \alpha_3)(s + \alpha_4)}.$$

As in the first case, the poles are now more widely separated since the pole at $-\alpha_2$ is taken out and instead a pole farther out, at $-\alpha_4$, substituted. Therefore the conclusions reached for the previous case apply here also.

Note that $H(s + \alpha_1)/(s + \alpha_0)$ can be realized by the network of Fig. 7-28a, and $(s + \alpha_2)/(s + \alpha_4)$ can be realized by the network of Fig. 7-28b.

These realizations produce the desired transfer function if the networks are driven from a zero-impedance source, and the outputs are not loaded.

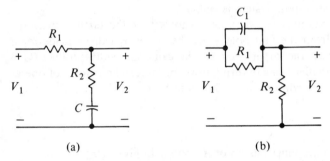

(a)   (b)

Fig. 7-28

EXAMPLE 7-8

An operational amplifier has a gain given by

$$A(s) = - \frac{A_0 \alpha_1 \alpha_2 \alpha_3}{(s + \alpha_1)(s + \alpha_2)(s + \alpha_3)},$$

where

$$A_0 = 10^5, \qquad \alpha_1 = 2\pi \times 20 \text{ rad/s}, \qquad \alpha_2 = 2\pi \times 10^6 \text{ rad/s}, \qquad \text{and}$$

$$\alpha_3 = 4\pi \times 10^6 \text{ rad/s}.$$

(a) Unity feedback is provided from the output to the negative input terminal. Show that a 3:2 gain margin is provided in A_0 before the network becomes an oscillator.

(b) Sketch the loci of the poles as β is varied from 0 to 1.

(c) The amplifier is connected in the noninverting mode. Sketch the frequency response for $\beta = 0, 0.001, 0.01,$ and 1. Discuss the results.

(d) If $A(s)$ is approximated as a one-pole rolloff amplifier, how would the responses differ from those in (c)?

SOLUTION

(a) Equation (7-58) gives the critical value of the gain which puts a pair of poles on the imaginary axis. For unity feedback, $\beta = 1$, the critical value is

$$A_{0c} = 2 + \frac{\alpha_1}{\alpha_2} + \frac{\alpha_1}{\alpha_3} + \frac{\alpha_2}{\alpha_1} + \frac{\alpha_2}{\alpha_3} + \frac{\alpha_3}{\alpha_1} + \frac{\alpha_3}{\alpha_2}$$

$$\cong \frac{\alpha_2}{\alpha_1} + \frac{\alpha_3}{\alpha_1} = 1.5 \times 10^5.$$

Since $A_0 = 10^5$, $3:2$ safety factor exists. This may be inadequate. However, if the noninverting amplifier is used in applications calling for a gain of 10 or more, then $\beta = 0.1$, and the critical value of the gain becomes 15×10^5, which results in a $15:1$ gain margin.

(b) The loci of the roots are found from

$$1 - \beta A(s) = 0,$$

$$(s + \alpha_1)(s + \alpha_2)(s + \alpha_3) + \alpha_1 \alpha_2 \alpha_3 \beta A_0 = 0,$$

$$s^3 + s^2(\alpha_1 + 6\pi \times 10^6) + s\left(\alpha_1 + \frac{4\pi}{3} \times 10^6\right)6\pi \times 10^6$$

$$+ \alpha_1 8\pi^2 \times 10^{12}(1 + 10^5\beta) = 0.$$

Since α_1 is so small, 40π, it hardly affects the coefficients of the s^2 and s terms. Like β, it has a pronounced effect only on the constant term of the equation. Substituting 40π for α_1, the equation for the poles becomes

$$s^3 + s^2 6\pi \times 10^6 + s8\pi^2 \times 10^{12} + 32\pi^3 \times 10^{13}(1 + 10^5\beta) = 0.$$

The roots of this equation are sketched as a function of β in Fig. 7-29.

For $\beta = 0$, the poles are the open-loop poles, all located on the negative real axis. As β is increased from 0, the leftmost pole moves farther away from the origin; the other two approach each other and for $\beta = 0.096$ they come together. For $\beta > 0.096$, they become complex and move in such a way that both the ω_o and the Q of the poles increase. For $\beta = 1$, the poles are very close to the imaginary axis. Since the poles are in the left half-plane even for $\beta = 1$, the system is stable. However, if the gain A_0 is increased from 10^5 to 1.5×10^5, then the complex poles follow the dotted lines and end up on the imaginary axis when $\beta = 1$.

(c) In the noninverting mode, the gain (the transfer function) is given by Eq. (7-48) as

$$\frac{V_o}{V_i} = -\frac{A(s)}{1 - \beta A(s)}$$

$$= \frac{A_0 \alpha_1 \alpha_2 \alpha_3}{(s + \alpha_1)(s + \alpha_2)(s + \alpha_3) + \alpha_1 \alpha_2 \alpha_3 A_0 \beta}$$

$$= \frac{32\pi^3 \times 10^{18}}{s^3 + s^2 6\pi \times 10^6 + s8\pi^2 \times 10^{12} + 32\pi^3 \times 10^{13}(1 + 10^5\beta)}.$$

For $s = j\omega$, (V_o/V_i) becomes $M(\omega)e^{j\theta(\omega)}$, where

$$M(\omega) =$$

$$\frac{32\pi^3 \times 10^{18}}{\sqrt{[32\pi^3 \times 10^{13}(1 + 10^5\beta) - 6\pi \times 10^6\omega^2]^2 + \omega^2(8\pi^2 \times 10^{12} - \omega^2)^2}};$$

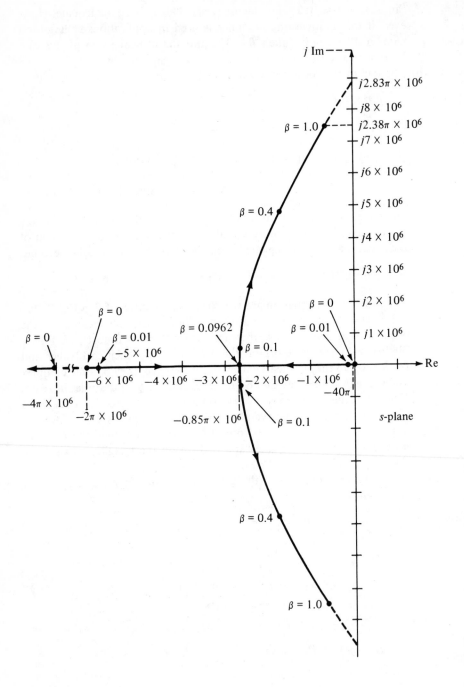

Fig. 7-29

$$\theta(\omega) = -\tan^{-1}\left[\frac{\omega(8\pi^2 \times 10^{12} - \omega^2)}{32\pi^3 \times 10^{13}(1 + 10^5\beta) - 6\pi \times 10^6\omega^2}\right].$$

The magnitude in dB, [20 log $M(\omega)$], and $\theta(\omega)$ are plotted as a function of f in Figs. 7-30a and 7-30b for the different values of β. In Fig. 7-30a, the $\beta = 0$ curve corresponds to the open-loop case with very high dc gain, (10^5), but small bandwidth, (20 Hz). For $\beta = 0.001$, the dc gain is 10^3 and the bandwidth 2000 Hz. Thus, the gain is reduced by the factor of 100 while the bandwidth is increased by the same factor keeping the gain-bandwidth product constant. This relationship is true practically down to the gain of 10 ($\beta = 0.1$), thereby indicating that the response is mainly dictated by the pole closest to the origin. As the pole-loci diagram shows, for $\beta = 1$, the poles are close to the imaginary axis, and therefore the response shows peaking at a frequency (in rad/s) very nearly equal to the value of the imaginary part of the pole. (Although not shown, the response will be peaked also for curves using values of β slightly less than 1; however, the peak value will not be as high and the frequency of peaking will be lower.) If for $\beta = 1$, A_0 is changed from 10^5 to 1.25×10^5, the poles will come closer to the imaginary axis, resulting in a magnitude characteristic with a much higher peak occurring at a slightly higher frequency than shown in Fig. 7-30a. If $A_0 = 1.5 \times 10^5$, the poles will be on the imaginary axis, causing the magni-

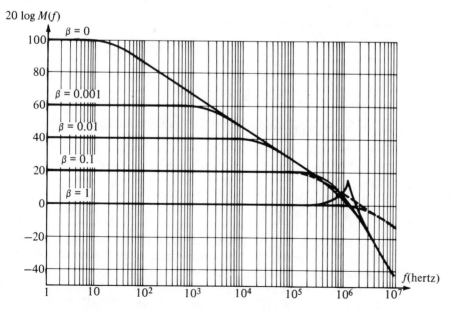

Fig. 7-30 (a) Magnitude characteristic as a function of β

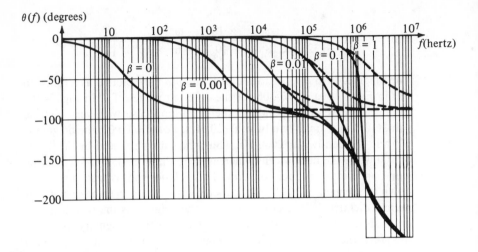

Fig. 7-30 (b) Phase characteristic as a function of β ($-\pi$ shift not shown)

tude characteristic to become infinite at $f = f_{os}$, which is given by Eq. (7-59) as

$$f_{os} = \frac{1}{2\pi}\sqrt{\alpha_1\alpha_2 + \alpha_1\alpha_3 + \alpha_2\alpha_3} \cong \frac{1}{2\pi}\sqrt{\alpha_2\alpha_3} = \sqrt{2} \times 10^6 \text{rad/s}.$$

The phase characteristic is shown without regard to the π-radian phase shift produced by the minus sign. Note the rather rapid change in phase, particularly for $\beta = 1$, occurring near f_{os}.

(d) When only the dominant pole is used, $A(s)$ becomes

$$A(s) = -\frac{A_0\alpha_1}{s + \alpha_1} = -\frac{4\pi \times 10^6}{s + 40\pi}.$$

With this $A(s)$, the expression for the transfer function becomes

$$\frac{V_o}{V_i} = -\frac{A(s)}{1 - \beta A(s)} = \frac{A_0\alpha_1}{s + \alpha_1(1 + \beta A_0)}$$

$$= \frac{4\pi \times 10^6}{s + 40\pi(1 + 10^5\beta)}.$$

As long as $\omega < 2\pi \times 10^5$, a tenth of α_2, the frequency-response curves for any β are practically indistinguishable from the curves obtained with the three-pole model. However, for higher frequencies, the departure becomes noticeable, as shown by the dotted magnitude curves in Fig. 7-30a. In particular, the simpler model does not show any peaking of the response and it falls at the rate of 6 dB/octave at high frequencies.

EXAMPLE 7-9

Under open-loop conditions, the output resistance of the operational amplifier is R_0 as shown in Fig. 7-31. Let

$$K_0 = 1 + \frac{R_2}{R_1} \quad \text{and} \quad A(s) = -\frac{GB}{s + \omega_a}.$$

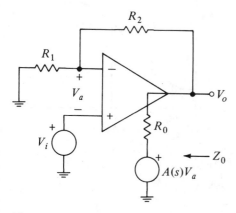

Fig. 7-31

(a) What is the closed-loop output impedance, $Z_0(s)$? Neglect loading of $(R_1 + R_2)$.
(b) Sketch $|Z_0(j\omega)|$ vs. ω.
(c) Obtain the Thévenin-equivalent circuit as seen from the output terminals.

SOLUTION

(a) Let $V_i = 0$, and drive the network with the external source V, as shown in Fig. 7-32a.
    Noting that

$$V_a = \frac{V R_1}{(R_1 + R_2)} = \frac{V}{K_0},$$

obtain the expression for the current I:

$$I = \frac{V}{R_1 + R_2} + \frac{V - \dfrac{AV}{K_0}}{R_0} = V\left(\frac{1}{R_1 + R_2} + \frac{1 - \dfrac{A}{K_0}}{R_0}\right).$$

The output admittance is

$$Y_0 = \frac{I}{V} = \frac{1}{R_1 + R_2} + \frac{1 - (A/K_0)}{R_0}.$$

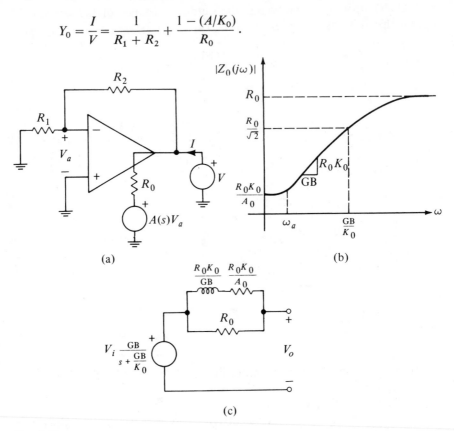

(a)   (b)

(c)

Fig. 7-32

Hence, the output impedance is the parallel combination of $(R_1 + R_2)$ with $[R_0/(1 - A/K_0)]$. If the $(R_1 + R_2)$ loading is neglected, the output impedance can be written as

$$Z_0(s) = \frac{R_0}{1 - \dfrac{A}{K_0}} = \frac{R_0}{1 + \dfrac{1}{K_0}\dfrac{GB}{s + \omega_a}} = R_0 \frac{s + \omega_a}{s + \omega_a(1 + A_0/K_0)}. \qquad Ans.$$

(b) Since $A_0/K_0 \gg 1$, simplify $Z_0(s)$ to

$$Z_0(s) = R_0 \frac{s + \omega_a}{s + \dfrac{\omega_a A_0}{K_0}} = R_0 \frac{s + \omega_a}{s + \dfrac{GB}{K_0}}.$$

Let $s = j\omega$ and obtain the magnitude of $Z_0(j\omega)$:

$$Z_0(j\omega) = R_0 \frac{j\omega + \omega_a}{j\omega + \dfrac{GB}{K_0}},$$

$$|Z_0(j\omega)| = R_0 \frac{\sqrt{\omega^2 + \omega_a^2}}{\sqrt{\omega^2 + \left(\dfrac{GB}{K_0}\right)^2}}.$$

The magnitude characteristic is sketched in Fig. 7-32b. It varies from $R_0 K_0/A_0$ (for $\omega \ll \omega_a$) to the open-loop value of R_0 (for $\omega \gg GB/K_0$). At mid-frequencies,

$$|Z_0(j\omega)| \cong R_0 \frac{K_0}{GB} \omega,$$

which represents inductive behavior with $L = R_0 K_0/GB$.

(c) As long as $R_0 \ll (R_1 + R_2)$—the usual case—the effect of the loading is negligible on the open-circuit voltage, V_{oc}. Therefore, use Eq. (7-17) to obtain V_{oc}:

$$V_{oc} = \frac{\dfrac{GB}{K_0}}{s + \dfrac{GB}{K_0}} V_i.$$

Because $Z_0(s)$ has a zero and then a pole on the negative real axis, it can be represented by an RL network. To obtain the Foster II representation, expand $Y_0(s)$ in partial fractions:

$$Y_0(s) = \frac{1}{R_0} \frac{s + \dfrac{GB}{K_0}}{s + \omega_a} = \frac{1}{R_0} + \frac{\dfrac{1}{R_0}\left(\dfrac{GB}{K_0} - \omega_a\right)}{s + \omega_a} \cong \frac{1}{R_0} + \frac{\dfrac{1}{R_0}\dfrac{A_0 \omega_a}{K_0}}{s + \omega_a}.$$

$Y_0(s)$ represents the parallel connection of a resistor of value R_0 with a branch composed of a resistor of value $R_0 K_0/A_0$ in series with an inductor of value $R_0 K_0/GB$.

The Thévenin-equivalent circuit is shown in Fig. 7-32c. Compare the low-, mid-, and high-frequency behavior of Z_0 presented in Fig. 7-32c with that of Fig. 7-32b.

7-6 SUMMARY

The operational amplifier is an extremely useful and versatile device. When an ideal operational amplifier is connected to a network, it produces two significant changes in the network (provided it operates in the linear mode). First, it forces the voltage to become zero between the two points in the circuit where its + and

— terminals are connected, and it does this without taking any current from these two points. Second, it introduces an ideal voltage source between ground and the point to which its output is connected. The value of the voltage source is determined by the external circuit.

While the model of the ideal operational amplifier is easy to apply, its usefulness is limited. Measurements done on actual circuits may differ considerably from the values determined by the ideal model. This is particularly noticeable if operation goes beyond low frequencies, in which case the one-pole rolloff model provides better agreement between theory and practice. The important feature of the one-pole rolloff model is the characterization of the operational amplifier in terms of its gain-bandwidth product, namely

$$A(s) = -\frac{GB}{(s + \omega_a)},$$

where ω_a is the open-loop bandwidth. In closed-loop operations where the closed-loop gain is much less than the open-loop gain, this model can be further simplified to

$$A(s) = -\frac{GB}{s},$$

which emphasizes the integrating action of the operational amplifier. An important relation to remember is

$$GB = \left(1 + \frac{R_2}{R_1}\right)\omega_{3\,dB},$$

where R_2 and R_1 represent the feedback resistors and $\omega_{3\,dB}$ the closed-loop bandwidth.

The stability of a network with amplifiers can be studied by observing the properties of the loop gain, which represents the transfer function around a particular loop of the network. It can be calculated by disconnecting the output lead of an amplifier from the rest of the network (thus breaking the feedback loop) and connecting this lead to an independent source, V_i. If all other independent signal sources in the network are made zero and if the output impedance of the amplifier is zero, then the loop gain is V_o/V_i where V_o represents the output voltage. The closed-loop poles are the roots of $(1 - \text{Loop gain}) = 0$. These roots must be in the left half-plane if the system is to be stable. Large loop gains make the system insensitive to amplifier-parameter variations, provided the system is stable.

The more widely separated the poles of an operational amplifier are, the less is the likelihood of oscillation under resistive feedback. For this reason, the operational amplifier is designed with a dominant pole. The dominant pole may produce, in combination with two distant poles, a response that peaks at high frequencies, particularly when unity feedback is used. The peaking can be reduced by reducing the loop gain.

PROBLEMS

7-1   Find the output voltage for the circuits shown in Fig. 7-33.

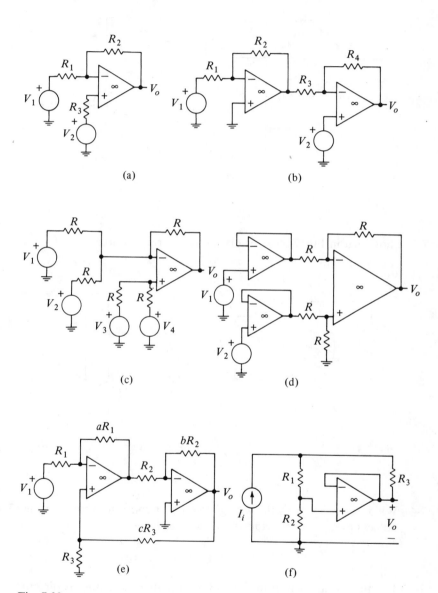

(a)

(b)

(c)

(d)

(e)

(f)

Fig. 7-33

7-2   Explain the operation of the ohmmeter circuit shown in Fig. 7-34.

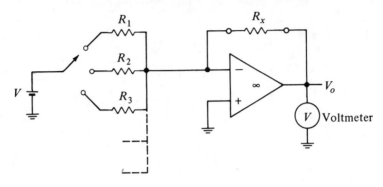

Fig. 7-34

7-3   Show that in Fig. 7-35 R_L is driven by a current source.

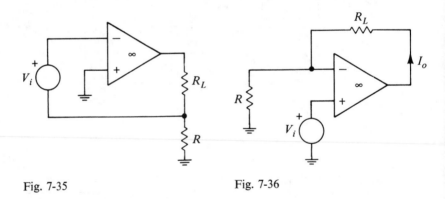

Fig. 7-35   Fig. 7-36

7-4   What advantage does the voltage-to-current converter shown in Fig. 7-36 possess over the circuit shown in Fig. 7-6a?

7-5   For the network given in Fig. 7-37, show that the output is directly proportional to changes in the feedback resistor.

7-6   Obtain the output voltage for the network shown in Fig. 7-38.

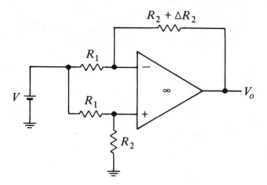

Fig. 7-37

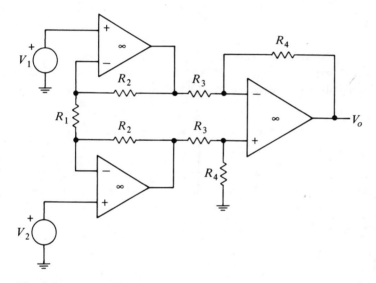

Fig. 7-38

7-7 In Fig. 7-39, find the output voltage $V_o(s)$. γ represents the initial voltage across the capacitor.

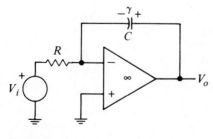

Fig. 7-39

7-8   Show that the circuit of Fig. 7-40 acts as an inverting integrator for the signal V_1 and as a noninverting integrator for the signal V_2.

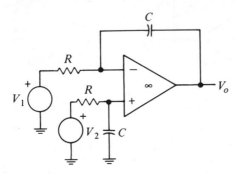

Fig. 7-40

7-9   (a) V_o may be written as shown in Fig. 7-41. Find R_a and R_b.
      (b) Show that large values of R_a and R_b may be obtained without using large values for the actual resistors.

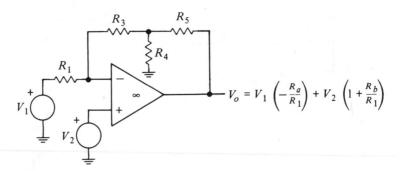

$$V_o = V_1 \left(-\frac{R_a}{R_1} \right) + V_2 \left(1 + \frac{R_b}{R_1} \right)$$

Fig. 7-41

7-10 What *passive* network has the same input impedance as the network shown in Fig. 7-42?

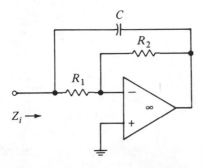

Fig. 7-42

7-11 Discuss the operation of the circuit shown in Fig. 7-43.

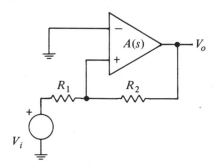

Fig. 7-43

7-12 Show that the roots of Eq. (7-51) are all on the nonpositive real axis if $\beta A(0) < 0$, and if the poles and zeros of $A(s)$ are simple and alternate on the negative real axis.

7-13 Show that Eq. (7-58) acquires its minimum value when $\alpha_1 = \alpha_2 = \alpha_3$.

7-14 Show that an operational amplifier with two negative real-axis poles and with zeros at infinity does not become unstable with any amount of resistive feedback to the negative input terminal. Draw the loci of the poles as a function of the loop gain.

7-15 Given

$$s^4 + a_3 s^3 + a_2 s^2 + a_1 s + a_0 = 0.$$

If this equation is to have a pair of imaginary-axis roots, what constraint must be applied to the a coefficients? What are the values of the imaginary roots?

7-16 The gain of an operational amplifier is given by

$$A(s) = \frac{-s A_0 \alpha_2 \alpha_3}{(s + \alpha_1)(s + \alpha_2)(s + \alpha_3)}.$$

Would this amplifier oscillate under resistive feedback applied to the negative terminals? Sketch the loci of the poles as a function of the dc loop gain.

7-17 An amplifier has poles at

$$-\alpha, \qquad -2\pi \times 10^6 \text{ rad/s}, \qquad -4\pi \times 10^6 \text{ rad/s};$$

the zeros are at infinity. $A_0 = 10^5$. Where should $-\alpha$ be placed to guarantee left half-plane poles if the amplifier is to be used as a unity inverter?

7-18 What would happen to the root loci of Fig. 7-26 if A_0 is negative, as is the case if feedback is applied to the positive rather than the negative terminal? Find the value of βA_0 which puts a pole in the right half-plane.

7-19 In Fig. 7-44, $A(s)$ is characterized by a dc gain of -10^4 and a 3-dB band-width of 100 rad/s. Find the closed-loop dc gain and bandwidth.

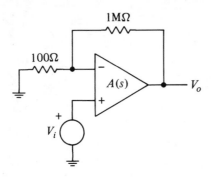

Fig. 7-44

7-20 In Example 7-6, let

$$A_0 = 10^5, \qquad \omega_a = 10, \qquad R = 1 \text{ M}\Omega, \qquad C = 10 \ \mu\text{F}.$$

Obtain the step response and compare it with the response of the ideal integrator.

7-21 In Fig. 7-45,

$$\beta_- = \frac{R_1}{R_1 + R_2} \qquad \text{and} \qquad \beta_+ = \frac{R_3}{R_3 + 4}.$$

When $\beta_+ = 0$, the stability-determining equation is $\beta_- A(s) - 1 = 0$. How is this equation modified if $\beta_+ \neq 0$?

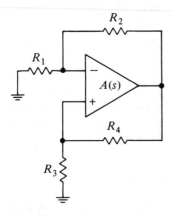

Fig. 7-45

7-22 Show that if α is properly chosen, the loop gains of the two networks shown in Fig. 7-46 are identical. What is the value of α?

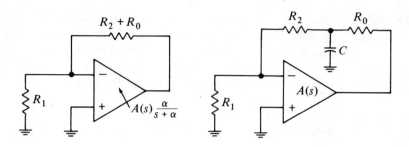

Fig. 7-46

7-23 Show that if C is properly chosen in Fig. 7-47, higher loop gains may be possible before the amplifier oscillates. Assume $A(s)$ is representable by three poles.

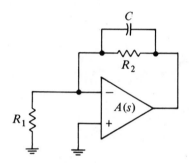

Fig. 7-47

7-24 Show that the larger the RC product, the larger the gain margin for the circuit shown in Fig. 7-48. Assume

$$A(s) = \frac{-A_0 \alpha_1 \alpha_2}{(s + \alpha_1)(s + \alpha_2)} .$$

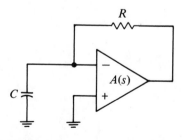

Fig. 7-48

7-25 The resistance values are adjusted so that the three circuits shown in Fig. 7-49 have the same dc gain, that is, $(V_o/V_i) = K_0$. All operational amplifiers have the same gain, namely $A(s) = -(GB/s)$. How much of an improvement in bandwidth is achieved by the circuits using two operational amplifiers over the circuit using one operational amplifier?

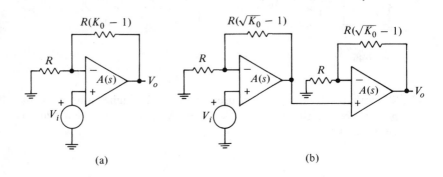

(a)   (b)

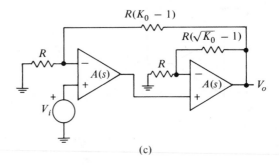

(c)

Fig. 7-49

7-26 A voltage, V_i, is injected in the feedback loop as shown in Fig. 7-50. Let $\beta = R_1/(R_1 + R_2)$. Obtain the expressions for V_{o1}/V_i and V_{o2}/V_i.

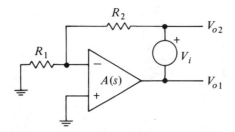

Fig. 7-50

7-27 (a) In Fig. 7-51, the switch is closed at $t = 0$. What should be the R_2/R_1 ratio so that a ramp is generated at the output?

(b) If the R_2/R_1 ratio is set slightly lower than the critical value calculated in (a), what would happen to the output waveform?

(c) Repeat (b) if the R_2/R_1 ratio is set slightly higher than the critical value.

(d) The R_2/R_1 ratio is adjusted to the value given in (a). The input is a voltage source v_i. The capacitor is replaced by an impedance Z. Show that this impedance is driven by a current source.

(e) A one-pole rolloff model is used for the operational amplifier. What impedance does the current source in (d) see?

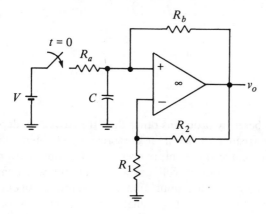

Fig. 7-51

The Operational Amplifier: Other Imperfections and Nonlinear Applications

The operational amplifier is beset by problems other than the frequency dependence of its gain, which is treated in detail in the previous chapter. Offset voltages and currents cause a voltage at the output even when the input terminals are grounded. This results in extraneous voltages at the output when input signals drive the amplifier. Common-mode input voltages also produce an output voltage. Thus, errors are introduced when two signals are subtracted. Another important consideration is the output swing. The dynamic range of the operational amplifier is governed by the output voltage and current limits. In addition to bandwidth limitations imposed by the poles of the operational amplifier, restrictions caused by slewing need to be considered also. All these imperfections are discussed in this chapter.

Although the principal aim of this book is to provide a firm foundation of linear network theory, it is also desirable to discuss some nonlinear applications involving the operational amplifier. By using diodes and feedback to the positive terminal, the amplifier can be operated as a comparator, bistable multivibrator, monostable multivibrator, or astable multivibrator. Thus timing, switching, and wave-shaping circuits can be constructed. A variety of examples is given in this chapter to show the immense versatility of the operational amplifier in this mode of operation.

8-1 OFFSET VOLTAGES AND CURRENTS

Actual operational amplifiers may produce a constant output voltage even when both inputs are grounded. This output voltage is primarily caused by the imperfections in the input circuitry of the operational amplifier. To permit

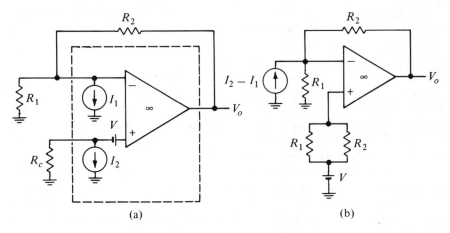

Fig. 8-1 Equivalent representations of offset sources

a quantitative discussion of these effects, equivalent current and voltage sources are introduced at the input of the amplifier, as shown in Fig. 8-1a, and the remaining portion of the amplifier is treated as ideal. (The actual amplifier is shown within the dashed block.) The designation of offset voltage is used only for the magnitude of V, and offset current for the magnitude of $I_1 - I_2$. The voltage $|V|$ can just as well be placed in series with the negative terminal of the amplifier.

To see the effect of the offset sources under closed-loop conditions, the circuit of Fig. 8-1a is used. Considering one source at a time, the output can be written by inspection as

$$V_o = I_1 R_2 - I_2 R_c\left(1 + \frac{R_2}{R_1}\right) + V\left(1 + \frac{R_2}{R_1}\right),$$

which can be rearranged as

$$V_o = (I_1 - I_2)R_2 + I_2\left[R_2 - R_c\left(1 + \frac{R_2}{R_1}\right)\right] + V\left(1 + \frac{R_2}{R_1}\right). \qquad (8\text{-}1)$$

The second term in Eq. (8-1) is made zero by adjusting the compensating resistor, R_c, to equal $(R_1 R_2)/(R_1 + R_2)$. Then, V_o becomes

$$V_o = (I_1 - I_2)R_2 + V\left(1 + \frac{R_2}{R_1}\right). \qquad (8\text{-}2)$$

The first term in Eq. (8-2) is the output voltage caused by the offset current. It can be minimized by selecting an operational amplifier with a small offset current and by using as small an R_2 as possible. The minimum size of R_2 is dependent upon the output-signal swing, the load to which the output is coupled, and the output-current capability of the operational amplifier.

The second term in Eq. (8-2) is the output voltage due to the offset voltage. It can be minimized by selecting an operational amplifier with a small offset voltage and by using as small an (R_2/R_1) ratio as possible. Since this ratio also determines the signal gain, it may not be desirable to make it too small.

With Eq. (8-2) as a guide, the equivalent circuit of Fig. 8-1b (using offset sources as excitations on an ideal amplifier) can be put together. A study of Fig. 8-1b shows that an external voltage source of $-V$ applied to the positive terminal of the amplifier nulls the voltage due to the offset voltage. Similarly, an external current source of $-(I_2 - I_1)$ applied to the negative terminal of the amplifier nulls the voltage due to the offset current.

It is possible to completely null the output voltage produced by the offset sources by introducing, either at input or at some internal point, an adjustable external dc voltage source. However, it should be kept in mind that the offset sources are temperature-dependent, and therefore the balance cannot be expected to hold over long periods of time or over large temperature ranges.

EXAMPLE 8-1

In Fig. 8-2, the input signal is a 100-mV dc source. The operational amplifier has an offset current of 0.05 μA and an offset voltage of 5 mV. The offset sources are not shown in the figure.

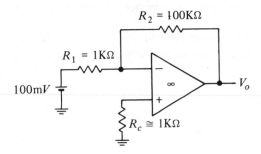

Fig. 8-2

(a) What is the dc level of the output due to the offset sources?
(b) Refer the output offset voltage to the input, and compare it with the desired signal.
(c) Would there be significant improvement in the desired-to-undesired signal ratio if R_2 is reduced to 10 KΩ and R_1 to 100 Ω? (The compensating resistance, R_c, is adjusted accordingly.)

SOLUTION

(a) Note that the compensating resistor is properly adjusted; that is, $R_c \cong R_1 || R_2$. Therefore, use Eq. (8-2) to calculate the output due to the offset sources. Since the polarities of these sources are not known, use worst-case analysis, i.e., assume that their effects are additive:

$$V_o\big|_{offset} = |I_1 - I_2|R_2 + |V|\left(1 + \frac{R_2}{R_1}\right)$$

$$= 0.05 \times 10^{-6} \times 10^5 + 5 \times 10^{-3} \times 101 = 0.005 + 0.505$$

$$\cong 0.5 \text{ V.} \quad Ans.$$

(b) The magnitude of the gain of the inverting circuit is 100. The offset voltage, when referred to the input, becomes

$$\frac{500 \text{ mV}}{100} = 5 \text{ mV.}$$

Hence, the undesired signal (the offset signal referred to the input) is 5% of the desired 100-mV input signal.

(c) Even with $R_2 = 100$ KΩ, the output voltage due to the offset current is 5 mV and therefore is negligible compared with the output due to the offset voltage (505 mV). Reducing R_2 by a factor of 10, while keeping the (R_2/R_1) ratio constant, will therefore not significantly alter the desired-to-undesired signal ratio.

EXAMPLE 8-2

The amplifier shown in Fig. 8-3 is to be used with sine-wave inputs only. Show that if C is properly chosen, the ac gains for the two input signals can be made large without at the same time increasing the output due to the offset voltage.

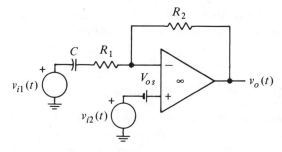

Fig. 8-3

SOLUTION

Let V_{os} represent the equivalent offset voltage as shown. Use the inverting and noninverting gain functions to obtain the output $V_o(s)$. Let $Z = R_1 + \dfrac{1}{sC}$.

$$V_o(s) = V_{i1}\left(-\frac{R_2}{Z}\right) + V_{i2}\left(1 + \frac{R_2}{Z}\right) + \frac{V_{os}}{s}\left(1 + \frac{R_2}{Z}\right)$$

$$= V_{i1}\left(-\frac{R_2}{R_1 + \dfrac{1}{sC}}\right) + V_{i2}\left(1 + \frac{R_2}{R_1 + \dfrac{1}{sC}}\right) + \frac{V_{os}}{s}\left(1 + \frac{R_2}{R_1 + \dfrac{1}{sC}}\right)$$

$$= V_{i1}\left(-\frac{R_2}{R_1}\right)\left(\frac{s}{s + \dfrac{1}{R_1 C}}\right) + V_{i2}\left(1 + \frac{R_2}{R_1}\right)\left[\frac{s + \dfrac{1}{(R_1 + R_2)C}}{s + \dfrac{1}{R_1 C}}\right]$$

$$+ \frac{V_{os}}{s}\left(1 + \frac{R_2}{R_1}\right)\left[\frac{s + \dfrac{1}{(R_1 + R_2)C}}{s + \dfrac{1}{R_1 C}}\right]. \qquad (8\text{-}3)$$

The natural response terms of Eq. (8-3) vanish with a time constant of $\tau = R_1 C$. If the input sine-wave frequencies, ω_1 and ω_2, are much larger than $(1/R_1 C)$, then the *steady-state output* is given by

$$v_{oss}(t) = \left(-\frac{R_2}{R_1}\right)v_{i1}(t) + \left(1 + \frac{R_2}{R_1}\right)v_{i2}(t) + V_{os}.$$

Thus, the input signals are amplified by

$$-(R_2/R_1) \qquad \text{and} \qquad \left(1 + \frac{R_2}{R_1}\right),$$

respectively, while the gain for the input-offset voltage is held at unity. These results can be seen directly from Fig. 8-3 if C is replaced by a short circuit for the ac signals and an open circuit for the dc source.

8-2 DIFFERENCE-MODE AND COMMON-MODE GAINS

In the ideal operational amplifier, the voltage between the positive and negative input terminals is zero, and the gain is infinite. In the actual operational amplifier, the voltage across the input terminals is nonzero. Furthermore, *the gain from the positive terminal to the output is slightly different in magnitude from the gain from the negative terminal to the output.* Consequently, when the same voltage is applied to both inputs, the output is not zero, i.e., perfect subtraction does not occur. This is indicated in Fig. 8-4a by expressing the output in terms of the gains (transfer functions) associated with each input signal.

Let A_1 represent the gain for the signal V_1 at the inverting input (while V_2 is held at 0), and A_2 represent the gain for the signal V_2 at the noninverting

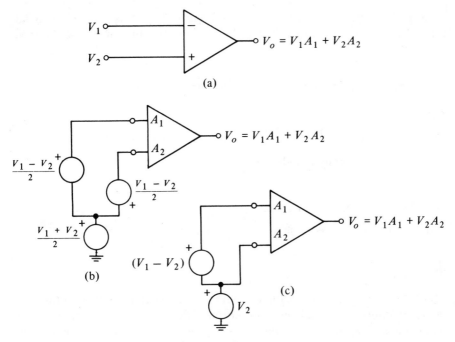

Fig. 8-4 Decomposition of the input signals

input (while V_1 is held at 0). Using the principle of superposition, the output can be expressed as the sum of outputs caused by each input signal acting alone; that is,

$$V_o = V_1 A_1 + V_2 A_2. \tag{8-4}$$

If $A_2 = -A_1$, then

$$V_o = (V_1 - V_2)A_1 = V_a(s)A(s),$$

where $V_a(s) = V_1 - V_2$ and $A(s) = A_1$ are in agreement with the model of the amplifier presented in Section 7-3. Consequently, the amplifier responds to the difference of the two input signals only. However, it is practically impossible to make one gain the *negative* of the other. As a result, when $V_1 = V_2$, the output is not zero but is given by

$$V_o = V_1(A_1 + A_2) = V_1 A_c,$$

where $A_c = A_1 + A_2$ represents the common-mode gain, as defined by Eq. (1-11a). Depending upon how the input signals are broken up, the general case ($V_2 \neq V_1$) is handled one of two ways. Either the *balanced decomposition* shown in Fig. 8-4b or the *unbalanced decomposition* shown in Fig. 8-4c is used. Regardless of how the input is decomposed, the output is the same, as all three presentations of Fig. 8-4 show.

In the balanced case of Fig. 8-4b, the output is written as the sum of signals arising from the common-mode and difference-mode components of the input signals:

$$V_o = \frac{(V_1 + V_2)}{2}(A_1 + A_2) + \frac{(V_1 - V_2)}{2}(A_1 - A_2)$$

$$= \frac{(V_1 + V_2)}{2}A_c + (V_1 - V_2)A_d, \tag{8-5}$$

where A_c is the common-mode gain,

$$A_c = \frac{V_o}{V_1}\bigg|_{V_2 = V_1},$$

and A_d is the difference-mode gain,

$$A_d = \frac{V_o}{V_1}\bigg|_{V_2 = -V_1}.$$

[Refer also to Fig. 1-11b and Eq. (1-12b) for further details.] Thus, as Eq. (8-5) shows, the output responds not only to the *difference* of the two input signals (desired response), but also to the *average value* of the two input signals (undesired response). Amplifiers are designed in such a way that $|A_c| \ll |A_d|$. However, the output may still contain an objectionable amount of undesirable signal if $\frac{1}{2}(V_1 + V_2)$ is large (as would be the case when both V_1 and V_2 are large and almost equal to each other). On the other hand, regardless of the value of A_c, there is no undesirable signal in the output if the input does not contain any common-mode component; that is, $V_2 = -V_1$. As a measure of the effectiveness of the amplifier in rejecting common-mode signals while amplifying difference-mode signals, the common-mode-rejection ratio, CMRR, is used. It is a function of s, and is defined by

$$\text{CMRR}(s) = \frac{A_d(s)}{A_c(s)}. \tag{8-6}$$

Stated in dB, CMRR is given by

$$\text{CMRR}_{\text{dB}} = 20 \log \left| \frac{A_d(j\omega)}{A_c(j\omega)} \right|. \tag{8-7}$$

CMRR is not only a function of frequency, but is also, in practice, a function of signal levels, thus indicating dependence upon *nonlinear* phenomena. Therefore, caution must be exercised in interpreting results.

In the unbalanced case of Fig. 8-4c, the output is written as the sum of signals arising from the V_2 and $(V_1 - V_2)$ components of the input signals,

$$V_o = V_2(A_1 + A_2) + (V_1 - V_2)A_1$$
$$= V_2 A_c + (V_1 - V_2)A_1, \tag{8-8}$$

where A_c is the common-mode gain, as in the previous case; that is,

$$A_c = \frac{V_o}{V_2}\bigg|_{V_1 = V} \qquad \text{and} \qquad A_1 = \frac{V_o}{V_1}\bigg|_{V_2 = 0}.$$

Because A_1 is defined as the gain with $V_2 = 0$ and A_d is defined as the gain with $V_2 = -V_1$,

$$A_1 \neq A_d.$$

To see the relationship between the two gains, V_o is eliminated between Eqs. (8-8) and (8-5). The result is

$$A_1 = A_d + \frac{A_c}{2}. \tag{8-9}$$

Thus, as long as

$$\tfrac{1}{2}|A_c(j\omega)| \ll |A_d(j\omega)|,$$

A_1 is practically equal to A_d.

As Eq. (8-8) shows, the output contains not only the desired signal, $(V_1 - V_2)A_1$, but also the undesired signal, $V_2 A_c$. Regardless of the value of A_c, the output contains only the desired signal if $V_2 = 0$. At first thought, this may seem inconsistent with the balanced case, where the desired signal is obtained only when $V_2 = -V_1$. However, a little thought shows that the outcome in both cases is the same; the interpretation, however, is different because of the way A_d and A_1 are defined. [Some authors prefer to use the decomposition of Fig. 8-4c and call V_2 the common-mode signal, whereas in Fig. 8-4b $(V_1 + V_2)/2$ is taken as the common-mode signal. These two signals are equal to each other if $V_1 = V_2$, as is the case when common-mode measurements are made.]

The output can be expressed in terms of the CMRR $= A_d/A_c$ by rearranging Eq. (8-5) as follows:

$$V_o = A_d\left(V_1 - V_2 + \frac{V_1 + V_2}{2}\frac{A_c}{A_d}\right) = A_d\left(V_1 - V_2 + \frac{V_1 + V_2}{2}\frac{1}{\text{CMRR}}\right).$$

As long as $V_1 \cong V_2$,

$$V_o \cong A_d\left(V_1 - V_2 + \frac{V_2}{\text{CMRR}}\right)$$

$$= A_d\left[V_1 - V_2\left(1 - \frac{1}{\text{CMRR}}\right)\right]. \tag{8-10}$$

Equation (8-10) shows that common-mode imperfections can be readily handled externally by decreasing the signal at the $+$ terminal from V_2 to

$$V_2\left(1 - \frac{1}{\text{CMRR}}\right),$$

and then using an ideal (as far as CMRR is concerned) amplifier. This equivalent presentation is diagrammatically illustrated in Fig. 8-5. Equation (8-10) is valid (under the assumption that $V_1 \cong V_2$) for both the balanced and unbalanced decompositions.

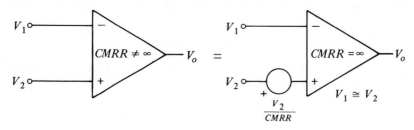

Fig. 8-5   Equivalent representation of CMRR

EXAMPLE 8-3

Show that the CMRR of an amplifier can be determined by connecting it as in Fig. 8-6, and measuring V_i and $(V_o - V_i)$.

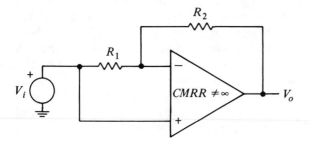

Fig. 8-6

SOLUTION

By inspecting the circuit and making use of the equivalence presented in Fig. 8-5, write the expression of the output voltage, and solve for the CMRR:

$$V_o = V_i\left(-\frac{R_2}{R_1}\right) + V_i\left(1 - \frac{1}{\text{CMRR}}\right)\left(1 + \frac{R_2}{R_1}\right),$$

$$V_o - V_i = -\frac{V_i}{\text{CMRR}}\left(1 + \frac{R_2}{R_1}\right);$$

$$\text{CMRR} = -\frac{V_i}{V_o - V_i}\left(1 + \frac{R_2}{R_1}\right). \qquad Ans.$$

EXAMPLE 8-4

The amplifier shown in Fig. 8-7 is used to measure the difference between two signals 1 mV apart. Assume that the resistance values are exact (see Problem 8-4 for the error introduced when the resistances are slightly different).

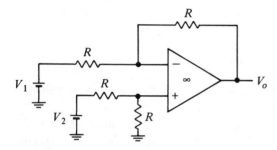

Fig. 8-7

(a) $V_1 = 10V$, $V_2 = 10.001V$, CMRR $= \infty$. Obtain V_o.
(b) Repeat (a), if CMRR is 60 dB.
(c) What should CMRR be if the difference of the two signals is to be measured with better than 10% accuracy?
(d) Repeat (c) if $V_1 = 100$ mV and $V_2 = 101$ mV.

SOLUTION

(a) Using the principle of superposition, write the output directly by inspection of Fig. 8-7:

$$V_o = V_1(-1) + \frac{V_2}{2} \times 2 = V_2 - V_1 = 1 \text{ mV.} \qquad Ans.$$

The circuit measures the difference between the two signals exactly.
(b) If CMRR is 60 dB (1000:1), then the signal at the positive input is

$$\frac{V_2}{2}\left(1 \pm \frac{1}{\text{CMRR}}\right) = \frac{10.001}{2}(1 \pm 0.001),$$

where both the plus and minus signs are used since the sign of the CMRR is not known. The output is

$$V_o = V_1(-1) + \frac{V_2}{2}\left(1 \pm \frac{1}{\text{CMRR}}\right)2 = (1 \pm 10) \text{ mV}$$

$$= 11 \text{ mV or } -9 \text{ mV.} \qquad Ans.$$

The 1-mV difference between the two input signals is swamped out in the output by the presence of the 11-mV (or -9-mV) common-mode signal. Hence, this amplifier cannot be used to measure the 1-mV difference between the two signals.

(c) In order to obtain the difference signal with better than 10% accuracy, the common-mode component in the output should be less than 0.1-mV. This means that

$$\frac{V_2}{\text{CMRR}} < 0.1 \text{ mV},$$

$$\text{CMRR} > \frac{10001}{0.1} \cong 10^5. \quad \textit{Ans.}$$

Better than 100-dB common-mode rejection is required in order to obtain the desired accuracy.

(d) The signals are still 1 mV apart; however, their common level is 100 times smaller. Therefore, with a much lower CMRR, the desired accuracy can be obtained:

$$\frac{V_2}{\text{CMRR}} < 0.1,$$

$$\text{CMRR} > \frac{101}{0.1} \cong 10^3. \quad \textit{Ans.}$$

With 60-dB CMRR, the output will indicate the desired difference of the two input signals with 10% accuracy. This example shows that it is easier to detect 1 mV in 100 mV than 1 mV in 10V, a result that makes physical sense.

8-3 DYNAMIC RANGE

If an actual operational amplifier is to operate in its linear range, certain precautions must be taken. Since the output can deliver only a specified amount of current and cannot exceed a specified voltage swing, the element values and the input signals should be chosen such that they do not force operation in the nonlinear regions. Consider, for example, the inverting network shown in Fig. 8-8a.

Suppose that the operational-amplifier output current, $i_o(t)$, is limited to $|i_o(t)| < I_{\max}$, where $I_{\max}$ is specified. Assuming that there is no load on the output, then as long as $|v_i(t)|/R_1 \leq I_{\max}$, the output can supply the current demanded by the input signal and the amplifier operates linearly. When the demand on the output current exceeds $\pm I_{\max}$, the amplifier is no longer in its linear range of operation; its output equivalent circuit may be represented by a constant current source, $\pm I_{\max}$. Then, as Fig. 8-8b indicates, the output is related to the input by

$$v_o(t) = v_i(t) - I_{\max}(R_1 + R_2), \qquad v_i(t) > I_{\max} R_1. \tag{8-11}$$

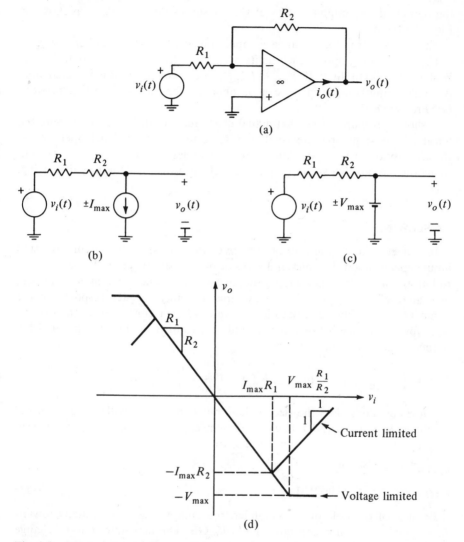

Fig. 8-8 Dynamic range of operation

This equation is plotted in Fig. 8-8d as a straight line with unity slope. A similar break in the characteristic occurs for negative values of $v_i(t)$. (In practice the breaks may not be as sharp as indicated in the figure.)

For operational amplifiers, a maximum swing of the output voltage, $\pm V_{max}$, is also specified. As long as the output does not exceed $\pm V_{max}$, the amplifier operates quite linearly (provided current limits are not exceeded). However, when

$$|v_i(t)| \frac{R_2}{R_1} > V_{max},$$

the output of the amplifier stays at the constant value of $\pm V_{max}$, as shown in Fig. 8-8c and d.

The limits of the linear range of operation may be set by either $\pm I_{max}$ or $\pm V_{max}$, depending upon which limits occur first. In Fig. 8-8d, current limiting is shown to occur first. If $V_{max} < I_{max} R_2$, then voltage limiting occurs first. A similar input–output curve can be obtained for the noninverting network (see Problem 8-8).

It should be emphasized that actual characteristics do not exhibit quite the linear relationships presented in Fig. 8-8d. Only for small-signal operation is the assumption of linearity valid. Also, the values of the voltage and current limits on the positive swing may not equal the values on the negative swing.

8-4 SLEWING

In a linear circuit, the rate of change of the output of the circuit to a step input depends upon the poles and zeros of the transfer function. For a one-pole rolloff amplifier, the open-loop gain-bandwidth product determines the rate of change of the output. For an example, consider the step response of the noninverting amplifier discussed in Example 7-5. For convenience, the expressions for the voltage between the $+$ and $-$ terminals of the amplifier and the output voltage are repeated here:

$$V_a(s) = \frac{-V}{s + GB/K_0}, \quad \text{and} \quad V_o(s) = V \frac{GB}{s(s + GB/K_0)}.$$

In these equations, V is the size of the input step and K_0 is the dc gain of the amplifier ($K_0 = 1 + R_2/R_1$). The time-domain expressions for the input (between $+$ and $-$ terminals) and output voltages are

$$v_a(t) = - Ve^{-t/(K_0/GB)}, \tag{8-12}$$

$$v_o(t) = K_0 V[1 - e^{-t/(K_0/GB)}]. \tag{8-13}$$

Because of the pole of the amplifier, the output rises exponentially toward $K_0 V$ volts with a time constant of $\tau = K_0/GB$. The maximum rate of change of the output occurs at $t = 0$ and is given by

$$\left. \frac{dv_o}{dt} \right|_{max} = VGB$$

$$= \text{Amplitude of input step} \times \text{gain-bandwidth product in rad/s.} \tag{8-14}$$

Because the output voltage cannot jump, the voltage between the $+$ and $-$ input terminals of the amplifier is forced to jump down V volts at $t = 0^+$. This is shown in Fig. 8-9.

Unless the step amplitude is extremely small, this sudden drop of V volts does overdrive the amplifier at $t = 0^+$. As a result, operation becomes nonlinear, and the output no longer behaves as expected. Instead of rising exponentially,

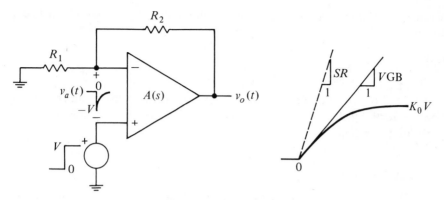

Fig. 8-9   Step-response waveforms $\left(K_0 = 1 + \dfrac{R_2}{R_1}\right)$

it may rise more or less linearly at a rate specified by the manufacturer. This rate, called slewing rate, SR, is a characteristic of the amplifier and is given in volts/μs. Whether the rate of rise of the output is limited by the slewing rate of the amplifier or its gain-bandwidth product, depends upon Eq. (8-14). If

$$\text{VGB} < \text{SR},\tag{8-15a}$$

the response is bandwidth-limited and the output rises exponentially, as shown in Fig. 8-9. On the other hand, if

$$\text{VGB} > \text{SR},\tag{8-15b}$$

the response is slewing-rate-limited, and the output rises more or less linearly until the amplifier recovers from overdriven conditions. Stated differently, either VGB or SR, whichever is *smaller*, determines the rate of rise of the output.

The slewing rate also affects the sinusoidal steady-state operation. Since it is impossible for the output to rise at a rate faster than the slewing rate, the constraint

$$\left.\frac{dv_o}{dt}\right|_{\text{max}} < \text{SR}\tag{8-16}$$

must be observed if linear operation is to result. For an output sine wave of maximum value V_m and frequency ω, the greatest rate of change occurs when the wave crosses zero; that is,

$$v_o = V_m \sin \omega t,$$

$$\left.\frac{dv_o}{dt}\right|_{\text{max}} = \omega V_m.$$

Thus, if operation is to be band-limited rather than slewing-rate-limited, the constraint

$$\omega V_m < SR \tag{8-17}$$

must apply.

Both Eq. (8-15a) and Eq. (8-17) indicate that for a given amplifier, limitations set forth by slewing rate can be avoided by reducing the maximum value of the output signal.

EXAMPLE 8-5

An operational amplifier has a $0.5 \text{V}/\mu s$ slewing rate and a gain-bandwidth product of 10^6 rad/s. It is connected in the noninverting mode for a gain of 10.
(a) The input is a 1-V step. Which will limit the rate of rise of the output, the slewing or the bandwidth?
(b) The input is a 0.5-V peak sine wave. Which will limit the upper frequency of operation, the slewing or the bandwidth?

SOLUTION

(a) With Eq. (8-15) in mind, obtain VGB and compare it with SR:

$$VGB = 1 \times 10^6 \text{ volts/s} = 1 \text{ volt}/\mu s,$$

$$SR = 0.5 \text{ volts}/\mu s.$$

Since SR is the smaller of the two, the rate of rise of the output is limited by the slewing rate. Therefore, the output changes linearly rather than exponentially.
(b) Obtain first the closed-loop 3-dB bandwidth of the amplifier, using Eq. (7-23) (the assumption is that this frequency is the highest useful frequency of operation for this circuit):

$$\omega_{3 \text{ dB}} = \frac{GB}{K_0} = \frac{10^6}{10} = 10^5 \text{ rad/s.}$$

Then, with Eq. (8-17) in mind, obtain $\omega_{3 \text{ dB}} V_m$ and compare it to SR. Make use of the facts that the peak value of the input is 0.5 V, $K_0 = 10$, and at the 3-dB frequency the output is down by 3 dB:

$$V_m = \frac{0.5 \times K_0}{\sqrt{2}} = \frac{5}{\sqrt{2}},$$

$$\omega_{3 \text{ dB}} V_m = \frac{10^5 \times 5}{\sqrt{2}} = \frac{0.5}{\sqrt{2}} \times 10^6 \text{ volts/s} = \frac{0.5}{\sqrt{2}} \text{ volts}/\mu s.$$

Since $\omega_{3\,dB}\,V_m$ is smaller than the SR, the rate of rise of the output is limited by the bandwidth. Therefore, slewing can be ignored. However, if the input-signal amplitude is made greater than $0.5\sqrt{2}$ volts (in peak value), then $\omega_{3\,dB}\,V_m$ will exceed the slewing rate, and the amplifier will not operate linearly at high frequencies; i.e., the output will no longer be a sine wave.

8-5 NONLINEAR APPLICATIONS

Although this is a book on linear circuits, a few applications of nonlinear circuits are given in this section merely to show the unlimited and exciting possibilities that arise when the operational amplifier is used in its nonlinear or piecewise-linear mode.

Comparator

A comparator is a device that produces an abrupt change in output when the input exceeds a predetermined voltage or current reference level. Such a circuit is shown in Fig. 8-10a.

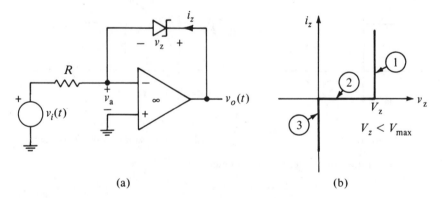

(a)   (b)

Fig. 8-10   Zero-crossing detector

To understand the operation of this circuit, consider first the three states of the zener shown in Fig. 8-10b.

When $i_z > 0$, operation is along ① and $v_z = V_z$;

when $i_z = 0$, operation is along ② and $0 \le v_z \le V_z$;

when $i_z < 0$, operation is along ③ and $v_z = 0$.

The output, v_o, depends upon the state of the zener, which, in turn, depends upon the input, v_i. Assume the zener is operating along ①. This means that

$v_z = V_z$, and the zener can be replaced by a battery, as shown in Fig. 8-11a. Since the output is fed back to the negative terminal of the amplifier, the network is stable; the amplifier is active and $v_a = 0$. Consequently, $v_o = V_z$. To assure that the zener is indeed operating along ①, i_z is calculated and set greater than zero:

$$i_z = -\frac{v_i(t)}{R} > 0.$$

Hence, $v_i(t) < 0$. The $v_o = V_z$, ($v_i < 0$)-curve is drawn as curve ④ in Fig. 8-11c.

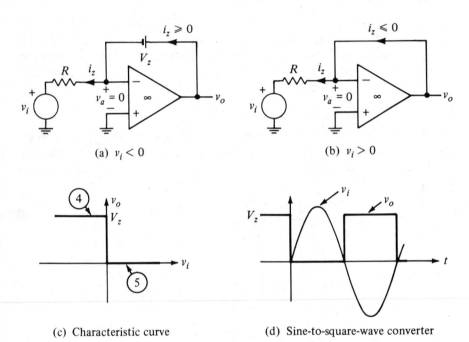

(a) $v_i < 0$ (b) $v_i > 0$

(c) Characteristic curve   (d) Sine-to-square-wave converter

Fig. 8-11   Operation of the zero-crossing detector

When the zener is operating along ③, $v_z = 0$. Thus, it can be replaced by a short circuit, as shown in Fig. 8-11b. The amplifier is under unity feedback and therefore stable; $v_a = 0$ and therefore $v_o = 0$. To assure that the zener is indeed operating along ③, i_z is calculated and set less than zero:

$$i_z = -\frac{v_i(t)}{R} < 0.$$

Hence, $v_i(t) > 0$. The $v_o = 0$, ($v_i > 0$)-curve is drawn as curve ⑤ in Fig. 8-11c.

When the zener is operating along ②, $i_z = 0$. The feedback loop is thus broken, and the amplifier operates in the open-loop mode. The input signal, $v_i(t)$, appears at the amplifier input terminals, that is, $v_a(t) = v_i(t)$. Since the amplifier is considered to be ideal, its gain is infinite. Consequently, $v_i(t)$ has to change only infinitesimally from zero to turn on the zener one way or the other, and thus close the feedback loop. Stated differently, operation along ② is not possible with an ideal amplifier. As a result, the output jumps from V_z to 0 as $v_i(t)$ crosses zero. The circuit, therefore, can be used as a zero-crossing detector.

This circuit can be used to convert a sine wave into a square wave, as shown in Fig. 8-11d. In practice, because of the imperfections of the amplifier, the output does not switch at the instant when the input crosses zero. Neither does it jump V_z volts in zero time.

Because the $+$ input of the amplifier is grounded, the input is compared to zero. It is possible to change the comparison point by inserting a reference source at the $+$ input. The comparison point can also be changed by connecting the reference source through a resistor to the $-$ input. (See Problem 8-15 for details.)

Circuit with Hysteresis

In Section 7-4, it is shown that feedback from the output to the positive terminal of the operational amplifier results in an unstable situation, driving the output to a constant voltage designated as $+V_{max}$ or $-V_{max}$.

Such a circuit is shown in Fig. 8-12a.

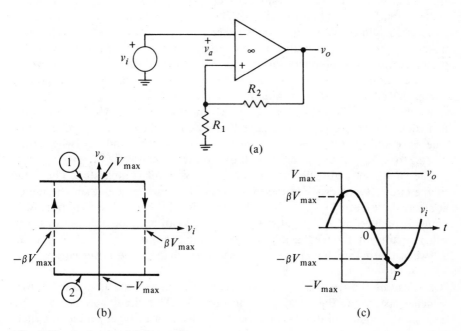

Fig. 8-12   A circuit with hysteresis

To see how this circuit operates, assume that the output is at $+V_{max}$ (resistance values are such that the output is voltage-limited rather than current-limited). The voltage at the $+$ input terminal is

$$v_{R_1} = \beta V_{max},$$

where $\beta = R_1/(R_1 + R_2)$. *Since the circuit is not operating in its linear mode, the voltage across the amplifier input terminals need not be zero.* As long as $v_i(t) < v_{R_1}$, the voltage $v_a(t)$ is negative, and the output is maintained at $+V_{max}$ since $v_a(t)$ is multiplied with a very large negative gain (ideally, *infinite*). The curve marked ① in Fig. 8-12b shows this region of operation. As soon as $v_i(t)$ exceeds the reference voltage,

$$v_{R_1} = \beta V_{max},$$

$v_a(t)$ becomes positive. A positive $v_a(t)$ multiplied with a large negative gain results in an output of $-V_{max}$. Thus, $v_o(t)$ switches from $+V_{max}$ to $-V_{max}$, as shown by the dotted line and arrow in Fig. 8-12b. (In practice, the switching is not instantaneous.) After switching, the voltage at the $+$ input terminal changes to

$$v_{R_1} = -\beta V_{max}.$$

As long as $v_i(t)$ exceeds this new reference level (curve ② in Fig. 8-12b), the voltage across the amplifier input, $v_a(t)$, is positive and therefore the output stays at $-V_{max}$. However, once $v_i(t)$ drops below v_{R_1}, $v_a(t)$ becomes negative and the output jumps to $+V_{max}$, as shown by the dotted line and arrow in the figure. Note that when v_i falls between $-\beta V_{max} < v_i < \beta V_{max}$, the v_o-vs.-v_i curve is double-valued (has two states). Therefore, in this range, knowledge of v_i alone is not sufficient to determine whether v_o is at $+V_{max}$ or $-V_{max}$. The past history of the circuit must also be given. For example, if $v_i(t)$ is initially highly negative, thus assuring operation on curve ①, and thereafter is raised to zero, then the output is at $+V_{max}$. Because of the double-valued nature of the v_o-vs.-v_i curve, this circuit is said to exhibit *hysteresis*.

When the input is a sine wave of peak value greater than βV_{max}, the output is a square wave, as shown in Fig. 8-12c. By controlling the amplitude of the sine wave and the (R_2/R_1) ratio, the square wave can be made to switch from $-V_{max}$ to $+V_{max}$ at any point from 0 to P on the sine wave.

The circuit may also be thought of as a two-level comparator. The state of the circuit switches when the input signal becomes greater than the upper comparison level of βV_{max} or when the input signal becomes less than the lower comparison level of $-\beta V_{max}$. Comparison levels can be made different by using other circuitry in the feedback loop. For example, a diode, which short-circuits R_1 when the output voltage becomes negative, changes the lower comparison level from $-V_{max}$ to 0.

Because of the feedback to the $+$ terminal, this circuit, as a comparator, switches faster than the circuit given in Fig. 8-10. This circuit can also be used as a bistable multivibrator, in which case v_i is used as a trigger source to switch the state of the output. For example, assume that $v_o = +V_{max}$ with $v_i = 0$. A positive spike of amplitude greater than βV_{max} triggers the circuit and the output switches to $-V_{max}$. It stays in this state until a negative spike of amplitude greater than βV_{max} is applied. The circuit can be switched back and forth by applying alternately positive and negative spikes. It can also be made to switch on a smaller positive spike than a negative spike by biasing v_i with a positive dc source.

Monostable Multivibrator

A monostable multivibrator is a device that possesses one stable state and one transitory (quasistable) stage. It remains in its stable state until triggered, at which time it goes into its transitory state. The duration of this state is controlled by the charging of a capacitor. At the termination of this period, the output switches and reverts to its stable state. Such a circuit is shown in Fig. 8-13a.

To understand the operation of the circuit, assume it is in its stable state. This means that all voltages and currents are in their steady state values. In particular, the current through the capacitor is zero and v_2 is at $-V$. Because the gain of the amplifier is infinite and v_2 is negative, the output is at $-V_{max}$. The stable state is therefore characterized by the following values:

$$v_t = 0, \qquad v_2 = -V, \qquad v_a = V, \qquad v_o = -V_{max}, \qquad v_{cap} = V_{max} - V.$$

Assume that at $t = 0$ the negative trigger is applied. If the trigger is to cause a transition of states, it should be of sufficient amplitude to cause a reversal of polarity in v_a. This requires that the trigger amplitude be at least V volts. As soon as v_a changes sign, the output switches to $+V_{max}$. The equivalent circuit of Fig. 8-13b can then be used to calculate the expression for $v_2(t)$.

Because the voltage across the capacitor cannot jump, the voltage at v_2 is $(2V_{max} - V)$ right after switching. This voltage tends toward $-V$ exponentially with a time constant of RC. Hence,

$$v_2(t) = -V + 2V_{max} e^{-t/RC}.$$

As long as $v_a(t) < 0$ (and therefore $v_o = +V_{max}$), the circuit remains in the transitory state. The instant v_a changes polarity, v_o becomes $-V_{max}$ and the transitory state is terminated. To determine this instant, let $v_a = 0$:

$$v_a(t) = [v_t(t) - v_2(t)]|_{t=T} = 0,$$

which simplifies to

$$v_2(T) = 0,$$

if it is assumed that at $t = T$ the trigger signal has already reached the zero level. Thus, the period-determining equation becomes

$$-V + 2V_{max} e^{-T/RC} = 0.$$

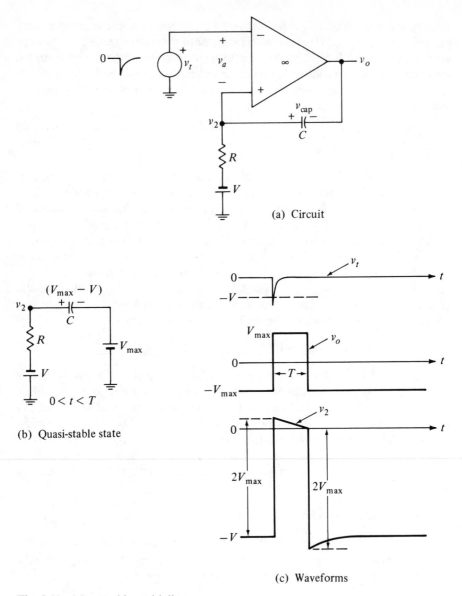

(a) Circuit

(b) Quasi-stable state

(c) Waveforms

Fig. 8-13   Monostable multivibrator

This equation is then solved for the transitory period T:

$$T = RC \ln \frac{2V_{max}}{V}.$$

The output reverts to the stable state the instant it switches to $-V_{max}$. However, the circuit does not recover completely until the voltage across the capacitor reaches the final value, that is,

$$v_{cap} = V_{max} - V.$$

This takes about three time constants, after which time the entire circuit is at steady state. The various waveforms of the monostable multivibrator (using an ideal amplifier) are presented in Fig. 8-13c. Monostable multivibrators are widely used to provide gating signals and delayed trigger signals.

Astable Multivibrator

An astable multivibrator is a device with no stable states. Instead it has two quasistable states, and it switches alternately from one to the other. Such a circuit is shown in Fig. 8-14a.

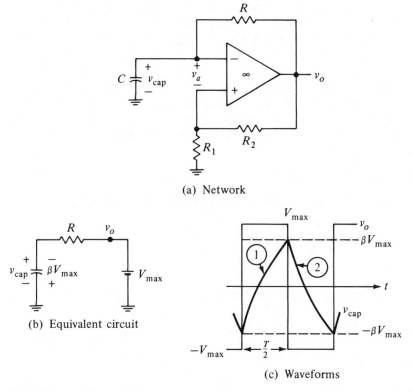

(a) Network

(b) Equivalent circuit

(c) Waveforms

Fig. 8-14 Astable multivibrator

Note that feedback is applied from the output to both input terminals. Since the circuit is not driven by a signal source, the operation of the circuit may not be obvious. For a start, assume the output is at $+V_{max}$ (or $-V_{max}$). There is reason to believe that this is the case because of the resistive feedback applied to the positive terminal. Let $\beta = R_1/(R_1 + R_2)$ represent the transfer function of this feedback network. The RC network can then be set aside and driven by the battery V_{max}, as shown in Fig. 8-14b. (Ignore for the moment that the capacitor has an initial voltage of $-\beta V_{\max}$, as shown in the diagram.) The capacitor voltage starts building up exponentially toward the battery voltage $V_{\max}$. It would reach $V_{\max}$ if the operational amplifier did not interfere. However, as soon as the capacitor voltage exceeds $\beta V_{\max}$ which is the dc comparison level established at the positive input terminal through the resistive feedback from the output, the voltage $v_a(t)$ across the amplifier input terminals changes sign; it becomes positive. This positive voltage, when multiplied by the infinite and negative gain of the operational amplifier, causes the output to switch to $-V_{\max}$. As a result, the equivalent circuit of Fig. 8-14b changes. The new equivalent circuit (not shown) differs from the previous one (Fig. 8-14b) in two respects. The polarity of the battery and the polarity of the initial condition are both reversed. (Merely change $V_{\max}$ to $-V_{\max}$ and $-\beta V_{\max}$ to $+\beta V_{\max}$ to get the new equivalent circuit.) The capacitor voltage, which has an initial value of $\beta V_{\max}$, tends now exponentially toward $-V_{\max}$. When it becomes less than the new comparison level, $-\beta V_{\max}$, $v_a(t)$ changes sign once more. It becomes negative and consequently causes the output to change back to $+V_{\max}$. A cycle of operation is thus completed with $-\beta V_{\max}$ volts left on the capacitor, as shown in Fig. 8-14b. From the equivalent circuit of Fig. 8-14b, the time for half a cycle of operation is calculated. Note that the capacitor starts with an initial voltage of $-\beta V_{\max}$ and tends toward $V_{\max}$ exponentially with a time constant of RC seconds:

$$v_{\text{cap}}(t) = V_{\max} - V_{\max}(1 + \beta)e^{-t/RC}.$$

This voltage is sketched as curve ① in Fig. 8-14c. After a half-cycle operation,

$$t = \frac{T}{2},$$

and

$v_{\text{cap}}(t)$ reaches the comparison level of $\beta V_{\max}$:

$$v_{\text{cap}}(t)\Big|_{t=T/2} = V_{\max} - V_{\max}(1 + \beta)e^{-T/2RC} = \beta V_{\max}.$$

Solving for T gives

$$T = 2RC \ln\left(\frac{1 + \beta}{1 - \beta}\right) = 2RC \ln\left(1 + \frac{2R_1}{R_2}\right).$$

As Fig. 8-14c shows, the output is a square wave with period T and peak-to-peak amplitude $2V_{\max}$.

EXAMPLE 8-6

Explain the operation of the circuit shown in Fig. 8-15. Assume b arbitrarily large.

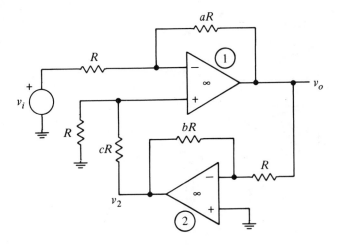

Fig. 8-15

SOLUTION

Assume that both amplifiers are active. Then, by inspection of the network, write the expressions for v_2 and v_o:

$$v_2 = v_o(-b),$$

$$v_o = v_i(-a) + v_o(-b)\left(\frac{1}{1+c}\right)(1+a).$$

Solve for v_o explicitly, and obtain:

$$v_o = \frac{-av_i}{1 + b\left(\dfrac{1+a}{1+c}\right)}.$$

The v_o-vs.-v_i curve is sketched as curve ① in Fig. 8-16. This curve is valid as long as both amplifiers operate in the linear region. Since b can be made arbitrarily large, amplifier #2 does become voltage-limited before amplifier #1. (Resistances are large enough so that voltage- rather than current-limiting occurs.) Amplifier #2 becomes voltage-limited in operation when

$$|(-b)v_o| = V_{max},$$

which occurs for

$$|v_i| = V_{max}\frac{\left[1 + b\left(\dfrac{1+a}{1+c}\right)\right]}{ab}.$$

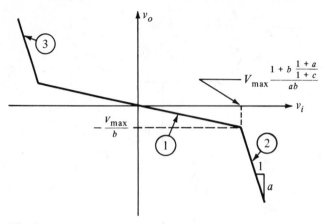

Fig. 8-16

When $|v_i|$ exceeds this value, amplifier #2 output stays at $\pm V_{max}$. Then the output v_o is given by

$$v_o = v_i(-a) \pm V_{max} \left(\frac{1}{1+c}\right)(1+a),$$

where the plus sign is used when the output is negative (curve ② in Fig. 8-16) and the minus sign is used when the output is positive (curve ③). As $b \to \infty$, curve ① becomes a horizontal line. Then, the v_o-vs.-v_i curve exhibits a dead zone, where the output stays zero over a wide range of input-voltage values. However, when the input goes beyond the dead zone, the full gain of $-a$ is restored in the circuit. Thus, signals above and below the thresholds, given by

$$\pm V_{max} \left. \frac{\left[1 + b\left(\dfrac{1+a}{1+c}\right)\right]}{ab}\right|_{b\to\infty} = \pm V_{max}\left[\frac{1}{a}\left(\frac{1+a}{1+c}\right)\right],$$

are amplified; those between are not passed.

8-6 SUMMARY

The actual operational amplifier is beset by a number of imperfections.

1. The offset voltage and current in an operational amplifier cause output voltages with no input signals. In high-gain dc applications, the offset effects may overcome the desired output unless special circuitry is used to balance them out. It may even be necessary to use chopper-stabilized amplifiers.

2. Operational amplifiers produce an output even when identical signals are applied to the plus and minus terminals. This is caused by the slight difference in the magnitudes of the two gains. As a result, perfect subtraction is not

possible. The common-mode rejection ratio is used as a figure of merit in determining the effectiveness of the amplifier in separating difference-mode and common-mode signals. In circuit applications, the common-mode effects are referred to the noninverting terminal of the operational amplifier by multiplying the signal there with $(1 - 1/\text{CMRR})$.

3. The output of the operational amplifier can deliver only a prescribed amount of current, which should not be exceeded if linear operation is desired. The output current is the sum of the currents taken by the feedback network and the external load. The output voltage swing is also limited. If the amplifier is driven into the voltage- or current-limited regions, operation is no longer linear. Consequently, the voltage between its plus and minus terminals may take any value.

4. Because of nonlinear effects arising from overdriven conditions, the output of the operational amplifier cannot rise any faster than the slewing rate. As a result, in large-signal and high-frequency applications, the steady-state output may not resemble the input. This phenomenon should not be confused with the one-pole rolloff characteristic of operational amplifiers, where the steady-state output amplitude is always directly proportional to the amplitude of the input signal.

The usefulness of the operational amplifier is vastly extended if nonlinear elements are admitted in the feedback circuit. A variety of wave-shaping and switching circuits can then be built. Square, triangular, and sawtooth waves can be generated if instability-causing feedback is applied. In this interesting and wide field, what can be said and done is limited only by one's imagination.

PROBLEMS

8-1   In Fig. 8-17,

$$R_c = R_3 - \frac{R_1 R_2}{R_1 + R_2}.$$

Find the output voltage caused by the offset current and voltage.

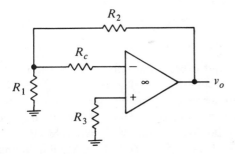

Fig. 8-17

8-2   Discuss the effect of the offset voltage on the output of the integrator shown in Fig. 8-18.

8-3   In Fig. 8-19, obtain the steady-state output voltage caused by the offset voltage source. Compare this answer with the one obtained with the positive terminal grounded. Assume $R_2 \gg R_1$.

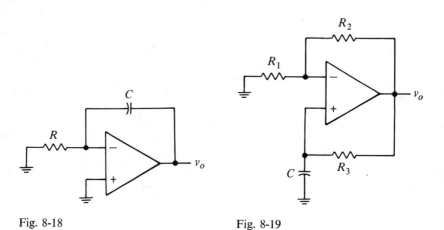

Fig. 8-18   Fig. 8-19

8-4   In the network of Fig. 8-20, the tolerance of the resistors is $\pm 1\%$. Therefore, the output is not exactly the difference of the two input signals.

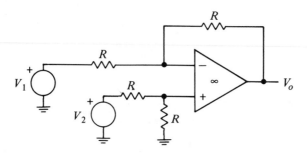

Fig. 8-20

(a) Consider the worst case, and express the output as the sum of the desired difference signal and an error signal based on V_2.

(b) The result of (a) may be interpreted as arising from a network using exact values for the resistors but employing an amplifier with imperfect common-mode rejection ratio. What is this equivalent CMRR?

8-5   In Fig. 8-21, the output is given by $V_o = V_2 G$ if the CMRR is infinite. What is the output if CMRR is not infinite?

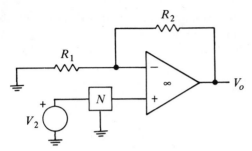

Fig. 8-21

8-6 The common-mode rejection ratio of each amplifier of Fig. 8-22 is the same. Show that, if the resistance ratios are properly chosen, the output can be made exactly proportional to the difference of the two input signals even if CMRR is not infinite. Find the correct ratio and output V_o.

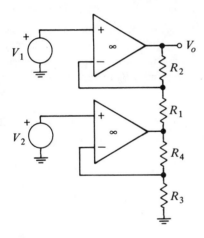

Fig. 8-22

8-7 In Fig. 8-23, find $v_o(t)$ if CMRR $= \infty$. What is the output if CMRR is 40 dB?

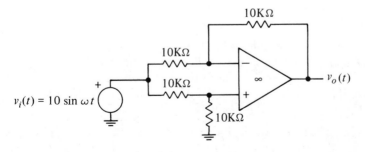

Fig. 8-23

8-8   Draw the v_o-vs.-v_i curve for the noninverting amplifier shown in Fig. 8-24. Consider both the maximum current and the maximum voltage limits.

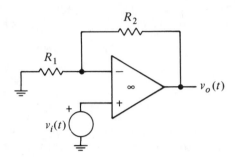

Fig. 8-24

8-9   Operational amplifiers are not exactly linear devices. They introduce distortion, so that under steady-state conditions the output contains terms that are not present in the input. These terms are lumped together and represented by the distortion signal, shown as V_d in Fig. 8-25.

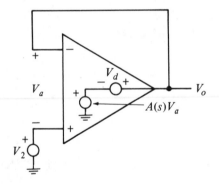

Fig. 8-25

(a) Obtain the expression for $V_o(s)$. Assume $|A| \gg 1$.
(b) Obtain the voltage V_a across the input terminals.
(c) Compare the desired-signal to the distortion-signal ratio at the amplifier input terminals with that at the amplifier output terminals. Explain the significance of this result.

8-10 A one-pole rolloff equivalent model $\left[A(s) = -\dfrac{GB}{s}\right]$ is used for anoperational amplifier. If it is used in the inverting mode, how fast does the output change at $t = 0^+$ if a V-volt step is applied to it? (Ignore slewing.)

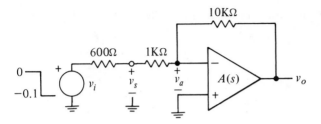

Fig. 8-26

8-11 The inverting amplifier shown in Fig. 8-26 is driven from a 600-Ω source. Sketch $v_o(t)$, $v_a(t)$, and $v_s(t)$, vs. t. Assume that the operational amplifier is modelled by $A(s) = -\dfrac{GB}{s}$.

8-12 A 10-V peak, undistorted, sine wave at 10 KHz is to be delivered at the output of an operational amplifier. What should be the slewing rate of the operational amplifier?

8-13 In the network of Fig. 8-27, the diode is ideal, i.e., it is a short circuit when it conducts and an open circuit otherwise. Sketch the v_o-vs.-v_i curve.

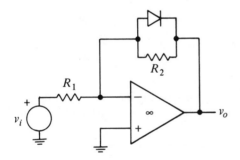

Fig. 8-27

8-14 Obtain the v_o-vs.-v_i characteristic of the network shown in Fig. 8-28. The back-to-back zener characteristic is as shown.

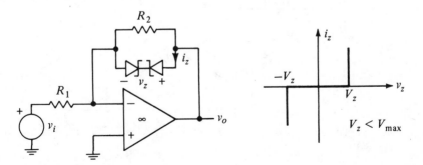

Fig. 8-28

8-15 For the two networks shown in Fig. 8-29, sketch the v_o-vs.-v_i curves.
When $i_z > 0$, $v_z = V_z$;
when $i_z < 0$, $v_z = 0$.

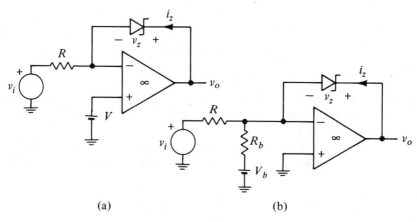

(a)   (b)

Fig. 8-29

8-16 Explain the operation of the two peak-holding circuits shown in Fig. 8-30.
Which circuit can be loaded without degrading the peak-holding capability?

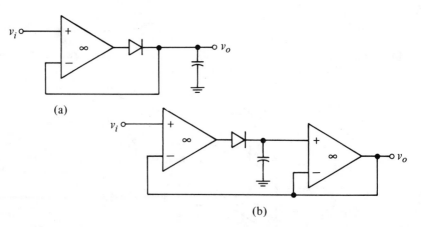

Fig. 8-30

8-17 (a) Obtain the v_o-vs.-v_i curve for the circuit shown in Fig. 8-31. Assume that $V_b = 0$. When forward-biased, the voltage across the diodes is 0.6 V.

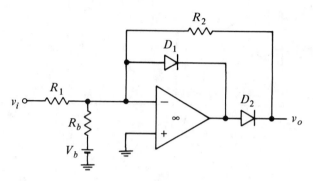

Fig. 8-31

(b) How does V_b alter the v_o-vs.-v_i curve obtained in (a)?
(c) What is the v_o-vs.-v_i curve if the diodes are reversed? $V_b = 0$.
(d) Show that is it possible to get an absolute-value circuit, that is,

$$v_o(t) = \pm|v_i(t)|,$$

by summing v_o with v_i. Draw a circuit with values.

8-18 (a) Explain the steady-state operation of the network shown in Fig. 8-32. Assume $R_2 > R_1$.
(b) Draw the waveforms for v_1 and v_2. Calculate the maximum and minimum values of the waveforms, and obtain the period.

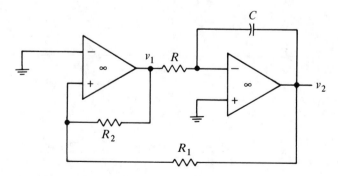

Fig. 8-32

8-19 Explain the operation of the network shown in Fig. 8-33 for both positions
of the switch. (The back-to-back zener characteristic is shown in Fig. 8-28.)

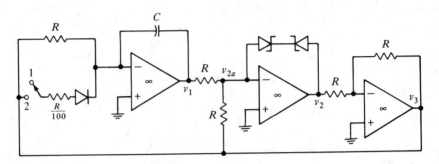

Fig. 8-33

9

Control of Poles
and Zeros
through Dependent Sources

Using dependent sources, the poles and zeros of system functions can be placed in any desired location in the s-plane. Pole–zero locations unattainable with passive networks become readily accessible when dependent sources are introduced. All this can be achieved without the use of inductors. This is significant, particularly in low-frequency work, where the use of large inductors may not be practical, because of size and weight limitations and the difficulty in providing negligible resistance and capacitance. On the other hand, RC-operational amplifier networks are small, light, and can be accurately tuned. They realize any function obtainable by RLC networks and also other functions that are not possible with passive networks. Furthermore, because of their low output impedances, amplifiers provide coupling between networks without much interaction. In the case of finite-gain amplifiers, the gain also provides an additional degree of freedom of adjustment.

This chapter deals mainly with techniques of moving the poles and zeros of second-order systems by means of amplifiers connected in the positive-, negative-, or infinite-gain mode. In particular, methods for generating constant-α and constant-ω_o root loci are given. Throughout the chapter, the emphasis is on knowing where the critical frequencies are and what to do to move them to new positions in the s-plane. Several examples are given to illustrate the principles involved.

9-1 CONTROL OF POLES AND ZEROS

To see how the poles and zeros can be controlled, consider Fig. 9-1a, where two amplifiers are used to change the transfer function of the RC network. The

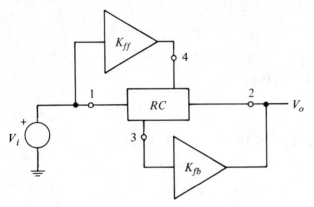

(a) Feedforward and feedback amplifiers

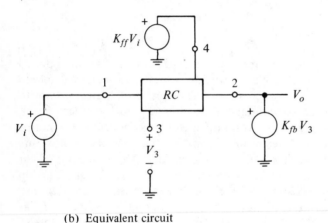

(b) Equivalent circuit

Fig. 9-1   *RC* network with dependent sources

upper amplifier takes the input signal, V_i, multiplies it with K_{ff}, and *feeds forward* the resulting signal to port 4 of the *RC* network. The lower amplifier takes the port 3 voltage of the *RC* network, multiplies it with K_{fb}, and *feeds* the resulting signal *back* to the *RC* network at port 2, which is also taken as the output. (Feedforward implies the feeding of the input signal, whereas feedback implies the feeding of a signal *other* than the input.) The amplifiers are considered ideal; that is, the inputs draw no current and the outputs act like voltage sources. Hence the network of Fig. 9-1a can be equivalently represented as in Fig. 9-1b. K_{ff} and K_{fb} are, in general, functions of s.

Since the output depends on V_3, it is calculated first. As Fig. 9-1b shows, three voltage sources act on the *RC* network to produce V_3. Considering one source at a time, V_3 is written as the sum of three responses:

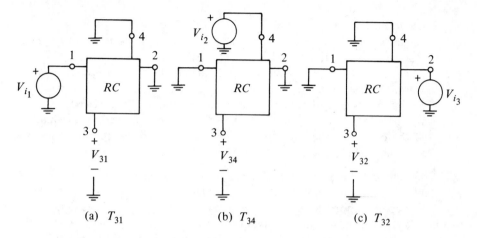

(a) T_{31} (b) T_{34} (c) T_{32}

Fig. 9-2   The three transfer functions associated with the *RC* network

$$V_3 = V_i T_{31} + K_{ff} V_i T_{34} + K_{fb} V_3 T_{32}. \tag{9-1}$$

The three transfer functions, T_{31}, T_{34}, and T_{32}, are associated with the various ports of the *RC* network. With reference to Fig. 9-2, they are defined as:

$$T_{31} = \left.\frac{V_{31}}{V_{i_1}}\right|_{V_2=V_4=0} = \frac{N_{31}}{D}, \quad T_{34} = \left.\frac{V_{34}}{V_{i_2}}\right|_{V_1=V_2=0} = \frac{N_{34}}{D}, \quad \text{and}$$

$$T_{32} = \left.\frac{V_{32}}{V_{i_3}}\right|_{V_1=V_4=0} = \frac{N_{32}}{D}, \quad (9\text{-}2)$$

The three transfer functions are assumed to have the same denominator polynomial D. Note that in each case, the voltage at port 3 is found while the signal is fed to one of the other ports; the remaining two ports are short-circuited. *Port 3 is significant because the feedback signal is sensed there.*

The voltage V_3 is obtained by solving Eq. (9-1) explicitly for V_3:

$$V_3 = \frac{V_i(T_{31} + K_{ff} T_{34})}{1 - K_{fb} T_{32}}. \tag{9-3}$$

Since the output is $K_{fb} V_3$, the system function is readily calculated.

$$\frac{V_o}{V_i} = \frac{K_{fb}(T_{31} + K_{ff} T_{34})}{1 - K_{fb} T_{32}}. \tag{9-4}$$

Because the transfer functions T_{31}, T_{34}, and T_{32} have the same poles, Eq. (9-4) simplifies to

$$\frac{V_o}{V_i} = \frac{K_{fb}(N_{31} + K_{ff} N_{34})}{D - K_{fb} N_{32}}. \tag{9-5}$$

Three important results are derived from Eq. (9-5).

1. The gain of the feedforward amplifier, K_{ff}, affects the positions of the *zeros* of the transfer function.
2. The gain of the feedback amplifier, K_{fb}, affects the positions of the *poles* of the transfer function.
3. The two effects are independent of each other; that is, K_{ff} appears only in the zero-defining equation, and K_{fb} in the pole-defining (characteristic) equation. (If the output of the circuit does not coincide with the output of the K_{fb} amplifier, then K_{fb} also affects the zeros.)

Consider first the effect of K_{fb}, which is assumed here to be a real number capable of taking any value between $-\infty$ and $+\infty$. How K_{fb} moves the poles of (V_o/V_i) depends upon the zeros of $D(s)$, the zeros of $N_{32}(s)$, and the sign of K_{fb}. Since $D(s)$ is the pole-defining polynomial arising from an *RC* network [see Eq. (9-2)], its zeros are simple and on the negative real axis. Since $N_{32}(s)$ is the numerator polynomial of a voltage ratio belonging to a grounded *RC* network, its zeros may be anywhere in the *s*-plane except on the positive real axis. As K_{fb} is varied from 0 to $\pm\infty$, the loci of the zeros of $(D - K_{fb}N_{32})$ start on the zeros of $D(s)$ and terminate on the zeros of $N_{32}(s)$. Consequently, by selecting D, N_{32}, and K_{fb} appropriately, the poles of (V_o/V_i) can be placed anywhere in the *s*-plane. (Refer to Section 6-2 for the proof of this statement.)

Consider next the effect of K_{ff}, which is also assumed to be a real number capable of taking on any value. How K_{ff} moves the zeros of (V_o/V_i) depends upon the zeros of the polynomials $N_{31}(s)$ and $N_{34}(s)$, and the sign of K_{ff}. Since both of these polynomials are associated with the transfer-function zeros of grounded *RC* networks, their zeros may be anywhere in the *s*-plane except the positive real axis. Therefore, by selecting N_{31}, N_{34}, and K_{ff} appropriately, the zeros of (V_o/V_i) can be placed anywhere in the *s*-plane.

The dependent sources used in Fig. 9-1 are voltage-controlled voltage sources. The voltage is sensed at some port of the *RC* network (without drawing any current from that port), and is fed at some other port as a voltage source. Current-dependent voltage sources can be used just as well to move the poles and zeros. For example, port 3 in Fig. 9-1a can be short-circuited; then the current through the short circuit can be sensed, converted to a voltage source, and fed back to port 2 (see Problem 9-1).

The important conclusion of this discussion is that any system function can be realized with *RC* networks and amplifiers.

EXAMPLE 9-1

The input impedance of the *RC* network shown in Fig. 9-3a has a pole and a zero on the negative real axis. By means of dependent sources, move the pole to infinity and the zero to the origin. Comment on the result.

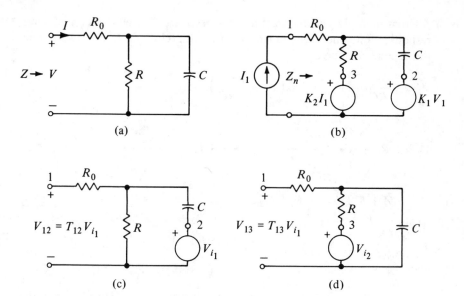

Fig. 9-3

SOLUTION

The input impedance of the passive network is

$$Z = \frac{V}{I} = R_0 \frac{s + \dfrac{R_0 + R}{R_0\,RC}}{s + \dfrac{1}{RC}}.$$ (9-6)

The pole of Z is at $s = -(1/RC)$. To move this pole to infinity requires that $-s$ be added to the denominator polynomial through a dependent source. The zero of Z is at

$$s = -\frac{(R_0 + R)}{R_0\,RC}.$$

To move this zero to the origin requires that $-(R_0 + R)/(R_0\,RC)$ be added to the numerator polynomial through a second dependent source. Thus, write the expression for the new impedance, indicating the desired changes in the numerator and denominator polynomials:

$$Z_n = \frac{V_1}{I_1} = R_0 \frac{\left(s + \dfrac{R_0 + R}{R_0\,RC}\right) - \dfrac{R_0 + R}{R_0\,RC}}{\left(s + \dfrac{1}{RC}\right) - s}.$$ (9-7)

Cross-multiply and then express the response, V_1, as the sum of three responses: one due to the independent input excitation, I_1, and two due to the dependent sources:

$$V_1\left(s + \frac{1}{RC}\right) - V_1 s = I_1 R_0\left(s + \frac{R_0 + R}{R_0 RC}\right) - I_1 \frac{R_0 + R}{RC},$$

$$V_1 = I_1 R_0 \frac{\left(s + \frac{R_0 + R}{R_0 RC}\right)}{s + \frac{1}{RC}} + V_1 \frac{s}{s + \frac{1}{RC}} + I_1 \frac{\left(-\frac{R_0 + R}{RC}\right)}{s + \frac{1}{RC}},$$

$$V_1 = I_1 Z + K_1 V_1 T_{12} + K_2 I_1 T_{13}, \qquad (9\text{-}8)$$

where

$$K_1 T_{12} = \frac{s}{s + \frac{1}{RC}}, \qquad K_2 T_{13} = -\frac{R_0 + R}{RC}\frac{1}{s + \frac{1}{RC}}.$$

As Eq. (9-8) shows, the first dependent source, $K_1 V_1$, depends upon the input voltage. The transfer function from this source to the input terminals, T_{12}, has a zero at the origin, which can be realized by introducing the source in series with the capacitor as shown in Fig. 9-3b. The second dependent source, $K_2 I_1$, depends upon the input current. The transfer function from this source to the input terminals, T_{13}, has a zero at infinity, which can be realized by introducing the source in series with resistor R, as shown in Fig. 9-3b.

By inspection of Fig. 9-3c, obtain

$$T_{12} = \frac{V_{12}}{V_{i_1}} = \frac{R}{R + \frac{1}{sC}} = \frac{s}{s + \frac{1}{RC}}.$$

Hence,

$$K_1 = 1. \qquad Ans.$$

By inspection of Fig. 9-3d, obtain

$$T_{13} = \frac{V_{13}}{V_{i_2}} = \frac{\frac{1}{sC}}{\frac{1}{sC} + R} = \frac{\frac{1}{RC}}{s + \frac{1}{RC}}.$$

Hence,

$$K_2 = -(R_0 + R). \qquad Ans.$$

To get a physical picture of the effects of the dependent sources, solve for Z_n in Fig. 9-3b without applying the constraints to K_1 and K_2:

$$Z_n = \frac{s R_0 RC + (K_2 + R_0 + R)}{s RC(1 - K_1) + 1}.$$

The pole of Z_n is at

$$s = -\frac{1}{RC(1 - K_1)}.$$

As K_1 increases from 0 to 1, the pole moves from $-(1/RC)$ to $-\infty$. Further increase of K_1 puts the pole of Z_n into the right half-plane.

The zero of Z_n is at

$$s = -\frac{K_2 + (R_0 + R)}{R_0 RC}.$$

As K_2 decreases from 0 to $-(R_0 + R)$, the zero moves from $-(R_0 + R)/(R_0 RC)$ to 0. Further decrease in K_2 puts the zero on the positive real axis.

When $K_1 = 1$ and $K_2 = -(R_0 + R)$, Z_n is given by

$$Z_n(s) = sR_0 RC.$$

Thus, the input impedance of this network becomes indistinguishable from the input impedance of a $R_0 RC$ henry inductor; yet the network itself does not contain any inductors. Instead, a capacitor, two resistors, and two dependent sources are used to simulate inductive behavior.

9-2 IMPLEMENTATION OF DEPENDENT SOURCES

A voltage-controlled voltage source or a current-controlled voltage source can be implemented with operational amplifiers. Figure 9–4 shows the equivalent

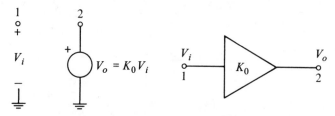

(a) Equivalent circuit   (b) Schematic diagram

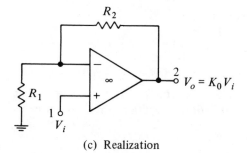

(c) Realization

Fig. 9-4   Amplifier with positive gain

circuit, the schematic diagram, and the realization of an ideal voltage-controlled voltage source *with positive gain*. The gain is greater than unity, and is given by

$$K_0 = 1 + \frac{R_2}{R_1}.$$

Note that the input does not draw any current.

If the operational amplifier has a one-pole rolloff characteristic, i.e., the open-loop gain is given by $-GB/(s + \omega_a)$ [see Eq. (7-11)], then, the closed-loop gain becomes frequency-dependent. The amplifier becomes a low-pass network with a bandwidth of $\omega_{3dB} = GB/K_0$ [see Eq. (7-22)]. To make the low-frequency gain as close to the ideal as possible, K_0 is usually limited to values between 1 and 100.

Figure 9-5 shows the equivalent circuit, the schematic diagram, and the realization of an ideal voltage-controlled voltage source *with negative gain*. The gain, $-K_0$, is given by $-(1 + R_2/R_1)(R_2/R_1)$. Note that the input does not draw any current.

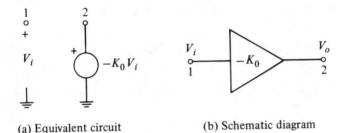

(a) Equivalent circuit   (b) Schematic diagram

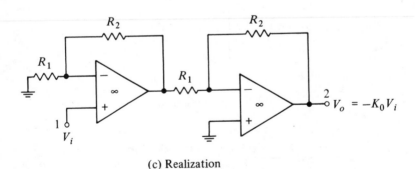

(c) Realization

Fig. 9-5   Amplifier with negative gain (2 stages)

If both operational amplifiers have the same one-pole rolloff characteristic, then for $K_0 > 100$, the overall 3-dB bandwidth is given very nearly by

$$\omega_{3dB} \cong \text{(Bandwidth of 1 stage)} \times \sqrt{\sqrt{2} - 1}$$
$$= 0.644 \times \text{(Bandwidth of 1 stage)}, \tag{9-9}$$

where the bandwidth of one stage is $(GB/\sqrt{K_0})$ rad/s. [See Eqs. (7-22) and (7-36).] $\sqrt{K_0}$ represents the magnitude of the dc gain of each stage. To make the low-frequency gain as close to the ideal as possible, K_0 is usually made no greater than 10,000.

It is also possible to obtain negative-gain amplifiers using one stage if the input-voltage sensing is done across an already existing grounded resistor, as shown in Fig. 9-6a.

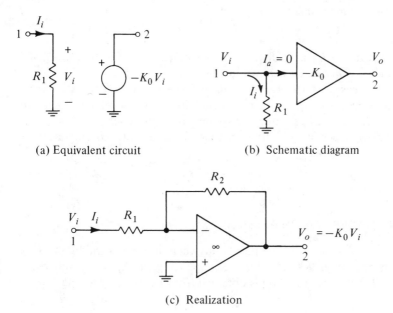

(a) Equivalent circuit   (b) Schematic diagram

(c) Realization

Fig. 9-6   Amplifier with negative gain (1 stage)

As shown in Fig. 9-6c, the resistor R_1 is made part of the amplifier, so that the gain is given by

$$-K_0 = -\frac{R_2}{R_1}.$$

Because the voltage between the $-$ and $+$ terminals of the amplifier is zero, one one end of R_1 is at virtual ground, thus simulating the equivalent circuit of Fig. 7-6a. In both cases the input current is $I_i = V_i/R_1$.

If the one-pole rolloff frequency dependence of the operational amplifier is taken into account, the closed-loop bandwidth of the circuit in Fig. 9-6c is given by $\omega_{3\,dB} = GB/(1 + K_0)$.

If I_i (rather than V_i) is considered as the input variable, the network of Fig. 9-6c can be considered a current-controlled voltage source. In that case,

$$V_o = -K_0 I_i \qquad \text{where} \qquad K_0 = R_2.$$

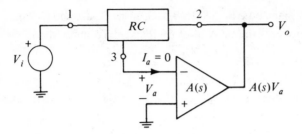

Fig. 9-7   Operation in the infinite-gain mode

Instead of using $\pm K_0$ amplifiers, *the operational amplifier can be directly imbedded in the RC network* as shown in Fig. 9-7.

For the ideal case, $|A(s)| \to \infty$, $V_a \to 0$, and the amplifier causes a virtual ground at port 3. This setup can be considered a special case of Fig. 9-1a with $K_{fb} \to -\infty$ while the $K_{fb} V_a$ product stays finite and equals V_o.

The frequency dependence of the amplifier is taken into consideration by letting $A(s) = -\text{GB}/(s + \omega_a)$.

9-3 SECOND-ORDER *KRC* REALIZATIONS

A pair of complex poles can be placed anywhere in the s-plane by the *KRC* network shown in Fig. 9-8. Ideally, the amplifier gain, K, is a positive real number, K_0. The transfer function is obtained from Eq. (9-4) by letting

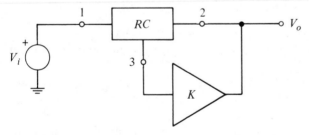

Fig. 9-8   *KRC* realization

$K_{ff} = 0$ and $K_{fb} = K,$

which results in

$$\frac{V_o}{V_i} = K \frac{T_{31}}{1 - KT_{32}}. \qquad (9\text{-}10)$$

$T_{31} = (N_{31}/D_{31})$ represents the transfer-voltage ratio from port 1 to port 3, with port 2 short-circuited. $T_{32} = (N_{32}/D_{32})$ represents the transfer-voltage ratio from port 2 to port 3, with port 1 short-circuited. KT_{32} is the loop gain. In general,

$$D_{31} = D_{32} = D,$$

in which case Eq. (9-10) becomes

$$\frac{V_o}{V_i} = \frac{KN_{31}}{D - KN_{32}}. \tag{9-11}$$

Because the output of the circuit coincides with the output of the amplifier, K does not affect the zeros of (V_o/V_i). The zeros of $D(s)$ are simple and on the negative real axis because $D(s)$ comes from the RC network. The zeros of N_{32} represent the transfer-function zeros from port 2 to port 3, while port 1 is short-circuited. Since KN_{32}/D represents the loop gain, N_{32} zeros correspond also to the loop-gain zeros. These zeros cannot be on the positive real axis because ports 2 and 3 have a common ground terminal. Except for this restriction, N_{32} may have zeros anywhere in the s-plane. The poles of (V_o/V_i) are given by the roots of

$$D(s) - KN_{32}(s) = 0. \tag{9-12}$$

If K is considered real and the system is of second order, Eq. (9-12) can be written as:

$$a_2(s + \alpha_1)(s + \alpha_2) - K(b_2 s^2 + b_1 s + b_0) = 0, \tag{9-13}$$

where $0 \le \alpha_1 < \alpha_2$ and the coefficients a_2 and b_2 are positive. The coefficients b_1 and b_0 are positive if the zeros of N_{32} are in the left half-plane; otherwise b_1, or b_0, or both are zero or negative.

To generate the $-\alpha_1$ and $-\alpha_2$ poles, an RC network with at least two capacitors and two resistors is required. Figure 9-9a shows a passive network with two poles using two capacitors and two resistors. (Depending upon where the input terminals are placed, this network may be viewed as a second-order Foster or Cauer network.)

To move the poles of the RC network off the negative real axis, feedback must be applied to effect the necessary changes in the denominator polynomial of the transfer function. Let

$$as^2 + bs + c = 0$$

represent the denominator polynomial. If complex poles are to result,

$$b^2 - 4ac < 0. \tag{9-14}$$

Hence, feedback must bring about one, or a combination, of the following changes in the denominator polynomial.
1. Reduce the coefficient of the s-term. (Make b smaller.)
2. Increase the coefficient of the s^2-term. (Make a larger.)
3. Increase the constant term. (Make c larger.)

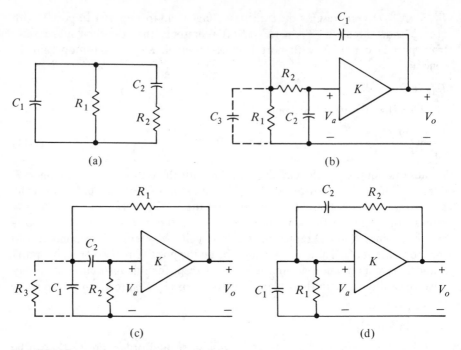

Fig. 9-9 Generating constant-ω_o pole loci

In the case of the positive-gain amplifier, case 1 applies, because K appears with a negative sign [see Eq. (9-13)]. Therefore, let $b_2 = b_0 = 0$, and obtain

$$a_2 (s + \alpha_1)(s + \alpha_2) - Kb_1 s = 0,$$

$$a_2 \left[s^2 + s\left(\alpha_1 + \alpha_2 - \frac{Kb_1}{a_2}\right) + \alpha_1 \alpha_2 \right] = 0. \tag{9-15}$$

Thus, the dependent source must be introduced such that it affects only the coefficient of the s-term. This requires that the zeros of the loop gain be at

$$s = 0 \quad \text{and} \quad s = \infty.$$

(There must be a zero at infinity because the system is second order.) Stated differently, the transfer function of the feedback network must be band-pass, that is, $Hs/[(s + \alpha_1)(s + \alpha_2)]$. Consequently, *in going from the output to the input terminals of the amplifier (via the RC network), one capacitor must appear in a series branch and the other in a shunt branch.* The amplifier can be introduced in the RC network of Fig. 9-9a in one of three ways, to achieve this. In Fig. 9-9b, the amplifier senses the voltage across C_2 and introduces it (after scaling by K) in series with C_1. The capacitor C_1 causes the zero at 0; the capacitor C_2 causes the zero at ∞. In Fig. 9-9c, the amplifier senses the voltage across R_2 and introduces it in series with R_1. The zero at the origin is produced by C_2 and the zero at infinity by C_1. In Fig. 9-9d, the amplifier senses the voltage across the $R_1 C_1$ circuit and injects it in series with the $R_2 C_2$ circuit. The zero at the origin is realized by C_2 and the zero at infinity by C_1.

Adding C_3 in Fig. 9-9b does not change the K dependence of the network. K still varies the coefficient of the s-term only. The attenuated output signal $V_o C_1/(C_1 + C_3)$, drives the $R_1 R_2 C_2$ circuit through the series capacitance of $(C_1 + C_3)$ (think in terms of the Thévenin equivalent). $(C_1 + C_3)$ causes the zero at the origin, and C_2 causes the zero at infinity. Note also that in the passive RC network, C_3 is in parallel with C_1; hence the order of the system is not changed. However, the inclusion of C_3 allows one more branch through which a grounded voltage source can be introduced for excitation. For the same reason, R_3 may be included in Fig. 9-9c. It allows an additional element through which the network may be excited, but it does not affect the K dependence of the poles.

As K is varied, the loci of the complex poles of the three active circuits shown in Fig. 9-9 trace circles centered at the origin because ω_o is independent of K $(\omega_o = \sqrt{\alpha_1 \alpha_2})$. Hence, the loci are constant-ω_o loci. In each case, if K is large enough, the poles can be placed in the right half-plane. The loci of the poles are shown in Fig. 9-10.

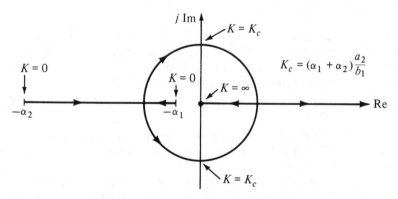

Fig. 9-10   Constant-ω_o control of poles through K

If $R_1 = R_2 = R$ and $C_1 = C_2 = C$, the poles of the three networks (without the C_3 and R_3 modifications) are given by the roots of

$$(sRC)^2 + (sRC)3 + 1 - K(sRC) = 0,$$
$$(sRC)^2 + (sRC)(3 - K) + 1 = 0. \tag{9-16}$$

Note that when K is adjusted to the critical value of $K_c = 3$, the poles are on the imaginary axis at $s = \pm j(1/RC)$. For $1 < K < 5$, the poles are complex and lie on a circle of radius $\omega_o = (1/RC)$. Therefore, by properly selecting the RC product and the value of K, the complex poles can be positioned anywhere in the s-plane.

9-4 SECOND-ORDER $-KRC$ REALIZATIONS

A pair of complex poles can be placed anywhere in the s-plane by the $-KRC$ network shown in Fig. 9-11. Ideally, the amplifier gain, $-K$, is a real negative number, $-K_0$.

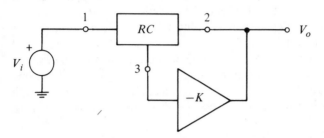

Fig. 9-11 $-KRC$ realization

The voltage ratio is obtained from Eq. (9-4) by letting $K_{ff} = 0$ and $K_{fb} = -K$:

$$\frac{V_o}{V_i} = -K \frac{T_{31}}{1 + KT_{32}}. \tag{9-17}$$

The transfer functions T_{31} and T_{32} are identical with those presented in Section 9-3. Because the output of the circuit coincides with the output of the amplifier, the amplifier gain, $-K$, does not affect the zeros of (V_o/V_i).

The poles of (V_o/V_i) are given by the zeros of

$$a_2(s + \alpha_1)(s + \alpha_2) + K(b_2 s^2 + b_1 s + b_0) = 0, \tag{9-18}$$

where $0 \le \alpha_1 < \alpha_2$ and the coefficients a_2 and b_2 are positive. The simplest two-pole RC network to which feedback can be applied to produce complex poles is shown in Fig. 9-12a.

The poles of the RC network can be moved off the real axis if the coefficient of the s^2-term or the constant term in Eq. (9-18) is made larger through feedback. Alternatively, both can be made larger, which, in effect, is equivalent to reducing the coefficient of the s-term.

Consider first the effect of increasing the coefficient of the s^2-term only:

$$a_2(s + \alpha_1)(s + \alpha_2) + Kb_2 s^2 = 0. \tag{9-19}$$

To produce the desired K dependence, the two transfer-function zeros of the feedback network (from the output to the input of the amplifier via the RC network) must be at the origin. Stated differently, the transfer function of the feedback network must be high-pass, that is, $Hs^2/(s + \alpha_1)(s + \alpha_2)$. This means that *the two capacitors, C_1 and C_2, must appear in the series branches of the feedback path*, as shown in Fig. 9-12b. As a function of K, the complex poles

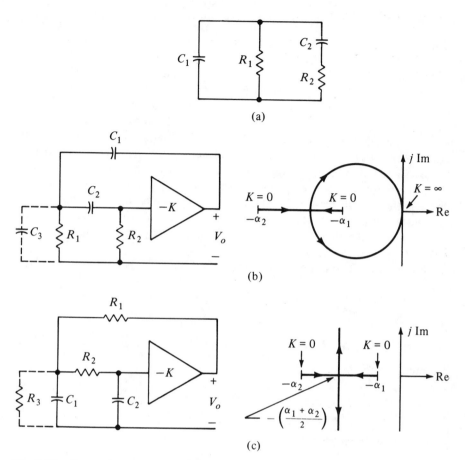

Fig. 9-12   Generating complex poles

describe a circle tangent to the imaginary axis, as shown. The center of the circle is at

$$s = -\frac{\alpha_1 \alpha_2}{\alpha_1 + \alpha_2},$$

and its radius is

$$r = \frac{\alpha_1 \alpha_2}{\alpha_1 + \alpha_2} \quad \text{(see Eq. 6-17)}.$$

Thus, the network of Fig. 9-12b can generate any left half-plane pole.

The inclusion of C_3 (shown dotted) does not change the circular nature (tangent to the imaginary axis) of the pole loci. However, C_3 provides another branch through which the network may be excited with a grounded voltage source.

Consider next the effect of increasing only the constant term. Equation (9-18) then reduces to

$$a_2(s + \alpha_1)(s + \alpha_2) + Kb_0 = 0. \tag{9-20}$$

To produce the desired K dependence given in Eq. (9-20), the two transfer-function zeros of the RC network (from the output to the input of the amplifier) must be at infinity. Stated differently, the transfer function of the feedback network must be low-pass, that is, $H/(s + \alpha_1)(s + \alpha_2)$. This means that *both capacitors must appear as shunt elements in the feedback path*, as shown in Fig. 9-12c. As a function of K, the complex poles describe a constant-α line as shown. The real part of the complex poles is $-(\alpha_1 + \alpha_2)/2$, and therefore the poles are always in the left half-plane. The addition of R_3 (shown dashed) does not change the constant-α nature of the pole loci; however, R_3 provides another branch through which the input may be introduced. (C_2 may also be shunted by a resistor.)

It is also possible to put poles in the right half-plane if T_{32} has right half-plane zeros.

EXAMPLE 9-2

Show that the feedback arrangement shown in Fig. 9-13 can produce complex poles.

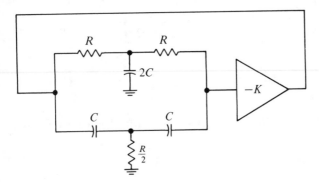

Fig. 9-13

SOLUTION

The characteristic equation is

$$1 - LG = 0,$$

where LG is the loop gain.

To obtain the LG, break the feedback loop and apply an external signal, as shown in Fig. 9-14a. Then, apply magnitude scaling (by R) and frequency

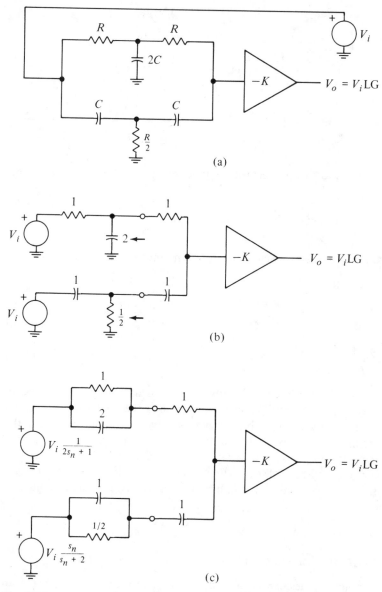

Fig. 9-14

scaling ($sRC = s_n$). To simplify further, apply separate excitations to the two T-networks and redraw the network as in Fig. 9-14b.

Next, obtain the Thévenin equivalent of the two circuits to the left of the arrows shown in Fig. 9-14b, and draw Fig. 9 14c. Then, use the principle of superposition and the voltage-divider rule, to obtain V_o:

$$V_o = -K\left[\left(V_i\frac{1}{2s_n+1}\right)\frac{\left(\dfrac{1}{s_n}+\dfrac{(1/s_n)\times\frac{1}{2}}{(1/s_n)+\frac{1}{2}}\right)}{\left(\dfrac{1}{s_n}+\dfrac{(1/s_n)\times\frac{1}{2}}{(1/s_n)+\frac{1}{2}}\right)+1+\left(\dfrac{(1/s_n2)\times1}{(1/s_n2)+1}\right)}\right.$$

$$\left.+\left(V_i\frac{s_n}{s_n+2}\right)\frac{\left(1+\dfrac{(1/s_n2)\times1}{(1/s_n2)+1}\right)}{\left(1+\dfrac{(1/s_n2)\times1}{(1/s_n2)+1}\right)+\dfrac{1}{s_n}+\left(\dfrac{(1/s_n)\times\frac{1}{2}}{(1/s_n)+\frac{1}{2}}\right)}\right].$$

Solve for the loop gain. Because of pole–zero cancellation, the LG simplifies to

$$LG = \frac{V_o}{V_i} = -K\frac{s_n^2+1}{s_n^2+4s_n+1}.$$

Substitute LG in the characteristic equation, and obtain

$$1+K\frac{s_n^2+1}{s_n^2+4s_n+1}=0,$$

$$s_n^2(1+K)+4s_n+(1+K)=0, \tag{9-21a}$$

$$s_n^2+\frac{4}{1+K}s_n+1=0. \tag{9-21b}$$

K increases the coefficients of both the s_n^2-term and the s_n^0-term, as shown by Eq. (9-21a). Viewed differently, K decreases the coefficient of the s_n-term, as shown by Eq. (9-21b). No matter which viewpoint is taken, the poles become complex when $K > 1$. Because K does not affect the ω_o of the poles, the complex poles move on a circle of radius 1 in the s_n-plane. [In the s-plane the radius is $(1/RC)$.] For $K = \infty$, the poles are on the imaginary axis. Thus, any value of left half-plane pole can be realized by adjusting RC and K.

9-5 SECOND-ORDER REALIZATIONS WITH INFINITE GAIN

A pair of complex poles can be placed anywhere in the s-plane by using an amplifier with negative and infinite gain (ideal operational amplifier). Figure 9-15 shows the connection diagram.

To obtain V_o/V_i, let $K \to \infty$ in Eq. (9-17):

$$\frac{V_o}{V_i} = -\frac{T_{31}}{T_{32}}. \tag{9-22}$$

As before, T_{31} represents the transfer function (voltage ratio) from port 1 to port 3, with port 3 open-circuited and port 2 short-circuited. T_{32} represents the transfer function (voltage ratio) from port 2 to port 3, with port 3 open-circuited

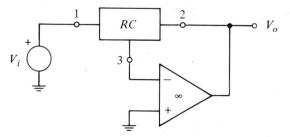

Fig. 9-15 Realization of complex poles with infinite gain

and port 1 short-circuited. The zeros of T_{31} or T_{32}, being transfer-function zeros of grounded RC networks, may be anywhere in the s-plane except the positive real axis.

Assuming that T_{31} and T_{32} have the same poles, V_o/V_i can be written as

$$\frac{V_o}{V_i} = -\frac{T_{31}}{T_{32}} = -\frac{\dfrac{N_{31}}{D}}{\dfrac{N_{32}}{D}} = -\frac{N_{31}}{N_{32}}.$$ (9-23)

The poles of V_o/V_i are, therefore, the zeros of N_{32}.

There are two grounded RC networks that produce, with two resistors and two capacitors, complex transfer-function zeros (zeros of N_{32}) in the left half-plane. These networks are discussed in Section 5-7 and are reproduced in Fig. 9-16. The (V_o/V_i) zeros of the network in Fig. 9-16a are complex if

$$\frac{C_1}{C_2} > \frac{1}{2} + \frac{1}{4}\left(\frac{R_1}{R_2} + \frac{R_2}{R_1}\right).$$ (9-24)

The (V_o/V_i) zeros of the network in Fig. 9-16b are complex if

$$\frac{R_2}{R_1} > \frac{1}{2} + \frac{1}{4}\left(\frac{C_1}{C_2} + \frac{C_2}{C_1}\right).$$ (9-25)

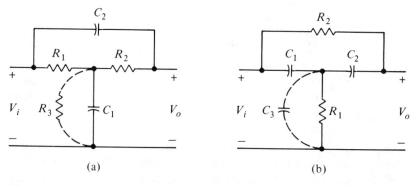

(a)   (b)

Fig. 9-16 Networks with complex transfer-function zeros

Two other networks with complex transfer-function zeros can readily be obtained by modifying the networks presented in Fig. 9-16. In Fig. 9-16a, R_3 (shown dashed) is added across C_1. For this case, complex transfer-function zeros result if

$$\sqrt{\frac{C_1}{C_2}} > \frac{1}{2}\left(\sqrt{\frac{R_1}{R_2}} + \sqrt{\frac{R_2}{R_1}} + \frac{\sqrt{R_1 R_2}}{R_3}\right). \tag{9-26}$$

In Fig. 9-16b, C_3 (shown dashed) is added across R_1. For this case, complex transfer-function zeros result if

$$\sqrt{\frac{R_2}{R_1}} > \frac{1}{2}\left(\sqrt{\frac{C_1}{C_2}} + \sqrt{\frac{C_2}{C_1}} + \frac{C_3}{\sqrt{C_1 C_2}}\right). \tag{9-27}$$

To obtain complex poles, these RC networks are connected between the output and the input of the operational amplifier, as shown in Fig. 9-17.

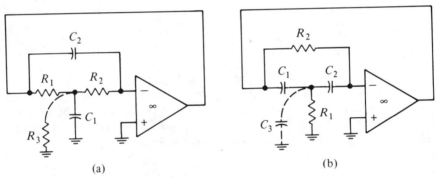

(a)   (b)

Fig. 9-17   Complex poles with operational amplifiers

In Fig. 9-17, a voltage excitation can be applied by breaking any of the ground leads and connecting the source between the broken lead and ground. Thus, in Fig. 9-17a, a voltage source can be inserted in series with R_3, C_1, or the $R_3 C_1$ parallel combination.

9-6 CONTROL OF ZEROS THROUGH FEEDFORWARD

It is shown in Section 9-1 that feeding the input signal forward through an amplifier alters the positions of the transfer-function zeros. The general case is illustrated in Fig. 9-18, where either a positive- or a negative-gain amplifier is employed to provide an additional input to the network.

The output can be obtained as the sum of two signals, one coming from port 1 while port 4 is held at zero, and the other coming from port 4 while port 1 is held at zero (this approach is somewhat different from the one used in Section 9-1).

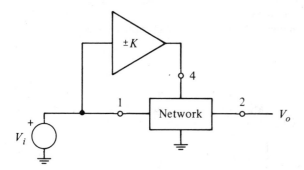

Fig. 9-18 Feedforward through $\pm K$ amplifier

$$V_o = V_i T_{21} \pm KV_i T_{24}.$$

The resulting transfer function is

$$\frac{V_o}{V_i} = T_{21} \pm KT_{24}. \tag{9-28}$$

In general, T_{21} and T_{24} have the same denominator polynomial. Therefore, Eq. (9-28) can be written as

$$\frac{V_o}{V_i} = \frac{N_{21} \pm KN_{24}}{D}. \tag{9-29}$$

Thus, K affects only the zeros of the transfer function. The precise nature of control exercised by the amplifier depends upon the gain of the amplifier and the positions of the two transfer-function zeros. In particular, if the network is of second order, the zero-defining equation can be written as

$$a_2 s^2 + a_1 s + a_0 \pm K(b_2 s^2 + b_1 s + b_0) = 0. \tag{9-30}$$

A special case of Eq. (9-30) is

$$a_2 s^2 + a_1 s + a_0 \pm Kb_1 s = 0, \tag{9-31}$$

in which case the real part of the complex zeros can be controlled by adjusting the gain of the amplifier. The resulting loci are constant-ω_o loci.

Another special case of Eq. (9-30) is

$$a_2 s^2 + a_1 s + a_0 \pm Kb_0 = 0, \tag{9-32}$$

in which case the imaginary part of the complex zeros can be controlled by adjusting the gain of the amplifier. The resulting loci are constant-α loci.

EXAMPLE 9-3

Construct a network with
(a) a pair of complex zeros which can be adjusted in a constant-ω_o manner;
(b) a pair of complex zeros which can be adjusted in a constant-α manner;

(c) a pair of variable imaginary-axis zeros;

(d) a pair of variable imaginary-axis zeros and complex poles.

SOLUTION

(a) Pick a second-order RC network and locate the input and output terminals such that a quadratic numerator results. An example of such a network is shown in Fig. 9-19a, where

$$\frac{V_{o1}}{V_i} = \frac{(sRC)^2 + 2(sRC) + 1}{(sRC)^2 + 3(sRC) + 1}. \tag{9-33}$$

To move the zeros of (V_o/V_i) in a constant-ω_o manner, use feedforward to reduce the coefficient of the s-term in the numerator polynomial. This requires that the amplifier gain be negative and the transfer function from the output of the feedforward amplifier to the output of the circuit (T_{24} of Fig. 9-18) be band-pass, that is, $Hs/(s + \alpha_1)(s + \alpha_2)$. Hence, the amplifier must be connected in such a way that one capacitor

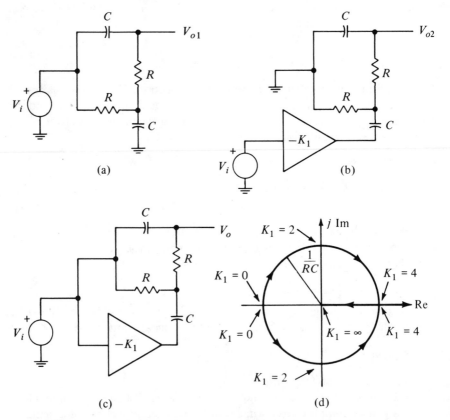

(a)   (b)   (c)   (d)

Fig. 9-19

causes a zero at the origin and the other at infinity. The result is Fig. 9-19b, where

$$\frac{V_{o2}}{V_i} = \frac{-K_1(sRC)}{(sRC)^2 + 3(sRC) + 1}.$$

To get the desired output, take the two responses together, as in Fig. 9-19c, and obtain

$$\frac{V_o}{V_i} = \frac{V_{o1} + V_{o2}}{V_i} = \frac{(sRC)^2 + 2(sRC) + 1 - K_1(sRC)}{(sRC)^2 + 3(sRC) + 1}$$

$$= \frac{(sRC)^2 + (sRC)(2 - K_1) + 1}{(sRC)^2 + 3(sRC) + 1}. \tag{9-34}$$

The zero-defining equation is

$$s^2 + \frac{s}{RC}(2 - K_1) + \frac{1}{R^2C^2} = 0.$$

When $K_1 = 0$, both zeros are on the real axis [at $s = -(1/RC)$]. When $0 < K_1 < 2$, the zeros are complex and in the left half-plane. When $K_1 = 2$, the zeros are on the imaginary axis at $s = \pm j(1/RC)$. When $2 < K_1 < 4$, the zeros are complex and in the right half-plane. When $4 < K_1$, the zeros are on the positive real axis. Figure 9-19d shows the zero loci as a function of K_1.

(b) Use the passive circuit of Fig. 9-20a as the basic circuit to obtain

$$\frac{V_{o1}}{V_i} = \frac{sRC(sRC + 2)}{(sRC)^2 + 3(sRC) + 1}. \tag{9-35}$$

To move the zeros of (V_o/V_i) in a constant-α manner, use feedforward to increase the value of the constant term in the numerator polynomial. This requires that the amplifier gain be positive and the transfer function from the output of the feedforward amplifier to the output of the circuit be low-pass, that is, $H/[(s + \alpha_1)(s + \alpha_2)]$. Hence, introduce the amplifier so that both capacitors produce zeros at infinity. The result is Fig. 9-20b, where

$$\frac{V_{o2}}{V_i} = \frac{K_2}{(sRC)^2 + 3(sRC) + 1}.$$

To get the desired output, put together the two responses as in Fig. 9-20c and obtain

$$\frac{V_o}{V_i} = \frac{V_{o1} + V_{o2}}{V_i} = \frac{(sRC)^2 + 2(sRC) + K_2}{(sRC)^2 + 3(sRC) + 1}. \tag{9-36}$$

The zero-defining equation is

$$s^2 + 2\frac{s}{RC} + \frac{K_2}{R^2C^2} = 0.$$

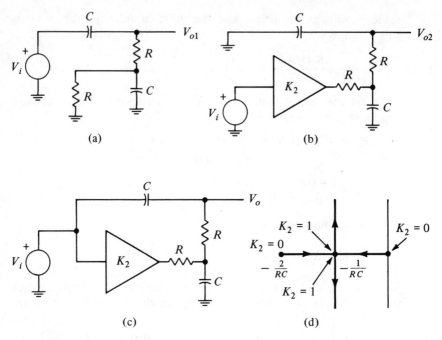

Fig. 9-20

When $K_2 > 1$, the zeros become complex and move vertically along the
$s = -(1/RC)$ line, as shown in Fig. 9-20d.

(c) Use the scheme presented in (a) and adjust K_1 to put the zeros on the
imaginary axis. Then, use the scheme presented in (b) and adjust K_2 to
move the zeros along the imaginary axis. The complete network is
shown in Fig. 9-21.

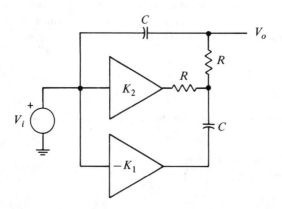

Fig. 9-21

The transfer function is given by

$$\frac{V_o}{V_i} = \frac{(sRC)^2 + (sRC)(2 - K_1) + K_2}{(sRC)^2 + 3(sRC) + 1}. \tag{9-37}$$

When $K_1 = 0$, Eq. (9-37) reduces to Eq. (9-36), as it must since the network of Fig. 9-21 becomes identical with the network of Fig. 9-20c. When $K_2 = 1$, Eq. (9-37) reduces to Eq. (9-34), as it must since the network of Fig. 9-21 reduces to the network of Fig. 9-19c (as far as transfer functions are concerned).

To get the desired control of the zeros, let $K_1 = 2$ in Eq. (9-37). The zeros are then on the imaginary axis at

$$s = \pm j\left(\frac{\sqrt{K_2}}{RC}\right).$$

Their position can be varied by changing K_2.

(d) The poles of the network given in Fig. 9-21 are on the negative real axis because the network is RC and the amplifiers are used only in the feedforward configuration. To make the poles complex, use feedback to reduce the coefficient of the s-term in the denominator polynomial. This requires a positive-gain amplifier and a band-pass feedback function. Since the network of Fig. 9-21 does not have any grounded passive elements left, replace the lower C by two $C/2$'s in parallel, and inject the feedforward signal through one $C/2$ and the feedback signal through the other. The result is the network of Fig. 9-22.

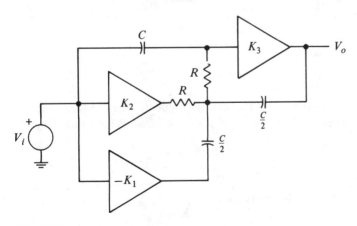

Fig. 9-22

The transfer function is given by

$$\frac{V_o}{V_i} = K_3 \frac{(sRC)^2 + (sRC)(2 - \frac{1}{2}K_1) + K_2}{(sRC)^2 + (sRC)(3 - \frac{1}{2}K_3) + 1}.$$

The factor $\frac{1}{2}$ appears before K_1 and K_3 because of the 2:1 attenuation introduced as a result of splitting the lower capacitor. The overall tuning procedure is as follows.

Adjust $K_1 = 4$ to put the zeros on the imaginary axis.

Adjust K_2 to move the zeros along the imaginary axis.

Adjust $2 < K_3 < 6$ to obtain complex left half-plane poles.

The resulting function is of the form

$$H\,\frac{s^2 + \omega_z^2}{s^2 + \dfrac{s\omega_o}{Q} + \omega_o^2},$$

and therefore can be used as a band-stop function. As shown in Chapter 13, there are other methods (using fewer amplifiers) of obtaining band-stop networks. However, this example emphasizes the usefulness of proper feedforward and feedback techniques for controlling the poles and zeros of a system function in a noninteractive manner.

EXAMPLE 9-4

In Fig. 9-23, the step generator has an internal resistance of R ohms and

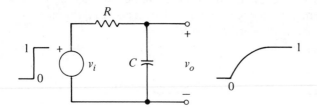

Fig. 9-23

its value cannot be altered. Neither can the value of C be changed. The rise time of the output waveform is found to be too large. What can be done to improve it?

SOLUTION

Obtain first the unit-step response:

$$v_o(t) = 1 - e^{-t/RC}.$$

Note that $-(1/RC)$ represents the pole of the system. Since the rise time is given by $t_r = 2.2RC$, express the pole position in terms of the rise time:

$$s = -\frac{1}{RC} = -\frac{2.2}{t_r}.$$

Thus, if the rise time is to be improved, i.e., made smaller, the system pole should be moved farther away from the origin. Three methods may be used to achieve this. (See also Problem 9-19.)

Method 1   Use a compensating network to cancel the system pole. Such a network is shown in Fig. 9-24.

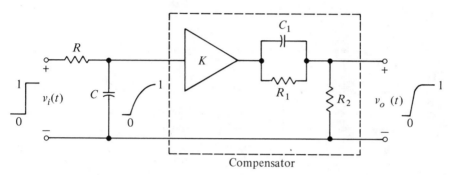

Fig. 9-24

The system is described by

$$\frac{V_o}{V_i} = \left(\frac{1}{RC} \frac{1}{s + \frac{1}{RC}}\right) K \left[\frac{s + \frac{1}{R_1 C_1}}{s + \left(\frac{R_1 + R_2}{R_1 R_2}\right)\frac{1}{C_1}} \right].$$

To cancel the old system pole, let $R_1 C_1 = RC$. Then, the new system is described by

$$\frac{V_o}{V_i} = \frac{K}{RC} \frac{1}{s + \frac{1 + (R_1/R_2)}{RC}}.$$

Adjust K to $1 + (R_1/R_2)$ to make up for the attenuation caused by the $R_1 R_2$ network:

$$\frac{V_o}{V_i} = \frac{1 + \left(\frac{R_1}{R_2}\right)}{RC} \frac{1}{s + \frac{1 + (R_1/R_2)}{RC}}.$$

The new system pole is at

$$s = -\frac{1}{RC}\left(1 + \frac{R_1}{R_2}\right),$$

which is $[1 + (R_1/R_2)]$ times farther out than the old system pole. Thus, the rise time is improved by a ratio of $[1 + (R_1/R_2)]$ to 1.

Method 2 Load the network with a resistor to move its pole away from the origin, and then amplify to make up for the attenuation caused by the loading. This scheme is shown in Fig. 9-25.

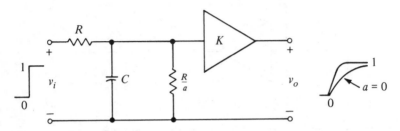

Fig. 9-25

The system is described by

$$\frac{V_o}{V_i} = \frac{K}{RC} \frac{1}{s + \dfrac{a+1}{RC}}.$$

Adjust K to $(a + 1)$ to produce a dc gain of 1:

$$\frac{V_o}{V_i} = \frac{a+1}{RC} \frac{1}{s + \dfrac{a+1}{RC}}.$$

The new system pole is at

$$s = -\frac{1}{RC}(a + 1),$$

which is $(a + 1)$ times farther out than the old system pole. Thus, the rise time is improved by the ratio of $(a + 1)$ to 1. Note that, in this instance, a resistor is used to control the pole position. This method of attenuation by loading, followed by amplification, is widely used in pulsed systems to improve rise time.

Method 3 Reduce the effective capacitance by shunting a negative capacitance across C. This is achieved by the circuit shown in Fig. 9-26.

To see how the negative capacitance is generated, calculate the input impedance of the network to the right of C. The amplifier senses the voltage $V_a(s)$, multiplies it by K, and introduces the result as a dependent voltage

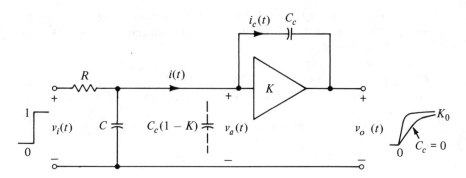

Fig. 9-26

source which drives the right-hand end of the capacitor C_c. Consequently, the voltage across C_c and the current through C_c are

$$V_c = V_a - KV_a,$$
$$I_c = V_c sC_c = V_a(1 - K)sC_c.$$

The input impedance of the network to the right of C is

$$Z = \frac{V_a}{I} = \frac{V_a}{I_c} = \frac{1}{sC_c(1 - K)},$$

which is equivalent to the impedance presented by a capacitance of value $C_c(1 - K)$. (See the dashed equivalent presentation shown in Fig. 9-26.) Thus, for $K > 1$, negative capacitance is generated. This negative capacitance, since it is in parallel with the undesired capacitance, reduces the total shunt capacitance, thereby increasing the bandwidth of operation.

Using Fig. 9-26 as a guide, calculate the output:

$$V_o = KV_a = K \frac{\dfrac{1}{s[C + C_c(1 - K)]}}{R + \dfrac{1}{s[C + C_c(1 - K)]}}$$

$$= \frac{K}{R[C + C_c(1 - K)]} \frac{1}{s + \dfrac{1}{R[C + C_c(1 - K)]}}.$$

The dc gain is K. The new system pole is at

$$s = -\frac{1}{RC\left[1 + \dfrac{C_c}{C}(1 - K)\right]},$$

which is $1/[1 + (C_c/C)(1 - K)]$ times farther out than the old system pole. Ideally, the pole can be placed at infinity by adjusting C_c/C to $1/(K - 1)$, and thus infinite bandwidth can be obtained.

In practice, the amplifier pole, which has been neglected in this analysis, prevents perfect compensation. The system is then of second order. When $C_c = 0$, the poles are at $-1/RC$ (input circuit pole) and $-GB/K_0$ (amplifier pole). As C_c is increased from zero, the poles come together on the negative real axis and, upon further increase, become complex. As long as

$$\frac{1}{RC} \leq 0.1 \frac{GB}{K_0},$$

the optimum bandwidth is achieved near

$$\frac{C_c}{C} = \frac{1}{K - 1}.$$

If (C_c/C) is adjusted to a value greater than this optimum value, the frequency response peaks (the step-response rings). Upon further increase, the circuit oscillates.

9-7 SUMMARY

Some networks are designed to realize specific system functions. Other networks are designed to modify the characteristics of existing systems. In either case, the aim of the circuit designer is to produce the prescribed pole–zero pattern. When the desired poles are complex, *RLC* networks or *RC*-active networks must be used. For low-frequency work, *RC*-active networks are superior to *RLC* networks.

To obtain complex poles, the poles of *RC* networks must be lifted off the real axis by feedback schemes involving $\pm K$ and infinite-gain amplifiers. These amplifiers are readily implemented with operational amplifiers. In the most commonly used second-order *KRC* realization, the amplifier gain is used to reduce the coefficient of the s-term in the denominator polynomial. As K is increased in value, constant-ω_o pole loci result. In the most commonly used second-order $-KRC$ realization, the amplifier gain is used to increase the value of the constant term in the denominator polynomial. As K is increased in value, constant-α pole loci result. In the second-order infinite-gain realizations, parallel *RC* networks are used to generate the complex poles.

Although complex zeros can be produced with *RC* networks alone, the adjustment procedure becomes simpler if feedforward amplifiers are used. With a positive-gain amplifier, the value of the constant term in the numerator polynomial can be readily increased, thus producing complex zeros along a constant-α line. With a negative-gain amplifier, the coefficient of the s-term in the numerator polynomial can be decreased, thus producing complex zeros along a constant-ω_o circle.

Thus, through the use of feedback and feedforward techniques, the poles and zeros of system functions can be readily moved about. A circuit designer who understands these techniques well has great versatility in generating new circuits and controlling their properties.

PROBLEMS

9-1   In Fig. 9-27, obtain the expression for V_o/V_i.

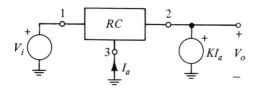

Fig. 9-27

9-2   The two RC networks shown in Fig. 9-28 are identical. Compare the poles

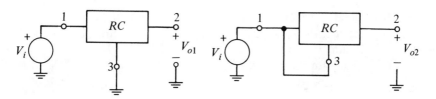

Fig. 9-28

and zeros of (V_{o1}/V_i) with the poles and zeros of (V_{o2}/V_i).

9-3   Derive Eq. (9-9).

9-4   (a) Obtain the input impedance of each circuit shown in Fig. 9-29.

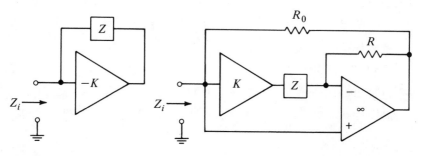

Fig. 9-29

(b) In each case, let $Z = (1/sC)$. Obtain an equivalent passive network that has the same input impedance. What is the effect of K?

9-5 Sketch, as a function of K, the loci of the poles for the network shown in Fig. 9-30.

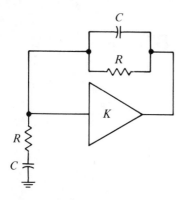

Fig. 9-30

9-6 Show that constant-α root loci result in Eq. (9-13) if $b_1/b_2 = \alpha_1 + \alpha_2$. Assume that the roots are complex and K is the variable.

9-7 (a) Obtain (V_o/V_i) for the network shown in Fig. 9-31.
     (b) What are the ω_o and Q of the poles?

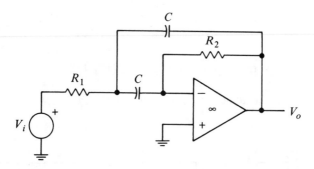

Fig. 9-31

9-8 For the network shown in Fig. 9-32, sketch the pole loci as a function of $(K_1 K_2)$. K_1 and K_2 are positive real numbers. What would change if R_3 is added as shown?

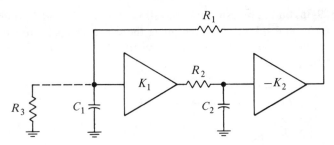

Fig. 9-32

9-9 For the network shown in Fig. 9-33, sketch the pole loci as a function of (K_1K_2). K_1 and K_2 are positive real numbers. What is the effect of adding C_3?

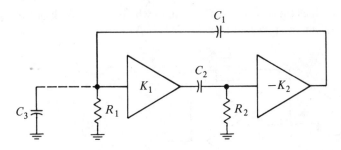

Fig. 9-33

9-10 Derive Eq. (9-26).

9-11 (a) Show that the dependent source shown in Fig. 9-34a affects both the poles and zeros of (V_o/V_i).

(b) For the network shown in Fig. 9-34b, calculate (V_o/V_i). How does K affect the poles and the zeros?

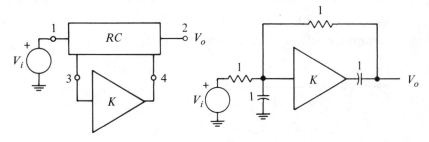

Fig. 9-34

9-12 (a) In Fig. 9-35, find I_a by using superposition from the V_i and V_o sources. Use appropriate subscripts to designate the various transfer functions.

(b) From (a) find (V_o/V_i). How does this result differ from the expression given by Eq. (9-22)?

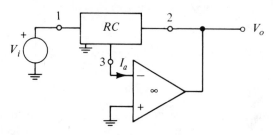

Fig. 9-35

9-13 In what way does the impedance Z, through which the network of Fig. 9-36 is excited, affect the poles and zeros of (V_o/V_i)?

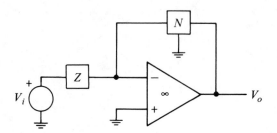

Fig. 9-36

9-14 Obtain a realization for

$$\frac{V_o}{V_i} = H\frac{s^2 - 3s + 1}{s^2 + 3s + 1}.$$

9-15 Find the input impedance of the network shown in Fig. 9-37. Assume

$$K_2 = -(R_0 + R) \quad \text{and} \quad K_1 = 1.$$

Comment on the significance of the result.

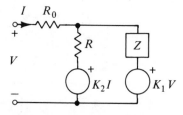

Fig. 9-37

9-16 (a) Obtain V_o/V_i for the block diagram shown in Fig. 9-38.

(b) Implement the block diagram with operational amplifiers. The transfer functions G and H represent voltage ratios and are defined on the basis of zero-source and infinite-load impedance.

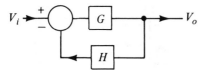

Fig. 9-38

9-17 The network shown in Fig. 9-39 is to be used to realize:

$$\frac{V_o}{V_i} = H \frac{s^2 + \omega_o^2}{s^2 + \dfrac{s\omega_o}{Q} + \omega_o^2}.$$

(a) Explain qualitatively the effect of $-K_1$ and K_2 amplifiers.

(b) Obtain the amplifier gains and the element values as a function of ω_o and Q.

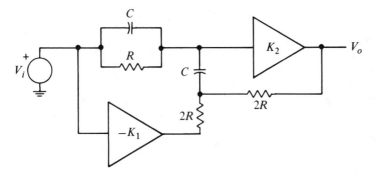

Fig. 9-39

9-18 Show that, in Fig. 9-40, K moves the zeros along the imaginary axis.

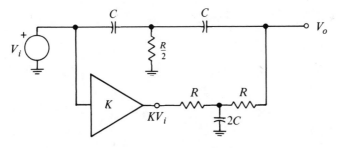

Fig. 9-40

9-19 If the amplifiers are considered ideal, the bandwidth of the system shown in in Fig. 9-41 can be made infinite. Show this.

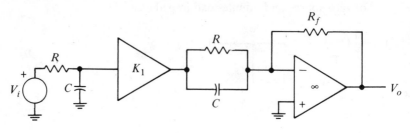

Fig. 9-41

10

Second-Order
Low-Pass Networks

In this chapter, several *RLC* and active *RC* realizations are given for the second-order low-pass function. The *RLC* networks use very few components, are simple to construct, and have excellent sensitivity characteristics. However, if the pass-band is narrow, i.e., covers only very low frequencies, the *RLC* realizations do become impractical because of the size requirements of the inductors and the capacitors. Furthermore, large inductors have characteristics that depart considerably from the characteristics of ideal inductors. This is where the use of active *RC* realizations becomes advantageous. They can be made small, light, and readily adjustable. However, if the passband is wide, i.e., extends into the high frequencies, the finite gain-bandwidth product of amplifiers does limit the usefulness of these active *RC* realizations. Thus, where the *RLC* realizations become impractical, the active *RC* realizations can be used.

In order to provide a basis for choosing between the various realizations, the ω_o-sensitivity and the Q-sensitivity functions of the active *RC* realizations are compared with the sensitivity functions of the passive realizations. In this way, a quantitative measure for the performance of the network is obtained. Techniques are presented for determining the pole, ω_o, and Q shifts caused by the finite gain-bandwidth product of the amplifiers. First-order correction terms are derived from which frequency limitations imposed by nonideal amplifiers are determined.

10-1 THE SECOND-ORDER LOW-PASS FUNCTION

An important second-order system function, which also serves as a building block for higher-order systems, is the transfer function with two zeros at infinity.

319

$$T(s) = \frac{H}{s^2 + s\dfrac{\omega_o}{Q} + \omega_o^2}.$$

Since H is merely a magnitude scale factor, let $H = \omega_o^2$.

$$T(s) = \frac{\omega_o^2}{s^2 + s\dfrac{\omega_o}{Q} + \omega_o^2} = \frac{1}{\left(\dfrac{s}{\omega_o}\right)^2 + \left(\dfrac{s}{\omega_o}\right)\dfrac{1}{Q} + 1}. \tag{10-1}$$

As Eq. (10-1) shows, the properties of $T(s)$ are dependent upon the constants ω_o and Q. The constant ω_o acts merely as a frequency scale factor. When s/ω_o rather than s is considered as the frequency variable, $T(s/\omega_o)$ becomes dependent only on the constant Q and can therefore be readily studied. In Fig. 10-1 the magnitude and phase of $T(s/\omega_o)$ are plotted against ω/ω_o.

Several important observations can be made from a study of the magnitude and phase curves and Eq. (10-1):

1. For $\omega/\omega_o \ll 1$, $|T(j\omega)| \cong 1$. Therefore, low frequencies are passed.

2. For $\omega/\omega_o \gg 1$, $|T(j\omega)| \cong (\omega_o/\omega)^2$. Therefore, high frequencies are attenuated.

3. For $Q > 1/\sqrt{2}$, the magnitude peaks at

$$\frac{\omega}{\omega_o} = \sqrt{1 - \frac{1}{2Q^2}} \quad \text{(see Problem 10-1).} \tag{10-2}$$

The peak occurs at a frequency lower than ω_o. For $Q > 5$, the frequency of peaking practically equals ω_o (within 1%).

4. For $Q > 5$, the 3-dB bandwidth is practically equal to ω_o/Q rad/s.

5. At $\omega/\omega_o = 1$, $|T(j\omega_o)| = Q$ and the phase is $-\pi/2$.

6. At $\omega/\omega_o \gg 1$, the phase is $-\pi$.

7. For $Q > 5$, the phase undergoes a rapid shift of π radians about ω_o.

As the magnitude curves indicate, $T(s)$, given by Eq. (10-1), *can be used as a low-pass function. For high Q values, it can also be used as a band-pass function.* Nonetheless, a second-order function with two zeros at infinity is usually labelled as a low-pass function. (In a high-order low-pass function, it is not uncommon to encounter some second-order factors with high Q's. However, the overall magnitude response does not peak appreciably because of the attenuation caused by other second-order factors.)

The design specifications for a second-order filter are usually given in terms of the allowable peaking in the pass band and the desired cutoff frequency, i.e., the frequency at which the magnitude characteristic is down a given number of dB's from the dc value. For example, if 10% peaking is acceptable and the 3-dB down frequency is to be 10^3 rad/s, then Fig. 10-1 shows that $Q \cong 1$ and $\omega/\omega_o = 1.3$. Hence, $\omega_o \cong 10^3/1.3 = 770$ rad/s.

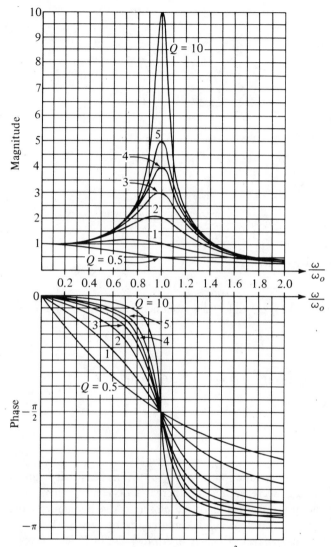

Fig. 10-1   Magnitude and phase of $\dfrac{\omega_o^2}{s^2 + s\dfrac{\omega_o}{Q} + \omega_o^2}$

10-2 PASSIVE REALIZATIONS

Two second-order *RLC* low-pass networks are given in Fig. 10-2. (Another network is given in Problem 10-3.) In each network, the inductor produces one of the zeros at infinity, and the capacitor the other.

The transfer function of the network in Fig. 10-2a is given by

$$\frac{V_o}{V_i} = \frac{1}{LC} \frac{1}{s^2 + s\dfrac{R}{L} + \dfrac{1}{LC}}. \tag{10-3}$$

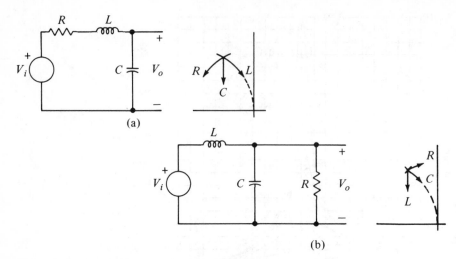

(a)

(b)

Fig. 10-2   Two passive, low-pass networks

By comparison with Eq. (10-1), ω_o and Q are obtained:

$$\omega_o = \frac{1}{\sqrt{LC}}, \qquad Q = \frac{1}{R}\sqrt{\frac{L}{C}}. \tag{10-4}$$

The three ω_o-sensitivity and the three Q-sensitivity functions are

$$S_L^{\omega_o} = S_C^{\omega_o} = -\tfrac{1}{2}, \qquad S_R^{\omega_o} = 0, \tag{10-5}$$

and

$$S_L^Q = -S_C^Q = -\tfrac{1}{2}S_R^Q = \tfrac{1}{2}. \tag{10-6}$$

The magnitudes of the sensitivity functions are 1, $\tfrac{1}{2}$, or 0. The sensitivity figures indicate that an increase in L decreases ω_o and increases Q. Hence, when complex, the poles move radially inward and rotate closer to the imaginary axis. The resultant motion is along a circle which is centered on the negative real axis and is tangent to the imaginary axis at the origin. An increase in C decreases ω_o as well as Q. Hence, when complex, the poles move radially inward and rotate away from the imaginary axis. The resultant motion is straight down, i.e., constant-α. An increase in R does not affect ω_o but decreases Q; the complex poles rotate away from the imaginary axis in a constant-ω_o manner. The direction of these changes (for positive changes in element values) is indicated for the upper half-plane pole in Fig. 10-2a.

The transfer function of the network in Fig. 10-2b is

$$\frac{V_o}{V_i} = \frac{1}{LC}\frac{1}{s^2 + s\dfrac{1}{RC} + \dfrac{1}{LC}}. \tag{10-7}$$

The resulting ω_o and Q are

$$\omega_o = \frac{1}{\sqrt{LC}}, \qquad Q = R\sqrt{\frac{C}{L}}. \tag{10-8}$$

The ω_o-sensitivity and Q-sensitivity functions are given by

$$S_L^{\omega_o} = S_C^{\omega_o} = -\tfrac{1}{2}, \qquad S_R^{\omega_o} = 0, \tag{10-9}$$

$$S_L^{Q} = -S_C^{Q} = -\tfrac{1}{2}S_R^{Q} = -\tfrac{1}{2}. \tag{10-10}$$

Figure 10-2b indicates the direction in which the upper half-plane pole moves if the element values are increased. As far as the sensitivity figures are concerned, both networks of Fig. 10-2 are considered to be good. These figures are used as a basis for comparing sensitivities of active realizations.

It should be stressed here that the results presented in this section are based on ideal inductors. The effect of nonideal inductors is treated in Problem 10-3.

10-3 *KRC* REALIZATION (IDEAL)

An *RC* network with transfer-function zeros at infinity (low-pass network with negative real-axis poles) is shown in Fig. 10-3a. If the voltage across C_2 is

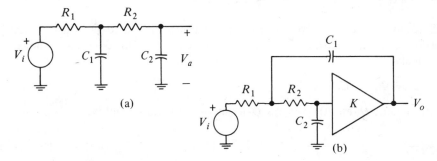

(a)

(b)

Fig. 10-3 Low-pass *KRC* network

sensed and fed back (without inverting) as a dependent voltage source in series with C_1, the *KRC* network shown in Fig. 10-3b results. As discussed in Section 9-3, this particular feedback scheme does not affect the transfer-function zeros. The zeros of (V_a/V_i) of Fig. 10-3a are the same as the zeros of (V_o/V_i) of Fig. 10-3b. If the amplifier is considered ideal, that is, if $K = K_0$, the poles of (V_o/V_i) can be made complex and moved in a constant-ω_o manner (because of band-pass type feedback) as K_0 is varied. See Fig. 9-10 for details.

For $K = K_0$, the transfer function of the network of Fig. 10-3b is given by

$$\frac{V_o}{V_i} = \frac{K_0}{R_1 R_2 C_1 C_2} \frac{1}{s^2 + s\left[\dfrac{1}{R_1 C_1} + \dfrac{1}{R_2 C_1} + \dfrac{1}{R_2 C_2}(1 - K_0)\right] + \dfrac{1}{R_1 R_2 C_1 C_2}}$$

$$= \frac{K_0 \omega_o^2}{s^2 + s\dfrac{\omega_o}{Q} + \omega_o^2}, \tag{10-11}$$

where K_0 represents the dc gain of the amplifier, and

$$\omega_o = \frac{1}{\sqrt{R_1 R_2 C_1 C_2}}, \qquad Q = \frac{1}{\sqrt{\dfrac{R_2 C_2}{R_1 C_1}} + \sqrt{\dfrac{R_1 C_2}{R_2 C_1}} + (1 - K_0)\sqrt{\dfrac{R_1 C_1}{R_2 C_2}}}. \tag{10-12}$$

The ω_o-sensitivity and the Q-sensitivity functions are given by

$$S_{R_1}^{\omega_o} = S_{R_2}^{\omega_o} = S_{C_1}^{\omega_o} = S_{C_2}^{\omega_o} = -\tfrac{1}{2}, \qquad S_{K_0}^{\omega_o} = 0, \tag{10-13}$$

$$S_{R_1}^{Q} = -S_{R_2}^{Q} = Q\sqrt{\frac{R_2 C_2}{R_1 C_1}} - \frac{1}{2}, \qquad S_{C_1}^{Q} = -S_{C_2}^{Q} = Q\sqrt{\frac{C_2}{C_1}}\left(\sqrt{\frac{R_1}{R_2}} + \sqrt{\frac{R_2}{R_1}}\right) - \frac{1}{2}, \tag{10-14}$$

$$S_{K_0}^{Q} = Q\left[\sqrt{\frac{R_1 C_1}{R_2 C_2}} + \sqrt{\frac{R_2 C_2}{R_1 C_1}} + \sqrt{\frac{R_1 C_2}{R_2 C_1}}\right] - 1.$$

Since five elements are available to set the values of ω_o and Q, considerable freedom exists in choosing the values of these elements. One convenient choice is to make

$$R_1 = R_2 = R, \qquad C_1 = C_2 = C.$$

Then, the ω_o, Q, and the various Q-sensitivity functions simplify to

$$\omega_o = \frac{1}{RC}, \qquad Q = \frac{1}{3 - K_0} \tag{10-15}$$

$$S_{R_1}^{Q} = -S_{R_2}^{Q} = Q - \tfrac{1}{2}, \qquad S_{C_1}^{Q} = -S_{C_2}^{Q} = 2Q - \tfrac{1}{2}, \qquad S_{K_0}^{Q} = 3Q - 1. \tag{10-16}$$

For a given ω_o and Q, the elements of the network of Fig. 10-3b are determined by using Eq. (10-15). For example, if $\omega_o = 1$ Krad/s and $Q = 5$, then

$$RC = \frac{1}{\omega_o} = 10^{-3}, \qquad K_0 = 3 - \frac{1}{Q} = 2.8.$$

If C is chosen as 0.1 μF, then $R = 10$ KΩ.

Another attractive choice is to make

$$R_1 = R_2 = R, \qquad K_0 = 1.$$

In this case, ω_o, Q, and the Q-sensitivity functions become

$$\omega_o = \frac{1}{R\sqrt{C_1 C_2}}, \qquad Q = \frac{1}{2}\sqrt{\frac{C_1}{C_2}}, \tag{10-17}$$

$$S_{R_1}^Q = S_{R_2}^Q = 0, \qquad S_{C_1}^Q = -S_{C_2}^Q = \tfrac{1}{2}, \qquad S_{K_0}^Q = 2Q^2. \tag{10-18}$$

For a given ω_o and Q, the elements of the network of Fig. 10-3b are determined by using Eq. (10-17). For example, if $\omega_o = 1$ Krad/s and $Q = 5$, then

$$\frac{C_1}{C_2} = 4Q^2 = 100, \qquad R = \frac{1}{\omega_o\sqrt{C_1 C_2}} = \frac{10^{-3}}{\sqrt{C_1 C_2}} = \frac{10^{-4}}{C_2}.$$

If C_2 is chosen as 0.01 μF, then $C_1 = 1$ μF and $R = 10$ KΩ.

A study of the Q-sensitivity functions given by Eqs. (10-14), (10-16), and (10-18) reveals that some of the sensitivity figures may be quite high. In particular, $S_{K_0}^Q$ shows a high degree of dependence on the Q of the poles. For example, if the Q of the poles is 10, then, for the equal-R and equal-C case, $S_{K_0}^Q = 29$; for the equal-R and unity-gain case, $S_{K_0}^Q = 200$. That these figures are so high for high-Q poles should not be surprising if the loci of the poles as a function of K_0 are considered. As Fig. 9-10 shows, the poles may even end up in the right half-plane if a sufficiently large positive change in K_0 occurs. As the following example indicates, in this KRC realization, no matter how the element values are chosen, $S_{K_0}^Q$ is always greater than $(2Q - 1)$.

It should be emphasized here that the sensitivity figure alone does not determine the change in Q. Recall that

$$\frac{\Delta Q}{Q} \cong S_{K_0}^Q \frac{\Delta K_0}{K_0}.$$

Thus, even though $S_{K_0}^Q$ may be high, $\Delta Q/Q$ may be kept low by keeping $\Delta K_0/K_0$ low. Since K_0 depends upon the ratio of two resistors, it can be maintained with almost no deviation from its nominal value. However, care should be exercised in the initial adjustment of K_0. (In the unity-gain case, $\Delta K_0 = 0$ for the ideal amplifier because the output of the amplifier is directly tied to the inverting terminal. However, in the nonideal amplifier, the dc value of the gain depends upon the open-loop gain of the operational amplifier, which may change.)

EXAMPLE 10-1

Show that in any second-order KRC system, where K_0 varies only the coefficient of the s-term of the denominator polynomial, the pole-Q sensitivity with respect to K_0 is greater than $(2Q - 1)$.

SOLUTION

Since K_0 varies only the coefficient of the s-term in the quadratic equation, write the equation for the poles as

$$a_2(s + \alpha_1)(s + \alpha_2) - K_0 b_1 s = 0, \qquad a_2 > 0, \quad b_1 > 0, \quad K_0 \geq 0.$$

For $K_0 = 0$, the system reduces to the passive RC network. Therefore, $-\alpha_1$ and $-\alpha_2$ represent the poles of the RC network. Rearrange the equation for the poles and obtain ω_o and Q:

$$s^2 + s(\alpha_1 + \alpha_2 - K_0 b_1/a_2) + \alpha_1 \alpha_2 = 0,$$

$$\omega_o = \sqrt{\alpha_1 \alpha_2}, \qquad Q = \frac{\sqrt{\alpha_1 \alpha_2}}{\alpha_1 + \alpha_2 - K_0 b_1/a_2}.$$

The sensitivity of Q with respect to K_0 is

$$S_{K_0}^Q = -\frac{-K_0 b_1/a_2}{\alpha_1 + \alpha_2 - K_0 b_1/a_2} = Q\left(\sqrt{\frac{\alpha_1}{\alpha_2}} + \sqrt{\frac{\alpha_2}{\alpha_1}}\right) - 1.$$

The minimum value of $\sqrt{(\alpha_1/\alpha_2)} + \sqrt{(\alpha_2/\alpha_1)}$ occurs when $\alpha_1 = \alpha_2$ and is equal to 2 [$y = x + (1/x)$ has a minimum of 2 at $x = 1$]. Since $-\alpha_1$ and $-\alpha_2$ represent the poles of the RC network, they can never be equal to each other. Hence, $\sqrt{(\alpha_1/\alpha_2)} + \sqrt{(\alpha_2/\alpha_1)} > 2$. Consequently,

$$S_{K_0}^Q > 2Q - 1. \qquad \textit{Ans.}$$

Note that this result is independent of the basic RC network to which feedback (band-pass type) is applied by the K_0 amplifier. As long as the system is of second order, the result is valid.

10-4 *KRC* REALIZATION (ONE-POLE ROLLOFF)

The poles of circuits designed for high-frequency filter applications have large magnitudes, i.e., the ω_o of the poles is large. The farther out these poles are, the less accurate become designs based on ideal amplifiers. To obtain better agreement between theory and practice, the small-signal, one-pole rolloff characteristics of the operational amplifier must be taken into account. When this is done, the ideal second-order system becomes a third-order system. If the operational amplifier is only slightly imperfect, i.e., its GB (gain-bandwidth) product is not infinite but quite large, then, the characteristics of the resulting third-order system will not depart appreciably from the second-order (ideal) system. Thus, the complex poles of the nonideal system will be slightly different from the complex poles of the ideal system. In addition, the nonideal system will have a negative real-axis pole which, being far out, has generally negligible effect on the frequency response. On the other hand, the higher the Q of the complex poles, the more significant are the changes in the frequency response due to complex-pole-position perturbations.

To derive a quantitative measure of the complex-pole displacements from their nominal positions, consider the general *KRC* realization shown in Fig. 10-4a. The RC network is driven by the independent source, V_i, and the dependent source, V_o. The output of the RC network, V_3, is constrained by the amplifier

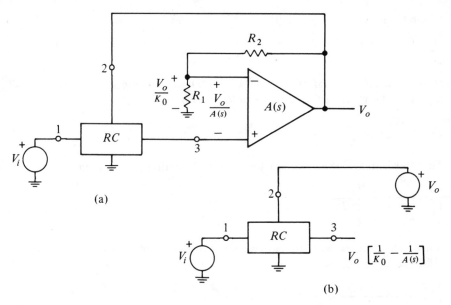

Fig. 10-4 General analysis of KRC network

to $V_o [1/K_0 - 1/A(s)]$, where $K_0 = 1 + (R_2/R_1)$. Consequently Fig. 10-4a can be redrawn as Fig. 10-4b.

Concentrating on the RC network of Fig. 10-4b, the voltage at port 3 is written as the superposition of voltages resulting from the excitations at ports 1 and 2:

$$V_o\left(\frac{1}{K_0} - \frac{1}{A(s)}\right) = V_i T_{31} = V_o T_{32}, \tag{10-19}$$

where

$$T_{31} = \frac{V_3}{V_1}\bigg|_{V_2=0} = \frac{N_{31}}{D}, \qquad T_{32} = \frac{V_3}{V_2}\bigg|_{V_1=0} = \frac{N_{32}}{D}.$$

Since T_{31} and T_{32} belong to the same RC network, the denominator polynomials are assumed to be identical. From Eq. (10-19), the transfer function is obtained as

$$\frac{V_o}{V_i} = \frac{T_{31}}{\dfrac{1}{K_0} - T_{32} - \dfrac{1}{A(s)}} = \frac{K_0 N_{31}}{D - K_0 N_{32} - \dfrac{K_0}{A(s)} D}. \tag{10-20}$$

Now let the amplifier be characterized by

$$A(s) = -\frac{GB}{s + \omega_a} \cong -\frac{GB}{s}, \tag{10-21}$$

where *the assumption*

$$\frac{\omega_o}{2Q} \gg \omega_a$$

is made to allow the simplification. (This assumption presupposes that s will represent the value of the complex poles, that is,

$$s = -\frac{\omega_o}{2Q} \pm j\omega_o \sqrt{1 - \frac{1}{4Q^2}}.$$

That this is indeed the case is justified later.) Equation (10-20) then becomes

$$\frac{V_o}{V_i} = \frac{K_0 N_{31}}{D - K_0 N_{32} \dfrac{K_0}{GB} sD}.$$ (10-22)

Since GB/K_0 represents the closed-loop 3-dB bandwidth of the amplifier, Eq. (10-22) can also be expressed as

$$\frac{V_o}{V_i} = \frac{K_0 N_{31}}{D - K_0 N_{32} + \dfrac{1}{\omega_{3dB}} sD}.$$ (10-23)

If the system is second order,

$$D = (s + \alpha_1)(s + \alpha_2) \qquad (0 \le \alpha_1 < \alpha_2),$$

where $-\alpha_1$ and $-\alpha_2$ represent the poles of the *RC* network with ports 1 and 2 grounded and port 3 open-circuited. Then, for the ideal case, that is, $GB = \infty$, Eq. (10-22) simplifies to

$$\frac{V_o}{V_i} = \frac{K_0 N_{31}}{D - K_0 N_{32}} = \frac{K_0 N_{31}}{(s + \alpha_1)(s + \alpha_2) - K_0(b_2 s^2 + b_1 s + b_0)}$$

$$= \frac{K_0 N_{31}}{(1 - K_0 b_2)s^2 + s(\alpha_1 + \alpha_2 - K_0 b_1) + (\alpha_1 \alpha_2 - K_0 b_0)}$$

$$= \frac{K_0 N_{31}}{(1 - K_0 b_2)\left(s^2 + s \dfrac{\omega_o}{Q} + \omega_o^2\right)}.$$

With ω_o and Q thus defined by the ideal system, the nonideal system, described by Eq. (10-22), can then be expressed as

$$\frac{V_o}{V_i} = \frac{K_0 N_{31}}{(1 - K_0 b_2)\left(s^2 + s \dfrac{\omega_o}{Q} + \omega_o^2\right) + \dfrac{K_0}{GB} s(s + \alpha_1)(s + \alpha_2)}.$$ (10-24)

The modified characteristic equation is a cubic, and is given by

$$(1 - K_0 b_2)\left(s^2 + s \dfrac{\omega_o}{Q} + \omega_o^2\right) + \dfrac{K_0}{GB} s(s + \alpha_1)(s + \alpha_2) = 0,$$ (10-25)

which can be written in factored form as

$$[(s + \alpha_a)^2 + \beta_a^2](s + \gamma) = \left(s^2 + s\frac{\omega_{oa}}{Q_a} + \omega_{oa}^2\right)(s + \gamma) = 0, \tag{10-26}$$

where (α_a, β_a) or (ω_{oa}, Q_a) represents the actual (modified) positions of the complex poles. Approximate expressions for ω_{oa} and Q_a can be found by first evaluating the derivative of the poles with respect to $1/GB$ at the nominal pole positions, and then displacing the poles along the line indicated by the derivative. (Refer to the discussion on the root-sensitivity function given in Section 6-3 for details.)

General Analysis of the Band-Pass Type Feedback

If the loop gain is a band-pass function, then $N_{32} = b_1 s$. Consequently, in the ideal case, K_0 varies the coefficient of the s-term only, which results in constant-ω_o pole loci with $\omega_o = \sqrt{\alpha_1 \alpha_2}$. (The loci don't quite form a circle because the nonideal system is of third order.) To determine the effect of finite amplifier GB on the complex-pole positions, let $b_2 = b_0 = 0$ and partition Eq. (10-25) as follows:

$$\underbrace{\left(s^2 + s\frac{\omega_o}{Q} + \omega_o^2\right)}_{P_1(s)} + \frac{1}{GB}\underbrace{\{K_0 s[s^2 + s(\alpha_1 + \alpha_2) + \omega_o^2]\}}_{P_2(s)} = 0, \tag{10-27}$$

where

$$\alpha_1 = \alpha_2 = \frac{\omega_o}{Q} + K_0 b_1.$$

The nominal pole positions $(GB = \infty)$ are the zeros of $P_1(s)$; that is,

$$p_{1,2} = -\frac{\omega_o}{2Q} \pm j\omega_o\sqrt{1 - \frac{1}{4Q^2}}. \tag{10-28}$$

From Eq. (6-34),

$$\frac{dp_{1,2}}{d\left(\dfrac{1}{GB}\right)} = -\left.\frac{P_2(s)}{d\dfrac{P_1(s)}{ds}}\right|_{s=p_{1,2}} \tag{10-29}$$

$$= -\left.\frac{K_0 s[s^2 + s(\alpha_1 + \alpha_2) + \omega_o^2]}{2s + \dfrac{\omega_o}{Q}}\right|_{s=p_{1,2}}$$

$$= -\left.\frac{K_0 s\left[s(\alpha_1 + \alpha_2) - s\dfrac{\omega_o}{Q}\right]}{2s + \dfrac{\omega_o}{Q}}\right|_{s=p_{1,2}}$$

[At this stage, note that the s in front of the brackets can be replaced by $(s + \omega_a)$, to take into consideration the bandwidth of the operational amplifier. Since s takes on the value of $p_{1,2}$,

$$p_{1,2} + \omega_a \cong p_{1,2} \qquad \text{as long as } \frac{\omega_o}{2Q} \gg \omega_a.]$$

Then

$$\frac{dp_{1,2}}{d\left(\dfrac{1}{\text{GB}}\right)} = -\frac{K_0\left(\alpha_1 + \alpha_2 - \dfrac{\omega_o}{Q}\right)\left(\dfrac{1}{2}\dfrac{\omega_o^2}{Q^2} - \omega_o^2 \mp j\dfrac{\omega_o^2}{Q}\sqrt{1 - \dfrac{1}{4Q^2}}\right)}{\pm j2\omega_o\sqrt{1 - \dfrac{1}{4Q^2}}}.$$

Hence,

$$\mathscr{S}^{p_{1,2}}_{\frac{1}{\text{GB}}} = \frac{1}{|p_{1,2}|}\frac{dp_{1,2}}{d\left(\dfrac{1}{\text{GB}}\right)} = \frac{K_0}{2}\left(\sqrt{\frac{\alpha_1}{\alpha_2}} + \sqrt{\frac{\alpha_2}{\alpha_1}} - \frac{1}{Q}\right)\left(\frac{1}{Q} \mp j\frac{1 - \dfrac{1}{2Q^2}}{\sqrt{1 - \dfrac{1}{4Q^2}}}\right)\omega_o. \qquad (10\text{-}30)$$

Using incremental rather than infinitesimal values and realizing that $\Delta(1/\text{GB}) \cong (1/\text{GB})$, the per-unit change in $p_{1,2}$ can be obtained from Eq (10-30):

$$\frac{\Delta p_{1,2}}{|p_{1,2}|} = \frac{-\Delta\alpha}{\omega_o} \pm j\frac{\Delta\beta}{\omega_o} \cong \mathscr{S}^{p_{1,2}}_{1/\text{GB}}\Delta\left(\frac{1}{\text{GB}}\right) = \mathscr{S}^{p_{1,2}}_{1/\text{GB}}\frac{1}{\text{GB}}; \qquad (10\text{-}31)$$

$$\frac{-\Delta\alpha}{\omega_o} \cong \frac{K_0}{2Q}\left(\sqrt{\frac{\alpha_1}{\alpha_2}} + \sqrt{\frac{\alpha_2}{\alpha_1}} - \frac{1}{Q}\right)\frac{\omega_o}{\text{GB}}; \qquad (10\text{-}32a)$$

$$\frac{\Delta\beta}{\omega_o} \cong \frac{K_0}{2}\left(\sqrt{\frac{\alpha_1}{\alpha_2}} + \sqrt{\frac{\alpha_2}{\alpha_1}} - \frac{1}{Q}\right)\frac{\left(1 - \dfrac{1}{2Q^2}\right)}{\sqrt{1 - \dfrac{1}{4Q^2}}}\frac{\omega_o}{\text{GB}}, \qquad (10\text{-}32b)$$

where the upper and lower signs are associated with the upper and lower half-plane poles respectively. The changes in ω_o and Q can be expressed in terms of the changes in α and β (see Prob. 10-5):

$$\frac{\Delta\omega_o}{\omega_o} \cong \frac{1}{2Q}\frac{\Delta\alpha}{\omega_o} + \sqrt{1 - \frac{1}{4Q^2}}\frac{\Delta\beta}{\omega_o}, \qquad (10\text{-}33a)$$

$$\frac{\Delta Q}{Q} \cong -2Q\left(1 - \frac{1}{4Q^2}\right)\frac{\Delta\alpha}{\omega_o} + \sqrt{1 - \frac{1}{4Q^2}}\frac{\Delta\beta}{\omega_o}. \qquad (10\text{-}33b)$$

Equation (10-33) is simplified by substituting the expressions of $(-\Delta\alpha/\omega_o)$ and $(\Delta\beta/\omega_o)$ given by Eq. (10-32):

$$\frac{\Delta\omega_o}{\omega_o} \cong Q\frac{\Delta\alpha}{\omega_o}, \tag{10-34a}$$

$$\frac{\Delta Q}{Q} \cong -\frac{\Delta\omega_o}{\omega_o}. \tag{10-34b}$$

Keeping in mind that the actual upper half-plane pole is given by

$$p_1 = -\alpha_a + j\beta_a = -(\alpha + \Delta\alpha) + j(\beta + \Delta\beta)$$

$$= \omega_{oa}\exp\left[j\left(\frac{\pi}{2} + \sin^{-1}\frac{1}{2Q_a}\right)\right] = (\omega_o + \Delta\omega_o)\exp\left[j\left(\frac{\pi}{2} + \sin^{-1}\frac{1}{2(Q + \Delta Q)}\right)\right],$$

the following important observations can be made about the changes in this pole caused by the nonideal amplifier.

1. The larger the ω_o/GB ratio, the more the pole becomes displaced from its nominal value.

2. The change from the nominal value is proportional to

$$\sqrt{\frac{\alpha_1}{\alpha_2}} + \sqrt{\frac{\alpha_2}{\alpha_1}} - \frac{1}{Q}.$$

Since

$$\sqrt{\frac{\alpha_1}{\alpha_2}} + \sqrt{\frac{\alpha_2}{\alpha_1}} > 2,$$

it is impossible to make the change zero by adjusting the poles of the RC network.

3. $-\dfrac{\Delta\alpha}{\omega_o} > 0$. Hence, the pole is always displaced to the right.

4. Depending on whether the nominal value of Q is less than or greater than $1/\sqrt{2}$, the pole is displaced up or down. For $Q = 1/\sqrt{2}$, the imaginary part does not change.

5. $\Delta\omega_o/\omega_o < 0$. Hence, the magnitude of the pole is always less than the nominal value.

6. $\Delta Q/Q > 0$. Hence, the Q of the pole always increases.

7. For high-Q poles, the magnitude characteristic peaks at a lower frequency.

8. The changes given by Eqs. (10-32) and (10-34) represent only first-order changes. The smaller ω_o/GB, the more accurate are these equations.

Pole, ω_o, and Q Changes in the Circuit of Fig. 10-3b.

The general results obtained so far can be readily applied to specific circuits. Consider, for example, the low-pass *KRC* network of Fig. 10-3b which uses band-pass type feedback. Since $(\alpha_1 + \alpha_2)$ represents the coefficient of the *s*-term in the denominator polynomial when $K_0 = 0$ (i.e., when the network is *RC*) and $\omega_o = \sqrt{\alpha_1 \alpha_2}$, the following relations are easily obtained from Eq. (10-11):

$$\sqrt{\frac{\alpha_1}{\alpha_2}} + \sqrt{\frac{\alpha_2}{\alpha_1}} = \frac{\alpha_1 + \alpha_2}{\omega_o} = \frac{\dfrac{1}{R_1 C_1} + \dfrac{1}{R_2 C_1} + \dfrac{1}{R_2 C_2}}{\sqrt{\dfrac{1}{R_1 R_2 C_1 C_2}}}$$

$$= \sqrt{\frac{R_2 C_2}{R_1 C_1}} + \sqrt{\frac{R_1 C_2}{R_2 C_1}} + \sqrt{\frac{R_1 C_1}{R_2 C_2}} = \frac{1}{Q} + K_0 \sqrt{\frac{R_1 C_1}{R_2 C_2}}.$$

Hence, the actual transfer function given by Eq. (10-24) simplifies to

$$\frac{V_o}{V_i} = \frac{K_0 \omega_o^2}{s^2 + s\dfrac{\omega_o}{Q} + \omega_o^2 + \dfrac{K_0}{GB} s \left[s^2 + s\dfrac{\omega_o}{Q}\left(1 + K_0 Q \sqrt{\dfrac{R_1 C_1}{R_2 C_2}}\right) + \omega_o^2 \right]}. \qquad (10\text{-}35)$$

Using Eqs. (10-32) and (10-34), the changes in the upper half-plane pole are obtained:

$$-\frac{\Delta\alpha}{\omega_o} \cong \frac{K_0^2}{2Q}\sqrt{\frac{R_1 C_1}{R_2 C_2}}\frac{\omega_o}{GB}, \qquad \frac{\Delta\beta}{\omega_o} \cong -\frac{K_0^2}{2}\sqrt{\frac{R_1 C_1}{R_2 C_2}}\frac{\left(1 - \dfrac{1}{2Q^2}\right)}{\sqrt{1 - \dfrac{1}{4Q^2}}}\frac{\omega_o}{GB}; \qquad (10\text{-}36)$$

$$\frac{\Delta\omega_o}{\omega_o} \cong -\frac{K_0^2}{2}\sqrt{\frac{R_1 C_1}{R_2 C_2}}\frac{\omega_o}{GB}, \qquad \frac{\Delta Q}{Q} \cong \frac{K_0^2}{2}\sqrt{\frac{R_1 C_1}{R_2 C_2}}\frac{\omega_o}{GB}. \qquad (10\text{-}37)$$

These relations can be simplified further by considering the equal-*R*, equal-*C* case and the unity-gain, equal-*R* case. A summary of all important results is provided in Table 10-1.

A comparison of the equal-*R*, equal-*C* case with the unity-gain, equal-*R* case shows that for high-*Q* circuits, the ω_o and *Q* changes are much greater for the latter case. Hence, as far as frequency limitations are concerned, the equal-*R*, equal-*C* circuit is preferable for high-*Q* circuits.

If (ω_o/GB) is not much less than 1, the equations in Table 10-1 predicting the changes in pole position from the nominal value are not accurate. It then becomes necessary to use the digital computer to solve directly for the roots of the characteristic equation, Eq. (10-35). Rearranged in descending powers of *s*, this equation is

$$s_n^3 + s_n^2\left(\frac{1}{Q} + K_0\sqrt{\frac{R_1 C_1}{R_2 C_2}} + \frac{GB_n}{K_0}\right) + s_n\left(1 + \frac{GB_n}{K_0 Q}\right) + \frac{GB_n}{K_0} = 0, \qquad (10\text{-}38)$$

Table 10-1 *KRC* Realization
(see Fig. 10-3b)

Equal-R, Equal-C

$$\omega_o = \frac{1}{RC}, \qquad Q = \frac{1}{3 - K_0}$$

$$\frac{V_o}{V_i} = \frac{\left(3 - \frac{1}{Q}\right)\omega_o^2}{s^2 + s\frac{\omega_o}{Q} + \omega_o^2 + \frac{\left(3 - \frac{1}{Q}\right)}{GB}s(s^2 + s3\omega_o + \omega_o^2)} \qquad \left(\omega_a \ll \frac{\omega_o}{2Q}\right)$$

$$-\frac{\Delta\alpha}{\omega_o} \cong \frac{1}{2Q}\left(3 - \frac{1}{Q}\right)^2 \frac{\omega_o}{GB}, \qquad \frac{\Delta\beta}{\omega_o} \cong -\frac{1}{2}\left(3 - \frac{1}{Q}\right)^2 \frac{\left(1 - \frac{1}{2Q^2}\right)}{\sqrt{1 - \frac{1}{4Q^2}}} \frac{\omega_o}{GB}$$

$$\frac{\Delta\omega_o}{\omega_o} \cong -\frac{1}{2}\left(3 - \frac{1}{Q}\right)^2 \frac{\omega_o}{GB}, \qquad \frac{\Delta Q}{Q} \cong \frac{1}{2}\left(3 - \frac{1}{Q}\right)^2 \frac{\omega_o}{GB}$$

Unity-gain, Equal-R

$$\omega_o = \frac{1}{R\sqrt{C_1 C_2}}, \qquad Q = \frac{1}{2}\sqrt{\frac{C_1}{C_2}}$$

$$\frac{V_o}{V_i} = \frac{\omega_o^2}{s^2 + s\frac{\omega_o}{Q} + \omega_o^2 + \frac{s}{GB}\left[s^2 + s\omega_o\left(2Q + \frac{1}{Q}\right) + \omega_o^2\right]} \qquad \left(\omega_a \ll \frac{\omega_o}{2Q}\right)$$

$$-\frac{\Delta\alpha}{\omega_o} \cong \frac{\omega_o}{GB}, \qquad \frac{\Delta\beta}{\omega_o} \cong -Q\frac{\left(1 - \frac{1}{2Q^2}\right)}{\sqrt{1 - \frac{1}{4Q^2}}} \frac{\omega_o}{GB}$$

$$\frac{\Delta\omega_o}{\omega_o} \cong -Q\frac{\omega_o}{GB}, \qquad \frac{\Delta Q}{Q} \cong Q\frac{\omega_o}{GB}$$

where

$$s_n = \frac{s}{\omega_o}, \qquad GB_n = \frac{GB}{\omega_o}.$$

The upper half-plane root of Eq. (10-38) is plotted as a function of Q and GB_n in Fig. 10-5. The solid lines represent constant nominal-Q lines. As the GB_n of the amplifier is decreased from infinity, the pole with a given nominal Q is displaced inward along the corresponding solid line. For example, when $Q = 1$ and $GB = \infty$, the pole in the equal-R, equal-C case is at $(-0.5 + j0.866)\omega_o$.

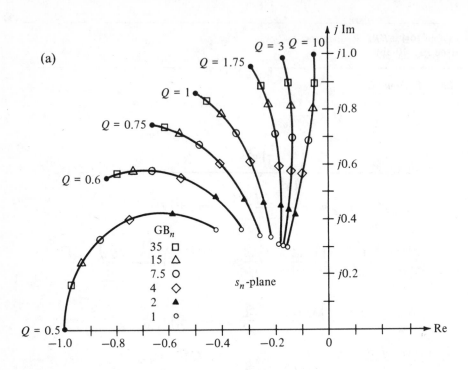

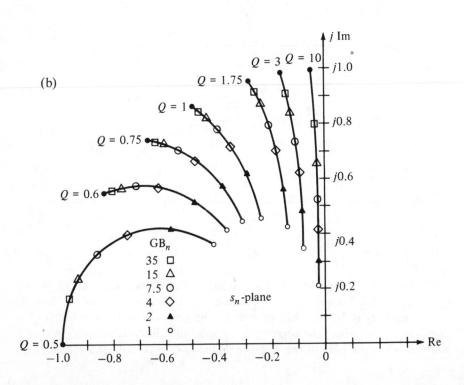

However, for $GB_n = 7.5$, the pole is at $(-0.34 + j0.71)\omega_o$, which shows considerable departure from the nominal position. The tangent drawn to any curve at the nominal-Q point, i.e., the outermost point, has a slope given by $[d(p_1/\omega_o)/d(\omega_o/GB)]$. The approximations presented in Table 10-1 predict changes in the pole position along this tangent line and therefore can give only first-order corrections to the pole position. The length of any line from the origin to a point (determined by GB_n) on a solid curve represents ω_{oa}/ω_o. The projection of this line on the negative real axis has a length of $(\omega_{oa}/\omega_o)/(2Q_a)$.

As the curves of Fig. 10-5 indicate, a second-order *KRC* network designed on the basis of an ideal amplifier will in practice yield a pole-ω_o that is lower in value. The network should therefore be designed originally with a higher ω_o to compensate for the reduction caused by the imperfect amplifier. A similar argument holds for the Q also. A trial-and-error procedure can be established to determine what ω_o and Q to use in the ideal circuit so that the desired ω_o and Q are obtained in the actual circuit. (See Problem 10-6).

EXAMPLE 10-2

Show that the general second-order *KRC* system using band-pass type feedback remains absolutely stable when a one-pole rolloff amplifier is substituted for the ideal amplifier.

SOLUTION

Rearrange Eq. (10-27) and obtain the characteristic equation:

$$s^3 + s^2 \underbrace{\left(\alpha_1 + \alpha_2 + \frac{GB}{K_0}\right)}_{a_2} + s\underbrace{\left(\omega_o^2 + \frac{\omega_o}{Q}\frac{GB}{K_0}\right)}_{a_1} + \underbrace{\omega_o^2\frac{GB}{K_0}}_{a_0} = 0.$$

Note that all the a coefficients are positive. If the system is stable, then it must satisfy

$$a_2 a_1 > a_0 \qquad \text{(Routh's criterion).}$$

Hence,

$$\left(\alpha_1 + \alpha_2 + \frac{GB}{K_0}\right)\left(\omega_o^2 + \frac{\omega_o}{Q}\frac{GB}{K_0}\right) > \omega_o^2 \frac{GB}{K_0},$$

◀Fig. 10-5a   Plot of upper half-plane root of

$$s_n^3 + s_n^2\left(3 + \frac{QGB_n}{3Q-1}\right) + s_n\left(1 + \frac{GB_n}{3Q-1}\right) + \frac{QGB_n}{3Q-1} = 0 \qquad \text{(Equal-}R\text{, equal-}C\text{)}$$

◀Fig. 10-5b   Plot of upper half plane root of

$$s_n^2 + s_n^2\left(2Q + \frac{1}{Q} + GB_n\right) + s_n\left(1 + \frac{GB_n}{Q}\right) + GB_n = 0 \qquad \text{(Unity-gain, equal-}R\text{)}$$

which simplifies to

$$(\alpha_1 + \alpha_2)\left(\omega_o^2 + \frac{\omega_o}{Q}\frac{GB}{K_0}\right) + \frac{\omega_o}{Q}\left(\frac{GB}{K_0}\right)^2 > 0. \tag{10-39}$$

As long as the ideal system has no right half-plane poles (Q not negative), the left side of Eq. (10-39) is positive. Therefore, the system is absolutely stable. Note that even for $Q = \infty$ (ideal system is oscillatory), the system is stable with the nonideal amplifier.

EXAMPLE 10-3

The transfer function of the ideal second-order *KRC* system with band-pass feedback is given by

$$T = \frac{H}{s^2 + s(\alpha_1 + \alpha_2 - K_0 b_1) + \omega_o^2}, \tag{10-40}$$

where $-\alpha_1$ and $-\alpha_2$ represent the poles of the *RC* network and $\omega_o = \sqrt{\alpha_1 \alpha_2}$.

To study the limitations caused by the nonideal amplifier, replace K_0 with the actual gain of the noninverting amplifier, that is, $[GB/(s + GB/K_0)]$ [see Eq. (7-17)]. Using the first two terms of the Maclaurin's series expansion of this gain, obtain the expressions for the actual ω_o and Q of the poles.

SOLUTION

Rearrange the expression of the gain and expand it in the Maclaurin's series:

$$K(s) = \frac{GB}{s + \dfrac{GB}{K_0}} = \frac{K_0}{1 + \dfrac{sK_0}{GB}}$$

$$\cong K_0\left[1 - \left(s\frac{K_0}{GB}\right) + \frac{1}{2}\left(s\frac{K_0}{GB}\right)^2 - \cdots\right] \qquad \left(\left|\frac{sK_0}{GB}\right| < 1\right).$$

As long as the first two terms of this expansion are used in Eq. (10-40), the characteristic polynomial remains quadratic with an added correction term resulting from the finite bandwidth of the amplifier:

$$s^2 + s\left[\alpha_1 + \alpha_2 - b_1 K_0\left(1 - s\frac{K_0}{GB}\right)\right] + \omega_o^2 = 0,$$

$$s^2\left(1 + b_1 K_0\frac{K_0}{GB}\right) + s(\alpha_1 + \alpha_2 - b_1 K_0) + \omega_o^2 = 0. \tag{10-41}$$

When $GB = \infty$, this equation reduces to the ideal case. Hence,

$$\alpha_1 + \alpha_2 - b_1 K_0 = \frac{\omega_o}{Q},$$

$$b_1 = \frac{\alpha_1 + \alpha_2 - \dfrac{\omega_o}{Q}}{K_0}.$$

Eliminate b_1 from Eq. (10-41) and put the quadratic equation in standard form:

$$s^2 + s \left[\underbrace{\frac{\dfrac{\omega_o}{Q}}{1 + \left(\alpha_1 + \alpha_2 - \dfrac{\omega_o}{Q}\right)\dfrac{K_0}{GB}}}_{\omega_{oa}/Q_a} \right] + \underbrace{\frac{\omega_o^2}{1 + \left(\alpha_1 + \alpha_2 - \dfrac{\omega_o}{Q}\right)\dfrac{K_0}{GB}}}_{\omega_{oa}^2} = 0.$$

Solve for ω_{oa} and Q_a:

$$\omega_{oa} = \frac{\omega_o}{\sqrt{1 + \left(\alpha_1 + \alpha_2 - \dfrac{\omega_o}{Q}\right)\dfrac{K_0}{GB}}}, \qquad Q_a = Q\sqrt{1 + \left(\alpha_1 + \alpha_2 - \frac{\omega_o}{Q}\right)\frac{K_0}{GB}}.$$

Since $[\alpha_1 + \alpha_2 - (\omega_o/Q)](K_0/GB) \ll 1$, simplify the expressions for ω_{oa} and Q_a by noting that

$$\frac{1}{\sqrt{1+x}} \cong 1 - \frac{x}{2} \qquad \text{and} \qquad \sqrt{1+x} \cong 1 + \frac{x}{2} \qquad \text{for} \quad |x| \ll 1.$$

Thus,

$$\omega_{oa} \cong \omega_o \left[1 - \frac{1}{2}\left(\alpha_1 + \alpha_2 - \frac{\omega_o}{Q}\right)\frac{K_0}{GB}\right],$$

$$Q_a \cong Q\left[1 + \frac{1}{2}\left(\alpha_1 + \alpha_2 - \frac{\omega_o}{Q}\right)\frac{K_0}{GB}\right]. \qquad \textit{Ans.}$$

Note that the changes in ω_o and Q are given by

$$\Delta\omega_o = -\frac{\omega_o}{2}\left(\alpha_1 + \alpha_2 - \frac{\omega_o}{Q}\right)\frac{K_0}{GB}, \qquad \Delta Q = \frac{Q}{2}\left(\alpha_1 + \alpha_2 - \frac{\omega_o}{Q}\right)\frac{K_0}{GB}.$$

These expressions for the changes agree with those given by Eq. (10-34) even though different approximation techniques are used to arrive at the results. Whenever applicable, the method presented in this example is simple and powerful.

EXAMPLE 10-4

In the *KRC* realization shown in Fig. 10-3, ω_o and Q are calculated by assuming an ideal amplifier.

(a) What constraint must be placed on the gain-bandwidth product of the amplifier if the resulting ω_{oa} and Q_a are to differ no more than 5% from ω_o and Q? Consider both the equal-R, equal-C and the unity-gain, equal-R case.

(b) Evaluate the constraint relations obtained in (a) for $GB = 10^6$ and $Q = 10$.

SOLUTION

(a) If ω_{oa} and Q_a are to be no more than 5% off from ω_o and Q, then, from Table 10-1,

$$\frac{1}{2}\left(3 - \frac{1}{Q}\right)^2 \frac{\omega_o}{\text{GB}} < 0.05, \quad \text{GB} > 10\left(3 - \frac{1}{Q}\right)^2 \omega_o \quad \begin{array}{l}\text{(equal-}R\text{, equal-}C \\ \text{case)}, \qquad Ans.\end{array}$$

$$\frac{Q\omega_o}{\text{GB}} < 0.05, \quad \text{GB} > 20Q\omega_o \quad \text{(equal-}R\text{, unity-gain case)}. \qquad Ans.$$

(b) $\omega_o < \dfrac{\text{GB}}{10\left(3 - \dfrac{1}{Q}\right)^2} = \dfrac{10^6}{10 \times 2.9^2}$,

$\omega_o < 11.9$ Krad/s   (equal-R, equal-C case),   *Ans.*

$$\omega_o < \frac{\text{GB}}{20Q} = \frac{10^6}{200},$$

$\omega_o < 5$ Krad/s   (equal-R, unity-gain case).   *Ans.*

Note that in this case it is possible to operate the equal-R, equal-C network with a much higher value of ω_o. As long as

$$\frac{1}{2}\left(3 - \frac{1}{Q}\right)^2 < Q,$$

the equal-R, equal-C network will allow higher ω_o realizations. (The inequality gives $Q > 3.85$.)

10-5 — KRC REALIZATION (IDEAL)

An RC network with transfer-function zeros at infinity is shown in Fig. 10-6. If the voltage across C_2 is sensed, inverted, and fed back as a dependent voltage source in series with R_3, the $-KRC$ realization shown in Fig. 10-6b results. As shown in Section 9-4, this particular feedback scheme does not affect the zeros of the transfer function. The zeros of (V_a/V_i) of Fig. 10-6a are the same as the zeros of (V_o/V_i) of Fig. 10-6b. However, the poles of (V_o/V_i) can be made complex and moved in a constant-α manner (because of low-pass type feedback) as $K = K_0$ is varied. (See Fig. 9-12c.) Note that both the input and feedback signals are applied through resistors. As shown in Fig. 10-6c, R_4 can be made part of the gain-producing network, that is,

$$K_0 = -\frac{R_5}{R_4}.$$

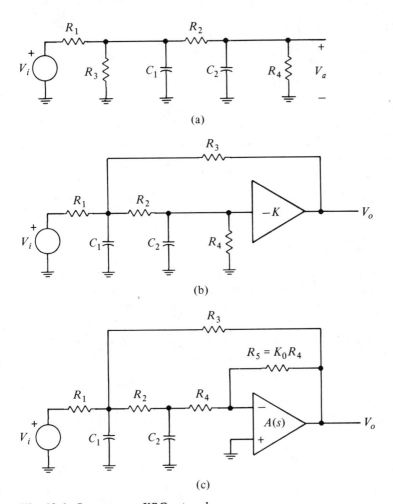

Fig. 10-6   Low-pass —*KRC* network

The transfer function for the network of Fig. 10-6b for $K = K_0 > 0$ is given by

$$\frac{V_o}{V_i} = \frac{-\dfrac{K_0}{R_1 R_2 C_1 C_2}}{s^2 + s\left[\left(1 + \dfrac{R_1}{R_3}\right)\dfrac{1}{R_1 C_1} + \left(1 + \dfrac{C_2}{C_1}\right)\dfrac{1}{R_2 C_2} + \dfrac{1}{R_4 C_2}\right]}$$

$$+ \frac{1 + (R_1/R_3)(1 + K_0) + (R_1/R_4)[1 + (R_2/R_3) + (R_2/R_1)]}{R_1 C_1 R_2 C_2}. \qquad (10\text{-}42)$$

The ω_o and Q of the poles are

$$\omega_o = \sqrt{\dfrac{1 + \dfrac{R_1}{R_3}(1 + K_0) + \dfrac{R_1}{R_4}\left(1 + \dfrac{R_2}{R_3} + \dfrac{R_2}{R_1}\right)}{R_1 R_2 C_1 C_2}},$$

$$Q = \dfrac{\sqrt{\dfrac{1 + (R_1/R_3)(1 + K_0) + (R_1/R_4)[1 + (R_2/R_3) + (R_2/R_1)]}{R_1 R_2 C_1 C_2}}}{\left(1 + \dfrac{R_1}{R_3}\right)\dfrac{1}{R_1 C_1} + \left(1 + \dfrac{C_2}{C_1}\right)\dfrac{1}{R_2 C_2} + \dfrac{1}{R_4 C_2}}. \tag{10-43}$$

Since seven elements (R_1, R_2, R_3, R_4, C_1, C_2, and K_0) can be adjusted to obtain a specified ω_o and Q, five of the elements can be arbitrarily chosen. A convenient choice is to make

$$R_1 = R_2 = R_3 = R_4 = R, \qquad C_1 = C_2 = C.$$

Then R, C, and K_0 are chosen to set ω_o and Q. (In this case, either R or C can still be picked arbitrarily.) Henceforth, only the equal-R, equal-C case is discussed. The corresponding V_o/V_i, ω_o, Q, and sensitivity functions are

$$\dfrac{V_o}{V_i} = -\dfrac{K_0}{R^2 C^2}\dfrac{1}{s^2 + s\dfrac{5}{RC} + \dfrac{5 + K_0}{R^2 C^2}}, \qquad \omega_o = \dfrac{\sqrt{5 + K_0}}{RC}, \qquad Q = \dfrac{\sqrt{5 + K_0}}{5};$$

$$\tag{10-44}$$

$$\left.\begin{array}{lll}
S_{R_1}^{\omega_o} = -\dfrac{1}{25Q^2}, & S_{R_2}^{\omega_o} = -\dfrac{1}{2} + \dfrac{1}{25Q^2}, & S_{R_3}^{\omega_o} = -\dfrac{1}{2} + \dfrac{3}{50Q^2} \\[2mm]
S_{R_4}^{\omega_o} = -\dfrac{3}{50Q^2}, & S_{C_1}^{\omega_o} = S_{C_2}^{\omega_o} = -\dfrac{1}{2}, & S_{K_0}^{\omega_o} = \dfrac{1}{2} - \dfrac{1}{10Q^2}
\end{array}\right\};$$

$$\tag{10-45}$$

$$\left.\begin{array}{lll}
S_{R_1}^{Q} = \dfrac{1}{5} - \dfrac{1}{25Q^2}, & S_{R_2}^{Q} = -\dfrac{1}{10} + \dfrac{1}{25Q^2}, & S_{R_3}^{Q} = -\dfrac{3}{10} + \dfrac{3}{50Q^2} \\[2mm]
S_{R_4}^{Q} = \dfrac{1}{5} - \dfrac{3}{50Q^2}, & S_{C_1}^{Q} = \dfrac{1}{10}, \quad S_{C_2}^{Q} = -\dfrac{1}{10}, & S_{K_0}^{Q} = \dfrac{1}{2} - \dfrac{1}{10Q^2}
\end{array}\right\}.$$

$$\tag{10-46}$$

For complex poles, $Q > \frac{1}{2}$. Hence all sensitivity figures are $\frac{1}{2}$ or less in magnitude. These values are equal to or better than the values of the passive realizations. The sensitivity values also indicate which way to change an element to achieve a desired change in ω_o or Q. For example, increasing K_0 increases both ω_o and Q (see $S_{K_0}^{\omega_o}$ and $S_{K_0}^{Q}$). On the other hand, increasing C_1 decreases ω_o and increases Q; as the $S_{C_1}^{\omega_o}$ and $S_{C_1}^{Q}$ figures indicate, the decrease in ω_o is much larger than the increase in Q.

For a given ω_o and Q, the elements of the network of Fig. 10-6c are determined by using Eq. (10-44). For example, if $\omega_o = 1$ Krad/s and $Q = 5$, then

$$K_0 = 25Q^2 - 5 = 620, \qquad RC = \frac{\sqrt{5 + K_0}}{\omega_o} = 25 \times 10^{-3}.$$

If C is chosen as 0.1 μF, then $R = 250$ KΩ.

10-6 — *KRC* REALIZATION (ONE-POLE ROLLOFF)

As the sensitivity figures indicate, the performance of the $-KRC$ network is excellent when the amplifier is considered ideal. However, when a one-pole roll-off model is used to characterize the operational amplifier, the results show a high degree of dependence on the gain-bandwidth product of the amplifier. What is worse, the circuit may even become unstable if the ω_o and Q are high enough.

To investigate the dependence of the pole locations on the GB of the operational amplifier, consider the general $-KRC$ realization shown in Fig. 10-7a.

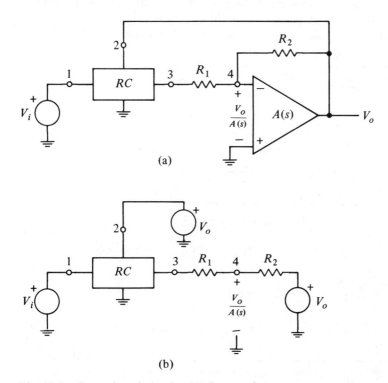

(a)

(b)

Fig. 10-7   General analysis of $-KRC$ network

The RC network is driven by the independent source, V_i, and the dependent source, V_o, which has access to the RC network through ports 2 and 3. For convenience, two separate V_o sources are shown in Fig. 10-7b for calculating the action of feedback.

The output of the RC network, V_3, is constrained by the amplifier to:

$$V_3 = \frac{V_o}{A(s)} - \left[\frac{V_o - \dfrac{V_o}{A(s)}}{R_2} \right] R_1 = -\frac{V_o}{K_0} + \frac{V_o}{A(s)} \left(1 + \frac{1}{K_0} \right), \tag{10-47}$$

where $K_0 = R_2/R_1$. But V_3 can also be written as the superposition of voltages resulting from excitations at ports 1, 2, and 4:

$$V_3 = V_i T_{31} + V_o T_{32} + \frac{V_o}{A(s)} T_{34}, \tag{10-48}$$

where

$$T_{31} = \frac{V_3}{V_1} \bigg|_{V_2 = V_4 = 0} = \frac{N_{31}}{D}, \qquad T_{32} = \frac{V_3}{V_2} \bigg|_{V_1 = V_4 = 0} = \frac{N_{32}}{D},$$

$$T_{34} = \frac{V_3}{V_4} \bigg|_{V_1 = V_2 = 0} = \frac{N_{34}}{D}.$$

Since all three transfer functions pertain to the same RCR_1 network, the denominator polynomials are assumed to be identical. The transfer function, (V_o/V_i), is obtained by eliminating V_3 from Eqs. (10-47) and (10-48):

$$\frac{V_o}{V_i} = -\frac{K_0 T_{31}}{1 + K_0 T_{32} - \dfrac{(1 + K_0) - K_0 T_{34}}{A(s)}}$$

$$= -\frac{K_0 N_{31}}{D + K_0 N_{32} - \dfrac{(1 + K_0)D - K_0 N_{34}}{A(s)}}. \tag{10-49}$$

With $A(s) \cong (-GB/s)$, Eq. (10-49) becomes

$$\frac{V_o}{V_i} = -\frac{K_0 N_{31}}{D + K_0 N_{32} + \dfrac{s}{GB}[(1 + K_0)D - K_0 N_{34}]}. \tag{10-50}$$

If the system is second order and $GB = \infty$, Eq. (10-50) simplifies to

$$\frac{V_o}{V_i} = -\frac{K_0 N_{31}}{D + K_0 N_{32}} = -\frac{K_0 N_{31}}{(s + \alpha_1)(s + \alpha_2) + K_0(b_2 s^2 + b_1 s + b_0)}$$

$$= -\frac{K_0 N_{31}}{(1 + K_0 b_2)s^2 + s(\alpha_1 + \alpha_2 + K_0 b_1) + (\alpha_1 \alpha_2 + K_0 b_0)}$$

$$= -\frac{K_0 N_{31}}{(1 + K_0 b_2)\left(s^2 + s\dfrac{\omega_o}{Q} + \omega_o^2 \right)}. \tag{10-51}$$

With ω_o and Q thus defined by the ideal system, the nonideal system can be expressed as

$$\frac{V_o}{V_i} = -\frac{K_0 N_{31}}{(1 + K_0 b_2)\left(s^2 + s\dfrac{\omega_o}{Q} + \omega_o^2\right) + \dfrac{s}{GB}[(1 + K_0)(s + \alpha_1)(s + \alpha_2) - K_0 N_{34}]}.$$
(10-52)

The effect of GB on the complex poles can be determined by studying the properties of

$$(1 + K_0 b_2)\left(s^2 + s\frac{\omega_o}{Q} + \omega_o^2\right) + \frac{s}{GB}[(1 + K_0)(s + \alpha_1)(s + \alpha_2) - K_0 N_{34}] = 0).$$
(10-53)

General Analysis of the Low-Pass Type Feedback

If the loop gain is a low-pass function, then $N_{32} = b_0$. Consequently, in the ideal case, K_0 varies the constant term in the denominator polynomial, which results in constant-α pole loci with

$$\alpha = \frac{\alpha_1 + \alpha_2}{2}.$$

(The loci don't quite form a vertical line because the nonideal system is of third order.) With $b_2 = b_1 = 0$, the characteristic equation, Eq. (10-53), simplifies to

$$\left(s^2 + s\frac{\omega_o}{Q} + \omega_o^2\right) + \frac{s}{GB}\left[(1 + K_0)\left(s^2 + s\frac{\omega_o}{Q} + \alpha_1\alpha_2\right) - K_0 N_{34}(s)\right] = 0.$$
(10-54)

To evaluate α_1, α_2, and N_{34}, ground ports 1 and 2 of the RC network shown in Fig. 10-7b, and drive port 4 with a voltage source. Calculate:

$$\frac{V_3}{V_4} = \frac{N_{34}}{(s + \alpha_1)(s + \alpha_2)},$$
(10-55)

and obtain α_1, α_2, and N_{34}. With all the values thus determined for a specific circuit, Eq. (10-54) can be used to determine the pole shifts caused by the GB of the amplifier.

Pole, ω_o, and Q Changes in the Circuit of Fig. 10-6c

As a specific example of the effect of the GB on the complex poles, consider the $-KRC$ realization given in Fig. 10-6c with

$$R_1 = R_2 = R_3 = R_4 = R, \qquad R_5 = K_0 R, \qquad \text{and} \qquad C_1 = C_2 = C.$$

If the amplifier is ideal, the transfer function is given by Eq. (10-44):

$$\frac{V_o}{V_i} = -\frac{K_0}{R^2 C^2} \frac{1}{s^2 + s \dfrac{5}{RC} + \dfrac{5 + K_0}{R^2 C^2}} = -\frac{K_0}{R^2 C^2} \frac{1}{s^2 + s \dfrac{\omega_o}{Q} + \omega_o^2}, \qquad (10\text{-}56)$$

where

$$\omega_o = \frac{\sqrt{5 + K_0}}{RC}, \qquad Q = \frac{\sqrt{5 + K_0}}{5}, \qquad \frac{\omega_o}{Q} = \frac{5}{RC}.$$

For $K_0 = 0$, the denominator of Eq. (10-56) describes the passive system. Hence, $D = (s + \alpha_1)(s + \alpha_2)$ and, consequently, $\alpha_1 \alpha_2$ can be readily obtained as

$$s^2 + s \frac{5}{RC} + \frac{5}{R^2 C^2} = s^2 + s \frac{\omega_o}{Q} + \frac{1}{5}\left(\frac{\omega_o}{Q}\right)^2 = (s + \alpha_1)(s + \alpha_2), \qquad (10\text{-}57)$$

$$\alpha_1 \alpha_2 = \frac{1}{5}\left(\frac{\omega_o}{Q}\right)^2.$$

To obtain N_{34}, the passive portion of the circuit is redrawn as shown in Fig. 10-8 and V_3/V_4 is calculated:

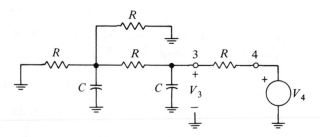

Fig. 10-8   Evaluation of N_{34}

$$\frac{V_3}{V_4} = \frac{1}{RC} \frac{s + \dfrac{3}{RC}}{s^2 + s \dfrac{5}{RC} + \dfrac{5}{R^2 C^2}}. \qquad (10\text{-}58)$$

Hence,

$$N_{34} = \frac{1}{RC}\left(s + \frac{3}{RC}\right) = \frac{1}{5}\frac{\omega_o}{Q}\left(s + \frac{3}{5}\frac{\omega_o}{Q}\right). \qquad (10\text{-}59)$$

Note that the denominator of V_3/V_4 can just as well be used to determine

$$D = (s + \alpha_1)(s + \alpha_2),$$

and hence $\alpha_1 \alpha_2$.

Equation (10-54) can now be written as

$$\underbrace{\left(s^2 + s\frac{\omega_o}{Q} + \omega_o^2\right)}_{P_1(s)}$$

$$+ \frac{1}{GB}\underbrace{s\left\{(1 + K_0)\left[s^2 + s\frac{\omega_o}{Q} + \frac{1}{5}\left(\frac{\omega_o}{Q}\right)^2\right] - K_0\frac{1}{5}\frac{\omega_o}{Q}\left(s + \frac{3}{5}\frac{\omega_o}{Q}\right)\right\}}_{P_2(s)} = 0.$$

At the nominal-pole position, the value of the derivative of the upper half-plane pole p_1 is given by

$$\frac{dp_1}{d\left(\dfrac{1}{GB}\right)} = -\left.\frac{P_2(s)}{\dfrac{dP_1(s)}{ds}}\right|_{s=p_1}$$

$$= -\left.\frac{s\left\{(1 + K_0)\left[s^2 + s\dfrac{\omega_o}{Q} + \dfrac{1}{5}\left(\dfrac{\omega_o}{Q}\right)^2\right] - \dfrac{K_0}{5}\dfrac{\omega_o}{Q}\left(s + \dfrac{3}{5}\dfrac{\omega_o}{Q}\right)\right\}}{2s + \dfrac{\omega_o}{Q}}\right|_{s=p_1}$$

$$= -\frac{p_1\left\{(1 + K_0)\left[-\omega_o^2 + \dfrac{1}{5}\left(\dfrac{\omega_o}{Q}\right)^2\right] - \dfrac{K_0}{5}\dfrac{\omega_o}{Q}\left[p_1 + \dfrac{3}{5}\dfrac{\omega_o}{Q}\right]\right\}}{2p_1 + \dfrac{\omega_o}{Q}}.$$

But $K_0 = 25Q^2 - 5$. Hence,

$$\frac{dp_1}{d\left(\dfrac{1}{GB}\right)} = \frac{p_1\left(5Q - \dfrac{1}{Q}\right)\left(5Q - \dfrac{1}{5Q} + \dfrac{p_1}{\omega_o}\right)\omega_o^2}{2p_1 + \dfrac{\omega_o}{Q}},$$

where

$$\frac{p_1}{\omega_o} = -\frac{1}{2Q} + j\sqrt{1 - \frac{1}{4Q^2}},$$

$$\frac{dp_1}{d\left(\dfrac{1}{GB}\right)} = \frac{25Q^2}{2}\left(1 - \frac{1}{5Q^2}\right)\left(1 - \frac{6}{25Q^2}\right)\omega_o^2 + j\frac{35Q}{4}\frac{\left(1 - \dfrac{1}{5Q^2}\right)\left(1 - \dfrac{6}{35Q^2}\right)\omega_o^2}{\sqrt{1 - \dfrac{1}{4Q^2}}}$$

$$= -\frac{d\alpha_1}{d\left(\dfrac{1}{GB}\right)} + j\frac{d\beta_1}{d\left(\dfrac{1}{GB}\right)}. \tag{10-60}$$

In terms of incremental changes,

$$-\frac{\Delta\alpha_1}{\omega_o} \cong \frac{25Q^2}{2}\left(1 - \frac{1}{5Q^2}\right)\left(1 - \frac{6}{25Q^2}\right)\frac{\omega_o}{GB}, \tag{10-61a}$$

$$\frac{\Delta\beta_1}{\omega_o} \cong \frac{35Q}{4}\frac{\left(1 - \frac{1}{5Q^2}\right)\left(1 - \frac{6}{35Q^2}\right)}{\sqrt{1 - \frac{1}{4Q^2}}}\frac{\omega_o}{GB}. \tag{10-61b}$$

Using Eq. (10-33), the corresponding changes in ω_o and Q are obtained as follows:

$$\frac{\Delta\omega_o}{\omega_o} \cong \frac{1}{2Q}\frac{\Delta\alpha_1}{\omega_o} + \sqrt{1 - \frac{1}{4Q^2}}\frac{\Delta\beta_1}{\omega_o}$$

$$= \frac{5Q}{2}\left(1 - \frac{1}{5Q^2}\right)\frac{\omega_o}{GB}, \tag{10-62a}$$

$$\frac{\Delta Q}{Q} \cong -2Q\left(1 - \frac{1}{4Q^2}\right)\frac{\Delta\alpha_1}{\omega_o} + \sqrt{1 - \frac{1}{4Q^2}}\frac{\Delta\beta_1}{\omega_o}$$

$$= 25Q^3\left(1 - \frac{1}{5Q^2}\right)\left(1 - \frac{7}{50Q^2}\right). \tag{10-62b}$$

A summary of these results is given in Table 10-2.

Table 10-2   — *KRC* Realization
(see Fig. 10-6c)

Equal-R, Equal-C

$$\omega_o = \frac{\sqrt{5 + K_0}}{RC}, \qquad Q = \frac{\sqrt{5 + K_0}}{5}$$

$$\frac{V_o}{V_i} = -\frac{\omega_o^2\left(1 - \frac{1}{5Q^2}\right)}{s^2 + s\dfrac{\omega_o}{Q} + \omega_o^2 + \dfrac{s}{GB}\left[s^2(25Q^2 - 4) + s\omega_o\left(20Q - \dfrac{3}{Q}\right) + \left(2 - \dfrac{1}{5Q^2}\right)\omega_o^2\right]}$$

$$\left(\omega_a \ll \frac{\omega_o}{2Q}\right)$$

$$-\frac{\Delta\alpha}{\omega_o} \cong \frac{25Q^2}{2}\left(1 - \frac{1}{5Q^2}\right)\left(1 - \frac{6}{25Q^2}\right)\frac{\omega_o}{GB}, \qquad \frac{\Delta\beta}{\omega_o} \cong \frac{35Q}{4}\frac{\left(1 - \frac{1}{5Q^2}\right)\left(1 - \frac{6}{35Q^2}\right)}{\sqrt{1 - \frac{1}{4Q^2}}}\frac{\omega_o}{GB}$$

$$\frac{\Delta\omega_o}{\omega_o} \cong \frac{5Q}{2}\left(1 - \frac{1}{5Q^2}\right)\frac{\omega_o}{GB}, \qquad \frac{\Delta Q}{Q} \cong 25Q^3\left(1 - \frac{1}{5Q^2}\right)\left(1 - \frac{7}{5Q^2}\right)\frac{\omega_o}{GB}$$

As Table 10-2 shows, the Q of the complex poles is highly dependent on the nominal Q of the realization. For high-Q poles, the change in Q is proportional to the third power of the nominal Q, that is,

$$\frac{\Delta Q}{Q} \cong 25Q^3 \frac{\omega_o}{GB}. \tag{10-63}$$

Consequently, unless ω_o/GB is very small, the frequency response will differ to a marked extent from the ideal.

If (ω_o/GB) is not small, the equations in Table 10-2 predicting the first-order changes in the pole positions are not accurate. The actual pole positions will then depart considerably from the nominal positions. In that case, the exact equation for the poles, obtained from Table 10-2, is solved for the roots using a digital computer. The resulting upper half-plane pole is plotted in Fig. 10-9, using $GB_n = GB/\omega_o$ and the nominal value of the Q as parameters. The solid lines are constant nominal-Q lines on which various values of GB_n are marked. All curves start at the nominal-Q positions on a unit circle ($GB_n = \infty$). As GB_n decreases from infinity, the actual Q of the poles becomes higher. For a given GB_n, the change in pole positions becomes progressively larger for high nominal-Q poles. In fact, the poles may even be displaced into the right half-plane, thus

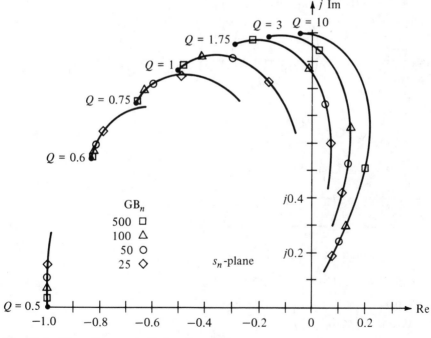

Fig. 10-9   Plot of upper half-plane root of

$$s_n^3(25Q^2 - 4) + s_n^2\left(20Q - \frac{3}{Q} + GB_n\right) + s_n\left(2 - \frac{1}{5Q^2} + \frac{GB_n}{Q}\right) + GB_n = 0$$

causing instability in the low-pass realization. Thus the circuit, which possesses excellent sensitivity characteristics when the amplifier is considered ideal, may become an oscillator when the one-pole rolloff characteristic of the operational amplifier is taken into account. This undesirable feature can be traced to the large dc gains that are required of the amplifier in order to produce high-Q poles. Consequently, the closed-loop bandwidth of the amplifier is quite narrow, and its effect on the poles is therefore more pronounced than for the KRC realization discussed in Section 10-4.

EXAMPLE 10-5

Obtain the condition for stability for the $-KRC$ (equal-R, equal-C) circuit of Fig. 10-6c. Assume that the amplifier is represented by the one-pole rolloff model and $Q \geq 5$.

SOLUTION

Obtain the characteristic equation from the denominator polynomial of (V_o/V_i) given in Table 10-2. Normalize the frequency with respect to ω_o, and rearrange the equation in descending powers of s:

$$s_n^3 \frac{(25Q^2 - 4)}{GB_n} + s_n^2 \left(\frac{20Q - \dfrac{3}{Q}}{GB_n} + 1 \right) + s_n \left(\frac{2 - \dfrac{1}{5Q^2}}{GB_n} + \frac{1}{Q} \right) + 1 = 0, \qquad (10\text{-}64)$$

where

$$s_n = \frac{s}{\omega_o}, \qquad GB_n = \frac{GB}{\omega_o}.$$

Since $Q \geq 5$, simplify Eq. (10-64) to

$$s_n^3 \underbrace{\left(\frac{25Q^2}{GB_n} \right)}_{a_3} + s_n^2 \underbrace{\left(\frac{20Q}{GB_n} + 1 \right)}_{a_2} + s_n \underbrace{\left(\frac{2}{GB_n} + \frac{1}{Q} \right)}_{a_1} + 1 = 0.$$

The condition for stability is

$$a_2 a_1 > a_3 \qquad \text{(Routh's criterion)},$$

$$\left(\frac{20Q}{GB_n} + 1 \right) \left(\frac{2}{GB_n} + \frac{1}{Q} \right) > \frac{25Q^2}{GB_n},$$

$$\frac{40Q}{GB_n^2} + \frac{22 - 25Q^2}{GB_n} + \frac{1}{Q} > 0.$$

Simplify further by neglecting insignificant terms $[(40Q/GB_n^2)$ and $(22/GB_n)]$:

$$-\frac{25Q^2}{GB_n} + \frac{1}{Q} > 0,$$

$$GB_n > 25Q^3,$$

$$\omega_o < \frac{GB}{25Q^3}. \qquad Ans.$$

If the $-KRC$ circuit of Fig. 10-6c is designed on the basis of an ideal amplifier for $\omega_o = 10^3$ rad/s and $Q = 10$, then the circuit will oscillate for any GB less than 25×10^6 rad/s. If $GB = 25 \times 10^6$, the circuit will oscillate at 10^3 rad/s.

EXAMPLE 10-6

In the $-KRC$ realization shown in Fig. 10-6c,

$$R_1 = R_2 = R_3 = R_4 = R, \qquad \text{and} \qquad C_1 = C_2 = C.$$

The ω_o and Q of the poles are calculated by assuming an ideal amplifier

(a) What constraint must be placed on the gain-bandwidth product of the amplifier if ω_{oa} and Q_a are to differ no more than 5% from ω_o and Q?
(b) Evaluate the constraint relation obtained in (a) for $GB = 10^6$ and $Q = 10$.

SOLUTION

(a) From Table 10-2,

$$\frac{\Delta\omega_o}{\omega_o} \cong \frac{5Q}{2}\left(1 - \frac{1}{5Q^2}\right)\frac{\omega_o}{GB}, \qquad \frac{\Delta Q}{Q} \cong 25Q^3\left(1 - \frac{7}{50Q^2}\right)\left(1 - \frac{1}{5Q^2}\right)\frac{\omega_o}{GB}.$$

Because

$$25Q^3\left(1 - \frac{7}{50Q^2}\right) > \frac{5Q}{2} \qquad \text{for any } Q \geq \frac{1}{2},$$

the constraint relation is set by the allowable deviation in Q. Therefore,

$$25Q^3\left(1 - \frac{7}{50Q^2}\right)\left(1 - \frac{1}{5Q^2}\right)\frac{\omega_o}{GB} < 0.05,$$

$$GB > 500Q^3\left(1 - \frac{7}{50Q^2}\right)\left(1 - \frac{1}{5Q^2}\right)\omega_o. \qquad Ans.$$

(b)

$$\omega_o < \frac{\text{GB}}{500Q^3 \left(1 - \dfrac{1}{5Q^2}\right)\left(1 - \dfrac{7}{50Q^2}\right)}$$

$$= \frac{10^6}{500 \times 10^3 \left(1 - \dfrac{1}{500}\right)\left(1 - \dfrac{7}{5000}\right)} \cong 2,$$

$$\omega_o < 2 \text{ rad/s.} \qquad Ans.$$

10-7 REALIZATION WITH INFINITE GAIN

An RC network with transfer function zeros at infinity is shown in Fig. 10-10a. In Section 9.5, it is demonstrated that complex poles result if an operational amplifier is introduced in this network as indicated in Fig. 10-10b. The zeros of the resulting (V_o/V_i) are the same as the zeros of (V_a/V_i). Note that two feedback loops are employed, one through R_1, the other through C_2. (For another interpretation of this feedback scheme, see Problem 10-11.)

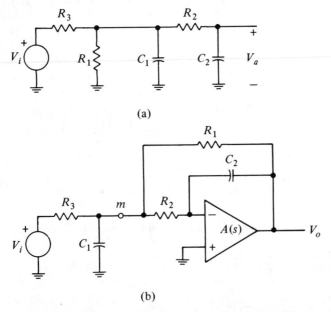

(a)

(b)

Fig. 10-10   Low-pass network with infinite-gain amplifier

For $|A(s)| = \infty$, the transfer function (V_o/V_i) is given by:

$$\frac{V_o}{V_i} = -\frac{1}{R_2 R_3 C_1 C_2} \frac{1}{s^2 + s\dfrac{1}{C_1}\left(\dfrac{1}{R_1} + \dfrac{1}{R_2} + \dfrac{1}{R_3}\right) + \dfrac{1}{R_1 R_2 C_1 C_2}}. \tag{10-65}$$

The ω_o and Q of the poles are:

$$\omega_o = \frac{1}{\sqrt{R_1 R_2 C_1 C_2}}, \qquad Q = \frac{\sqrt{\dfrac{C_1}{C_2}}}{\sqrt{\dfrac{R_1}{R_2}} + \sqrt{\dfrac{R_2}{R_1}} + \dfrac{\sqrt{R_1 R_2}}{R_3}}. \tag{10-66}$$

Since R_3 affects the Q and not the ω_o, it can be used for the fine tuning of the Q.

To fix the ω_o and Q of a complex pair of poles, five elements are available. Therefore, three of them can be chosen arbitrarily. Let $R_1 = R_2 = R_3 = R$, and pick C_1 or C_2 arbitrarily. Then, ω_o, Q, and the sensitivity functions become

$$\omega_o = \frac{1}{R\sqrt{C_1 C_2}}, \qquad Q = \frac{1}{3}\sqrt{\frac{C_1}{C_2}}, \tag{10-67}$$

$$S_{R_1}^{\omega_o} = S_{R_2}^{\omega_o} = S_{C_1}^{\omega_o} = S_{C_2}^{\omega_o} = -\tfrac{1}{2}, \qquad S_{R_3}^{\omega_o} = 0, \tag{10-68}$$

$$S_{R_1}^{Q} = S_{R_2}^{Q} = -\tfrac{1}{6}, \qquad S_{R_3}^{Q} = \tfrac{1}{3}, \qquad S_{C_1}^{Q} = -S_{C_2}^{Q} = \tfrac{1}{2}. \tag{10-69}$$

All the sensitivity figures are no greater than $\tfrac{1}{2}$ in magnitude. Therefore, this active network compares favorably with the passive networks given in Fig. 10-2.

It is interesting to show that the input impedance of the circuit to the right of the terminal m to ground is an inductor in parallel with a resistor (see Problem 10-12). Therefore, if the output is taken from m to ground, the active network is equivalent to an *RLC* network.

For a given ω_o and Q, the elements of the network of Fig. 10-10b are determined by using Eq. (10-67). For example, if $\omega_o = 1$ Krad/s and $Q = 5$, then

$$\frac{C_1}{C_2} = 9Q^2 = 225, \qquad R = \frac{1}{\omega_o\sqrt{C_1 C_2}} = \frac{10^{-3}}{\sqrt{C_1 C_2}} = \frac{10^{-3}}{15 C_2}.$$

If C_2 is chosen as $0.01\ \mu F$, then $C_1 = 2.25\ \mu F$ and $R = 6.67\ K\Omega$.

10-8 REALIZATION WITH INFINITE GAIN (ONE-POLE ROLLOFF)

How sensitive are infinite-gain realizations to changes in the GB of an imperfect operational amplifier? To obtain a quantitative measure of the dependence of the transfer function on the GB, refer to the general case shown in Fig. 10-11a.

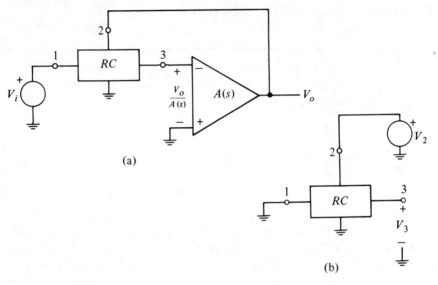

(a)

(b)

Fig. 10-11   General analysis of infinite-gain network

The voltage at port 3 can be written as the superposition of voltages arising from port 1 and port 2 inputs, that is,

$$\frac{V_o}{A(s)} = V_i T_{31} + V_o T_{32}.$$ (10-70)

The transfer functions T_{31} and T_{32} are defined as

$$T_{31} = \left.\frac{V_3}{V_1}\right|_{V_2=0} = \frac{N_{31}}{D},$$ (10-71a)

and

$$T_{32} = \left.\frac{V_3}{V_2}\right|_{V_1=0} = \frac{N_{32}}{D} = \frac{b_2 s^2 + b_1 s + b_0}{(s + \alpha_1)(s + \alpha_2)},$$ (10-71b)

where b_2 represents the high-frequency gain of T_{32}, and $-\alpha_1$, $-\alpha_2$ are the poles of the RC network with ports 1 and 2 grounded and port 3 open-circuited. When solved for V_o/V_i, Eq. (10-70) gives

$$\frac{V_o}{V_i} = -\frac{T_{31}}{T_{32} - \dfrac{1}{A(s)}} = -\frac{N_{31}}{N_{32} - \dfrac{D}{A(s)}}$$

$$= -\frac{N_{31}}{(b_2 s^2 + b_1 s + b_0) + \dfrac{s}{\text{GB}}(s + \alpha_1)(s + \alpha_2)}.$$ (10-72)

For an ideal amplifier, Eq. (10-72) reduces to

$$\frac{V_o}{V_i} = -\frac{N_{31}}{b_2 s^2 + b_1 s + b_0} = -\frac{N_{31}}{b_2\left(s^2 + s\dfrac{\omega_o}{Q} + \omega_o^2\right)}.$$

With ω_o and Q thus defined, the transfer function of the nonideal system can be written as

$$\frac{V_o}{V_i} = -\frac{N_{31}}{b_2\left(s^2 + s\dfrac{\omega_o}{Q} + \omega_o^2\right) + \dfrac{s}{GB}(s + \alpha_1)(s + \alpha_2)}. \tag{10-73}$$

It is interesting to note that this same result can be obtained directly from Eq. (10-24) if $K_0 \to \infty$ and $A(s)$ is replaced by $-A(s)$. That these operations are valid can be seen by comparing Fig. 10-11a with Fig. 10-4a.

To evaluate b_2, α_1, and α_2, drive the RC circuit as shown in Fig. 10-11b and obtain:

$$\frac{V_3}{V_2} = \frac{b_2 s^2 + \cdots}{(s + \alpha_1)(s + \alpha_2)}. \tag{10-74}$$

As Eq. (10-73) shows, the GB of the operational amplifier affects the position of the poles. The sensitivity of the poles to variations in GB can be readily calculated for a given circuit.

Pole, ω_o, and Q Changes in the Circuit of Fig. 10-10b

When Eqs. (10-73) and (10-74) are applied to the network of Fig. 10-10b, the following transfer function (see Problem 10-15) results for the equal-R case:

$$\frac{V_o}{V_i} = -\frac{\omega_o^2}{s^2 + s\dfrac{\omega_o}{Q} + \omega_o^2 + \dfrac{s}{GB}\left[s^2 + s\omega_o\left(3Q + \dfrac{1}{Q}\right) + 2\omega_o^2\right]}, \tag{10-75}$$

where

$$\omega_o = \frac{1}{R\sqrt{C_1 C_2}}, \qquad Q = \frac{1}{3}\sqrt{\frac{C_1}{C_2}}.$$

The nominal complex-pole positions are determined by ω_o and Q. The actual complex-pole positions are given by the complex roots of

$$\underbrace{s^2 + s\frac{\omega_o}{Q} + \omega_o^2}_{P_1(s)} + \underbrace{\frac{1}{GB}\left\{s\left[s^2 + s\omega_o\left(3Q + \frac{1}{Q}\right) + 2\omega_o^2\right]\right\}}_{P_2(s)} = 0. \tag{10-76}$$

The value of the derivative of the upper half-plane pole p_1, with respect to $(1/GB)$ at the nominal position of the pole, is

$$\frac{dp_1}{d\left(\frac{1}{GB}\right)} = -\frac{P_2(s)}{\frac{dP_1(s)}{ds}}\Bigg|_{s=p_1} = -\frac{s\left[s^2 + s\omega_o\left(3Q + \frac{1}{Q}\right) + 2\omega_o^2\right]}{2s + \frac{\omega_o}{Q}}\Bigg|_{s=p_1}$$

$$= \omega_o^2 + j\frac{\omega_o^2}{2}\frac{3Q - \frac{1}{Q}}{\sqrt{1 - \frac{1}{4Q^2}}} \tag{10-77}$$

$$= -\frac{d\alpha_1}{d\left(\frac{1}{GB}\right)} + j\frac{d\beta_1}{d\left(\frac{1}{GB}\right)},$$

where $-\alpha_1 + j\beta_1$ represents the pole. In terms of incremental changes, the normalized displacements are

$$-\frac{\Delta\alpha_1}{\omega_o} \cong \frac{\omega_o}{GB} \tag{10-78a}$$

$$\frac{\Delta\beta_1}{\omega_o} = -\frac{1}{2}\frac{3Q - \frac{1}{Q}}{\sqrt{1 - \frac{1}{4Q^2}}}\frac{\omega_o}{GB}. \tag{10-78b}$$

Using Eq. (10-33), the corresponding changes in ω_o and Q are obtained:

$$\frac{\Delta\omega_o}{\omega_o} \cong \frac{1}{2Q}\frac{\Delta\alpha_1}{\omega_o} + \sqrt{1 - \frac{1}{4Q^2}}\frac{\Delta\beta_1}{\omega_o}$$

$$= -\frac{3Q}{2}\frac{\omega_o}{GB}; \tag{10-79a}$$

$$\frac{\Delta Q}{Q} \cong -2Q\left(1 - \frac{1}{4Q^2}\right)\frac{\Delta\alpha_1}{\omega_o} + \sqrt{1 - \frac{1}{4Q^2}}\frac{\Delta\beta_1}{\omega_o}$$

$$= \frac{Q}{2}\frac{\omega_o}{GB}. \tag{10-79b}$$

A summary of these results is given in Table 10-3.

If (ω_o/GB) is not much less than 1, then the equations in Table 10-3 do not give accurate results. In that case, the cubic equation of the poles, given by Eq. (10-76), is solved with the aid of a digital computer. The results obtained this way are plotted in Fig. 10-12. The solid lines indicate constant nominal-Q curves,

Table 10-3   Infinite-gain Realization
(see Fig. 10-10b)

Equal-R

$$\omega_o = \frac{1}{R\sqrt{C_1 C_2}}; \qquad Q = \frac{1}{3}\sqrt{\frac{C_1}{C_2}}$$

$$\frac{V_o}{V_i} = -\frac{\omega_o^2}{s^2 + s\frac{\omega_o}{Q} + \omega_o^2 + \frac{s}{GB}\left[s^2 + s\omega_o\left(3Q + \frac{1}{Q}\right) + 2\omega_o^2\right]} \qquad \left(\omega_a \ll \frac{\omega_o}{2Q}\right)$$

$$-\frac{\Delta\alpha}{\omega_o} \cong \frac{\omega_o}{GB}, \qquad \frac{\Delta\beta}{\omega_o} \cong -\frac{1}{2}\frac{3Q - \frac{1}{Q}}{\sqrt{1 - \frac{1}{4Q^2}}}\frac{\omega_o}{GB}$$

$$\frac{\Delta\omega_o}{\omega_o} \cong -\frac{3Q}{2}\frac{\omega_o}{GB}, \qquad \frac{\Delta Q}{Q} \cong \frac{Q}{2}\frac{\omega_o}{GB}$$

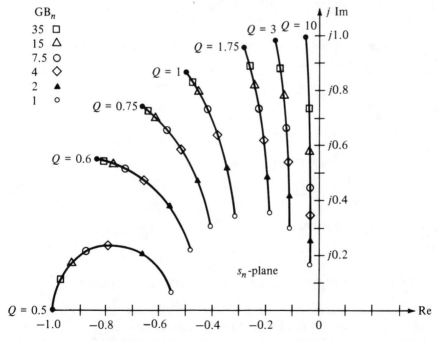

Fig. 10-12   Plot of upper half-plane root of

$$s_n^3 + s_n^2\left(3Q + \frac{1}{Q} + GB_n\right) + s_n\left(2 + \frac{GB_n}{Q}\right) + GB_n = 0$$

on which $GB_n = (GB/\omega_o)$ values are marked. Note that as GB_n is decreased from infinity, Q_a increases first, becomes invariant, then decreases, whereas ω_{oa} always decreases. Since the pole does not enter the right half-plane, the circuit is absolutely stable.

EXAMPLE 10-7

In the low-pass, infinite-gain realization shown in Fig. 10-10b, $R_1 = R_2 = R_3$. The ω_o and Q of the poles are calculated by assuming the amplifier is ideal.
(a) What constraint must be placed on the gain-bandwidth product of the operational amplifier if ω_{oa} and Q_a are to differ no more than 5% from ω_o and Q?
(b) Evaluate the constraint relation obtained in (a) for $GB = 10^6$ and $Q = 10$.

SOLUTION

(a) From Table 10-3,

$$\frac{\Delta\omega_o}{\omega_o} \cong -\frac{3Q}{2}\frac{\omega_o}{GB}, \qquad \frac{\Delta Q}{Q} \cong \frac{Q}{2}\frac{\omega_o}{GB}.$$

Since ω_o changes more than Q, $\Delta\omega_o/\omega_o$ determines the maximum allowable deviation.

$$\frac{3Q}{2}\frac{\omega_o}{GB} < 0.05,$$

$$GB > 30Q\omega_o. \qquad Ans.$$

(b)

$$\omega_o < \frac{GB}{30Q} = \frac{10^6}{300},$$

$$\omega_o < 3.33 \text{ Krad/s.} \qquad Ans.$$

10-9 REALIZATION WITH THREE OPERATIONAL AMPLIFIERS

Second-order systems can also be generated by using three active building blocks: the inverter, the integrator, and the lossy integrator.

The inverter is shown in Fig. 10-13a. It realizes the transfer function

$$\frac{V_o}{V_i} = -\frac{R_2}{R_1}.$$

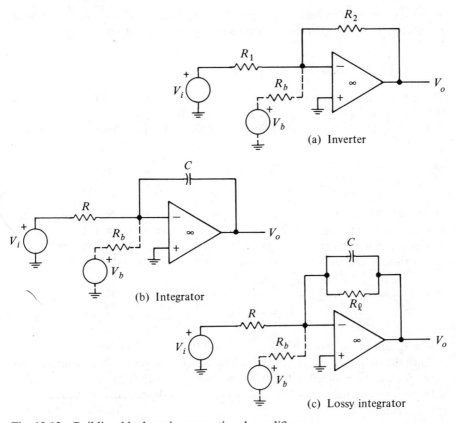

Fig. 10-13   Building blocks using operational amplifiers

It can also be used with more than one input. For example, when the second source V_b is added (shown dashed in the diagram), the output becomes the inverted and weighted sum of the two input signals:

$$V_o = -\frac{R_2}{R_1} V_i - \frac{R_2}{R_b} V_b.$$

The two transfer functions, one for each input while the other input is held at zero, are

$$-\frac{R_2}{R_1} \quad \text{and} \quad -\frac{R_2}{R_b}.$$

The integrator is shown in Fig. 10-13b. It realizes the transfer function

$$\frac{V_o}{V_i} = -\frac{1}{sRC}.$$

It can also be used to integrate more than one input. For example, if the input signals are V_i and V_b (shown dashed in the figure), the output is

$$V_o = -\frac{1}{sRC}V_i - \frac{1}{sR_bC}V_b.$$

The transfer function, associated with the signal V_i, is $-(1/sRC)$. The transfer function associated with the signal V_b is $-(1/sR_bC)$.

The lossy integrator is shown in Fig. 10-13c. It realizes the transfer function

$$\frac{V_o}{V_i} = -\frac{1}{\left(s + \dfrac{1}{R_lC}\right)RC}.$$

Note that the transfer function of the lossy integrator can be readily obtained from the transfer function of the ideal integrator by replacing s with $[s + (1/R_lC)]$. The product R_lC may be thought of as the loss-time-constant. When two (or more) signals are used as inputs, the output of the lossy integrator becomes

$$V_o = -\frac{1}{\left(s + \dfrac{1}{R_lC}\right)RC}V_i - \frac{1}{\left(s + \dfrac{1}{R_lC}\right)R_bC}V_b,$$

where V_b and R_b (shown dashed in Fig. 10-13) represent the second source.

Since the output of each network in Fig. 10-13 is an ideal voltage source, these networks can be coupled without interaction. The transfer function associated with two or more networks connected in cascade is the product of the individual transfer functions.

To see how these building blocks can be used to obtain second-order realizations, consider the low-pass function

$$\frac{V_o}{V_i} = \frac{1}{s^2 + s\dfrac{\omega_o}{Q} + \omega_o^2},$$

which can be written as

$$V_o\left(s^2 + s\frac{\omega_o}{Q} + \omega_o^2\right) = V_i. \tag{10-80}$$

To obtain a V_o term by itself, divide both sides of Eq. (10-80) by $[s^2 + s(\omega_o/Q)]$, and rearrange the result:

$$V_o\left[1 + \frac{\omega_o^2}{s\left(s + \dfrac{\omega_o}{Q}\right)}\right] = \frac{V_i}{s\left(s + \dfrac{\omega_o}{Q}\right)},$$

$$V_o = V_i\left(-\frac{1}{s}\right)\left(-\frac{1}{s + \dfrac{\omega_o}{Q}}\right) + V_o(-\omega_o^2)\left(-\frac{1}{s}\right)\left(-\frac{1}{s + \dfrac{\omega_o}{Q}}\right). \tag{10-81}$$

It should be clear that the arrangement indicated in Eq. (10-81) is not unique. The terms could be grouped together differently. The various parenthetical terms in Eq. (10-81) are readily recognized as the transfer function of an inverter, integrator, or lossy integrator. The output V_o is then interpreted as the superposition of two signals. One is due to the input signal V_i, which is multiplied by the transfer functions

$$\left(-\frac{1}{s}\right) \quad \text{and} \quad \left(-\frac{1}{s + \dfrac{\omega_o}{Q}}\right).$$

The other is due to the output signal (thought of as a dependent-source signal), which is multiplied by the transfer functions

$$(-\omega_o^2), \quad \cdot \left(-\frac{1}{s}\right), \quad \text{and} \quad \left(-\frac{1}{s + \dfrac{\omega_o}{Q}}\right);$$

the product of these three transfer functions represents the loop gain of the system. With these points in mind, Eq. (10-81) is implemented as the network shown in Fig. 10-14 (with the switch in position 1). An inverter, integrator, and lossy integrator serve as the building blocks for this realization.

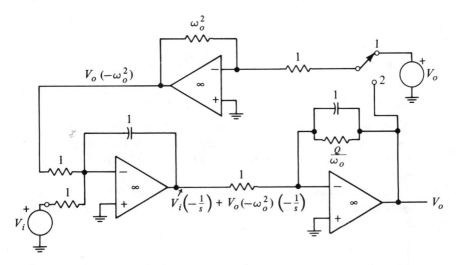

Fig. 10-14   Generation of a low-pass network

Since the dependent source V_o is the same as the output V_o, the switch can be thrown to position 2 to obtain the desired low-pass network. Because each transfer function is the ratio of two impedances, the impedances associated with each building block can be scaled up (or down) without affecting (V_o/V_i). Thus, the two resistances in the inverter can be multiplied by 10^4, or the two input resistances of the integrator can be multiplied by 10^6 and the capacitance divided by 10^6 without changing the transfer function, and so on. It should also be clear that the inverter gain can be changed from $(-\omega_o^2)$ to (-1) provided the input resistance of the integrator (corresponding to the feedback input) is changed to $(1/\omega_o^2)$. (Think in terms of the transfer-function products.)

The approach used in this development is not unique, as the following example indicates.

EXAMPLE 10-8

Starting with the low-pass function, arrange the output voltage in a form different from Eq. (10-81) and obtain a network realization.

SOLUTION

From Eq. (10-80) obtain the rearranged low-pass function:

$$V_o\left(s^2 + s\frac{\omega_o}{Q} + \omega_o^2\right) = V_i.$$

To obtain a different decomposition for the output, divide both sides of the equation by s^2 and regroup the result:

$$V_o\left(1 + \frac{1}{s}\frac{\omega_o}{Q} + \frac{\omega_o^2}{s^2}\right) = \frac{V_i}{s^2},$$

$$V_o = V_i\left(-\frac{1}{s}\right)\left(-\frac{1}{s}\right) + V_o\left(-\frac{\frac{\omega_o}{Q}}{s}\right) + V_o(-\omega_o^2)\left(-\frac{1}{s}\right)\left(-\frac{1}{s}\right). \tag{10-82}$$

The output is formed by the superposition of three signals, one due to the input and the other two due to the output itself. On the righthand side of Eq. (10-82), there are two terms dependent on V_o. Therefore, there are two feedback loops.

Keeping in mind the principle of superposition, Eq. (10-82) can be implemented as shown in Fig. 10-15. First, the input, V_i, is operated on by the two integrators to obtain

$$V_i\left(-\frac{1}{s}\right)\left(-\frac{1}{s}\right).$$

Second, the dependent source, V_o, is operated on by the second integrator to obtain

$$V_o\left(-\frac{\omega_o/Q}{s}\right).$$

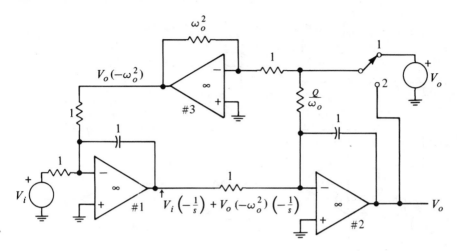

Fig. 10-15

Third, the dependent source, V_o, is passed through the inverter and the two integrators to form

$$V_o(-\omega_o^2)\left(-\frac{1}{s}\right)\left(-\frac{1}{s}\right).$$

Thus, in all, two integrators and one inverter are needed. Since the dependent source V_o is available at the output of the second integrator, the switch can be thrown to position 2 to obtain the desired network.

At first it may appear that this is a new realization for the low-pass function. Upon closer examination, it is seen that the network of Fig. 10-15 is the same as the network of Fig. 10-14 if the switches are in position 2. However, two different methods are used to arrive at the same network.

The $V_o[-(\omega_o/Q)/s]$ term in Eq. (10-82) can also be generated by the scheme shown in Problem 10-17.

Sensitivity Functions

The network of Fig. 10-14 is redrawn in Fig. 10-16 with each element labelled for identification so that the effect of each element on the ω_o- and Q-sensitivities can be determined.

The transfer function (V_o/V_i), the ω_o, and the Q of the poles are given by

$$\frac{V_o}{V_i} = \frac{1}{R_0 R_2 C_1 C_2} \frac{1}{s^2 + s\dfrac{1}{R_Q C_2} + \dfrac{R_4/R_3}{R_1 R_2 C_1 C_2}}; \tag{10-83}$$

$$\omega_o = \sqrt{\frac{R_4/R_3}{R_1 R_2 C_1 C_2}}, \qquad Q = \sqrt{\frac{R_4}{R_3}}\frac{R_Q C_2}{\sqrt{R_1 R_2 C_1 C_2}}. \tag{10-84}$$

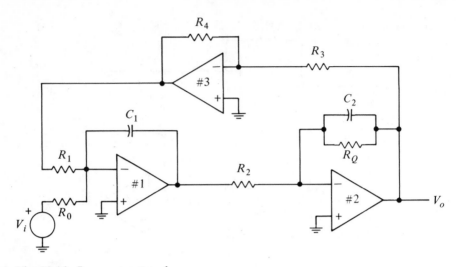

Fig. 10-16   Low-pass network

Since R_Q affects Q but not ω_o, it can be used to adjust the Q of the poles. The locus of the poles as a function of R_Q is a circle centered at the origin.

The ω_o-sensitivity and the Q-sensitivity functions are given by

$$S_{R_1}^{\omega_o} = S_{R_2}^{\omega_o} = S_{R_3}^{\omega_o} = -S_{R_4}^{\omega_o} = S_{C_1}^{\omega_o} = S_{C_2}^{\omega_o} = -\tfrac{1}{2}, \qquad S_{R_Q}^{\omega_o} = S_{R_0}^{\omega_o} = 0,$$

$$S_{R_1}^{Q} = S_{R_2}^{Q} = S_{R_3}^{Q} = -S_{R_4}^{Q} = -\tfrac{1}{2}S_{R_Q}^{Q} = S_{C_1}^{Q} = -S_{C_2}^{Q} = -\tfrac{1}{2}, \qquad S_{R_0}^{Q} = 0. \quad (10\text{-}85)$$

The sensitivity of the Q with respect to R_Q is 1. All other sensitivity figures are either zero or equal to $\tfrac{1}{2}$ in magnitude. Therefore, when the amplifiers are considered ideal, this network possesses good sensitivity figures.

For a given ω_o and Q, the elements of the network of Fig. 10-16 are determined by using Eq. (10-84), which for

$$R_1 = R_2 = R_3 = R_4 = R \qquad \text{and} \qquad C_1 = C_2 = C$$

becomes

$$\omega_o = \frac{1}{RC}, \qquad Q = \frac{R_Q}{R}.$$

For example, if $\omega_o = 1$ Krad/s and $Q = 5$, then

$$RC = \frac{1}{\omega_o} = 10^{-3}, \qquad R_Q = RQ = 5R = \frac{5 \times 10^{-3}}{C}.$$

If C is chosen as 0.01 μF, then $R = 100$ KΩ, $R_Q = 500$ KΩ.

Realization with Three Operational Amplifiers (One-Pole Rolloff Amplifiers)

When the one-pole rolloff model rather than the ideal amplifier model is used for the operational amplifiers in the three-amplifier realization, the poles of the system are affected by the GB of the operational amplifiers. To avoid too many variables, let all operational amplifiers be represented by the same $A(s)$, that is,

$$A(s) = -\frac{GB}{s + \omega_a} \cong -\frac{GB}{s}. \tag{10-86}$$

Also, let $R_3 = R_4$. The characteristic equation of the system shown in Fig. 10-16 is

$$1 - T_1(s)T_2(s)T_3(s) = 0, \tag{10-87}$$

where $T_1(s)T_2(s)T_3(s)$ represents the loop gain. The transfer functions $T_1(s)$, $T_2(s)$, and $T_3(s)$ are associated with the integrator, lossy integrator, and inverter, respectively. They are given by (see Problem 10-18):

$$T_1(s) = -\frac{1}{sR_1C_1}\frac{GB}{s + GB + \dfrac{1}{R_1C_1}}, \tag{10-88a}$$

$$T_2(s) = -\frac{1}{R_2C_2}\frac{GB}{s^2 + s\left(GB + \dfrac{1}{R_QC_2} + \dfrac{1}{R_2C_2}\right) + \dfrac{GB}{R_QC_2}}, \tag{10-88b}$$

$$T_3(s) = -\frac{\dfrac{GB}{2}}{s + \dfrac{GB}{2}}. \tag{10-88c}$$

When the results of Eq. (10-88) are substituted in Eq. (10-87), a fifth-order equation is obtained. For

$$R_1 = R_2 = R_3 = R_4 = R, \qquad C_1 = C_2 = C, \qquad \omega_o = \frac{1}{RC}, \qquad \text{and} \qquad Q = \frac{R_Q}{Q},$$

this equation can be rearranged as:

$$
s^2 + s\frac{\omega_o}{Q} + \omega_o^2
$$

$$
+ \frac{2}{GB}\left\{s\left[\frac{s^4}{GB^2} + s^3\frac{1}{GB}\left(\frac{5}{2} + \frac{\omega_o}{QGB} + \frac{2\omega_o}{GB}\right)\right.\right.
$$

$$
+ s^2\left(2 + \frac{5}{2}\frac{\omega_o}{QGB} + \frac{3\omega_o}{GB} + \frac{\omega_o^2}{GB^2} + \frac{\omega_o^2}{QGB^2}\right)
$$

$$
\left.\left. + s\left(\frac{2\omega_o}{Q} + \frac{3}{2}\frac{\omega_o^2}{QGB} + \omega_o + \frac{\omega_o^2}{2GB}\right) + \frac{1}{2}\frac{\omega_o^2}{Q}\right]\right\} = 0. \tag{10-89}
$$

Without much loss of accuracy, the terms divided by GB and GB² within the brackets can be neglected, thereby reducing the order of the equation to three. [Digital-computer solutions of the roots of Eq. (10-89) justify these simplifications.]

$$\underbrace{\left(s^2 + s\frac{\omega_o}{Q} + \omega_o^2\right)}_{P_1(s)} + \frac{1}{GB}\underbrace{\left\{4s\left[s^2 + s\omega_o\left(\frac{1}{2} + \frac{1}{Q}\right) + \frac{\omega_o^2}{4Q}\right]\right\}}_{P_2(s)} = 0. \tag{10-90}$$

With $P_1(s)$ and $P_2(s)$ thus defined, the changes in the upper half-plane pole can be calculated (see Problem 10-20). The results are summarized in Table 10-4.

Table 10-4   Three-Amplifier Realization
(see Fig. 10-16)

Equal-R (except R_Q) and Equal-C

$$\omega_o = \frac{1}{RC}, \qquad Q = \frac{R_Q}{R}$$

$$\frac{V_0}{V_i} \simeq \frac{\omega_o^2\left(\dfrac{2}{GB}s + 1\right)}{s^2 + s\dfrac{\omega_o}{Q} + \omega_o^2 + \dfrac{1}{GB}\left\{4s\left[s^2 + s\omega_o\left(\dfrac{1}{2} + \dfrac{1}{Q}\right) + \dfrac{\omega_o^2}{4Q}\right]\right\}} \qquad \left(\omega_a \ll \frac{\omega_o}{2Q}\right)$$

$$-\frac{\Delta\alpha}{\omega_o} \simeq 2\left(1 + \frac{1}{4Q}\right)\frac{\omega_o}{GB}, \qquad \frac{\Delta\beta}{\omega_o} \simeq -\frac{\left(1 - \dfrac{1}{Q} - \dfrac{1}{4Q^2}\right)}{\sqrt{1 - \dfrac{1}{4Q^2}}}\frac{\omega_o}{GB}$$

$$\frac{\Delta\omega_o}{\omega_o} \simeq -\frac{\omega_o}{GB}, \qquad \frac{\Delta Q}{Q} \simeq 4Q\frac{\omega_o}{GB}$$

As Table 10-4 indicates, the higher the Q, the larger is the change in Q. As the GB of the operational amplifiers decreases from infinity, high-Q poles move to the right and down.

If ω_o/GB is not small enough to make the approximate expressions in Table 10-4 valid, the equation of the poles, given by Eq. (10-90), can be solved using a digital computer. The results are displayed in Fig. 10-17.

EXAMPLE 10-9

In the three-amplifier realization shown in Fig. 10-16,

$$R_1 = R_2 = R, \qquad R_3 = R_4, \qquad C_1 = C_2 = C.$$

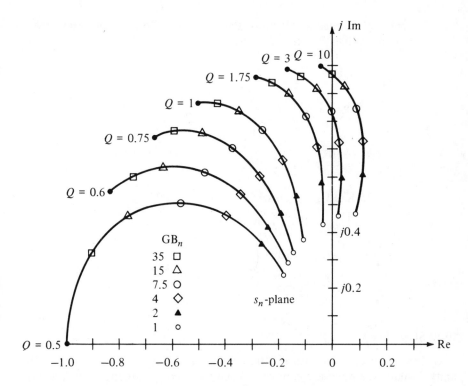

Fig. 10-17 Plot of upper half-plane root of

$$s_n^3 + s_n^2\left(\frac{1}{2} + \frac{1}{Q} + \frac{GB_n}{4}\right) + s_n\frac{1}{4Q}\left(1 + GB_n\right) + \frac{GB_n}{4} = 0$$

The ω_o and Q of the poles are calculated by assuming the amplifiers are ideal.

(a) What constraint must be placed on the gain-bandwidth product of the operational amplifiers if ω_{oa} and Q_a are to differ no more than 5% from ω_o and Q?

(b) Evaluate the constraint relation obtained in (a) for $GB = 10^6$ and $Q = 10$.

(c) $Q = 10$ and $GB = 10^6$. For what nominal ω_o would the system oscillate?

SOLUTION

(a) From Table 10-4,

$$\frac{\Delta\omega_o}{\omega_o} = -\frac{\omega_o}{GB}, \qquad \frac{\Delta Q}{Q} = 4Q\frac{\omega_o}{GB}.$$

Since Q changes much more than ω_o, the allowable deviation from nominal values is determined by

$$4Q \frac{\omega_o}{GB} < 0.05,$$

$$GB > 80Q\omega_o. \qquad Ans.$$

(b)

$$\omega_o < \frac{GB}{80Q} = \frac{10^6}{800},$$

$$\omega_o < 1.25 \text{ Krad/s}. \qquad Ans.$$

(c) From Problem 10-19c, the system oscillates if:

$$\omega_o \geq \frac{GB}{4Q} = \frac{10^6}{40}, \qquad GB \gg \left(2 + \frac{4}{Q}\right)\omega_o,$$

$$\omega_o \geq 25 \text{ Krad/s}. \qquad Ans.$$

10-10 SUMMARY

The second-order low-pass function is characterized with two zeros at infinity. Usually the poles are complex. When the pole Q's are high, the low-pass function can just as well be used as a band-pass function. The passive realizations involve a series inductor and a shunt capacitor in a resistively-terminated ladder structure. The complex-pole ω_o- and Q-sensitivity figures of the passive circuits are used as a basis for comparing the sensitivities of the RC active circuits.

Four active realization methods are used to implement the low-pass function. In the KRC realization, a positive-gain amplifier is used to reduce the coefficient of the s-term in the denominator polynomial, thereby moving the poles in a constant-ω_o manner. The sensitivity of the Q with respect to the dc gain of the amplifier is quite large for high-Q poles. Therefore, care should be exercised in adjusting the value of the dc gain when the poles are close to the imaginary axis. First-order correction terms show a decrease in ω_o and an increase in Q from the nominal values, as the GB of the amplifier decreases from the nominal value of infinity. The circuit is stable even for large departures of GB from infinity.

In the $-KRC$ realization, a negative-gain amplifier is used to increase the constant term in the denominator polynomial, thereby moving the poles in a constant-α manner. Although the various sensitivity figures (based on an ideal amplifier) show this circuit to be excellent, in practice, because of nonideal amplifiers, it may oscillate even for low values of ω_o if the Q is high.

In the infinite-gain realization, the Q of the poles depends upon the square root of the ratio of two capacitors. Therefore, to realize high Q's, large capacitor ratios must be used. When there is such a large spread in capacitor values, it may not be easy to maintain constant Q as the environmental conditions change. First-order correction terms show a decrease in ω_o and an increase in Q as the GB of the amplifier decreases from infinity. The circuit is stable even for large departures of GB from infinity.

In the three-amplifier realization, an integrator, lossy integrator, and inverter are connected in a loop to generate complex poles. The Q of the poles is controlled by the feedback resistor in the lossy integrator. As GB decreases from infinity, ω_o decreases and Q increases in value. The higher the $\omega_o Q$ product of the poles, the larger must be the gain-bandwidth product of the operational amplifiers to prevent the circuit from oscillating.

PROBLEMS

10-1 Derive Eq. (10-2).

10-2 The poles of the network shown in Fig. 10-2b are complex. Show that the poles move:
(a) in a constant-α manner if L is changed;
(b) in a constant-ω_o manner if R is changed;
(c) on a circle tangent to the imaginary axis if C is changed.

10-3 Figure 10-18 shows a low-pass network with resistive terminations at the source as well as at the load end. (R_1, or part of R_1, may be taken as the coil's resistance.)

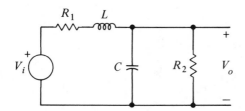

Fig. 10-18

(a) Obtain (V_o/V_i).
(b) Calculate the ω_o-sensitivity and the Q-sensitivity functions.

10-4 (a) Derive the Q-sensitivity functions given by Eq. (10-14).
(b) Show that these functions reduce to the expressions given by Eq. (10-18), if $K_0 = 1$ and $R_1 = R_2 = R$.
(c) In (b), $Q = 10$. If Q is to change no more than 10%, to what tolerance must K_0 be held?

10-5 Derive Eq. (10-33).

10-6 The KRC realization shown in Fig. 10-19 is used to obtain a pair of poles with $\omega_o = 250$ Krad/s and $Q = 1$. The gain-bandwidth product of the amplifier is 10^6 rad/s.

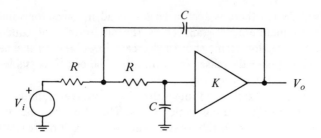

Fig. 10-19

(a) K_0, R, and C are chosen to obtain the desired ω_o and Q by assuming the amplifier to be ideal. What will be the actual ω_{oa} and Q_a if the ideal amplifier is subsequently replaced with a one-pole rolloff amplifier?
(b) What must be the nominal values of ω_o and Q in order that the desired ω_o and Q are obtained with the nonideal amplifier?

10-7   Derive the ω_o-sensitivity functions given in Eq. (10-46).
10-8   Derive the expression for (V_o/V_i) given in Table 10-2.
10-9   In Fig. 10-20,

$$\frac{V_o}{V_m} = K(s) = -\frac{K_0 \omega_a}{s + \omega_a}.$$

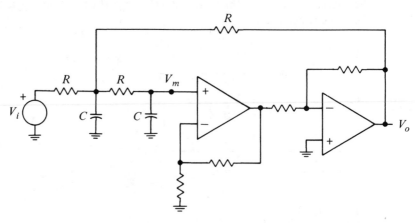

Fig. 10-20

(a) Obtain (V_o/V_i), ω_o, and Q, if $\omega_a = \infty$.
(b) Let

$$K(s) = -\frac{K_0}{1 + \dfrac{s}{\omega_a}} \cong -K_0\left[1 - \left(\frac{s}{\omega_a}\right) + \left(\frac{s}{\omega_a}\right)^2\right].$$

Then find ω_{oa} and Q_a.

10-10 (a) Obtain (V_o/V_i) for the network shown in Fig. 10-21. Assume the amplifiers are ideal.
  (b) Find ω_o and Q of the poles for $R_1 = R_2 = R_3 = R$ and $C_1 = C_2 = C$.

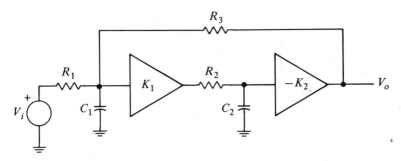

Fig. 10-21

10-11 In Fig. 10-10b, treat R_2, C_2, and the operational amplifier as a unit. Thus, the circuit can be equivalently redrawn by using a dependent source that is a function of s. Show how this is done. Assume the operational amplifier is ideal.

10-12 (a) Find the input admittance of the network shown in Fig. 10-22.
  (b) If $Z = (1/sC)$, show that the input impedance can equivalently be represented by the parallel connection of a resistor, of value $(R_1 R_2)/(R_1 + R_2)$ with an inductor, of value $R_1 R_2 C$.
  (c) Explain the operation of the circuit of Fig. 10-10b in terms of the equivalent circuit obtained in part b.

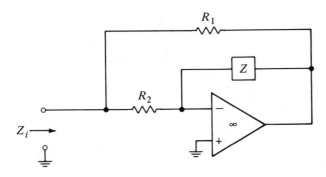

Fig. 10-22

10-13 (a) Obtain the input impedance of the network shown in Fig. 10-23.
  (b) If $Z = (1/sC)$, show that Z_i can be represented by the series connection of a resistor [of value $(R_1 + R_2)$] with an inductor [of value $(R_1 R_2 C)$].

(c) A capacitor, C_2, is placed across the input of this circuit and $Z = (1/sC_1)$. Show that the resulting network is identical with that of Fig. 10-3b, if the excitation is introduced at the proper location.

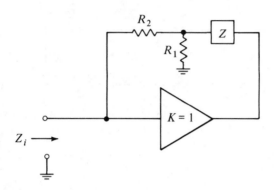

Fig. 10-23

10-14 Using Routh's criterion for stability, show that the low-pass infinite-gain realization given in Fig. 10-10b (equal-R) is absolutely stable.

10-15 Derive Eq. (10-75).

10-16 The initial voltage γ on the capacitor of the integrator shown in Fig. 10-24 can be referred to the input as an additional signal. What is the value of this signal?

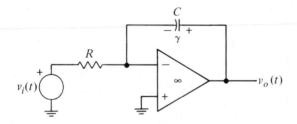

Fig. 10-24

10-17 (a) In Fig. 10-25,

$$\frac{R_2}{R_1 + R_2} = \frac{\omega_o}{Q}\frac{1}{1 + \omega_o^2}.$$

Show that the poles of this network are the roots of

$$s^2 + s\frac{\omega_o}{Q} + \omega_o^2 = 0.$$

(b) Indicate how this network can be excited without altering its pole locations.

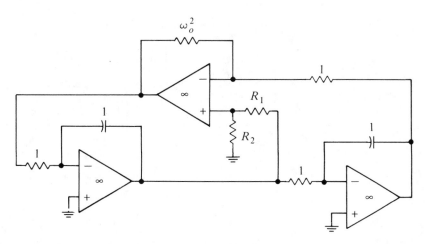

Fig. 10-25

10-18 (a) Show that (V_o/V_i) in Fig. 10-26 is given by

$$\frac{V_o}{V_i} = -\frac{Z}{R}\frac{1}{1 - \frac{1 + (Z/R)}{A(s)}}.$$

(b) Using (a), derive Eq. (10-88).

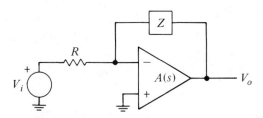

Fig. 10-26

10-19 (a) Derive Eq. (10-89).
    (b) Obtain V_o/V_i as given in Table 10-4.
    (c) Under what condition is the system stable?
10-20 Obtain the expressions for $\Delta\omega_o/\omega_o$ and $\Delta Q/Q$ given in Table 10-4.
10-21 (a) For the network shown in Fig. 10-27, obtain the transfer function (V_n/V_i). (Let $\tau_i = R_i C_i$.)

(b) Show that the coefficients of the denominator polynomial can be independently adjusted.

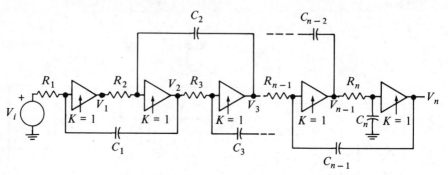

Fig. 10-27

11

Second-Order
Band-Pass Networks

In this chapter several RLC and active RC realizations are given for the second-order band-pass function. The active realizations are important for low-frequency applications, whereas passive realizations yield simple circuits for high-frequency filter work. For each circuit that is presented, the ω_o- and Q-sensitivity functions are obtained. These functions can be used for fine-tuning the network and for comparing the sensitivity characteristics of different realizations. Also discussed are the frequency limitations imposed by nonideal amplifiers. In particular, first-order correction terms are given to evaluate the changes in the ω_o and Q of the complex poles as a function of the gain-bandwidth product of the operational amplifier.

11-1 THE SECOND-ORDER BAND-PASS FUNCTION

The second-order band-pass function has one zero at the origin and another at infinity:

$$T(s) = \frac{Hs}{s^2 + s\dfrac{\omega_o}{Q} + \omega_o^2}.$$

To normalize the peak value of the magnitude function to unity, let $H = (\omega_o/Q)$:

$$T(s) = \frac{\dfrac{\omega_o}{Q}s}{s^2 + s\dfrac{\omega_o}{Q} + \omega_o^2} = \frac{\dfrac{1}{Q}\dfrac{s}{\omega_o}}{\left(\dfrac{s}{\omega_o}\right)^2 + \left(\dfrac{s}{\omega_o}\right)\dfrac{1}{Q} + 1}. \tag{11-1}$$

The characteristics of $T(s)$ depend upon the ω_o and Q of the poles. The effect of ω_o is to scale the frequency variable, as shown in Eq. (11-1). The effect of Q is displayed in the magnitude and phase curves shown in Fig. 11-1.

A study of Eq. (11-1) and Fig. 11-1 reveals the following important features concerning the band-pass function:

1. For $\omega/\omega_o \ll 1$ and for $\omega/\omega_o \gg 1$,

$$|T(j\omega)| \cong 0.$$

Therefore, low and high frequencies are attenuated.

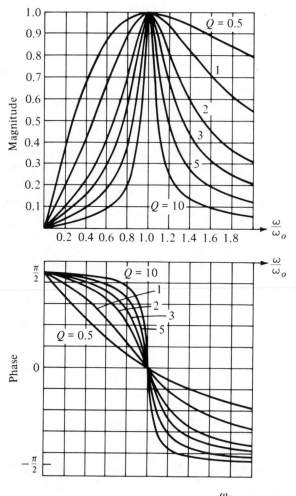

Fig. 11-1   Magnitude and phase of $\dfrac{\dfrac{\omega_o}{Q} s}{s^2 + s \dfrac{\omega_o}{Q} + \omega_o^2}$

2. Regardless of the value of the Q, the magnitude curve peaks exactly at $\omega = \omega_o$ and the 3-dB bandwidth is exactly ω_o/Q. (See Example 6-3.)

3. At $\omega/\omega_o = 1$, the phase is 0. At $\omega/\omega_o \ll 1$, the phase is $\pi/2$. At $\omega/\omega_o \gg 1$, the phase is $-\pi/2$. When the Q is high, the phase undergoes a rapid shift of π radians about ω_o.

4. $|T(j\omega)|$ has the same value at two frequencies. The geometric average of these two frequencies is ω_o. Thus,

$$|T(j\omega)| = \left| T\left(j\frac{\omega_o^2}{\omega} \right) \right|.$$

Furthermore, the phase at the higher frequency is the *negative* of the phase at the lower frequency, that is,

$$\theta(\omega) = -\theta\left(\frac{\omega_o^2}{\omega} \right).$$

For a second-order band-pass network, it is easy to determine the complex-pole positions experimentally. The network is driven with a sine wave and the steady-state output is observed. Then, three measurements are taken, namely, the frequency at which the maximum output occurs (f_{max}) and the two frequencies at which the output is down 3 dB from the maximum value (f_1 and f_2). The ω_o and Q of the poles are calculated from

$$\omega_o = 2\pi f_{max}, \qquad Q = \frac{f_{max}}{f_2 - f_1}. \tag{11-2}$$

11-2 PASSIVE REALIZATIONS

Two second-order *RLC* band-pass networks are given in Fig. 11-2. (Another network is given in Problem 11-1.)

In the band-pass network of Fig. 11-2a, the inductor causes the transfer function zero at the origin; the zero at infinity is caused by the capacitor. The voltage ratio is given by:

$$\frac{V_o}{V_i} = \frac{1}{RC} \frac{s}{s^2 + s\dfrac{1}{RC} + \dfrac{1}{LC}}. \tag{11-3}$$

The ω_o and Q of the poles are related to the element values by

$$\omega_o = \frac{1}{\sqrt{LC}}, \qquad Q = R\sqrt{\frac{C}{L}}. \tag{11-4}$$

The magnitude response peaks at ω_o and has a 3-dB bandwidth of

$$\mathrm{BW} = \frac{\omega_o}{Q} = \frac{1}{RC}. \tag{11-5}$$

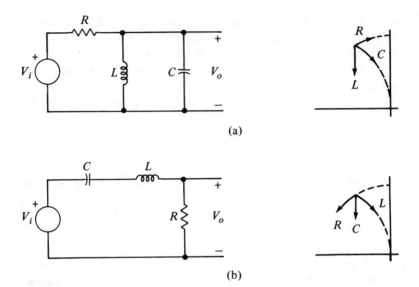

(a)

(b)

Fig. 11-2   Passive band-pass networks

By changing L, the center frequency $(1/\sqrt{LC})$ can be varied without altering the bandwidth $(1/RC)$. Stated pictorially, constant-bandwidth, variable center-frequency operation results if the poles are moved in a constant-α manner (by changing L). On the other hand, the bandwidth can be varied without altering the center frequency by changing R. The resulting motion of the poles is along the constant-ω_o circle. The direction of these changes, for positive changes in element values, is given in Fig. 11-2a. A quantitative measure of these changes is obtained by calculating the ω_o- and Q-sensitivity functions.

$$S_L^{\omega_o} = S_C^{\omega_o} = -\tfrac{1}{2}, \qquad S_R^{\omega_o} = 0; \tag{11-6a}$$

$$S_L^Q = -S_C^Q = -\tfrac{1}{2}S_R^Q = -\tfrac{1}{2}. \tag{11-6b}$$

The magnitudes of the sensitivity functions are 1, $\tfrac{1}{2}$, or 0. These figures serve also as a figure of merit for comparing the sensitivity functions of the active realizations.

In the band-pass network of Fig. 11-2b, the capacitor causes the transfer-function zero at the origin; the inductor causes the zero at infinity. The transfer function, ω_o, Q, 3-dB bandwidth, and the various sensitivity functions are given by

$$\frac{V_o}{V_i} = \frac{R}{L} \frac{s}{s^2 + s\dfrac{R}{L} + \dfrac{1}{LC}}, \tag{11-7}$$

$$\omega_o = \frac{1}{\sqrt{LC}}, \qquad Q = \frac{1}{R}\sqrt{\frac{L}{C}}, \tag{11-8}$$

$$BW = \frac{R}{L}, \tag{11-9}$$

$$S_L^{\omega_o} = S_C^{\omega_o} = -\tfrac{1}{2}, \qquad S_R^{\omega_o} = 0, \tag{11-10a}$$

$$S_L^Q = -S_C^Q = -\tfrac{1}{2}S_R^Q = \tfrac{1}{2}. \tag{11-10b}$$

The resistor R changes the poles in a constant-ω_o manner, thereby varying the bandwidth but not the center frequency. The capacitor C changes the poles in a constant-α manner, thereby varying the center frequency but not the bandwidth. The inductor L changes the poles along a circle tangent to the imaginary axis at the origin, thereby changing the center frequency and the bandwidth. The direction of these changes (for positive changes in element values) is shown for the upper half-plane pole in Fig. 11-2b.

The results presented in this section are based on ideal inductors. For high-Q circuits, the losses in the inductors cause departures from the ideal characteristics.

11-3 *KRC* REALIZATION

An RC network with two transfer-function zeros, one at the origin, the other at infinity, is shown in Fig. 11-3a. (See Problem 11-3 for another arrangement of the RC network.) Although this network possesses band-pass characteristics, the bandwidth is more than twice the frequency of peaking (see Problem 11-2). In order to obtain narrower bandwidths, the poles of the network must be made complex. This can be done by sensing the voltage across R_3 and feeding it back

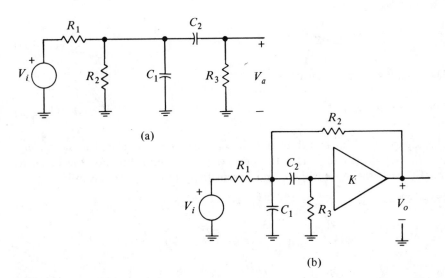

(a)

(b)

Fig. 11-3 Band-pass *KRC* network

(without inverting) in series with the resistor R_2, as shown in Fig. 11-3b. As discussed in Section 9-3, this kind of feedback (band-pass) does produce complex poles without affecting the zeros of the transfer function.

When the amplifier is considered ideal, that is, when $K = K_0$, the transfer function is given by

$$\frac{V_o}{V_i} = \frac{K_0}{R_1 C_1} \frac{s}{s^2 + s\left(\dfrac{1}{R_1 C_1} + \dfrac{1}{R_3 C_2} + \dfrac{1}{R_3 C_1} + \dfrac{1 - K_0}{R_2 C_1}\right) + \dfrac{R_1 + R_2}{R_1 R_2 R_3 C_1 C_2}}. \quad (11\text{-}11)$$

The ω_o and Q of the poles and the 3-dB bandwidth, BW, are given by

$$\omega_o = \sqrt{\frac{R_1 + R_2}{R_1 R_2 R_3 C_1 C_2}}, \qquad Q = \frac{\sqrt{\dfrac{R_2 C_1(R_1 + R_2)}{R_1 R_3 C_2}}}{1 + \dfrac{R_2}{R_1} + \dfrac{R_2}{R_3}\left(1 + \dfrac{C_1}{C_2}\right) - K_0},$$

$$BW = \frac{1}{R_1 C_1} + \frac{1}{R_3 C_2} + \frac{1}{R_3 C_1} + \frac{1 - K_0}{R_2 C_1} = \frac{\omega_o}{Q}. \quad (11\text{-}12)$$

With the exception of K_0, every element of the network affects both the ω_o and the Q of the poles. The amplifier gain, K_0, affects only the Q, and therefore it can be used to vary the bandwidth without altering the center frequency.

Since six elements (R_1, R_2, R_3, C_1, C_2, and K_0) are available to set the ω_o and Q of the poles, four of them can be arbitrarily chosen. Therefore, let

$$R_1 = R_2 = R_3 = R, \qquad C_1 = C_2 = C.$$

This leaves three elements (R, C, and K_0) to fix ω_o and Q, which are given by the simplified expressions

$$\omega_o = \frac{\sqrt{2}}{RC}, \qquad Q = \frac{\sqrt{2}}{4 - K_0}. \quad (11\text{-}13a)$$

The resulting 3-dB bandwidth is

$$BW = \frac{\omega_o}{\sqrt{2}}(4 - K_0). \quad (11\text{-}13b)$$

The capacitor C can be arbitrarily chosen; then R is calculated to satisfy ω_o, and K_0 is adjusted to obtain the desired Q or bandwidth. For the equal-R, equal-C case, the ω_o-sensitivity and Q-sensitivity functions are given by

$$S_{R_1}^{\omega_o} = S_{R_2}^{\omega_o} = -\tfrac{1}{4}, \qquad S_{R_3}^{\omega_o} = S_{C_1}^{\omega_o} = S_{C_2}^{\omega_o} = -\tfrac{1}{2}, \qquad S_{K_0}^{\omega_o} = 0, \quad (11\text{-}14a)$$

$$S_{R_1}^{Q} = -\frac{1}{4} + \frac{Q}{\sqrt{2}}, \qquad S_{R_2}^{Q} = \frac{3}{4} - \frac{3Q}{\sqrt{2}}, \qquad S_{R_3}^{Q} = -\frac{1}{2} + Q\sqrt{2},$$

$$S_{C_1}^{Q} = \frac{1}{2} - \frac{Q}{\sqrt{2}}, \qquad S_{C_2}^{Q} = -\frac{1}{2} + \frac{Q}{\sqrt{2}}, \qquad S_{K_0}^{Q} = \frac{4Q}{\sqrt{2}} - 1. \quad (11\text{-}14b)$$

Since all the Q-sensitivity functions depend directly on the Q of the poles, this network, like the *KRC* low-pass realization, needs to be constructed rather carefully for high-Q circuits.

The effect of the finite gain-bandwidth product of the operational amplifier on the performance of this band-pass circuit can be obtained by the methods presented in Section 10-4. Without going into the details of derivation, the important results are summarized in Table 11-1.

Table 11-1 *KRC* Realization
(see Fig. 11-3b)

Equal-R, Equal-C

$$\omega_o = \frac{\sqrt{2}}{RC}, \qquad Q = \frac{\sqrt{2}}{4 - K_0}, \qquad BW = \frac{\omega_o}{\sqrt{2}}(4 - K_0)$$

$$\frac{V_o}{V_i} = \frac{\frac{1}{\sqrt{2}}\left(4 - \frac{\sqrt{2}}{Q}\right)\omega_o s}{s^2 + s\frac{\omega_o}{Q} + \omega_o^2 + \frac{1}{GB}\left(4 - \frac{\sqrt{2}}{Q}\right)s\left(s^2 + s2\sqrt{2}\omega_o + \omega_o^2\right)} \qquad \left(\omega_a \ll \frac{\omega_o}{2Q}\right)$$

$$-\frac{\Delta\alpha}{\omega_o} \cong \frac{4\sqrt{2}}{Q}\left(1 - \frac{1}{2\sqrt{2}Q}\right)^2 \frac{\omega_o}{GB}, \qquad \frac{\Delta\beta}{\omega_o} \cong -4\sqrt{2}\frac{\left(1 - \frac{1}{2Q^2}\right)\left(1 - \frac{1}{2\sqrt{2}Q}\right)^2}{\sqrt{1 - \frac{1}{4Q^2}}}\frac{\omega_o}{GB}$$

$$\frac{\Delta\omega_o}{\omega_o} \cong -4\sqrt{2}\left(1 - \frac{1}{2\sqrt{2}Q}\right)^2 \frac{\omega_o}{GB}, \qquad \frac{\Delta Q}{Q} \cong 4\sqrt{2}\left(1 - \frac{1}{2\sqrt{2}Q}\right)^2 \frac{\omega_o}{GB}$$

$$\frac{\Delta BW}{BW} \cong -8\sqrt{2}\left(1 - \frac{1}{2\sqrt{2}Q}\right)^2 \frac{\omega_o}{GB}$$

The denominator polynomial of (V_o/V_i) given in Table 11-1 is very similar to the denominator polynomial of the *KRC* low-pass realization presented in Table 10-1. Hence, Fig. 10-5a can be used to determine the approximate location of the upper half-plane pole if it departs considerably from its nominal position.

The effect of the nonideal amplifier can be demonstrated in still another way. In Fig. 11-4, the magnitude of V_o/V_i (obtained from Table 11-1) is plotted for $Q = 10$ and for three values of GB, namely ∞, $100\omega_o$, and $50\omega_o$. The magnitude is normalized so that the peak value for the ideal case is unity. Note the reduction of the center frequency as the GB becomes progressively smaller. For $GB/\omega_o = 100$, the center frequency is about 5% lower. This result is in close agreement with the $\Delta\omega_o/\omega_o$ correction term presented in Table 11-1.

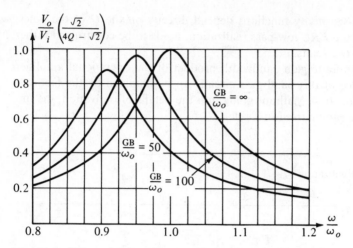

Fig. 11-4 Effect of GB on the magnitude curve for $Q = 10$

11-4 $-KRC$ REALIZATION

An RC band-pass network is shown in Fig. 11-5a. This network is a simpler version of the network shown in Fig. 11-3a. If the voltage across R_2 is sensed and fed back in series with C_1, the poles can be made complex without affecting the transfer-function zeros. See Section 9-4 for details. The resistor R_2 can be made part of the gain-producing network of the amplifier, as shown in Fig. 11-5c.

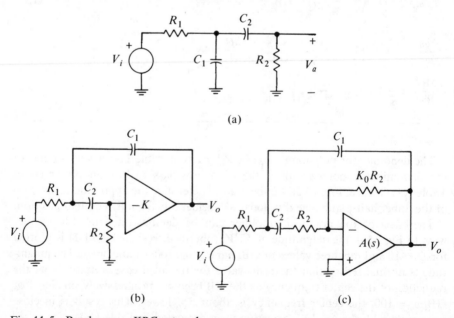

Fig. 11-5 Band-pass $-KRC$ network

When the amplifier is considered ideal, that is, when $K = K_0$, the transfer function is given by

$$\frac{V_o}{V_i} = -\frac{K_0}{1 + K_0} \frac{1}{R_1 C_1}$$

$$\times \frac{s}{s^2 + s\frac{1}{(1 + K_0)}\left(\frac{1}{R_1 C_1} + \frac{1}{R_2 C_2} + \frac{1}{R_2 C_1}\right) + \frac{1}{(1 + K_0)R_1 R_2 C_1 C_2}}.$$

$$(11\text{-}15)$$

The ω_o and Q of the poles and the 3-dB bandwidth are given by

$$\omega_o = \frac{1}{\sqrt{(1 + K_0)R_1 R_2 C_1 C_2}}, \qquad Q = \frac{\sqrt{1 + K_0}}{\sqrt{\frac{R_2 C_2}{R_1 C_1}} + \sqrt{\frac{R_1 C_1}{R_2 C_2}} + \sqrt{\frac{R_1 C_2}{R_2 C_1}}},$$

$$BW = \frac{1}{1 + K_0}\left(\frac{1}{R_1 C_1} + \frac{1}{R_2 C_2} + \frac{1}{R_2 C_1}\right). \qquad (11\text{-}16)$$

To simplify the network, let $R_1 = R_2 = R$ and $C_1 = C_2 = C$. Then,

$$\omega_o = \frac{1}{RC\sqrt{1 + K_0}}, \qquad Q = \frac{\sqrt{1 + K_0}}{3}, \qquad BW = \frac{3\omega_o}{\sqrt{1 + K_0}}. \qquad (11\text{-}17)$$

The Q of the pole is set by adjusting K_0. Large gains are required to obtain large Q's. For example, $Q = 10$ requires $K_0 = 899$. The ω_o of the pole is set by choosing the RC product properly (either R or C can be arbitrarily chosen). For the equal-R, equal-C case, the sensitivity functions are given by

$$S_{R_1}^{\omega_o} = S_{R_2}^{\omega_o} = S_{C_1}^{\omega_o} = S_{C_2}^{\omega_o} = -\frac{1}{2}, \qquad S_{K_0}^{\omega_o} = -\frac{1}{2}\left(1 - \frac{1}{9Q^2}\right); \qquad (11\text{-}18a)$$

$$S_{R_1}^{Q} = -S_{R_2}^{Q} = -S_{C_1}^{Q} = S_{C_2}^{Q} = -\frac{1}{6}, \qquad S_{K_0}^{Q} = \frac{1}{2}\left(1 - \frac{1}{9Q^2}\right). \qquad (11\text{-}18b)$$

As Eq. (11-18) shows, all the sensitivity figures are low regardless of the Q of the poles.

An analysis similar to the one given in Section 10-6 can be undertaken to obtain first-order correction terms to the complex-pole positions caused by the finite gain-bandwidth product of the operational amplifier. A summary of important results is given in Table 11-2.

Application of Routh's criterion of stability to this system shows that the poles are in the left half-plane regardless of the value of GB. However, for high-Q poles, the shifts in pole locations are much larger than the shifts in the KRC realization given in Table 11-1. For example, the change in the ω_o is $[3Q/(8\sqrt{2})]$ times more in the $-KRC$ realization, as can be seen by comparing corresponding first-order correction terms from Tables 11-1 and 11-2.

Table 11-2 $-KRC$ Realization
(see Fig. 11-5c)

Equal-R, Equal-C

$$\omega_o = \frac{1}{RC\sqrt{1+K_0}}, \qquad Q = \frac{\sqrt{1+K_0}}{3}, \qquad BW = \frac{3\omega_o}{\sqrt{1+K_0}}$$

$$\frac{V_o}{V_i} = \frac{3Q\left(1 - \frac{1}{9Q^2}\right)\omega_o s}{s^2 + s\frac{\omega_o}{Q} + \omega_o^2 + \frac{s}{GB}\left[s^2 + s\frac{\omega_o}{3Q}(2 + 9Q^2) + \omega_o^2\right]} \qquad \left(\omega_a \ll \frac{\omega_o}{2Q}\right)$$

$$-\frac{\Delta\alpha}{\omega_o} \cong \frac{3}{2}\left(1 - \frac{1}{9Q^2}\right)\frac{\omega_o}{GB}, \qquad \frac{\Delta\beta}{\omega_o} \cong -\frac{3Q\left(1 - \frac{1}{9Q^2}\right)\left(1 - \frac{1}{2Q^2}\right)}{2\sqrt{1 - \frac{1}{4Q^2}}}\frac{\omega_o}{GB}$$

$$\frac{\Delta\omega_o}{\omega_o} \cong -\frac{3Q}{2}\left(1 - \frac{1}{9Q^2}\right)\frac{\omega_o}{GB}, \qquad \frac{\Delta Q}{Q} \cong \frac{3Q}{2}\left(1 - \frac{1}{9Q^2}\right)\frac{\omega_o}{GB}$$

$$\frac{\Delta BW}{BW} \cong -3Q\left(1 - \frac{1}{9Q^2}\right)\frac{\omega_o}{GB}$$

11-5 REALIZATION WITH INFINITE GAIN

An *RC* network with transfer-function zeros at the origin and infinity is shown in Fig. 11-5a. As demonstrated in Section 9-5, the poles of this network can be moved off the negative real axis and made complex if an operational amplifier is introduced as shown in Fig. 11-6. Two feedback signals are applied, one through C_1, the other through R_2. (For another interpretation of this feedback scheme, see Problem 11-7.)

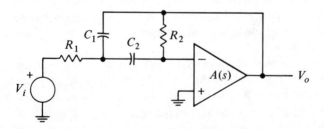

Fig. 11-6   Band-pass network

For $|A(s)| = \infty$, the transfer function is given by

$$\frac{V_o}{V_i} = \frac{-1}{R_1 C_1} \frac{s}{s^2 + \dfrac{1}{R_2}\left(\dfrac{1}{C_1} + \dfrac{1}{C_2}\right)s + \dfrac{1}{R_1 R_2 C_1 C_2}}. \tag{11-19}$$

The ω_o and Q of the poles and the bandwidth are given by

$$\omega_o = \frac{1}{\sqrt{R_1 R_2 C_1 C_2}}, \qquad Q = \frac{\sqrt{\dfrac{R_2}{R_1}}}{\sqrt{\dfrac{C_1}{C_2}} + \sqrt{\dfrac{C_2}{C_1}}}, \qquad \mathrm{BW} = \frac{1}{R_2}\left(\frac{1}{C_1} + \frac{1}{C_2}\right). \tag{11-20}$$

To simplify, let $C_1 = C_2 = C$. Then,

$$\frac{V_o}{V_i} = -\frac{(2Q^2 \mathrm{BW})s}{s^2 + s\mathrm{BW} + \omega_o^2}, \tag{11-21}$$

where

$$\omega_o = \frac{1}{C\sqrt{R_1 R_2}}, \qquad Q = \frac{1}{2}\sqrt{\frac{R_2}{R_1}}, \qquad \mathrm{BW} = \frac{2}{R_2 C}.$$

Either the capacitance C or one of the resistances can be arbitrarily chosen. The other element values are then determined by Eq. (11-21).

At the frequency of peaking, $s = j\omega_o$, the magnitude of the transfer function is $V_o/V_i = 2Q^2$. Therefore, large output voltages may result for high-Q poles. In an actual amplifier, this may cause distortion. To avoid overdriving the amplifier, the input signal is sometimes attenuated before applying it to the circuit (see Example 11-1).

As Eq. (11-21) shows, R_1 appears in ω_o but not in BW. Therefore, R_1 can be used to vary the center frequency while maintaining a constant bandwidth.

The various sensitivity functions for the equal-C case are

$$S_{R_1}^{\omega_o} = S_{R_2}^{\omega_o} = S_{C_1}^{\omega_o} = S_{C_2}^{\omega_o} = -\tfrac{1}{2}, \tag{11-22a}$$

$$S_{R_1}^{Q} = -S_{R_2}^{Q} = -\tfrac{1}{2}, \qquad S_{C_1}^{Q} = S_{C_2}^{Q} = 0. \tag{11-22b}$$

The sensitivity figures given by Eq. (11-22) have all low values.

By studying the results presented in Table 11-3, the nonideal amplifier's effect on the performance of this circuit can be determined.

The denominator polynomial of (V_o/V_i) given in Table 11-3 is the same as the denominator polynomial of (V_o/V_i) given in Table 10-1 (unity-gain, equal-R). In other words, the infinite-gain band-pass realization has the same dependence of the poles on the GB of the operational amplifier as the KRC low-pass realization. Therefore, Fig. 10-5b can be used for both circuits to determine the complex-pole positions as a function of GB and nominal Q.

Table 11-3   Realization with Infinite Gain
(see Fig. 11-6)

Equal-C

$$\omega_o = \frac{1}{C\sqrt{R_1 R_2}}, \qquad Q = \frac{1}{2}\sqrt{\frac{R_2}{R_1}}, \qquad BW = \frac{2}{R_2 C}$$

$$\frac{V_o}{V_i} = -\frac{2Q\omega_o s}{s^2 + s\dfrac{\omega_o}{Q} + \omega_o^2 + \dfrac{s}{GB}\left[s^2 + s\omega_o\left(2Q + \dfrac{1}{Q}\right) + \omega_o^2\right]} \qquad \left(\omega_a \ll \frac{\omega_o}{2Q}\right)$$

$$-\frac{\Delta a}{\omega_o} \cong \frac{\omega_o}{GB}, \qquad \frac{\Delta\beta}{\omega_o} \cong -\frac{Q\left(1 - \dfrac{1}{2Q^2}\right)}{\sqrt{1 - \dfrac{1}{4Q^2}}}\frac{\omega_o}{GB}$$

$$\frac{\Delta\omega_o}{\omega_o} \cong -Q\frac{\omega_o}{GB}, \qquad \frac{\Delta Q}{Q} \cong Q\frac{\omega_o}{GB}$$

$$\frac{\Delta BW}{BW} \cong -2Q\frac{\omega_o}{GB}$$

EXAMPLE 11-1

(a) Obtain the transfer function of the network shown in Fig. 11-7. The amplifier gain is given by $A(s)$.

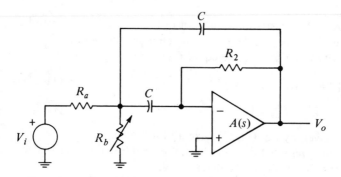

Fig. 11-7

(b) Let $|A(s)| = \infty$. Plot the magnitude characteristics as a function of the setting of R_b. Discuss the results.

(c) Repeat (b) for $A(s) = -(GB/s)$.

(d) Repeat (b) for $A(s) = -GB/(s + \omega_a)$.

SOLUTION

(a) Convert the $V_i R_a R_b$ network into its Thévenin equivalent and then use the transfer function of Table 11-3 with appropriate changes:

$$\frac{V_o}{V_i} = -\frac{R_b}{R_a + R_b} \frac{2Q\omega_o s}{s^2 + s\dfrac{\omega_o}{Q} + \omega_o^2 - \dfrac{1}{A(s)}\left[s^2 + s\omega_o\left(2Q + \dfrac{1}{Q}\right) + \omega_o^2\right]},$$

Ans. (11-23)

where

$$\omega_o = \frac{1}{C\sqrt{R_1 R_2}}, \qquad Q = \frac{1}{2}\sqrt{\frac{R_2}{R_1}}, \qquad R_1 = \frac{R_a R_b}{R_a + R_b}.$$

(b) In Eq. (11-23), let $|A(s)| = \infty$ and obtain

$$\frac{V_o}{V_i} = -\frac{R_b}{R_a + R_b} \frac{2Q\omega_o s}{s^2 + s\dfrac{\omega_o}{Q} + \omega_o^2}.$$

(11-24)

The bandwidth and the maximum value of $|V_o/V_i|$ are

$$BW = \frac{\omega_o}{Q} = \frac{2}{R_2 C},$$

(11-25)

$$\left|\frac{V_o(j\omega_o)}{V_i(j\omega_o)}\right| = \frac{2R_b}{R_a + R_b} \qquad Q^2 = \frac{1}{2}\frac{R_2}{R_a}.$$

(11-26)

Neither the BW nor $|V_o/V_i|_{max}$ is a function of R_b. Hence, as R_b changes, the center frequency of the band-pass curve varies but the bandwidth and the peak value remain constant. When Eq. (11-24) is normalized with respect to its peak value, it becomes:

$$\frac{V_o}{V_i}\frac{2R_a}{R_2} = -\frac{BWs}{s^2 + sBW + \omega_o^2}.$$

(11-27)

The solid curves of Fig. 11-8a are plots of the magnitude of Eq. (11-27) for $BW = 40\pi$ rad/s and for four values of ω_o, namely,

$$\omega_o = 500\pi, \quad 1000\pi, \quad 1500\pi, \quad \text{and} \quad 2000\pi.$$

Note the constant nature of the bandwidth and the peak value as ω_o is varied. The ($\omega_o = 500\pi$)-curve corresponds to a Q of 12.5; the ($\omega_o = 2000\pi$)-curve corresponds to a Q of 50.

(c) In Eq. (11-23), let $A(s) = -GB/s$ and multiply V_o/V_i by $2R_a/R_2$ to normalize its magnitude:

$$\frac{V_o}{V_i}\frac{2R_a}{R_2} = -\frac{BWs}{s^2 + sBW + \omega_o^2 + \dfrac{s}{GB}\left[s^2 + s\omega_o\left(2Q + \dfrac{1}{Q}\right) + \omega_o^2\right]}.$$

(11-28)

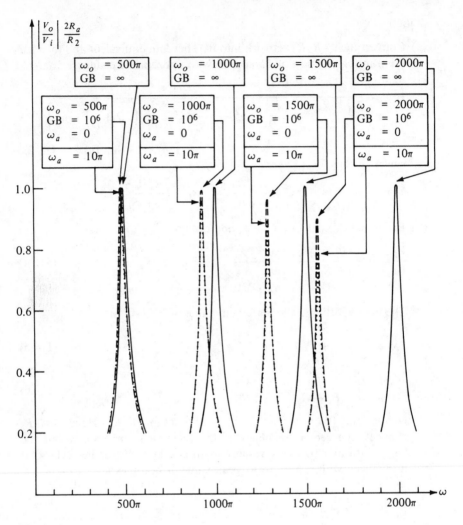

Fig. 11-8a

Substitute $Q = \omega_o/\mathrm{BW}$ and obtain

$$\frac{V_o}{V_i}\frac{2R_a}{R_2} = -\frac{\mathrm{BW}s}{s^2 + s\mathrm{BW} + \omega_o^2 + \dfrac{s}{\mathrm{GB}}\left[s^2 + s\mathrm{BW}\left(1 + 2\dfrac{\omega_o^2}{\mathrm{BW}^2}\right) + \omega_o^2\right]}.$$

$$(11\text{-}29)$$

The dashed curves of Fig. 11-8a represent plots of the magnitude of Eq. (11-29) for the same set of ω_o values as in (a) above and for $\mathrm{GB} = 10^6$ rad/s. A comparison of the dashed and solid curves shows that the peak value of the response does not remain constant; it decreases with the nominal ω_o of the poles. More importantly, the center frequency shifts to the left and therefore no longer coincides with ω_o. The higher the ω_o, the more is the departure of the center frequency from ω_o. A first-order measure of this shift can be obtained from the $\Delta\omega_o/\omega_o$ expression given in Table 11-3:

$$\frac{\Delta\omega_o}{\omega_o} \cong -Q\frac{\omega_o}{GB} = -\frac{\omega_o}{BW}\frac{\omega_o}{GB};$$

$$\Delta\omega_o = -\frac{1}{BW}\frac{\omega_o^3}{GB}. \tag{11-30}$$

For $BW = 40\pi$, $\omega_o = 1000\pi$, and $GB = 10^6$, Eq. (11-30) becomes

$$\Delta\omega_o = -\frac{(1000\pi)^3}{40\pi \times 10^6} = -78.5\pi,$$

which is in close agreement with the shift shown in Fig. 11-8a. Further study of the two sets of curves indicates that the bandwidth does not stay constant, but decreases. A quantitative but approximate measure of the reduction in bandwidth is obtained from the expression given in Table 11-3:

$$\frac{\Delta BW}{BW} \cong -2Q\frac{\omega_o}{GB} = -\frac{2\omega_o^2}{BWGB}. \tag{11-31}$$

For $BW = 40\pi$, $\omega_o = 1000\pi$, and $GB = 10^6$, the reduction is $5\pi\%$. These results indicate that as ω_o is increased, the upper half-plane pole moves to the right instead of following the constant-α ($\alpha = BW/2$) line. This is graphically illustrated in Fig. 11-8b by the pole-locus diagram (dashed curve).

(d) In Eq. (11-29), replace s/GB with $(s + \omega_a)/GB$ and obtain

$$\frac{V_o}{V_i}\frac{2R_a}{R_2} = -\frac{BWs}{s^2 + sBW + \omega_o^2 + \dfrac{(s + \omega_a)}{GB}\left[s^2 + sBW\left(1 + \dfrac{2\omega_o^2}{BW^2}\right) + \omega_o^2\right]}. \tag{11-32}$$

The dotted curves (inside the dashed curves) of Fig. 11-8a represent plots of Eq. (11-32) with $\omega_a = 10\pi$. All other parameters are the same as in (c). A study of the curves indicates that the open-loop bandwidth of the operational amplifier, ω_a, causes no further shift in the peaking frequency. However, compared to (c) above, it increases the bandwidth and causes a further reduction in the peak amplitude. These effects are particularly noticeable at large ω_o values. These results predict a shift of the upper half-plane pole [from its location given in (c)] to the left at constant-ω_o. The dotted pole-locus curve shown in Fig. 11-8b shows the combined effect of GB and ω_a.

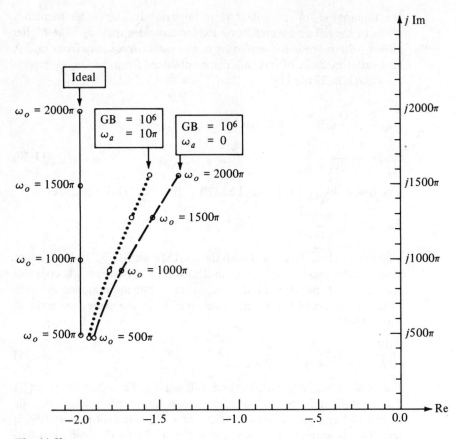

Fig. 11-8b

By the methods described in detail in Section 10-4, the first-order changes in ω_o, Q, and BW can be obtained as a function of GB and ω_a:

$$\frac{\Delta\omega_o}{\omega_o} \cong -\frac{\omega_o}{BW}\frac{\omega_o}{GB},$$ (11-33a)

$$\frac{\Delta Q}{Q} \cong \frac{\omega_o}{BW}\left(1 - 2\frac{\omega_a}{BW}\right)\frac{\omega_o}{GB},$$ (11-33b)

$$\frac{\Delta BW}{BW} \cong -2\frac{\omega_o}{BW}\left(1 - \frac{\omega_a}{BW}\right)\frac{\omega_o}{GB}.$$ (11-33c)

Note that ω_o does not depend on ω_a (at least to a first-order approximation), a result that is confirmed by the plots of Fig. 11-8. However, when the band-pass bandwidth approaches the open-loop bandwidth of the operational amplifier, significant changes occur in $\Delta Q/Q$ and $\Delta BW/BW$. As Eq. (11-33c) shows, the bandwidth decreases if $\omega_a < BW$; it increases if $\omega_a > BW$.

The various stages of this example progressively show the tradeoff that exists between complexity of analysis and accuracy of actual results. The engineer quite often chooses a simple model for the circuit elements so that he develops a good physical understanding of the operation of the circuit. As a consequence, he may be disappointed at the actual performance of the circuit. On the other hand, it may not be desirable or practical to use very complicated models, either. For example, it may not be easy to cope in an orderly manner with the complex inter-action of the many variables involved. In order to find an optimum solution for a successful design, patience, observation, and good judg-ment are necessary. Experience combined with sound theoretical under-standing helps the process.

11-6 REALIZATION WITH THREE OPERATIONAL AMPLIFIERS

The network shown in Fig. 11-9 is a rearranged version of the network of Fig. 10-16, with the excitation applied to the lossy integrator rather than to the integrator.

The transfer function is given by

$$\frac{V_o}{V_i} = -\frac{1}{R_0 C_2} \frac{s}{s^2 + s\,\dfrac{1}{R_Q C_2} + \dfrac{R_4/R_3}{R_1 R_2 C_1 C_2}}. \qquad (11\text{-}34)$$

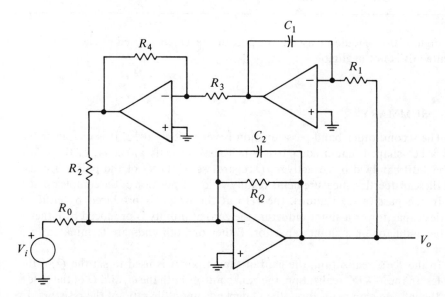

Fig. 11-9   Band-pass network

Since the poles of Eq. (11-34) are identical with the poles of the network given in Fig. 10-16, the discussions involving sensitivity and the limitation imposed by the amplifier's finite gain-bandwidth product given there apply here, too. For convenience, some of the important results are repeated in Table 11-4.

Table 11-4   Three-Amplifier Realization
(see Fig. 11-9)

Equal-R (except R_Q) and Equal-C

$$\omega_o = \frac{1}{RC}, \qquad Q = \frac{R_Q}{R}$$

$$\frac{V_o}{V_i} \cong - \frac{s\omega_o \left(\frac{2}{GB}s+1\right)\left(\frac{1}{GB}s+1\right)}{s^2 + s\frac{\omega_o}{Q} + \omega_o^2 + \frac{1}{GB}\left\{4s\left[s^2 + s\omega_o\left(\frac{1}{2}+\frac{1}{Q}\right)+\frac{\omega_o^2}{4Q}\right]\right\}} \qquad \left(\omega_a \ll \frac{\omega_o}{2Q}\right)$$

$$\frac{-\Delta\alpha}{\omega_o} \cong 2\left(1+\frac{1}{4Q}\right)\frac{\omega_o}{GB}, \qquad \frac{\Delta\beta}{\omega_o} \cong -\frac{\left(1-\frac{1}{Q}-\frac{1}{4Q^2}\right)}{\sqrt{1-\frac{1}{4Q^2}}}\frac{\omega_o}{GB}$$

$$\frac{\Delta\omega_o}{\omega_o} \cong -\frac{\omega_o}{GB}, \qquad \frac{\Delta Q}{Q} \cong 4Q\frac{\omega_o}{GB}$$

$$\frac{\Delta BW}{BW} \cong -4Q\frac{\omega_o}{GB}$$

Figure 10-17 should be used if the pole displacement caused by the amplifiers' finite GB is not negligible.

11-7 SUMMARY

The second-order band-pass function is very widely used. It is characterized by a bell-shaped magnitude curve which peaks exactly at the ω_o of the poles. The 3-dB bandwidth is exactly ω_o/Q regardless of the Q of the poles. Because of these properties, measurements readily yield the position of the complex poles.

In the passive realizations, the zero at the origin is produced by either a series capacitor or a shunt inductor; the zero at infinity is produced by either a series inductor or a shunt capacitor. Either or both ends are terminated in a resistor.

In the KRC realization, the gain of the amplifier is used to set the Q of the poles. In the $-KRC$ realization, the gain controls both the ω_o and Q of the poles. In the infinite-gain realization, the Q depends upon the ratio of the resistors. In the three-amplifier realization, the resistor of the lossy integrator determines the value of the Q. In all cases, the RC product can be used to adjust the value of the ω_o. As far as the finite value of the gain-bandwidth product of the operational amplifier is concerned, the low-pass and band-pass realizations have identical or similar frequency limitations, which set an upper bound for the ω_o of the active realizations.

PROBLEMS

11-1 Figure 11-10 shows a band-pass network with resistive terminations at both ends.

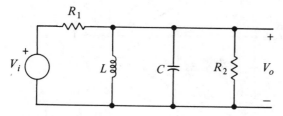

Fig. 11-10

(a) Obtain V_o/V_i.

(b) Calculate the ω_o-sensitivity and the Q-sensitivity functions.

(c) How do the poles change as a function of each element in the network?

11-2 Show that any RC band-pass network has a bandwidth that is always greater than twice the frequency of peaking.

11-3 The capacitor C_2 in Fig. 11-11 produces a transfer-function zero at $s = 0$, whereas the capacitor C_1 produces a zero at $s = \infty$. Hence, the network possesses band-pass characteristics. Let

$$R_1 = R_2 = R_3 = R \quad \text{and} \quad C_1 = C_2 = C.$$

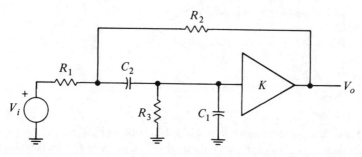

Fig. 11-11

(a) Obtain V_o/V_i, ω_o, and Q. Assume the amplifier is ideal.

(b) Obtain the denominator polynomial if

$$K = \frac{K_0 \omega_2}{s + \omega_2} .$$

(ω_2 represents the closed-loop bandwidth of the amplifier.)

11-4 What is the locus of the poles as a function of K_0 for the network shown in Fig. 11-3b? For the network in Fig. 11-5b?

11-5 Obtain V_o/V_i, ω_o, and Q for the network shown in Fig. 11-12.

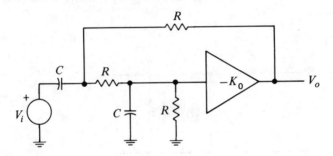

Fig. 11-12

11-6 Derive all the results of Table 11-2.

11-7 In Fig. 11-6, treat R_2, C_2, and the operational amplifier as a unit. Draw the resulting network in terms of a dependent source that is a differentiator.

11-8 (a) In Fig. 11-13, obtain V_o/V_i.

(b) What effect does C_1 have on the ω_o and BW?

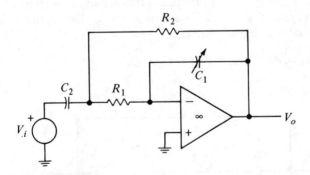

Fig. 11-13

11-9 (a) For the circuit shown in Fig. 11-14, obtain (V_o/V_i).

(b) Sketch $|V_o/V_i|$ for two values of C_3. Compare the peak values and the bandwidths of the two curves.

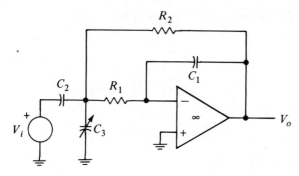

Fig. 11-14

11-10 In Fig. 11-15, if r is appropriately chosen, the real part of the complex poles does not vary with GB. Let $A(s) = -\mathrm{GB}/s$ and consider only first-order variation of the poles from the nominal positions. What is the value of r?

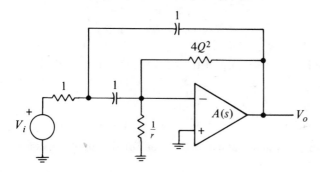

Fig. 11-15

11-11 Show that the output of the integrator in the network of Fig. 11-9 has a low-pass characteristic.

11-12 Obtain the (V_o/V_i) for the network shown in Fig. 11-16.

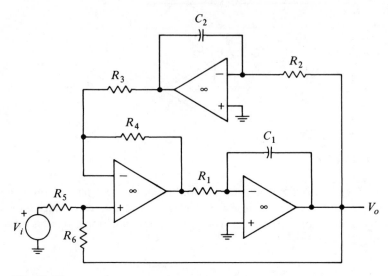

Fig. 11-16

11-13 (a) Find the input impedance of the network shown in Fig. 11-17.
(b) For

$$K = 1 + \frac{R_1}{R_2} \qquad \text{and} \qquad Z = \frac{1}{sC},$$

obtain an equivalent passive realization for the input impedance.

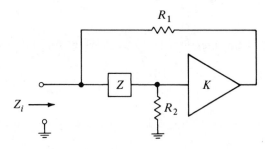

Fig. 11-17

Second-Order
High-Pass Networks

In this chapter, several *RLC* and active *RC* realizations are given for the second-order high-pass function. The passive realizations find applications in high-frequency filter work. The active realizations are used in low-frequency applications. The frequency limitation imposed by the amplifier becomes an important consideration when the band of frequencies to be passed extends into the high frequencies. For each circuit that is presented, the ω_o- and Q-sensitivity functions are developed, and first-order correction terms are given to evaluate the limitations imposed by the finite gain-bandwidth product of the operational amplifiers.

12-1 THE SECOND-ORDER HIGH-PASS FUNCTION

The second-order high-pass function has two zeros at the origin:

$$T(s) = \frac{Hs^2}{s^2 + s\dfrac{\omega_o}{Q} + \omega_o^2}.$$

To normalize the high-frequency magnitude of $T(j\omega)$ to unity, let $H = 1$:

$$T(s) = \frac{s^2}{s^2 + s\dfrac{\omega_o}{Q} + \omega_o^2} = \frac{\left(\dfrac{s}{\omega_o}\right)^2}{\left(\dfrac{s}{\omega_o}\right)^2 + \left(\dfrac{s}{\omega_o}\right)\dfrac{1}{Q} + 1}. \tag{12-1}$$

The characteristics of $T(s)$ depend upon ω_o and Q. The effect of ω_o is to scale the frequency variable. The effect of Q is displayed in the magnitude and phase curves shown in Fig. 12-1.

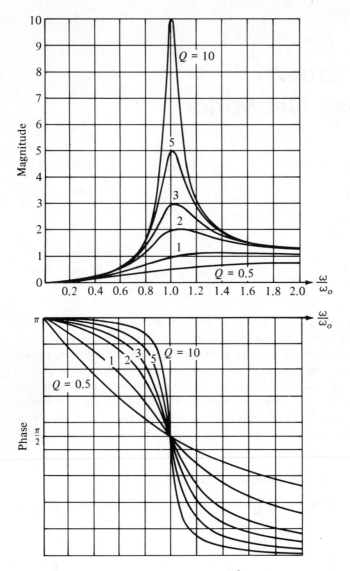

Fig. 12-1   Magnitude and phase of $\dfrac{s^2}{s^2 + s\dfrac{\omega_o}{Q} + \omega_o^2}$

A study of Eq. (12-1) and Fig. 12-1 reveals the following important features concerning the high-pass function.

1. For $\omega/\omega_o \ll 1$, $|T(j\omega)| \cong 0$. Therefore, low frequencies are attenuated.

2. For $\omega/\omega_o \gg 1$, $|T(j\omega)| \cong 1$. Therefore, high frequencies are passed.

3. For $Q > 1/\sqrt{2}$, the magnitude peaks at (see Problem 12-2)

$$\frac{\omega}{\omega_o} = \frac{1}{\sqrt{1 - \dfrac{1}{2Q^2}}}. \tag{12-2}$$

The peak occurs at a frequency higher than ω_o. For $Q > 5$, the frequency of peaking occurs practically at $\omega = \omega_o$ (within 1%).

4. For $Q > 5$, the 3-dB bandwidth (band-pass) is practically equal to ω_o/Q rad/s.

5. At $\omega = \omega_o$, $|T(j\omega_o)| = Q$ and the phase is $\pi/2$.

6. At $\omega/\omega_o \gg 1$, the phase is 0.

7. For $Q > 5$, the phase undergoes a rapid shift of nearly π radians about ω_o.

As the magnitude curves indicate, $T(s)$, given by Eq. (12-1), *can be used as a high-pass function. For high-Q values, it can also be used as a band-pass function.* Nonetheless, a second-order function with two zeros at the origin is usually labelled as a high-pass function. (In a high-order high-pass function, it is not uncommon to encounter some second-order factors with high Q's. However, the overall magnitude response may not peak appreciably because other second-order factors pull the overall response down.)

12-2 PASSIVE REALIZATIONS

Two second-order *RLC* high-pass networks are given in Fig. 12-2. (Another network is given in Problem 12-3.) In each network, the inductor causes one transfer-function zero at the origin and the capacitor the other.

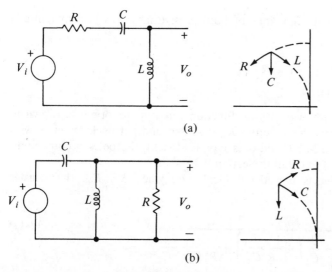

(a)

(b)

Fig. 12-2   Passive high-pass networks

For the circuit shown in Fig. 12-2a, the transfer function, ω_o, Q, and the various sensitivity functions are

$$\frac{V_o}{V_i} = \frac{s^2}{s^2 + s\dfrac{R}{L} + \dfrac{1}{LC}}; \tag{12-3}$$

$$\omega_o = \frac{1}{\sqrt{LC}}, \qquad Q = \frac{1}{R}\sqrt{\frac{L}{C}}; \tag{12-4}$$

$$S_L^{\omega_o} = S_C^{\omega_o} = -\tfrac{1}{2}, \qquad S_R^{\omega_o} = 0, \tag{12-5}$$

$$S_L^Q = -S_C^Q = \tfrac{1}{2}, \qquad S_R^Q = -1. \tag{12-6}$$

As each element is increased from its nominal value, the upper half-plane pole moves as shown in Fig. 12-2a.

For the circuit shown in Fig. 12-2b, the transfer function, ω_o, Q, and the various sensitivity functions are

$$\frac{V_o}{V_i} = \frac{s^2}{s^2 + s\dfrac{1}{RC} + \dfrac{1}{LC}}; \tag{12-7}$$

$$\omega_o = \frac{1}{\sqrt{LC}}, \qquad Q = R\sqrt{\frac{C}{L}}; \tag{12-8}$$

$$S_L^{\omega_o} = S_C^{\omega_o} = -\tfrac{1}{2}, \qquad S_R^{\omega_o} = 0, \tag{12-9}$$

$$S_L^Q = -S_C^Q = -\tfrac{1}{2}S_R^Q = -\tfrac{1}{2}. \tag{12-10}$$

The effect of a positive increment in each element is shown in Fig. 12-2b for the upper half-plane pole.

All the pole-sensitivity functions of both passive networks have magnitude values of one or less.

12-3 *KRC* REALIZATION

An *RC* network with two transfer-function zeros at the origin is shown in Fig. 12-3a. When the voltage across R_2 is sensed and fed back in series with R_1 as shown in Fig. 12-3b (band-pass type feedback), the poles become complex. This is discussed in detail in Section 9-3.

When the amplifier is considered ideal, i.e., when $K = K_0$, the transfer function is given by

$$\frac{V_o}{V_i} = \frac{K_0 s^2}{s^2 + s\left[\dfrac{1}{R_2}\left(\dfrac{1}{C_1} + \dfrac{1}{C_2}\right) + \dfrac{1 - K_0}{R_1 C_1}\right] + \dfrac{1}{R_1 R_2 C_1 C_2}}. \tag{12-11}$$

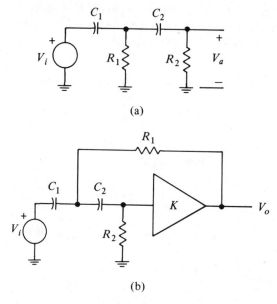

(a)

(b)

Fig. 12-3　High-pass *KRC* network

The ω_o and Q of the poles are

$$\omega_o = \frac{1}{\sqrt{R_1 R_2 C_1 C_2}}, \qquad Q = \frac{1}{\sqrt{\dfrac{R_1}{R_2}}\left(\sqrt{\dfrac{C_2}{C_1}} + \sqrt{\dfrac{C_1}{C_2}}\right) + \sqrt{\dfrac{R_2 C_2}{R_1 C_1}}(1 - K_0)}.$$

$$(12\text{-}12)$$

To simplify, let $R_1 = R_2 = R$, $C_1 = C_2 = C$ (see Problem 12-5 for the equal-C, unity-gain case):

$$\omega_o = \frac{1}{RC}, \qquad Q = \frac{1}{3 - K_0}. \tag{12-13}$$

Either R or C is chosen arbitrarily. The remaining element values are fixed by Eq. (12-13), i.e., the RC product fixes the ω_o whereas K_0 fixes the Q. The two adjustments do not interact.

For the equal-R, equal-C case, the ω_o-sensitivity and Q-sensitivity functions are

$$S_{R_1}^{\omega_o} = S_{R_2}^{\omega_o} = S_{C_1}^{\omega_o} = S_{C_2}^{\omega_o} = -\tfrac{1}{2}, \qquad S_{K_0}^{\omega_o} = 0, \tag{12-14a}$$

$$S_{R_1}^{Q} = -S_{R_2}^{Q} = \tfrac{1}{2} - 2Q, \qquad S_{C_1}^{Q} = -S_{C_2}^{Q} = \tfrac{1}{2} - Q, \qquad S_{K_0}^{Q} = 3Q - 1. \tag{12-14b}$$

Since the Q-sensitivity functions are all directly proportional to the Q of the poles, care should be exercised when high-Q circuits are constructed.

The results obtained so far are based on an ideal amplifier. If a one-pole rolloff model is used for the amplifier, the transfer function, and hence the poles, change from their nominal positions. The first-order changes are summarized in Table 12-1.

Table 12-1   *KRC* Realization
(see Fig. 12-3b)

Equal-R, Equal-C

$$\omega_o = \frac{1}{RC}, \qquad Q = \frac{1}{3 - K_0}$$

$$\frac{V_o}{V_i} = \frac{\left(3 - \frac{1}{Q}\right)s^2}{s^2 + s\frac{\omega_o}{Q} + \omega_o^2 + \frac{\left(3 - \frac{1}{Q}\right)s}{GB}(s^2 + s3\omega_o + \omega_o^2)} \qquad \left(\omega_a \ll \frac{\omega_o}{2Q}\right)$$

$$-\frac{\Delta\alpha}{\omega_o} \cong \frac{\left(3 - \frac{1}{Q}\right)^2}{2Q}\frac{\omega_o}{GB}, \qquad \frac{\Delta\beta}{\omega_o} \cong -\frac{\left(1 - \frac{1}{2Q^2}\right)\left(3 - \frac{1}{Q}\right)^2}{2\sqrt{1 - \frac{1}{4Q^2}}}\frac{\omega_o}{GB}$$

$$\frac{\Delta\omega_o}{\omega_o} \cong -\frac{\left(3 - \frac{1}{Q}\right)^2}{2}\frac{\omega_o}{GB}, \qquad \frac{\Delta Q}{Q} \cong \frac{\left(3 - \frac{1}{Q}\right)^2}{2}\frac{\omega_o}{GB}$$

Since the characteristic equation is identical with that of the *KRC* low-pass realization (equal-*R*, equal-*C*), Fig. 10-5a can be used to determine the position of the poles if large departures from nominal positions occur.

Because of nonideal amplifiers, active *RC* realizations produce a response that falls off at high frequencies. Hence, strictly speaking, it is not possible to realize the high-pass function. The resulting response is actually band-pass. However, as long as frequencies of interest do not extend into the very high frequencies, the desired filtering can be accomplished.

EXAMPLE 12-1

In the *KRC* high-pass circuit given in Fig. 12-3b, let

$$R_1 = R_2 = R, \qquad C_1 = C_2 = C, \qquad \text{and} \qquad K = GB/(s + GB/K_0).$$

(a) Up to what frequency can this circuit be used as a high-pass circuit?
(b) Plot the magnitude response as a function of Q for $GB = 1000\omega_o$. Discuss the results.

SOLUTION

(a) The transfer function of this network is identical with that given in Table 12-1. Therefore, obtain V_o/V_i directly from Table 12-1:

$$\frac{V_o}{V_i} = \frac{\left(3 - \frac{1}{Q}\right)s^2}{s^2 + s\frac{\omega_o}{Q} + \omega_o^2 + \frac{s}{GB}\left[\left(3 - \frac{1}{Q}\right)(s^2 + s3\omega_o + \omega_o^2)\right]}. \tag{12-15}$$

Rearrange this equation to put the denominator polynomial in monic form:

$$\frac{V_o}{V_i} = \frac{GBs^2}{s^3 + s^2\left[3\omega_o + \frac{GB}{3 - (1/Q)}\right] + s\left[\omega_o^2 + \frac{\omega_o}{Q}\frac{GB}{3 - (1/Q)}\right] + \omega_o^2\left[\frac{GB}{3 - (1/Q)}\right]}. \tag{12-16}$$

When GB is infinite, Eq. (12-15) reduces to the high-pass function; that is,

$$\frac{V_o}{V_i} = \frac{\left(3 - \frac{1}{Q}\right)s^2}{s^2 + s\frac{\omega_o}{Q} + \omega_o^2}. \tag{12-17}$$

When GB is not infinite but very large, a slight change occurs in the nominal-pole positions given by Eq. (12-17). More importantly, a real-axis pole appears. It is this pole that causes the decrease in the magnitude characteristic at high frequencies. Let its position be given by $s_3 = -\gamma$. For all practical purposes, the other two poles can still be considered at their nominal positions; that is,

$$s_{1,2} \cong -\frac{\omega_o}{2Q} \pm j\omega_o\sqrt{1 - \frac{1}{4Q^2}}.$$

In a cubic equation in monic form, the coefficient of the s^2-term is the negative sum of the three roots. Therefore,

$$3\omega_o + \frac{GB}{3 - \frac{1}{Q}} = -(s_1 + s_2 + s_3)$$

$$= -\left(-\frac{\omega_o}{2Q} + j\omega_o\sqrt{1 - \frac{1}{4Q^2}} - \frac{\omega_o}{2Q} - j\omega_o\sqrt{1 - \frac{1}{4Q^2}} - \gamma\right)$$

$$= \frac{\omega_o}{Q} + \gamma.$$

Solve for γ and obtain

$$\gamma = \frac{GB}{3 - \dfrac{1}{Q}} + \omega_o\left(3 - \frac{1}{Q}\right), \tag{12-18}$$

which, for $(GB/\omega_o) \gg [3 - (1/Q)]^2$, reduces to

$$\gamma \cong \frac{GB}{3 - \dfrac{1}{Q}} = \frac{GB}{K_0}. \tag{12-19}$$

Note that the magnitude of the real-axis pole is identical with the closed-loop bandwidth of the amplifier, that is, $\omega_{3\,dB} = \gamma$. With γ thus evaluated, Eq. (12-16) can be written in factored form:

$$\frac{V_o}{V_i} \cong \frac{GBs^2}{\left(s^2 + s\dfrac{\omega_o}{Q} + \omega_o^2\right)\left[s + \dfrac{GB}{3 - (1/Q)}\right]}$$

$$= \frac{\left(3 - \dfrac{1}{Q}\right)s^2}{\left(s^2 + s\dfrac{\omega_o}{Q} + \omega_o^2\right)\left[s\dfrac{3 - (1/Q)}{GB} + 1\right]}. \tag{12-20}$$

As long as $\omega \ll GB/[3 - (1/Q)]$, the magnitude of Eq. (12-20) is identical with the magnitude of the high-pass function given in Eq. (12-17). At $\omega = GB/[3 - (1/Q)]$, the magnitude is down 3 dB. Therefore, this *KRC* circuit can be considered functioning as a high-pass circuit for frequencies less than $GB/[3 - (1/Q)]$.

(b) To plot the magnitude response, use magnitude and frequency normalizations on Eq. (12-16) and obtain

$$\frac{V_o}{V_i}\frac{1}{\left(3 - \dfrac{1}{Q}\right)} = \frac{\dfrac{GB}{\omega_o}}{3 - \dfrac{1}{Q}} \times$$

$$\frac{\left(\dfrac{s}{\omega_o}\right)^2}{\left(\dfrac{s}{\omega_o}\right)^3 + \left(\dfrac{s}{\omega_o}\right)^2\left[3 + \dfrac{(GB/\omega_o)}{3 - (1/Q)}\right] + \left(\dfrac{s}{\omega_o}\right)\left[1 + \dfrac{1}{Q}\dfrac{(GB/\omega_o)}{3 - (1/Q)}\right] + \dfrac{(GB/\omega_o)}{3 - (1/Q)}}. \tag{12-21}$$

Equation (12-21) is plotted in Fig. 12-4 for three values of Q and for $GB/\omega_o = 1000$. As the curves show, the lower 3-dB frequency is determined to a large extent by ω_o and to a lesser extent by Q. For example, for $Q = 1$,

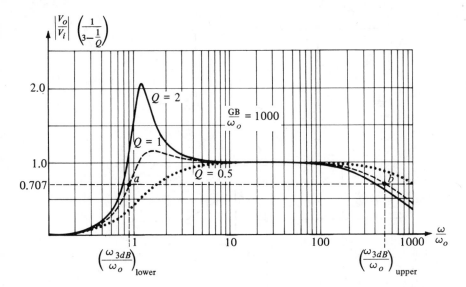

Fig. 12-4

$$\left(\frac{\omega_{3\,dB}}{\omega_o}\right)_{lower} \cong 0.8 \qquad \text{(point } a \text{ in Fig. 12-4).}$$

The upper 3-dB frequency is determined by ω_o, Q, and GB. For example, for $Q = 1$ and $\text{GB}/\omega_o = 1000$,

$$\left(\frac{\omega_{3\,dB}}{\omega_o}\right)_{upper} \cong 500 \qquad \text{(point } b \text{ in Fig. 12-4).}$$

Note that this 3-dB frequency is in agreement with that obtained from the approximate expression given by Eq. (12-19); that is,

$$\left(\frac{\omega_{3\,dB}}{\omega_o}\right)_{upper} \cong \frac{1}{3 - \dfrac{1}{Q}}\frac{\text{GB}}{\omega_o} = \left(\frac{1}{3 - 1}\right)1000 = 500.$$

Had a larger value of GB been used, the upper 3-dB frequency and hence the bandwidth would have been proportionally higher. The peaking that occurs near ω_o for high-Q poles would be objectionable for a second-order system. However, if used in cascade with one or more second-order sections with lower Q's, the overall response can be made to exhibit no peaking.

12-4 —*KRC* REALIZATION

An *RC* network with two transfer-function zeros at the origin is shown in Fig. 12-5a. This network is slightly different from that given in Fig. 12-3a since it contains an additional capacitor C_3. When the voltage across R_2 is sensed

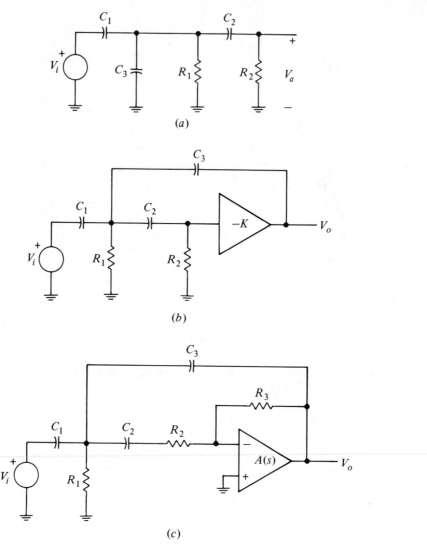

Fig. 12-5 High-pass $-KRC$ networks

and fed back in series with capacitor C_3 (high-pass feedback), as shown in Fig. 12-5b, the poles can be made complex. See Section 9-4 for details. The resistor R_2 can be used as part of the amplifier, as shown in Fig. 12-5c.

When the amplifier is considered ideal, that is, when $K = K_0$, the transfer function is given by

$$\frac{V_o}{V_i} = \frac{-\dfrac{K_0}{1 + (C_3/C_1) + K_0}\, s^2}{s^2 + s\,\dfrac{(R_1 + R_2)C_2 + R_1(C_1 + C_3)}{R_1 R_2 C_2[C_1 + C_3(1 + K_0)]} + \dfrac{1}{R_1 R_2 C_2[C_1 + C_3(1 + K_0)]}}.$$

$$(12\text{-}22)$$

The ω_o and Q of the poles are

$$\omega_o = \frac{1}{\sqrt{R_1 R_2 C_2 [C_1 + C_3(1 + K_0)]}}, \qquad Q = \frac{\sqrt{R_1 R_2 C_2 [C_1 + C_3(1 + K_0)]}}{(R_1 + R_2)C_2 + R_1(C_1 + C_3)}.$$

$$(12\text{-}23)$$

To simplify, let $R_1 = R_2 = R$ and $C_1 = C_2 = C_3 = C$. Then

$$\omega_o = \frac{1}{RC\sqrt{2 + K_0}}, \qquad Q = \frac{\sqrt{2 + K_0}}{4}. \qquad (12\text{-}24)$$

The corresponding sensitivity functions are

$$S_{R_1}^{\omega_o} = S_{R_2}^{\omega_o} = S_{C_2}^{\omega_o} = -\frac{1}{2}, \qquad S_{C_1}^{\omega_o} = -\frac{1}{32Q^2}, \qquad S_{C_3}^{\omega_o} = -\frac{1}{2} + \frac{1}{16Q^2}, \qquad (12\text{-}25a)$$

$$S_{R_1}^{Q} = -S_{R_2}^{Q} = -\frac{1}{4}, \qquad S_{C_1}^{Q} = \frac{1}{32Q^2} - \frac{1}{4}, \qquad S_{C_2}^{Q} = 0,$$

$$S_{C_3}^{Q} = \frac{1}{4} - \frac{1}{32Q^2}, \qquad S_{K_0}^{Q} = \frac{1}{2} - \frac{1}{16Q^2}. \qquad (12\text{-}25b)$$

All the sensitivity figures are low, being $\frac{1}{2}$ or smaller in magnitude.

The first-order effects resulting from the finite value of the gain-bandwidth product of the operational amplifier are summarized in Table 12-2.

Table 12-2   − KRC Realization
(see Fig. 12-5c)

Equal-R, Equal-C

$$\omega_o = \frac{1}{RC\sqrt{2 + K_0}}, \qquad Q = \frac{\sqrt{2 + K_0}}{4}$$

$$\frac{V_o}{V_i} = \frac{-\left(1 - \dfrac{1}{8Q^2}\right)s^2}{s^2 + s\dfrac{\omega_o}{Q} + \omega_o^2 + \dfrac{s}{GB}\left[2\left(1 - \dfrac{1}{16Q^2}\right)s^2 + s\left(4Q + \dfrac{1}{2Q}\right)\omega_o + \omega_o^2\right]} \qquad \left(\omega_a \ll \dfrac{\omega_o}{2Q}\right)$$

$$-\frac{\Delta\alpha}{\omega_o} \cong \left(\frac{5}{2} - \frac{13}{16Q^2} + \frac{1}{16Q^4}\right)\frac{\omega_o}{GB}, \qquad \frac{\Delta\beta}{\omega_o} \cong -\frac{\left(2Q - \dfrac{2}{Q} + \dfrac{15}{32Q^3} - \dfrac{1}{32Q^5}\right)}{\sqrt{1 - \dfrac{1}{4Q^2}}}\frac{\omega_o}{GB}$$

$$\frac{\Delta\omega_o}{\omega_o} \cong -\left(2Q - \frac{3}{4Q} + \frac{1}{16Q^3}\right)\frac{\omega_o}{GB}, \qquad \frac{\Delta Q}{Q} \cong \left(3Q - \frac{7}{8Q} + \frac{1}{16Q^3}\right)\frac{\omega_o}{GB}$$

12-5 REALIZATION WITH INFINITE GAIN

An RC network with two transfer-function zeros at the origin is shown in Fig. 12-6a. As discussed in Section 9-5, the poles of this network can be moved off the real axis and made complex if an operational amplifier is introduced, as shown in Fig. 12-6b.

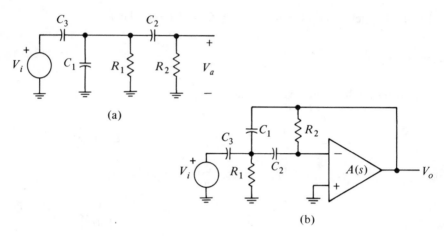

(a)

(b)

Fig. 12-6 High-pass network

For $|A(s)| = \infty$, the transfer function becomes

$$\frac{V_o}{V_i} = -\frac{C_3}{C_1} \frac{s^2}{s^2 + s\dfrac{1}{R_2}\left(\dfrac{1}{C_1} + \dfrac{1}{C_2} + \dfrac{C_3}{C_1 C_2}\right) + \dfrac{1}{R_1 R_2 C_1 C_2}}. \tag{12-26}$$

The ω_o and Q of the poles are

$$\omega_o = \frac{1}{\sqrt{R_1 R_2 C_1 C_2}}, \qquad Q = \frac{\sqrt{\dfrac{R_2}{R_1}}}{\sqrt{\dfrac{C_1}{C_2}} + \sqrt{\dfrac{C_2}{C_1}} + \dfrac{C_3}{\sqrt{C_1 C_2}}}. \tag{12-27}$$

To simplify, let $C_1 = C_2 = C_3 = C$. Then,

$$\omega_o = \frac{1}{C\sqrt{R_1 R_2}}, \qquad Q = \frac{1}{3}\sqrt{\frac{R_2}{R_1}}. \tag{12-28}$$

The various sensitivity functions for the equal-C case are

$$S_{R_1}^{\omega_o} = S_{R_2}^{\omega_o} = S_{C_1}^{\omega_o} = S_{C_2}^{\omega_o} = -\tfrac{1}{2}, \qquad S_{C_3}^{\omega_o} = 0; \tag{12-29a}$$

$$S_{R_1}^{Q} = -S_{R_2}^{Q} = -\tfrac{1}{2}, \qquad S_{C_1}^{Q} = S_{C_2}^{Q} = \tfrac{1}{6}, \qquad S_{C_3}^{Q} = -\tfrac{1}{3}. \tag{12-29b}$$

All the sensitivity figures are low.

The first-order effects resulting from the finite value of the gain-bandwidth product of the operational amplifier are summarized in Table 12-3.

Table 12-3   Realization with Infinite Gain
(see Fig. 12-6b)

Equal-C

$$\omega_o = \frac{1}{C\sqrt{R_1 R_2}}, \qquad Q = \frac{1}{3}\sqrt{\frac{R_2}{R_1}}$$

$$\frac{V_o}{V_i} = -\frac{s^2}{s^2 + s\dfrac{\omega_o}{Q} + \omega_o^2 + \dfrac{s}{GB}\left[2s^2 + s\omega_o\left(3Q + \dfrac{1}{Q}\right) + \omega_o^2\right]} \qquad \left(\omega_a \ll \frac{\omega_o}{2Q}\right)$$

$$-\frac{\Delta\alpha}{\omega_o} \cong \left(1 - \frac{1}{4Q^2}\right)\frac{\omega_o}{GB}, \qquad \frac{\Delta\beta}{\omega_o} \cong \frac{\left(3Q - \dfrac{3}{Q} + \dfrac{1}{2Q^3}\right)}{4\sqrt{1 - \dfrac{1}{4Q^2}}}\frac{\omega_o}{GB}$$

$$\frac{\Delta\omega_o}{\omega_o} \cong -\frac{3Q}{4}\left(1 - \frac{1}{3Q^2}\right)\frac{\omega_o}{GB}, \qquad \frac{\Delta Q}{Q} \cong \frac{5Q}{4}\left(1 - \frac{1}{25Q^2}\right)\frac{\omega_o}{GB}$$

12-6 REALIZATION WITH FOUR OPERATIONAL AMPLIFIERS

The network shown in Fig. 12-7 uses the network of Fig. 11-9 to generate a pair of complex poles. A fourth amplifier is then used to combine the appropriate output signals to obtain the desired high-pass response. (Problem 12-6

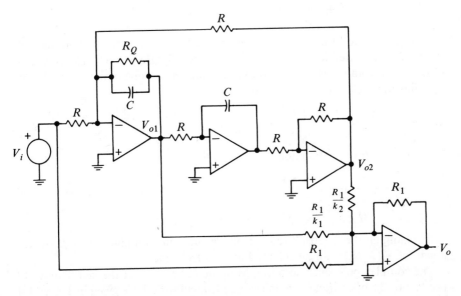

Fig. 12-7   Generation of a high-pass function

gives another multioperational amplifier scheme for generating the high-pass function.)

The transfer function V_{o1}/V_i is obtained from Eq. (11-34). The transfer function V_{o2}/V_i is obtained by multiplying V_{o1}/V_i with $(1/sRC)$.

$$\frac{V_{o1}}{V_i} = -\frac{s\omega_o}{s^2 + s\dfrac{\omega_o}{Q} + \omega_o^2},\tag{12-30}$$

$$\frac{V_{o2}}{V_i} = -\frac{\omega_o^2}{s^2 + s\dfrac{\omega_o}{Q} + \omega_o^2},\tag{12-31}$$

where

$$\omega_o = \frac{1}{RC}, \qquad Q = \frac{R_Q}{R}.$$

The output voltage is obtained by summing V_i, V_{o1}, and V_{o2}:

$$V_o = V_i(-1) + V_{o1}(-k_1) + V_{o2}(-k_2),$$

$$\frac{V_o}{V_i} = -1 + k_1 \frac{s\omega_o}{s^2 + s\dfrac{\omega_o}{Q} + \omega_o^2} + k_2 \frac{\omega_o^2}{s^2 + s\dfrac{\omega_o}{Q} + \omega_o^2}$$

$$= -\frac{s^2 + s\dfrac{\omega_o}{Q}(1 - k_1 Q) + \omega_o^2(1 - k_2)}{s^2 + s\dfrac{\omega_o}{Q} + \omega_o^2}.\tag{12-32}$$

After the ω_o and the Q of the poles are set by adjusting the RC product and the R_Q/R ratio, attention is directed to V_o/V_i given by Eq. (12-32). The coefficient of the s-term and the constant term in the numerator polynomial are made zero by adjusting the gains k_1 and k_2 to

$$k_1 = \frac{1}{Q}, \qquad k_2 = 1.\tag{12-33}$$

The resulting transfer function is

$$\frac{V_o}{V_i} = -\frac{s^2}{s^2 + s\dfrac{\omega_o}{Q} + \omega_o^2}.\tag{12-34}$$

By changing the feedback resistor of the summing amplifier from R_1 to R_f, a gain of $-(R_f/R_1)$ can be achieved at high frequencies.

The complex-pole shifts resulting from the finite value of the gain-bandwidth product of the operational amplifiers are identical with those given in Table 11-4. When the frequency dependence of the amplifiers is taken into account, and some simplifying assumptions are made (see disussion in Section 10-9), a four-pole system with three finite zeros is obtained. Nonetheless, it is still possible to null the coefficient of the *s*-term and the constant term [though for slightly different values than given by Eq. (12-33)]. The frequency limitation imposed by the summing amplifier comes as a separate one-pole rolloff factor. As Example 7-4 illustrates, the 3-dB bandwidth (high-frequency) associated with this factor is given by

$$\omega_{3\,dB} = \frac{GB}{1 + \left(1 + \dfrac{1}{k_1} + \dfrac{1}{k_2}\right)}. \tag{12-35}$$

Thus, for frequencies greater than $\omega_{3\,dB}$ given by Eq. (12-35), the magnitude response starts falling off at 6 dB/octave.

12-7 SUMMARY

The second-order high-pass function is characterized by two zeros at the origin. As a result, low frequencies are attenuated and high frequencies are passed. The transition from attenuation to passing occurs near ω_o.

The passive realizations contain a series capacitor and a shunt inductor which produce the desired zeros. Resistive terminations are used at either or both ends.

The active realizations contain two capacitors in the series path between input and output, to achieve the desired transfer-function zeros. Through feedback the poles are made complex. When the nonideal nature of the operational amplifiers is taken into account, the system order becomes higher than two. As a result, the complex poles are displaced from their nominal positions. Moreover, the response falls off at high frequencies, thereby terminating the high-pass type response.

PROBLEMS

12-1 Obtain two first-order high-pass networks. What is the 3-dB frequency in each case?

12-2 Derive Eq. (12-2).

12-3 Figure 12-8 shows a high-pass network with resistive terminations at both ends.

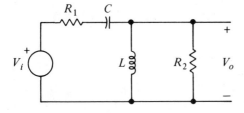

Fig. 12-8

(a) Obtain V_o/V_i.

(b) Calculate the ω_o-sensitivity and the Q-sensitivity functions.

12-4 Assume $R_2 = R_1 = R$ and draw, as a function of C_2, the loci of the poles of the network shown in Fig. 12-3b. Mark the $C_2 = C_1$ points on the loci.

12-5 In Fig. 12-3b, let

$$C_1 = C_2 = C \quad \text{and} \quad K = \frac{GB}{s + GB}.$$

(a) Find ω_o and Q if $GB = \infty$.

(b) Obtain the first-order effects caused by the amplifier's gain-bandwidth product on the ω_o and Q of the poles.

12-6 Show that the network given in Fig. 12-9 does generate all three of the second-order functions, namely low-pass, band-pass, and high-pass.

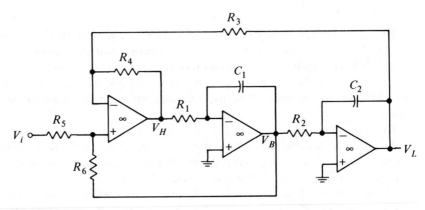

Fig. 12-9

13

Second-Order
Band-Stop Networks

There are applications which require the elimination of an undesired sinusoidal signal of a known frequency from a given waveform. This can be accomplished by passing the waveform through a system with a pair of imaginary-axis zeros which are coincident with the poles of the undesired sine wave. A second-order band-stop network can be used for this purpose. If a band of frequencies is to be rejected, additional poles and imaginary-axis zeros must be introduced to broaden the rejection characteristic.

In this chapter, several second-order RLC and RC band-stop networks are given and their characteristics are discussed. The RLC band-stop networks are widely used at high frequencies to trap or reject unwanted frequencies. Rejection can also be achieved by RC networks. However, the rejection characteristics of RC networks are rather broad. To achieve sharper rejection, the poles of the networks are made complex by employing feedback amplifiers. Several schemes for accomplishing this are given. In particular, noninteractive adjustment procedures are given by which the rejection frequency and the bandwidth of rejection can be independently controlled.

13-1 THE SECOND-ORDER BAND-STOP FUNCTION

A pair of imaginary-axis zeros is necessary to stop the steady-state transmission of a sinusoidal signal of a given frequency. Hence, the numerator polynomial must be of the form $(s^2 + \omega_z^2)$ where $\pm j\omega_z$ represent the zero locations. The corresponding second-order function with unity scale factor is

$$T(s) = \frac{s^2 + \omega_z^2}{s^2 + s\dfrac{\omega_o}{Q} + \omega_o^2}. \tag{13-1}$$

The frequency ω_z is called the rejection or notch frequency.

In general, $\omega_z = \omega_o$. [The $\omega_z \neq \omega_o$ case may arise when $T(s)$ represents a second-order factor in a higher-order system.] Consequently, Eq. (13-1) can be simplified to

$$T(s) = \frac{s^2 + \omega_o^2}{s^2 + s\dfrac{\omega_o}{Q} + \omega_o^2} = \frac{\left(\dfrac{s}{\omega_o}\right)^2 + 1}{\left(\dfrac{s}{\omega_o}\right)^2 + \left(\dfrac{s}{\omega_o}\right)\dfrac{1}{Q} + 1}. \tag{13-2}$$

$T(s)$ as given by Eq. (13-2) is called the band-stop (band-reject, band-eliminate) function. In Fig. 13-1, the magnitude and phase of $T(s)$ are plotted vs. ω/ω_o for

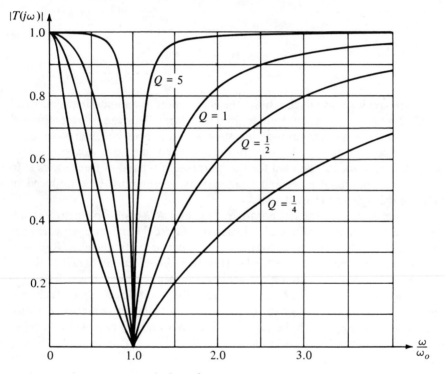

Fig. 13-1a   Magnitude of $\dfrac{s^2 + \omega_o^2}{s^2 + s\dfrac{\omega_o}{Q} + \omega_o^2}$

several values of the variable Q. By studying these curves and Eq. (13-2), the following observations can be made regarding the second-order band-stop function:

1. For $\omega/\omega_o \ll 1$ and for $\omega/\omega_o \gg 1$,

$$|T(j\omega)| \cong 1.$$

Therefore, high and low frequencies are passed.

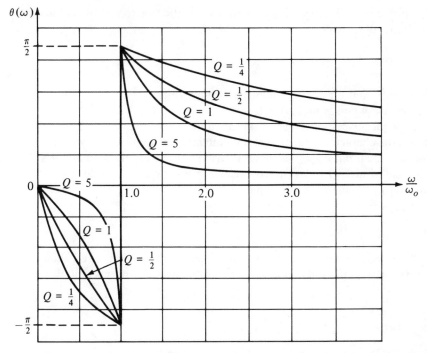

Fig. 13-1b   Phase of $\dfrac{s^2 + \omega_o^2}{s^2 + s\dfrac{\omega_o}{Q} + \omega_o^2}$

2. For $\omega = \omega_o$,

$$|T(j\omega)| = 0.$$

Therefore, a sine wave of frequency ω_o is stopped in the steady state. [In practice, because of imperfections, $|T(j\omega_o)|$ is not zero but very small. Consequently, the notch depth is specified in terms of "so many dB's down" from 0.]

3. If the bandwidth of rejection, BW, is defined as the band of frequencies over which the magnitude is down 3 dB or more, then the BW is given by (see Problem 13-1):

$$\text{BW} = \frac{\omega_o}{Q}. \tag{13-3}$$

Thus, the higher the Q of the poles, the narrower is the rejection characteristic.

4. $|T(j\omega)|$ has the same value at two frequencies. The geometric average of these two frequencies is ω_o. Thus,

$$|T(j\omega)| = \left| T\!\left(j\frac{\omega_o^2}{\omega}\right) \right|.$$

The phase of $T(j\omega)$ at the higher frequency is the negative of the phase at the lower frequency, that is,

$$\theta(\omega) = -\theta\left(\frac{\omega_o^2}{\omega}\right).$$

5. For $\omega \ll 1$ and for $\omega \gg 1$, the phase is practically 0.
6. The phase jumps π radians at ω_o.

The frequency-response curves given in Fig. 13-1 represent the sinusoidal steady-state characteristics of the BS function. In applications where abrupt changes in the input occur, the natural term arising from the system poles may, in the case of high-Q poles (narrow rejection bandwidths), produce a lingering though damped sine wave of frequency

$$\omega_o\sqrt{1 - \frac{1}{4Q^2}} \cong \omega_o.$$

This "sine-wave ringing" is the transient part of the response and should not be confused with the input sine wave which the network rejects completely.

13-2 PASSIVE REALIZATIONS: *RLC*

Two *RLC* networks which realize the second-order BS function are given in Fig. 13-2. Two others are given in Problem 13-2.

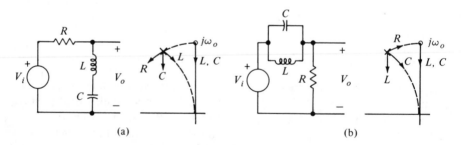

Fig. 13-2   Two *RLC* band-stop networks

In Fig. 13-2a, the transfer-function zeros are produced by the series *LC* circuit located in the shunt arm of the ladder. [The impedance of this *LC* circuit becomes zero for $s = \pm j(1/\sqrt{LC})$.] The transfer function is given by

$$\frac{V_o}{V_i} = \frac{s^2 + \dfrac{1}{LC}}{s^2 + s\dfrac{R}{L} + \dfrac{1}{LC}}. \tag{13-4}$$

By comparison with Eq. (13-2), it is seen that

$$\omega_o = \frac{1}{\sqrt{LC}}, \qquad Q = \frac{1}{R}\sqrt{\frac{L}{C}}. \tag{13-5}$$

The various sensitivity functions have magnitudes equal to 1, $\frac{1}{2}$, or 0. They are given by:

$$S_L^{\omega_o} = S_C^{\omega_o} = -\tfrac{1}{2}, \qquad S_R^{\omega_o} = 0, \tag{13-6a}$$

$$S_L^Q = -S_C^Q = \tfrac{1}{2}, \qquad S_R^Q = -1. \tag{13-6b}$$

An increase in L will move the upper half-plane zero down and the upper half-plane pole along a circular path passing through the origin. An increase in C moves the zero and the pole down. An increase in R decreases the Q of the poles but does not affect the zeros. These changes are indicated in Fig. 13-2a. Regardless of the changes in the elements, the poles and zeros are always equidistant from the origin, i.e., they fall on the circle of radius ω_o.

The transfer-function zeros of the network of Fig. 13-2b are produced by the parallel LC circuit located in the series arm of the ladder. [The impedance of this LC circuit becomes infinite for $s = \pm j(1/\sqrt{LC})$]. The transfer function is given by

$$\frac{V_o}{V_i} = \frac{s^2 + \dfrac{1}{LC}}{s^2 + s\dfrac{1}{RC} + \dfrac{1}{LC}}. \tag{13-7}$$

By comparison with Eq. (13-2) it is seen that

$$\omega_o = \frac{1}{\sqrt{LC}}, \qquad Q = R\sqrt{\frac{C}{L}}. \tag{13-8}$$

The various sensitivity functions all have low magnitudes, equal to or less than 1. They are given by

$$S_L^{\omega_o} = S_C^{\omega_o} = -\tfrac{1}{2}, \qquad S_R^{\omega_o} = 0, \tag{13-9a}$$

$$S_L^Q = -S_C^Q = -\tfrac{1}{2}S_R^Q = +\tfrac{1}{2}. \tag{13-9b}$$

The direction of the changes for the upper half-plane zero and pole (for a positive change in the element values) is given in Fig. 13-2b. Regardless of the values of the elements, the poles and zeros lie on a circle of radius $(1/\sqrt{LC})$.

EXAMPLE 13-1

Modify the network of Fig. 13-2a such that:
(a) the magnitude of the zero is larger than the magnitude of the pole;
(b) the magnitude of the zero is smaller than the magnitude of the pole.

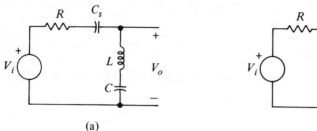

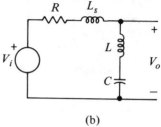

(a)   (b)

Fig. 13-3

SOLUTION

(a) As shown in Fig. 13-3a, add a capacitor, C_s, in the series branch of the ladder. C_s does not affect the transfer-function zeros, but it alters the poles of the network. Calculate the transfer function and obtain ω_o and ω_z:

$$\frac{V_o}{V_i} = \frac{s^2 + \dfrac{1}{LC}}{s^2 + s\dfrac{R}{L} + \dfrac{1}{L}\left(\dfrac{1}{C} + \dfrac{1}{C_s}\right)} = \frac{s^2 + \omega_z^2}{s^2 + s\dfrac{\omega_o}{Q} + \omega_o^2}, \tag{13-10}$$

$$\omega_z = \frac{1}{\sqrt{LC}}, \qquad \omega_o = \frac{1}{\sqrt{L\dfrac{CC_s}{C + C_s}}} > \omega_z. \qquad Ans.$$

(b) As shown in Fig. 13-3b, add an inductor, L_s, in the series branch of the ladder. L_s does not affect the transfer-function zeros, but it alters the poles of the network. Calculate the transfer function and obtain ω_o and ω_z:

$$\frac{V_o}{V_i} = \frac{L}{L + L_s} \frac{s^2 + \dfrac{1}{LC}}{s^2 + s\dfrac{R}{L + L_s} + \dfrac{1}{(L + L_s)C}} = \frac{L}{L + L_s} \frac{s^2 + \omega_z^2}{s^2 + s\dfrac{\omega_o}{Q} + \omega_o^2}, \tag{13-11}$$

$$\omega_z = \frac{1}{\sqrt{LC}}, \qquad \omega_o = \frac{1}{\sqrt{(L + L_s)C}} < \omega_z. \qquad Ans.$$

Obtaining a Null with a Lossy Inductor

The networks given in Fig. 13-2 are rather idealized since inductors are assumed to be free of resistance. Any losses in the inductors degrade the notch depth by displacing the zeros from the imaginary axis into the left half-plane. At the expense of having an ungrounded output, the effect of losses can be

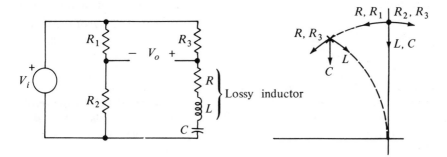

Fig. 13-4   Band-stop bridge network

balanced out by the network shown in Fig. 13-4. The resistor R represents the effect of the losses in the inductor. (See Problem 13-4 for a network with a grounded output.)

Elements R_3, R, L, and C produce the desired poles. The (R_2/R_1) ratio can be adjusted to put a pair of zeros on the imaginary axis. The transfer function for this bridge network is

$$
\frac{V_o}{V_i} = \frac{sL + R + \dfrac{1}{sC}}{sL + R + R_3 + \dfrac{1}{sC}} - \frac{R_2}{R_1 + R_2}
$$

$$
= \frac{R_1}{R_1 + R_2} \frac{s^2 + s\dfrac{R_3}{L}\left(\dfrac{R}{R_3} - \dfrac{R_2}{R_1}\right) + \dfrac{1}{LC}}{s^2 + s\dfrac{(R + R_3)}{L} + \dfrac{1}{LC}} = \frac{R_1}{R_1 + R_2} \frac{s^2 + s\dfrac{\omega_o}{Q_z} + \omega_o^2}{s^2 + s\dfrac{\omega_o}{Q} + \omega_o^2}, \qquad (13\text{-}12)
$$

where

$$
\omega_o = \frac{1}{\sqrt{LC}}, \qquad Q = \frac{1}{R + R_3}\sqrt{\frac{L}{C}}, \qquad Q_z = \frac{1}{R_3}\sqrt{\frac{L}{C}} \frac{1}{\dfrac{R}{R_3} - \dfrac{R_2}{R_1}}.
$$

When

$$
\frac{R_2}{R_1} = \frac{R}{R_3}, \qquad\qquad\qquad (13\text{-}13)
$$

Eq. (13-12) becomes

$$
\frac{V_o}{V_i} = \frac{R_3}{R + R_3} \frac{s^2 + \dfrac{1}{LC}}{s^2 + s\dfrac{R + R_3}{L} + \dfrac{1}{LC}} = \frac{R_3}{R + R_3} \frac{s^2 + \omega_o^2}{s^2 + s\dfrac{\omega_o}{Q} + \omega_o^2}. \qquad (13\text{-}14)
$$

The condition for rejection, Eq. (13-13), can also be obtained by inspection of the network of Fig. 13-4. When $s = \pm j(1/\sqrt{LC})$, the impedance of the series-LC circuit is zero, and hence the steady-state voltage across it is zero. The network then looks like a resistive bridge network which is balanced by making $R_1 R = R_2 R_3$.

When $R_1 R$ is not quite equal to $R_2 R_3$, the transfer-function zeros are not on the imaginary axis. It then becomes necessary to talk about the Q_z of the zeros. If $R_1 R > R_2 R_3$, Q_z is positive and the zeros are in the left half-plane. If $R_1 R < R_2 R_3$, Q_z is negative and the zeros are in the right half-plane.

The various sensitivity functions associated with the poles and zeros are given by

$$\text{Poles}\begin{cases} S_L^{\omega_o} = S_C^{\omega_o} = -\tfrac{1}{2}, & S_{R_1}^{\omega_o} = S_{R_2}^{\omega_o} = S_{R_3}^{\omega_o} = S_R^{\omega_o} = 0, & (13\text{-}15a) \\[2mm] S_L^Q = -S_C^Q = \dfrac{1}{2}, & S_R^Q = -\dfrac{R}{R + R_3}, \quad S_{R_3}^Q = -\dfrac{R_3}{R + R_3}, \\[3mm] & S_{R_1}^Q = S_{R_2}^Q = 0, & (13\text{-}15b) \end{cases}$$

$$\text{Zeros} \; \{ S_L^{\omega_z} = S_C^{\omega_z} = -\tfrac{1}{2}, \quad S_{R_1}^{\omega_z} = S_{R_2}^{\omega_z} = S_{R_3}^{\omega_z} = S_R^{\omega_z} = 0. \qquad (13\text{-}16)$$

When $R_1 R = R_2 R_3$, the sensitivity of Q_z with respect to any element is infinite because the nominal value of Q_z is infinite. As soon as the zeros move off the imaginary axis, Q_z changes from an infinite to a finite value, thus resulting in an infinite change. Therefore, in this case, it is misleading to talk about the sensitivity of Q_z. Instead, the sensitivity of the transfer-function zeros is calculated to obtain a quantitative measure of their displacement from the nominal positions as an element of the network departs from its nominal value. For example, to find the effect of R_1 on the upper half-plane zero, the numerator polynomial of Eq. (13-12) is rearranged to show the explicit dependence on R_1 as follows:

$$P(s) = \underbrace{-sR_2 R_3 C}_{P_1(s)} + \underbrace{R_1(LCs^2 + sRC + 1)}_{P_2(s)}$$

Then, Eq. (6-30) is used to find the sensitivity of the upper half-plane zero with respect to R_1 (at its nominal position):

$$S_{R_1}^z = \frac{R_1}{|z|} \frac{dz}{dR_1} = -\frac{R_1}{|z|} \frac{P_2(s)}{\dfrac{dP(s)}{ds}}\Bigg|_{s=j\omega_o}$$

$$= -\frac{R_1}{\dfrac{1}{\sqrt{LC}}} \frac{LCs^2 + sRC + 1}{-R_2 R_3 C + 2R_1 LCs + R_1 RC}\Bigg|_{s=j\omega_o}$$

$$= -R_1\sqrt{LC}\, \frac{LCs^2 + sRC + 1}{2R_1 LCs}\Bigg|_{s=j\omega_o} = -\frac{R}{2}\sqrt{\frac{C}{L}}. \qquad (13\text{-}17)$$

Thus, if R_1 becomes $R_1 + \Delta R$, then $z = j(1/\sqrt{LC})$ becomes

$$z + \Delta z \cong z + |z| S_{R_1}^z \frac{\Delta R_1}{R_1} = \left(-\frac{R}{2L}\right) \frac{\Delta R_1}{R_1} + j \frac{1}{\sqrt{LC}}. \tag{13-18}$$

The result is approximate because incremental rather than infinitesimal changes are considered. As Eq. (13-18) shows, for $\Delta R_1 > 0$, the zero moves away from the imaginary axis into the left half-plane.

Similarly, the sensitivity of the upper half-plane zero with respect to the other elements can be obtained:

$$S_{R_2}^z = S_{R_3}^z = -S_R^z = \frac{R}{2} \sqrt{\frac{C}{L}}, \qquad S_L^z = S_C^z = -j\frac{1}{2}. \tag{13-19}$$

The direction of the movement of the zero as a function of a positive change in the element values is shown in Fig. 13-4. Note that the poles and zeros have a common ω_o, that is, they fall on the circle of radius $(1/\sqrt{LC})$.

13-3 PASSIVE REALIZATIONS: *RC*

It is also possible to get imaginary-axis transfer-function zeros with *RC* networks. A widely used network for this purpose is the twin-*T* network shown in Fig. 13-5.

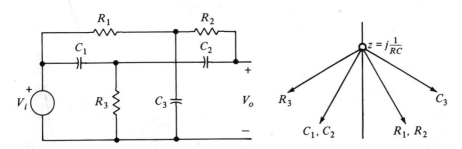

Fig. 13-5　The twin-*T* network

The transfer function of this network is:

$$\frac{V_o}{V_i} = \frac{R_1 R_2 R_3 C_1 C_2 C_3 s^3 + (R_1 + R_2) R_3 C_1 C_2 s^2 + R_3 (C_1 + C_2)s + 1}{\begin{array}{l} R_1 R_2 R_3 C_1 C_2 C_3 s^3 + (R_1 R_2 C_2 C_3 + R_1 R_3 C_1 C_2 + R_1 R_3 C_1 C_3 \\ \quad + R_1 R_3 C_2 C_3 + R_2 R_3 C_1 C_2)s^2 + (R_1 C_2 + R_1 C_3 + R_2 C_2 \\ \quad\quad\quad\quad\quad\quad\quad\quad\quad\quad\quad + R_3 C_1 + R_3 C_2)s + 1 \end{array}}.$$

$$\tag{13-20}$$

If (V_o/V_i) is to have imaginary-axis zeros, the following condition (see Problem 13-5) must be met:

$$\frac{\dfrac{R_1 R_2}{R_1 + R_2}}{R_3} = \frac{C_1 + C_2}{C_3}, \tag{13-21}$$

in which case Eq. (13-20) simplifies to

$$\frac{V_o}{V_i} = \frac{[sR_3(C_1 + C_2) + 1][s^2(R_1 + R_2)R_3 C_1 C_2 + 1]}{[sR_3(C_1 + C_2) + 1][s^2(R_1 + R_2)R_3 C_1 C_2 + s(R_1 C_2 + R_1 C_3 + R_2 C_2) + 1]}. \tag{13-22}$$

The significant aspect of Eq. (13-22) is not the cancellation of a zero with a pole but rather the appearance of a pair of imaginary-axis zeros. By comparison with Eq. (13-1), it is seen that

$$\omega_z = \frac{1}{\sqrt{(R_1 + R_2)R_3 C_1 C_2}} = \frac{1}{\sqrt{R_1 R_2 \dfrac{C_1 C_2}{C_1 + C_2} C_3}} = \omega_o, \tag{13-23}$$

$$Q = \frac{\sqrt{(R_1 + R_2)R_3 C_1 C_2}}{R_1 C_2 + R_1 C_3 + R_2 C_2}. \tag{13-24}$$

In order to simplify the network, let

$$R = R_1 = R_2 = 2R_3, \qquad C = C_1 = C_2 = \tfrac{1}{2}C_3. \tag{13-25}$$

Equation (13-22) then becomes

$$\frac{V_o}{V_i} = \frac{s^2 + \omega_o^2}{s^2 + s\dfrac{\omega_o}{Q} + \omega_o^2}, \tag{13-26}$$

where

$$\omega_o = \frac{1}{RC}, \qquad Q = \frac{1}{4}.$$

The poles are on the negative real axis and therefore hardly affect the band-stop characteristic even when they are displaced somewhat (due to element tolerances, for example) from their nominal positions. On the other hand, the rejection frequency and the notch depth are highly affected if the zeros are not placed exactly in their nominal positions on the imaginary axis. Since the numerator polynomial, Eq. (13-20), is cubic, it is practically impossible to obtain a general expression for the ω_o and Q for the complex zeros. Therefore, the ω_o-sensitivity and the Q-sensitivity functions for the complex zeros cannot be calculated. Rather, the sensitivity of the upper half-plane zero is calculated when it is on the imaginary axis. For example, the sensitivity of the zero with respect to R_1 is obtained as follows:

First, the numerator polynomial of Eq. (13-20) is rearranged as

$$P(s) = \underbrace{[R_2 R_3 C_1 C_2 s^2 + R_3(C_1 + C_2)s + 1]}_{P_1(s)} + \underbrace{R_1[R_3 C_1 C_2 s^2(R_2 C_3 s + 1)]}_{P_2(s)}.$$

Then, using Eq. (6-30), $S_{R_1}^z$ is obtained:

$$
S_{R_1}^z = \frac{-R_1}{|z|} \frac{P_2(s)}{\dfrac{\delta P(s)}{\delta s}} \Bigg|_{s=z}
$$

$$
= \frac{-R_1}{|z|} \frac{R_3 C_1 C_2 z^2 (R_2 C_3 z + 1)}{3 R_1 R_2 R_3 C_1 C_2 C_3 z^2 + 2(R_1 + R_2)R_3 C_1 C_2 z + R_3(C_1 + C_2)},
$$

(13-27)

where

$$z = j \frac{1}{\sqrt{R_1 R_2 \dfrac{C_1 C_2}{C_1 + C_2} C_3}}.$$

When the conditions expressed by Eq. (13-25) are used to simplify the results, Eq. (13-27) reduces to

$$S_{R_1}^z = \tfrac{1}{8}(1 - j3).$$ (13-28)

Thus, if R_1 is increased, the upper half-plane zero moves to the right and downward, as shown in Fig. 13-5. If R_1 is increased by 10%, the zero will be approximately at

$$\frac{|z|}{80} + j|z|\left(1 - \frac{3}{80}\right).$$

Similarly, the sensitivity of the zero with respect to the other elements can be obtained:

$$S_{R_2}^z = S_{R_1}^z = \tfrac{1}{8}(1 - j3), \qquad S_{R_3}^z = \tfrac{1}{4}(-1 - j), \qquad S_{C_1}^z = S_{C_2}^z = \tfrac{1}{8}(-1 - j3),$$
$$S_{C_3}^z = \tfrac{1}{4}(1 - j). \quad (13\text{-}29)$$

The resulting motions of the upper half-plane zero (for positive changes in the element values) are shown in Fig. 13-5. As Eq. (13-21) shows, the zeros are kept on the imaginary axis if

$$\frac{(C_1 + C_2)}{C_3}, \qquad \left(\frac{\dfrac{R_1 R_2}{R_1 + R_2}}{R_3}\right), \qquad \text{or} \qquad \frac{R_3}{C_3}$$

is kept constant. This can be accomplished by ganging all three capacitors, all three resistors, or R_3 with C_3.

An Easily Adjustable RC-Null Network

Figure 13-6 shows an *RC* network that can be readily adjusted to produce imaginary-axis zeros. It also uses fewer capacitors than the twin-*T* network. However, the input and the output do not possess a common terminal.

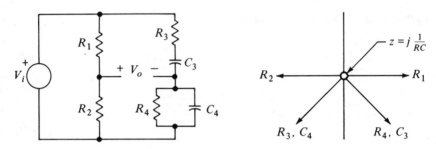

Fig. 13-6   Wien-bridge network

Since the network is *RC*, the poles of V_o/V_i are on the negative real axis. The transfer function is given by

$$\frac{V_o}{V_i} = \frac{R_2}{R_1 + R_2} \frac{s^2 + \left(\dfrac{1}{R_3 C_3} + \dfrac{1}{R_4 C_4} - \dfrac{1}{R_3 C_4} \dfrac{R_1}{R_2}\right)s + \dfrac{1}{R_3 R_4 C_3 C_4}}{s^2 + \left(\dfrac{1}{R_3 C_3} + \dfrac{1}{R_4 C_4} + \dfrac{1}{R_3 C_4}\right)s + \dfrac{1}{R_3 R_4 C_3 C_4}}. \tag{13-30}$$

By adjusting the resistor ratio to

$$\frac{R_1}{R_2} = \frac{R_3}{R_4} + \frac{C_4}{C_3}, \tag{13-31}$$

the zeros are placed on the imaginary axis at

$$s = z = \pm j \frac{1}{\sqrt{R_3 R_4 C_3 C_4}}. \tag{13-32}$$

The ω_o and Q_z of the zeros are given by

$$\omega_o = \frac{1}{\sqrt{R_3 R_4 C_3 C_4}}, \qquad Q_z = \frac{1}{\sqrt{\dfrac{R_4 C_4}{R_3 C_3}} + \sqrt{\dfrac{R_3 C_3}{R_4 C_4}} - \dfrac{R_1}{R_2}\sqrt{\dfrac{R_4 C_3}{R_3 C_4}}}. \tag{13-33}$$

When $Q_z > 0$, the zeros are in the left half-plane; when $Q_z < 0$, the zeros are in the right half-plane. To simplify the results, let

$$R_3 = R_4 = R \qquad \text{and} \qquad C_3 = C_4 = C.$$

Then Eq. (13-30) becomes

$$\frac{V_o}{V_i} = \frac{1}{3} \frac{s^2 + \frac{1}{RC}\left(2 - \frac{R_1}{R_2}\right)s + \frac{1}{R^2C^2}}{s^2 + \frac{3}{RC}s + \frac{1}{R^2C^2}}. \tag{13-34}$$

When $(R_1/R_2) = 2$, the upper half-plane zero is at $z = j(1/RC)$. The various sensitivity functions associated with this zero are

$$S_{R_1}^z = -S_{R_2}^z = 1, \qquad S_{R_3}^z = S_{C_4}^z = \tfrac{1}{2}(-1-j), \qquad S_{R_4}^z = S_{C_3}^z = \tfrac{1}{2}(1-j). \tag{13-35}$$

The effect of each element on the upper half-plane zero is shown in Fig. 13-6. The zero can be moved along the imaginary axis by ganging R_3 with R_4, or by ganging C_3 with C_4 (so that the R_1/R_2 ratio is kept constant). See Eq. (13-32).

A Variable RC Null Network Requiring No Ganging of Elements

Figure 13-7 shows another *RC* network which is capable of producing imaginary-axis zeros. Furthermore, these zeros can be moved along the imaginary axis by changing the potentiometer setting. Thus, no ganging is required for varying the zeros.

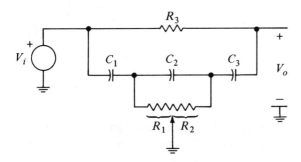

Fig. 13-7   Network with variable zeros

The transfer function of this network is given by

$$\frac{V_o}{V_i} = \frac{\begin{aligned}&s^3 R_1 R_2 R_3 C_1 C_2 C_3 + s^2 R_1 R_2 (C_1 C_2 + C_1 C_3 + C_2 C_3)\\&\qquad + s[R_1(C_1 + C_2) + R_2(C_2 + C_3)] + 1\end{aligned}}{\begin{aligned}&s^3 R_1 R_2 R_3 C_1 C_2 C_3 + s^2 (R_1 R_2 C_1 C_2 + R_1 R_2 C_1 C_3\\&\quad + R_1 R_3 C_1 C_3 + R_1 R_2 C_2 C_3 + R_1 R_3 C_2 C_3 + R_2 R_3 C_2 C_3)\\&\qquad + s(R_1 C_1 + R_1 C_2 + R_2 C_2 + R_2 C_3 + R_3 C_3) + 1\end{aligned}}. \tag{13-36}$$

To simplify Eq. (13-36), let $C_1 = C_2 = C_3 = C$. Then,

$$\frac{V_o}{V_i} = \frac{s^3 + \dfrac{3}{R_3 C} s^2 + \dfrac{2}{R_3 C^2}\left(\dfrac{1}{R_1} + \dfrac{1}{R_2}\right)s + \dfrac{1}{R_1 R_2 R_3 C^3}}{s^3 + \dfrac{1}{C}\left(\dfrac{1}{R_1} + \dfrac{2}{R_2} + \dfrac{3}{R_3}\right)s^2 + s\dfrac{1}{C^2}\left(\dfrac{2}{R_2 R_3} + \dfrac{2}{R_1 R_3} + \dfrac{1}{R_1 R_2}\right) + \dfrac{1}{R_1 R_2 R_3 C^3}},$$

$$(13\text{-}37)$$

If V_o/V_i is to have zeros on the imaginary axis, the coefficients of the numerator polynomial must satisfy

$$\left(\frac{3}{R_3 C}\right)\left[\frac{2}{R_3 C^2}\left(\frac{1}{R_1} + \frac{1}{R_2}\right)\right] = \frac{1}{R_1 R_2 R_3 C^3},$$

which simplifies to

$$R_3 = 6(R_1 + R_2). \qquad\qquad (13\text{-}38)$$

This is an easy condition to meet because $(R_1 + R_2)$ is the total resistance of the potentiometer; thus, R_3 is constant and independent of the potentiometer setting. The transfer function can then be written as

$$\frac{V_o}{V_i} = \frac{\left[s + \dfrac{1}{2(R_1 + R_2)C}\right]\left(s^2 + \dfrac{1}{3R_1 R_2 C^2}\right)}{s^3 + \left[\dfrac{4R_1^2 + 7R_1 R_2 + 2R_2^2}{2R_1 R_2(R_1 + R_2)C}\right]s^2 + \dfrac{4}{3R_1 R_2 C^2}s + \dfrac{1}{6R_1 R_2(R_1 + R_2)C^3}},$$

$$(13\text{-}39)$$

which exposes the upper half-plane transfer-function zero at

$$z = j\,\frac{1}{C\sqrt{3R_1 R_2}}. \qquad\qquad (13\text{-}40)$$

Using the numerator of Eq. (13-36) and the root-sensitivity expression given in Eq. (6-30), the sensitivity of the upper half-plane zero with respect to each element is obtained for the equal-C case:

$$S_{R_1}^z = \frac{-\dfrac{1}{\sqrt{3}}\sqrt{\dfrac{R_1}{R_2}} - j\dfrac{1}{6}\left(7 + 4\dfrac{R_2}{R_1}\right)}{\dfrac{4}{3}\left(\sqrt{\dfrac{R_1}{R_2}} + \sqrt{\dfrac{R_2}{R_1}}\right)^2 + 1}, \qquad\qquad (13\text{-}41a)$$

$$S_{R_2}^z = \frac{-\dfrac{1}{\sqrt{3}}\sqrt{\dfrac{R_2}{R_1}} - j\dfrac{1}{6}\left(7 + 4\dfrac{R_1}{R_2}\right)}{\dfrac{4}{3}\left(\sqrt{\dfrac{R_1}{R_2}} + \sqrt{\dfrac{R_2}{R_1}}\right)^2 + 1}, \qquad\qquad (13\text{-}41b)$$

$$S^z_{R_3} = \frac{\dfrac{1}{\sqrt{3}}\left(\sqrt{\dfrac{R_1}{R_2}} + \sqrt{\dfrac{R_2}{R_1}}\right)\left[1 - j\dfrac{2}{\sqrt{3}}\left(\sqrt{\dfrac{R_1}{R_2}} + \sqrt{\dfrac{R_2}{R_1}}\right)\right]}{\dfrac{4}{3}\left(\sqrt{\dfrac{R_1}{R_2}} + \sqrt{\dfrac{R_2}{R_1}}\right)^2 + 1};$$

(13-41c)

$$S^z_{C_1} = \frac{\dfrac{1}{6\sqrt{3}}\left(2\sqrt{\dfrac{R_2}{R_1}} - \sqrt{\dfrac{R_1}{R_2}}\right) - j\dfrac{1}{3}\left(\dfrac{R_1}{R_2} + 4 + 2\dfrac{R_2}{R_1}\right)}{\dfrac{4}{3}\left(\sqrt{\dfrac{R_1}{R_2}} + \sqrt{\dfrac{R_2}{R_1}}\right)^2 + 1},$$

(13-41d)

$$S^z_{C_2} = \frac{-\dfrac{1}{6\sqrt{3}}\left(\sqrt{\dfrac{R_1}{R_2}} + \sqrt{\dfrac{R_2}{R_1}}\right) - j\dfrac{1}{3}\left(\dfrac{R_1}{R_2} + 3 + \dfrac{R_2}{R_1}\right)}{\dfrac{4}{3}\left(\sqrt{\dfrac{R_1}{R_2}} + \sqrt{\dfrac{R_2}{R_1}}\right)^2 + 1},$$

(13-41e)

$$S^z_{C_3} = \frac{\dfrac{1}{6\sqrt{3}}\left(2\sqrt{\dfrac{R_1}{R_2}} - \sqrt{\dfrac{R_2}{R_1}}\right) - j\dfrac{1}{3}\left(2\dfrac{R_1}{R_2} + 4 + \dfrac{R_2}{R_1}\right)}{\dfrac{4}{3}\left(\sqrt{\dfrac{R_1}{R_2}} + \sqrt{\dfrac{R_2}{R_1}}\right)^2 + 1}.$$

(13-41f)

These sensitivity functions can be used as an aid in the tuning of the zeros. For example, when $R_1 = R_2$, the upper half-plane zero moves to the left and downward for positive increments in R_1, R_2, or C_2. On the other hand, the zero moves to the right and downward if R_3, C_1, or C_3 is increased.

The three RC networks introduced in this section have one common property: They all produce a pair of imaginary-axis zeros. However, since the networks are RC, the poles are on the negative real axis; consequently, the rejection band is very wide. A narrow rejection band requires complex poles. The poles of the RC networks can be moved and made complex by means of dependent sources, as discussed in Chapter 9. Several such schemes for RC null networks are considered in the following sections.

13-4 KRC REALIZATION

The poles of the RC twin-T network shown in Fig. 13-5 can be made complex, and thereby the notch narrowed, by sensing the output voltage and feeding it back in series with either R_3 or C_3 or both. In Problem 13-11, the dependent source is fed in series with R_3. In Fig. 13-8 it is fed in series with $C_3 = 2C$.

The transfer function (V_o/V_i) is given by

$$\frac{V_o}{V_i} = \frac{K\left(s^2 + \dfrac{1}{R^2C^2}\right)}{s^2 + 2\dfrac{s}{RC}(2 - K) + \dfrac{1}{R^2C^2}}.$$

(13-42)

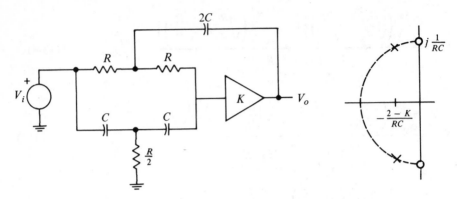

Fig. 13-8 Sharpening the notch

The zeros are independent of the amplifier gain K. Therefore, the twin-T sensitivity functions given by Eqs. (13-28) and (13-29) apply here also. The zero movement diagram given in Fig. 13-5 can be used to tune for a null. If K is considered to be constant, that is, $K = K_0$, the ω_o of the complex poles is $(1/RC)$; hence the poles and zeros lie on the same circle. The Q of the poles is $1/[2(2 - K_0)]$, and therefore it can be set or altered by adjusting K_0.

If the frequency dependence of the amplifier is considered, that is,

$$K = \frac{GB}{s + GB/K_0},$$

the poles and the zeros do not fall on the same circle. The pole ω_o will be lower. As a result, for narrow rejection bandwidths, the magnitude response peaks before the null frequency. The narrower the bandwidth, the more pronounced becomes this peaking. (Refer to Section 10-6 for the details of the pole shifts caused by the imperfect amplifier.)

EXAMPLE 13-2

(a) A second-order band-stop function can always be constructed by subtracting a properly scaled second-order band-pass function from unity. Show this.

(b) Using the *KRC* band-pass realization, implement the $BS = 1 - BP$ scheme presented in (a).

SOLUTION

(a) Represent the BP function by

$$BP = \frac{Hs}{s^2 + s\dfrac{\omega_o}{Q} + \omega_o^2}.$$

Subtract BP from unity and obtain

$$1 - BP = 1 - \frac{Hs}{s^2 + s\dfrac{\omega_o}{Q} + \omega_o^2} = \frac{s^2 + s\left(\dfrac{\omega_o}{Q} - H\right) + \omega_o^2}{s^2 + s\dfrac{\omega_o}{Q} + \omega_o^2}. \tag{13-43}$$

Adjust the scale factor H to equal ω_o/Q and thus generate the BS function:

$$BS = \frac{s^2 + \omega_o^2}{s^2 + s\dfrac{\omega_o}{Q} + \omega_o^2}. \qquad Ans. \tag{13-44}$$

(b) Use the *KRC* network given in Section 11-3 to generate the BP function. As shown in Fig. 13-9, combine its output with the input using a second amplifier.

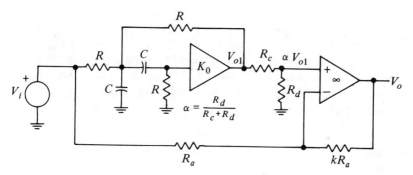

Fig. 13-9

Using Eq. (11-11), obtain V_{o1} :

$$V_{o1} = V_i \frac{\left(2\sqrt{2} - \dfrac{1}{Q}\right)\omega_o s}{s^2 + s\dfrac{\omega_o}{Q} + \omega_o^2}, \tag{13-45}$$

where

$$\omega_o = \frac{\sqrt{2}}{RC}, \qquad Q = \frac{\sqrt{2}}{4 - K_0}.$$

By inspection of Fig. 13-9, obtain V_o and then V_o/V_i :

$$V_o = \alpha V_{o1}(1 + k) + V_i(-k), \tag{13-46}$$

$$\frac{V_o}{V_i} = \alpha(1+k) \frac{\left(2\sqrt{2} - \frac{1}{Q}\right)\omega_o s}{s^2 + s\frac{\omega_o}{Q} + \omega_o^2} - k$$

$$= -k \frac{s^2 + s\left[\frac{1}{Q} - \alpha\frac{(1+k)}{k}\left(2\sqrt{2} - \frac{1}{Q}\right)\right]\omega_o + \omega_o^2}{s^2 + s\frac{\omega_o}{Q} + \omega_o^2}. \tag{13-47}$$

To generate the BS function, make the coefficient of the s-term zero by adjusting α:

$$\alpha = \frac{k}{1+k}\frac{1}{Q}\frac{1}{2\sqrt{2} - \frac{1}{Q}}. \qquad Ans. \tag{13-48}$$

EXAMPLE 13-3

Measurements taken on a second-order BS network show that the zeros are not quite on the imaginary axis; i.e., the notch depth is not what it should be. Without adjusting the network, determine whether the zeros are in the left or right half-plane.

SOLUTION

Write the BS function as

$$T(s) = \frac{s^2 + s\delta + \omega_o^2}{s^2 + s\frac{\omega_o}{Q} + \omega_o^2}, \tag{13-49}$$

where $\delta/2$ represents the real part of the zeros. When $\delta = 0$, the zeros are on the imaginary axis. When $\delta > 0$, the zeros are in the left half-plane. When $\delta < 0$, the zeros are in the right half-plane. Examine the properties of $T(s)$ for s near $j\omega_o$ by letting $s = j(\omega_o + \Delta\omega)$:

$$T(j\Delta\omega) \cong \frac{-2\Delta\omega + j\delta}{-2\Delta\omega + j\frac{\omega_o}{Q}}, \tag{13-50a}$$

$$|T(j\Delta\omega)| \cong \frac{Q}{\omega_o}\sqrt{4(\Delta\omega)^2 + \delta^2}, \tag{13-50b}$$

$$\theta(\Delta\omega) \cong -\tan^{-1}\left(\frac{\delta}{2\Delta\omega}\right) + \tan^{-1}\left(\frac{\omega_o}{2Q\Delta\omega}\right) \tag{13-50c}$$

$$\left.\begin{aligned} &\cong 2\left(\frac{1}{\delta} - \frac{Q}{\omega_o}\right)\Delta\omega, && \delta > 0 \\[2mm] &\cong -\pi - 2\left(\frac{1}{|\delta|} + \frac{Q}{\omega_o}\right)\Delta\omega, && \delta < 0 \end{aligned}\right\} \quad \left|\frac{\delta}{\Delta\omega}\right| > 2, \quad \left|\frac{\omega_o}{\Delta\omega}\right| > 2Q. \tag{13-50d}$$

It is impossible to tell the position of the zeros (except when they are on the imaginary axis) from the magnitude characteristic because it is a function of δ^2. On the other hand, the phase, being dependent on δ, produces markedly different characteristics (near $\omega = \omega_o$) for $\delta > 0$ and $\delta < 0$. As ω approaches ω_o and goes past it, the phase increases for $\delta > 0$ and decreases for $\delta < 0$. [See Eq. (13-50d).] Near ω_o, the phase characteristic has a positive slope for $\delta > 0$ and a negative slope for $\delta < 0$. This is clearly shown in Fig. 13-10, which represents a plot of Eq. (13-49) for five values of δ, namely

$$\delta = 0.02, \quad 0.01, \quad 0, \quad -0.01, \quad -0.02.$$

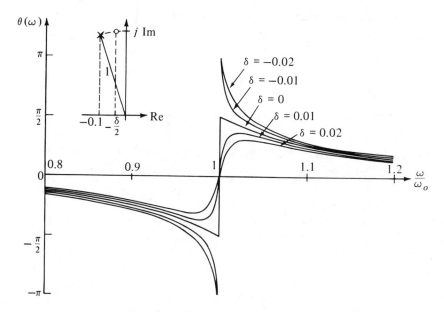

Fig. 13-10   Phase of $\dfrac{s^2 + s\delta + \omega_o^2}{s^2 + s\dfrac{\omega_o}{Q} + \omega_o^2}$ for $Q = 5$ and for various values of δ.

For all the curves, the pole Q is chosen as 5. For higher pole-Q values, the changes near ω_o occur more rapidly. Thus, by observing the phase characteristic near ω_o, the position of the zeros can be determined. The zeros are in the left half-plane if $(d\theta/d\omega)$ is positive near ω_o. The zeros are in the right half-plane if $(d\theta/d\omega)$ is negative. Note that the phase curves for $\delta < 0$ can be made continuous by subtracting 2π radians from $\theta(\omega)$ for $\omega > \omega_o$. Similarly, the $(\delta = 0)$-curve can be drawn with 2π radians subtracted from $\theta(\omega)$ for $\omega > \omega_o$, in which case the phase jumps $-\pi$ instead of $+\pi$ radians at ω_o.

13-5 $-KRC$ REALIZATION

Using a second-order band-pass $-KRC$ network and a summing amplifier, a second-order band-stop network can be constructed. This is demonstrated in Fig. 13-11.

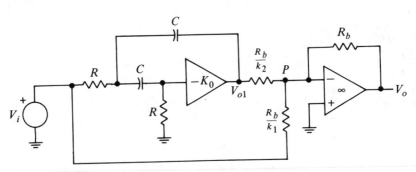

Fig. 13-11   BS network employing $-KRC$ BP network

The $-KRC$ band-pass network is identical with the network presented in Section 11-5. Therefore, Eq. (11-15) can be used to evaluate V_{o1}:

$$V_{o1} = -V_i \frac{\left(3Q - \dfrac{1}{3Q}\right)\omega_o s}{s^2 + s\dfrac{\omega_o}{Q} + \omega_o^2}, \tag{13-51}$$

where

$$\omega_o = \frac{1}{RC\sqrt{1 + K_0}}, \qquad Q = \frac{\sqrt{1 + K_0}}{3}.$$

As shown in Fig. 13-11, V_o is obtained by summing V_{o1} with V_i:

$$V_o = -(V_{o1}k_2 + V_i k_1). \tag{13-52}$$

Hence,

$$\frac{V_o}{V_i} = \frac{k_2\left(3Q - \dfrac{1}{3Q}\right)\omega_o s}{s^2 + s\dfrac{\omega_o}{Q} + \omega_o^2} - k_1$$

$$= -k_1 \frac{s^2 + s\left[\dfrac{1}{Q} - \dfrac{k_2}{k_1}\left(3Q - \dfrac{1}{3Q}\right)\right]\omega_o + \omega_o^2}{s^2 + s\dfrac{\omega_o}{Q} + \omega_o^2}. \tag{13-53}$$

In order to obtain the BS function, the coefficient of the s-term in the numerator polynomial must be made zero. This requires that

$$\frac{k_2}{k_1} = \frac{3}{9Q^2 - 1}. \tag{13-54}$$

The overall adjustment procedure for a given frequency of rejection (ω_o) and bandwidth of rejection (BW) is as follows.

1. Calculate Q using $Q = \omega_o/\text{BW}$.
2. Determine the value of K_0 using $K_0 = 9Q^2 - 1$.
3. Determine the RC product using $RC = [1/(3Q\omega_o)]$. Choose either R or C arbitrarily.
4. Decide on the desired low- (or high-) frequency gain $(-k_1)$.
5. Determine the value of k_2 using $k_2 = 3k_1/(9Q^2 - 1)$.

It is interesting to observe that at the rejection frequency the output of the first amplifier is exactly 180° out of phase with the input. The k_2/k_1 adjustment merely controls the relative amplitudes of the two signals so that perfect cancellation occurs at the output of the second amplifier. In fact, if gain- and low-output impedance are not required, the second amplifier can be dispensed with, and the output taken at point P in Fig. 13-11. The summing is performed by the two resistors, and the k_2/k_1 ratio is still given by Eq. (13-54).

Nonideal amplifiers affect the performance of the BS network given in Fig. 13-11 in two ways. First, the poles of the BP network are displaced from their nominal positions. (Refer to Section 11-4 for details.) As a result, for narrow-band-rejection filters, the magnitude characteristic peaks at a frequency slightly lower than the rejection frequency. Second, the third pole of the BP network and the finite bandwidth of the summing amplifier cause attenuation at high frequencies.

EXAMPLE 13-4

Discuss the feasibility of obtaining a narrow-band BS network with the arrangement shown in Fig. 13-12.

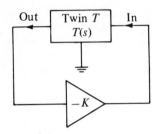

Fig. 13-12

SOLUTION

To obtain a narrow-band BS network, the poles must be complex. Using the characteristic equation, check to see whether the network of Fig. 13-12 can produce complex poles. Use Eq. (13-26) for the transfer function of the twin-T network.

$$1 - LG = 0 \qquad (LG = \text{loop gain}),$$

$$1 + KT(s) = 0,$$

$$1 + K\frac{s^2 + \omega_o^2}{s^2 + s4\omega_o + \omega_o^2} = 0,$$

$$s^2 + s\frac{4\omega_o}{1 + K} + \omega_o^2 = 0.$$

The Q of the poles is given by

$$Q = \frac{1 + K}{4}.$$

For $K > 1$, Q becomes greater than $\frac{1}{2}$, and therefore complex poles result.

To produce the desired transfer-function zeros, the input signal must be introduced at the amplifier input through a summing circuit and the band-stop output must be taken at the output of the twin-T network.

13-6 REALIZATION WITH INFINITE GAIN

Band-stop networks can also be constructed by using a single amplifier which serves in the generation of the band-pass function as well as in the subtraction of this function from unity. An example of this scheme is illustrated in Fig. 13-13.

For $\alpha = 0$, the network of Fig. 13-13 reduces to the BP network realization with infinite gain, which is presented in Section 11-5. The BP function is obtained from Eq. (11-19):

$$\text{BP} = -\frac{2\omega_o Q s}{s^2 + s\dfrac{\omega_o}{Q} + \omega_o^2}, \qquad (13\text{-}55)$$

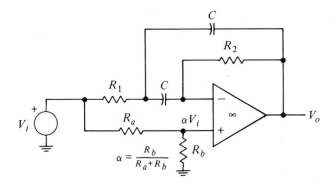

Fig. 13-13   Band-stop network

where

$$\omega_o = \frac{1}{C\sqrt{R_1 R_2}}, \qquad Q = \frac{1}{2}\sqrt{\frac{R_2}{R_1}}, \qquad \frac{\omega_o}{Q} = \frac{2}{R_2 C}.$$

Using the principle of superposition, the output for $\alpha \neq 0$ is obtained by inspection of Fig. 13-13:

$$V_o = V_i \,\mathrm{BP} + \alpha V_i (1 - \mathrm{BP}).$$

Hence,

$$\frac{V_o}{V_i} = \alpha + (1 - \alpha)\mathrm{BP} = \alpha - (1 - \alpha)\,\frac{2\omega_o Q s}{s^2 + s\dfrac{\omega_o}{Q} + \omega_o^2}$$

$$= \alpha \, \frac{s^2 + s\left(\dfrac{1}{Q} - \dfrac{1 - \alpha}{\alpha}\,2Q\right)\omega_o + \omega_o^2}{s^2 + s\dfrac{\omega_o}{Q} + \omega_o^2}. \qquad (13\text{-}56)$$

Equation (13-56) reduces to the BS function if α is adjusted to

$$\alpha = \frac{2Q^2}{2Q^2 + 1}. \qquad (13\text{-}57)$$

For a given frequency and bandwidth of rejection (ω_o and BW), the following adjustment procedure can be used to obtain the BS realization:

1. Ground the positive terminal of the amplifier, i.e., make $\alpha = 0$. The resulting network is band-pass and peaks at $\omega = \omega_o$, and has a 3-dB bandwidth of BW $= (\omega_o/Q)$. Choose a convenient value for C and use $R_2 C = (2/\mathrm{BW})$ to set the bandwidth and $R_1 C = \mathrm{BW}/(2\omega_o^2)$ to set the ω_o.
2. Disconnect the ground. Use Eq. (13-57) to adjust α to the correct value and obtain the desired notch.

EXAMPLE 13-5

In Fig. 13-13, assume that the operational amplifier has a one-pole rolloff characteristic, that is,

$$A(s) = -\frac{GB}{s + \omega_a} \cong -\frac{GB}{s}.$$

What effect does the finite value of GB have on the band-stop magnitude characteristic?

SOLUTION

With the nonideal amplifier and with $\alpha = 2Q^2/(2Q^2 + 1)$, Eq. (13-56) becomes

$$\frac{V_o}{V_i} = \frac{2Q^2}{2Q^2 + 1} \frac{s^2 + \omega_o^2}{s^2 + s\dfrac{\omega_o}{Q} + \omega_o^2 + \dfrac{s}{GB}\left[s^2 + s\left(2Q + \dfrac{1}{Q}\right)\omega_o + \omega_o^2\right]}. \tag{13-58}$$

The effect of the imperfect amplifier is to displace the poles of the BS network from the nominal positions and to cause a 6-dB/octave rolloff at high frequencies. Refer to the expressions given in Table 11-3 for a quantitative measure of the first-order changes in the pole positions. To obtain the position of the real-axis pole, put the denominator polynomial of Eq. (13-58) in monic form:

$$s^3 + s^2\left[GB + \left(2Q + \frac{1}{Q}\right)\omega_o\right] + \cdots = 0.$$

The coefficient of the s^2-term is the negative of the sum of the three poles of the network. Assume that the displacement of the complex poles from their nominal positions is small, so that the sum of the two complex poles is still $-(\omega_o/Q)$. Hence, to a first-order approximation,

$$GB + \left(2Q + \frac{1}{Q}\right)\omega_o \cong \frac{\omega_o}{Q} + \gamma,$$

where $-\gamma$ represents the real-axis pole. Solve for γ and obtain

$$\gamma \cong GB + 2Q\omega_o \cong GB \qquad (2Q\omega_o \ll GB).$$

Hence, Eq. (13-58) can be approximated to

$$\frac{V_o}{V_i} \cong \frac{2Q^2}{2Q^2 + 1} \frac{s^2 + \omega_o^2}{\left(s^2 + s\dfrac{\omega_o}{Q} + \omega_o^2\right)\left(\dfrac{s}{GB} + 1\right)} \qquad \left(\omega_o \ll \frac{GB}{2Q}\right). \tag{13-59}$$

The high-frequency 3-dB point is

$$\omega_{3\,dB} \cong GB. \tag{13-60}$$

To plot the magnitude characteristic, normalize Eq. (13-58):

$$\left(\frac{2Q^2+1}{2Q^2}\right)\frac{V_o}{V_i}$$

$$=\frac{\left(\dfrac{s}{\omega_o}\right)^2+1}{\left(\dfrac{s}{\omega_o}\right)^2+\left(\dfrac{s}{\omega_o}\right)\dfrac{1}{Q}+1+\left(\dfrac{s}{\omega_o}\right)\left(\dfrac{\omega_o}{GB}\right)\left[\left(\dfrac{s}{\omega_o}\right)^2+\left(\dfrac{s}{\omega_o}\right)\left(2Q+\dfrac{1}{Q}\right)+1\right]}.$$

$$\text{(13-61)}$$

Equation (13-61) is plotted in Fig. 13-14 for $(GB/\omega_o) = 200$ and for three values of rejection bandwidth, namely,

$$\frac{BW}{\omega_o}=\frac{1}{Q}=1,\quad 0.143,\quad 0.1.$$

Note the prerejection peaking for narrow bandwidths (indicating a lowering of the ω_o of the poles) and the high-frequency 3-dB point ($\omega_{3\,dB}/\omega_o = 200$), which is in close agreement with Eq. (13-60). For the same ω_o and BW, a higher GB would result in a prerejection peak that is lower in value, in a postrejection gain that is closer to unity, and in a 3-dB bandwidth (high frequency) that is wider.

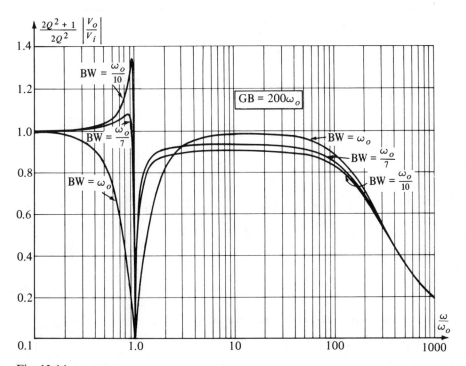

Fig. 13-14

13-7 REALIZATION WITH FOUR AMPLIFIERS

When the magnitude of the zeros is different from the magnitude of the poles, the network of Fig. 13-15 can be used to generate the second-order band-stop function. The portion of the network consisting of amplifiers #1, #2, and #3 is identical with the network presented in Section 11-6. It is used for generating the second-order BP, LP, and −LP functions. Amplifier #4 combines the various outputs to produce the desired BS function.

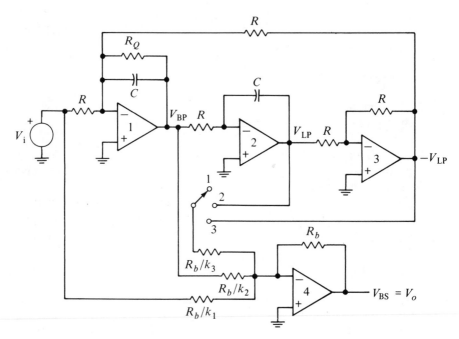

Fig. 13-15   Generation of band-stop function

The V_{BP} [see Eq. (11-34)] and V_{LP} signals are given by

$$V_{BP} = -V_i \frac{\omega_o s}{s^2 + s\dfrac{\omega_o}{Q} + \omega_o^2}, \qquad V_{LP} = V_i \frac{\omega_o^2}{s^2 + s\dfrac{\omega_o}{Q} + \omega_o^2}, \qquad (13\text{-}62)$$

where

$$\omega_o = \frac{1}{RC}, \qquad Q = \frac{R_Q}{R}, \qquad \frac{\omega_o}{Q} = \frac{1}{R_Q C}.$$

The output is obtained by inspection of Fig. 13-15:

$$V_o = V_i(-k_1) + V_{BP}(-k_2) + qV_{LP}(-k_3), \qquad (13\text{-}63)$$

where $q = 0$, 1, or −1, depending upon the position of the switch (1, 2, or 3).

The transfer function is obtained by using Eqs. (13-62) and (13-63):

$$\frac{V_o}{V_i} = -k_1 \frac{s^2 + s\frac{\omega_o}{Q}\left(1 - \frac{k_2}{k_1}Q\right) + \omega_o^2\left(1 + q\frac{k_3}{k_1}\right)}{s^2 + s\frac{\omega_o}{Q} + \omega_o^2}. \tag{13-64}$$

The BS function requires that the coefficient of the s-term be zero in the numerator polynomial. Hence, the k_2/k_1 ratio must be adjusted to

$$\frac{k_2}{k_1} = \frac{1}{Q}. \tag{13-65}$$

The zeros are located at

$$s = \pm j\omega_o\sqrt{1 + q\frac{k_3}{k_1}}, \qquad (q = 0, 1, -1). \tag{13-66}$$

The magnitude of the zeros is equal to, greater than, or smaller than the magnitude of the poles depending upon the value of q, which in turn is determined by the position of the switch.

The following adjustment procedure can be used to tune the BS network:

1. With the oscilloscope or some other measuring instrument connected to the BP output (output of amplifier #1), adjust for a response peak at $\omega = \omega_o$ by setting $RC = (1/\omega_o)$. Adjust for the 3-dB bandwidth, BW $= (\omega_o/Q)$, by setting

$$\frac{R_Q}{R} = \frac{\omega_o}{\text{BW}}.$$

These adjustments are noninteractive if the center frequency is set first and then the bandwidth controlled by R_Q:

2. Decide on the desired high-frequency gain and set k_1 to achieve it.
3. Set the k_3/k_1 ratio for the desired value of the zeros using the equation

$$\omega_z = \omega_o\sqrt{1 + q\frac{k_3}{k_1}}.$$

4. With the oscilloscope connected to the BS output (output of amplifier #4), tune for a null at $\omega = \omega_z$ by adjusting $k_2 = k_1(\text{BW}/\omega_o)$.

Refer to Section 11-6 for a discussion of the effect of nonideal amplifiers on the pole displacements from the nominal positions.

13-8 SUMMARY

The second-order band-stop function is characterized by two imaginary-axis zeros. If the poles are on the negative real axis, the bandwidth of rejection is more than twice the frequency of rejection. If the poles are complex and are on a circle passing through the zero locations, then the bandwidth of rejection is given by ω_o/Q. The magnitude characteristic has the same value at two frequencies which have a geometric average of ω_o. At these same two frequencies, the phases are equal in value but opposite in sign.

A very narrow-band, second-order rejection filter is undesirable for several reasons. First, it is difficult to set and maintain exactly the frequency of rejection. A slight departure of component values from the nominal values causes drastic degradation of the notch depth. (If a narrow rejection band is required, it is better to use a fourth-order filter which has a steeper rejection characteristic but a broader rejection base.) Second, in passive realizations, lossy inductors and, in active realizations, nonideal amplifiers affect the notch characteristics. The narrower the notch, the more pronounced become the resulting changes. Third, the transient response causes undue ringing.

A useful technique for generating the band-stop function is to subtract a band-pass function from unity. In passive realizations, a bridge structure is used to implement the subtraction. In active realizations, a summing or difference amplifier is used depending upon the sign of the band-pass output. The rejection frequency and the bandwidth of rejection are determined by the band-pass circuit. The rejection depth is controlled by the subtraction network. To achieve a good notch depth (ideally infinite attenuation), the zeros should be exactly on the imaginary axis. In making the notch-depth adjustment, an oscillator with low distortion should be used to prevent extraneous signals from appearing at the output.

PROBLEMS

13-1 (a) The second-order BS function [Eq. (13-2)] has the same magnitude value at two frequencies. Show that the geometric average of these two frequencies is ω_o.

   (b) Show that at these same two frequencies the phases are equal in magnitude but opposite in sign.

   (c) Show that the 3-dB rejection bandwidth is ω_o/Q.

13-2 For the networks shown in Fig. 13-16, obtain V_o/V_i and the ω_o- and Q-sensitivity functions associated with the poles and zeros.

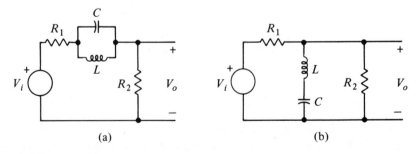

(a)   (b)

Fig. 13-16

13-3 Modify the network of Fig. 13-2b such that:
   (a) $\omega_o < \omega_z$,
   (b) $\omega_o > \omega_z$.

13-4 (a) For the network shown in Fig. 13-17, obtain the transfer function.
   (b) What value of R does produce a pair of imaginary-axis zeros? What are the values of these zeros?
   (c) Let $Q_c = (\omega_o L/r)$ represent the quality factor of the coil and let Q represent the pole Q of the transfer function. Show that $Q < Q_c$.

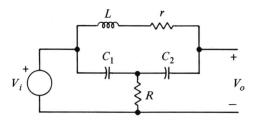

Fig. 13-17

13-5 Derive Eqs. (13-21) and (13-22). Also show that right half-plane zeros result when

$$\frac{C_1 + C_2}{C_3} < \frac{R_1 R_2}{R_1 + R_2} \frac{1}{R_3}.$$

13-6 Derive the sensitivity functions given by Eq. (13-29).

13-7 As shown in Fig. 13-18, the RC twin-T network is terminated in resistors at both ends. Discuss the effect of the terminations on the magnitude characteristic. Is the null frequency affected? How do the low- and high-frequency characteristics differ from the case with $R_s = 0$ and $R_L = \infty$?

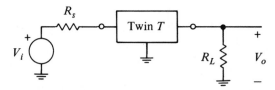

Fig. 13-18

13-8 (a) Sketch the null frequency vs. the potentiometer setting for the network shown in Fig. 13-7. Let $C_1 = C_2 = C_3 = C$.
   (b) Modify the circuit so that a 2:1 change in null frequency is obtained from one end of the potentiometer setting to the other.

13-9 Derive the sensitivity functions given by Eq. (13-41).

13-10 If k is chosen properly in Fig. 13-19, (V_o/V_i) has a pair of imaginary-axis zeros. Find k and the null frequency.

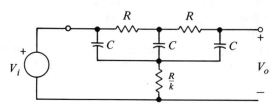

Fig. 13-19

13-11 Obtain (V_o/V_i) for the network shown in Fig. 13-20.

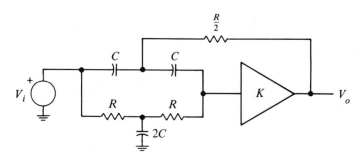

Fig. 13-20

13-12 (a) Calculate the input impedance Z_i shown in Fig. 13-21.
       (b) Show that (V_o/V_i) can be a band-stop function with complex poles.

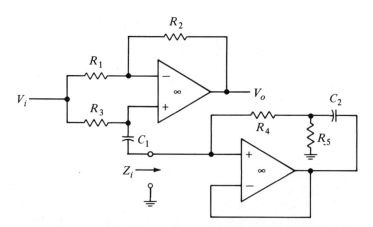

Fig. 13-21

13-13 (a) For the network shown in Fig. 13-22, obtain V_o/V_i.
   (b) For what k_1/k_2 ratio does (V_o/V_i) becomes BS?

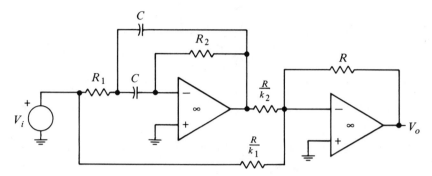

Fig. 13-22

13-14 (a) Obtain (V_o/V_i) for the circuit shown in Fig. 13-23.
   (b) Show that the zeros can be placed on the imaginary axis by adjusting R_5.
   (c) What is controlled by R_1? By R_3?

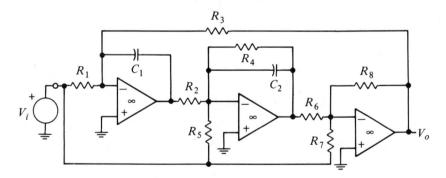

Fig. 13-23

13-15 (a) In Fig. 13-24, the network N is not grounded, i.e., it has only the three terminals shown. When $V_2 = 0$, $V_o = V_1 G$. Find V_o when $V_2 \neq 0$.
   (b) Is the answer in (a) valid if N has a grounded terminal? Explain.

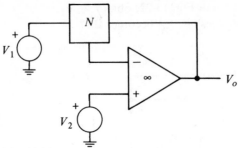

Fig. 13-24

13-16 In Fig. 13-25, the output can be written as

$$V_o = V_i T_1 + \alpha V_i T_2.$$

(a) Express T_2 in terms of T_1. (The network N does not have a grounded terminal.)

(b) Let T_1 be band-pass. Show that V_o/V_i can be made band-stop.

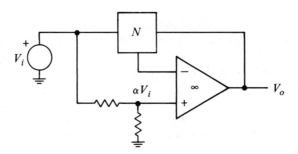

Fig. 13-25

14

First- and Second-Order All-Pass Networks

All-pass networks are used to alter the phase characteristics of systems without affecting the magnitude characteristics. First-order all-pass networks are designed to vary the phase of a sine wave at constant amplitude over a range of angle values, generally between 0 and $-\pi$. Second- and higher-order all-pass networks are used as phase correction networks.

In this chapter, several RC, RLC, and active RC all-pass networks are presented, and their characteristics are discussed.

14-1 THE FIRST- AND SECOND-ORDER ALL-PASS FUNCTION

In order for a function to be all-pass, (AP), its poles and zeros must be symmetrically placed with respect to the imaginary axis. Thus, a pole at $s = -\alpha$ is matched with a zero at $s = \alpha$; a pair of complex-conjugate poles at $-\alpha \pm j\beta$ are matched by a pair of complex-conjugate zeros at $\alpha \pm j\beta$. Graphically speaking, each pole-to-$j\omega$-axis distance is matched by an equal zero-to-$j\omega$-axis distance.

The first-order all-pass function is given by

$$T(s) = -\frac{s - \alpha}{s + \alpha}. \tag{14-1}$$

The minus sign is not a necessary part of the all-pass function. However, it is included here to show that the numerator polynomial can be obtained from the denominator polynomial by substituting $(-s)$ for s. Also, the function may have a scale factor other than unity.

The $s = j\omega$ characteristics of $T(s)$ are

$$T(j\omega) =$$

$$-\frac{j\omega - \alpha}{j\omega + \alpha} = \exp\left\{j\left[\tan^{-1}\left(\frac{-\omega}{\alpha}\right) - \tan^{-1}\left(\frac{\omega}{\alpha}\right)\right]\right\} = \exp\left[-j2\tan^{-1}\left(\frac{\omega}{\alpha}\right)\right],$$

$$(14\text{-}2\text{a})$$

$$|T(j\omega)| = 1, \qquad \theta(\omega) = -2\tan^{-1}\left(\frac{\omega}{\alpha}\right).$$

$$(14\text{-}2\text{b})$$

In Fig. 14-1, the phase is plotted vs. ω/α. Note that the phase curve always bends the same way, i.e., its second derivative is positive for all ω.

The second-order all-pass function is given by

$$T(s) = \frac{(s - \alpha)^2 + \beta^2}{(s + \alpha)^2 + \beta^2} = \frac{s^2 - s\dfrac{\omega_o}{Q} + \omega_o^2}{s^2 + s\dfrac{\omega_o}{Q} + \omega_o^2}.$$

$$(14\text{-}3)$$

The $s = j\omega$ characteristics of $T(s)$ are

$$T(j\omega) = \frac{(\omega_o^2 - \omega^2) - j\omega\dfrac{\omega_o}{Q}}{(\omega_o^2 - \omega^2) + j\omega\dfrac{\omega_o}{Q}} = \exp j\left[\tan^{-1}\left(\frac{-\omega\omega_o/Q}{\omega_o^2 - \omega^2}\right) - \tan^{-1}\left(\frac{\omega\omega_o/Q}{\omega_o^2 - \omega^2}\right)\right],$$

$$(14\text{-}4\text{a})$$

$$|T(j\omega)| = 1, \qquad \theta(\omega) = -2\tan^{-1}\left[\frac{\dfrac{\omega}{\omega_o}\dfrac{1}{Q}}{1 - \left(\dfrac{\omega}{\omega_o}\right)^2}\right].$$

$$(14\text{-}4\text{b})$$

In Fig. 14-1, using Q as a parameter, the phase is plotted vs. (ω/ω_o). When $Q = \frac{1}{2}$, both poles are on the negative real axis at $s = -\omega_o$. For $Q > \frac{1}{2}$, the poles, and hence the zeros, are complex. Note that for low-Q values the phase curve always bends the same way for all values of ω, whereas for high-Q values it bends first one way and then the opposite way, i.e., its second derivative changes sign. To obtain the point of inflection, $\theta(\omega/\omega_o)$ is differentiated twice with respect to (ω/ω_o) and the result is set equal to zero:

$$\frac{d\theta}{d\left(\dfrac{\omega}{\omega_o}\right)} = -\frac{2}{Q}\frac{\left(\dfrac{\omega}{\omega_o}\right)^2 + 1}{\left[1 - \left(\dfrac{\omega}{\omega_o}\right)^2\right]^2 + \dfrac{1}{Q^2}\left(\dfrac{\omega}{\omega_o}\right)^2},$$

$$(14\text{-}5)$$

$$\frac{d^2\theta}{d\left(\dfrac{\omega}{\omega_o}\right)^2} = 0 = \left(\frac{\omega}{\omega_o}\right)^4 + 2\left(\frac{\omega}{\omega_o}\right)^2 + \left(\frac{1}{Q^2} - 3\right).$$

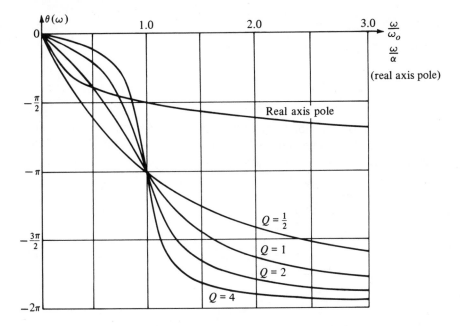

Fig. 14-1 All-pass phase characteristics

The point of inflection is obtained by solving for (ω/ω_o):

$$\left(\frac{\omega}{\omega_o}\right) = \sqrt{\sqrt{4 - \frac{1}{Q^2}} - 1},\tag{14-6}$$

which is positive provided $Q > (1/\sqrt{3})$. Hence, for $Q > (1/\sqrt{3}) = 0.578$, the phase characteristic changes the direction of bending at the frequency given by Eq. (14-6). For $Q \gg 1$, this occurs practically at $\omega = \omega_o$. This reversal in bending can be utilized in phase-equalizing applications. Thus, by choosing ω_o and Q appropriately, the right amount of phase can be added over a given band of frequencies by the all-pass function in order to bring the overall phase characteristic of a system closer to the desired characteristic. Note further that the all-pass function provides phase lag.

The effect of the all-pass function can best be summarized by comparing the steady-state sinusoidal output with the input:

Input $= V_m \sin \omega t$;

Steady-state output $= V_m \sin[\omega t + \theta(\omega)]$.

EXAMPLE 14-1

Show that a second-order AP can be constructed by subtracting an appropriately scaled second-order BP function from unity.

SOLUTION

Represent the BP function by

$$BP = \frac{Hs}{s^2 + s\dfrac{\omega_o}{Q} + \omega_o^2}.$$

Subtract this BP function from unity and choose H to make the coefficient of the s-term in the numerator polynomial the opposite of the corresponding coefficient in the denominator polynomial:

$$1 - BP = 1 - \frac{Hs}{s^2 + s\dfrac{\omega_o}{Q} + \omega_o^2}$$

$$= \frac{s^2 + s\left(\dfrac{\omega_o}{Q} - H\right) + \omega_o^2}{s^2 + s\dfrac{\omega_o}{Q} + \omega_o^2}.$$

Let $H = (2\omega_o/Q)$ and obtain

$$AP = \frac{s^2 - s\dfrac{\omega_o}{Q} + \omega_o^2}{s^2 + s\dfrac{\omega_o}{Q} + \omega_o^2}. \qquad Ans.$$

14-2 FIRST- AND SECOND-ORDER PASSIVE REALIZATIONS

A first-order all-pass realization can be obtained by rearranging the first-order function and using a scale factor of $\frac{1}{2}$:

$$\frac{V_o}{V_i} = -\frac{1}{2}\frac{s - \alpha}{s + \alpha} = \frac{1}{2} - \frac{s}{s + \alpha}. \tag{14-7}$$

The bridge network shown in Fig. 14-2 implements Eq. (14-7) with $\alpha = (1/RC)$. Note that the input and the output do not have a common terminal. (See also Problem 14-9.)

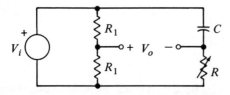

Fig. 14-2   First-order RC all-pass network

The amplitude of the output sine wave is one-half the amplitude of the input sine wave at all frequencies. This network is used generally in one of two ways. Either the input frequency is fixed and R is varied to change the phase of the output signal, or R is held constant and the phase of the output signal is made dependent upon the input frequency. In applications where R is varied, the phase of the output is 0 when $R = 0$ (being the voltage across the lower R_1). It is $-\pi$ when $R = \infty$ (being the negative of the voltage across the upper R_1). It is $-(\pi/2)$ when $R = (1/\omega C)$. As R is continuously varied from 0 to ∞, the phase is continuously shifted from 0 to $-\pi$. In applications where the input frequency is varied, the phase of the output is 0 when $\omega = 0$. It is $-\pi$ when $\omega = \infty$. It is $-(\pi/2)$ when $\omega = (1/RC)$. In either kind of application, the amount of phase shift for a given R or given frequency can be determined from Eq. (14-2b) or from the curve given in Fig. 14-1.

A second-order RLC all-pass network is shown in Fig. 14-3. The resistance r represents the coil's resistance.

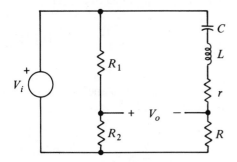

Fig. 14-3 A second-order all-pass network

The transfer function is

$$\frac{V_o}{V_i} = \frac{R_2}{R_1 + R_2} - \frac{s\left(\dfrac{R}{L}\right)}{s^2 + s\left[\dfrac{(R + r)}{L}\right] + \dfrac{1}{LC}}$$

$$= \frac{R_2}{R_1 + R_2}\left\{1 - \frac{\left(1 + \dfrac{R_1}{R_2}\right)\left(\dfrac{R}{L}\right)s}{s^2 + s\left[\dfrac{(R + r)}{L}\right] + \dfrac{1}{LC}}\right\}$$

$$= \frac{R_2}{R_1 + R_2}\left\{\frac{s^2 - s\left(\dfrac{1}{L}\right)\left(\dfrac{RR_1}{R_2} - r\right) + \dfrac{1}{LC}}{s^2 + s\left[\dfrac{(R + r)}{L}\right] + \dfrac{1}{LC}}\right\},$$

which for $(R_1/R_2) = 1 + (2r/R)$ becomes the all-pass function

$$\frac{V_o}{V_i} = \frac{R_2}{R_1 + R_2} \left[\frac{s^2 - s\frac{(R+r)}{L} + \frac{1}{LC}}{s^2 + s\frac{(R+r)}{L} + \frac{1}{LC}} \right].$$

$(14\text{-}8)$

The ω_o and Q of the poles and the associated sensitivity functions are

$$\omega_{op} = \frac{1}{\sqrt{LC}}, \qquad Q_p = \frac{1}{R+r}\sqrt{\frac{L}{C}},$$

$(14\text{-}9)$

$$S_L^{\omega_{op}} = S_C^{\omega_{op}} = -\tfrac{1}{2}, \qquad S_R^{\omega_{op}} = S_r^{\omega_{op}} = 0,$$

$(14\text{-}10a)$

$$S_L^{Q_p} = -S_C^{Q_p} = \frac{1}{2}, \qquad S_R^{Q_p} = -\frac{R}{R+r}, \qquad S_r^{Q_p} = -\frac{r}{R+r}.$$

$(14\text{-}10b)$

The ω_o of the zeros is the same as the ω_o of the poles. The Q of the zeros is

$$Q_z = -\sqrt{\frac{L}{C}} \frac{1}{\dfrac{RR_1}{R_2} - r}.$$

$(14\text{-}11)$

The Q_z is negative because the zeros are in the right half-plane. The various Q_z-sensitivity functions at the nominal position of the zeros, $(R_1/R_2) = 1 + (2r/R)$, are given by

$$-S_L^{Q_z} = S_C^{Q_z} = \frac{1}{2}, \qquad S_R^{Q_z} = S_{R_1}^{Q_z} = -S_{R_2}^{Q_z} = -\frac{1 + \dfrac{2r}{R}}{1 + \dfrac{r}{R}}, \qquad S_r^{Q_z} = \frac{\dfrac{r}{R}}{1 + \dfrac{r}{R}}.$$

$(14\text{-}12)$

Since r/R is generally much less than 1, all the sensitivity functions (for the poles as well as for the zeros) have magnitudes that are of the order 1 or less.

By selecting the ω_o and Q of the poles, a variety of phase characteristics can be obtained from this second-order all-pass network. Refer to Fig. 14-1 for the resulting phase curves.

14-3 ACTIVE *RC* REALIZATIONS

In this section four active *RC* realizations are given. In each case a band-pass realization is used in conjunction with a summing amplifier to produce the all-pass characteristics. The band-pass networks are obtained from Chapter 11. These networks use *KRC*, −*KRC*, infinite-gain *RC* or three-amplifier *RC* realization techniques to generate the BP functions. Using the BP functions, either the band-stop or the all-pass functions are constructed as follows:

$$1 - BP = 1 - \frac{Hs}{s^2 + s\dfrac{\omega_o}{Q} + \omega_o^2} \tag{14-13}$$

$$= \frac{s^2 + s\left(\dfrac{\omega_o}{Q} - H\right) + \omega_o^2}{s^2 + s\dfrac{\omega_o}{Q} + \omega_o^2} =$$

$$H = \frac{\omega_o}{Q} \longrightarrow \frac{s^2 + \omega_o^2}{s^2 + s\dfrac{\omega_o}{Q} + \omega_o^2} \quad \text{(BS)}, \tag{14-14a}$$

$$H = 2\frac{\omega_o}{Q} \longrightarrow \frac{s^2 - s\dfrac{\omega_o}{Q} + \omega_o^2}{s^2 + s\dfrac{\omega_o}{Q} + \omega_o^2} \quad \text{(AP)}. \tag{14-14b}$$

Whether a BS or an AP function is obtained depends solely on the adjusted value of the scale factor, H. Hence, by doubling the gain of the BP output, all the BS realizations of Chapter 13 (using the $1 - BP$ technique) can be converted to AP realizations. For convenience, these networks are reproduced here, and a summary of adjustment procedures is given.

KRC AP Network

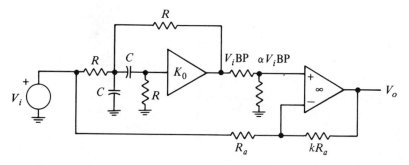

Fig. 14-4   *KRC AP network*

$$\frac{V_o}{V_i} = -k \frac{s^2 + s\left[\dfrac{1}{Q} - \alpha\dfrac{(1+k)}{k}\left(2\sqrt{2} - \dfrac{1}{Q}\right)\right]\omega_o + \omega_o^2}{s^2 + s\dfrac{\omega_o}{Q} + \omega_o^2}, \tag{14-15}$$

$$\omega_o = \frac{\sqrt{2}}{RC}, \qquad Q = \frac{\sqrt{2}}{4 - K_0}, \qquad BW = \frac{\omega_o}{Q}. \tag{14-16}$$

To obtain the AP function, set α to

$$\alpha = \frac{2k}{1+k}\frac{1}{Q}\frac{1}{2\sqrt{2}-\dfrac{1}{Q}} \qquad (\alpha \leq 1). \tag{14-17}$$

To realize a given ω_o and Q, use the following adjustment procedure.

1. Set the ω_o and the Q of the poles by monitoring the BP output. Use the RC product to set the frequency of peaking,

$$RC = \frac{\sqrt{2}}{\omega_o},$$

and adjust K_0 to produce the desired bandwidth,

$$K_0 = 4 - \sqrt{2}\,\frac{BW}{\omega_o}.$$

2. Using Eq. (14-17), determine α and put the zeros in their correct location by monitoring V_o. If ω_o and Q are not too high (so that frequency limitations imposed by imperfect amplifiers can be ignored), the magnitude response at V_o should be flat when α is adjusted correctly.

Refer to Section 11-3 for a discussion of frequency limitations concerning the band-pass portion of the network.

$-KRC$ *AP Network*

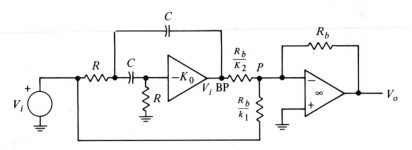

Fig. 14-5 $-KRC$ AP network

$$\frac{V_o}{V_i} = -k_1\,\frac{s^2 + s\left[\dfrac{1}{Q} - \dfrac{k_2}{k_1}\left(3Q - \dfrac{1}{3Q}\right)\right]\omega_o + \omega_o^2}{s^2 + s\dfrac{\omega_o}{Q} + \omega_o^2}, \tag{14-18}$$

$$\omega_o = \frac{1}{RC\sqrt{1+K_0}}, \qquad Q = \frac{\sqrt{1+K_0}}{3}, \qquad BW = \frac{\omega_o}{Q}. \tag{14-19}$$

To obtain the AP function, set (k_2/k_1) to

$$\frac{k_2}{k_1} = \frac{6}{9Q^2 - 1}.$$ (14-20)

To realize a given ω_o and Q, use the following adjustment procedure.

1. Set the ω_o and the Q of the poles by monitoring the BP output. Use the RC product to set the frequency of peaking,

$$RC = \frac{1}{3Q\omega_o},$$

and adjust K_0 to produce the desired bandwith,

$$K_0 = 9\left(\frac{\omega_o}{BW}\right)^2 - 1.$$

2. Using Eq. (14-20) determine the k_2/k_1 ratio. Monitor V_o for a flat magnitude response (which is obtained when ω_o is much smaller than the gain-bandwidth product of the operational amplifier).

   Refer to Section 11-4 for a discussion of the frequency limitations concerning the band-pass portion of the network.

AP *Network with Infinite Gain*

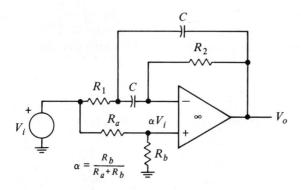

Fig. 14-6   AP network with infinite gain

$$\frac{V_o}{V_i} = \alpha \frac{s^2 + s\left(\frac{1}{Q} - \frac{1-\alpha}{\alpha} 2Q\right)\omega_o + \omega_o^2}{s^2 + s\frac{\omega_o}{Q} + \omega_o^2},$$ (14-21)

$$\omega_o = \frac{1}{C\sqrt{R_1 R_2}}, \qquad Q = \frac{1}{2}\sqrt{\frac{R_2}{R_1}}, \qquad BW = \frac{\omega_o}{Q}.$$ (14-22)

To obtain the AP function, set α to

$$\alpha = \frac{Q^2}{Q^2 + 1}. \tag{14-23}$$

To realize a given ω_o and Q, use the following adjustment procedure.

1. Ground the positive terminal of the amplifier. This converts the network to BP. Use the $R_1 C$ product to set the frequency of peaking,

$$R_1 C = \frac{BW}{2\omega_o^2},$$

and the $R_2 C$ product to set the bandwidth,

$$R_2 C = \frac{2}{BW}.$$

2. Disconnect the ground and use Eq. (14-23) to determine α. When α is adjusted correctly, the magnitude response at the output should be flat (provided ω_o is much less than the gain-bandwidth product of the operational amplifier).

EXAMPLE 14-2

In Fig. 14-6, assume that the operational amplifier has a one-pole rolloff characteristic, that is,

$$A(s) = -\frac{GB}{s + \omega_a} \cong -\frac{GB}{s}.$$

What effect does the finite value of GB have on the all-pass magnitude characteristic?

SOLUTION

With the nonideal amplifier and with $\alpha = Q^2/(Q^2 + 1)$, Eq. (14-21) becomes

$$\frac{V_o}{V_i} = \frac{Q^2}{Q^2 + 1} \frac{s^2 - s\dfrac{\omega_o}{Q} + \omega_o^2}{s^2 + s\dfrac{\omega_o}{Q} + \omega_o^2 + \dfrac{s}{GB}\left[s^2 + s\left(2Q + \dfrac{1}{Q}\right)\omega_o + \omega_o^2\right]}. \tag{14-24}$$

As Eq. (14-24) shows, the nonideal amplifier affects the AP function in two ways:

First, the poles are displaced from their nominal positions. As discussed in Section 11-5, the ω_o of the poles decreases and the Q increases. As a result, for high-Q AP realizations, the magnitude response, instead of being flat, peaks at a frequency slightly lower than ω_o. This is followed by a dip as the frequency of operation is swept past the upper half-plane zero.

Second, the magnitude response falls off at the rate of 6 dB/octave at high frequencies because the transfer function has two zeros and three poles. As discussed in Example 13-5, the high-frequency 3-dB point is at $\omega_{3\,\mathrm{dB}} \cong \mathrm{GB}$.

To study these effects further, magnitude- and frequency-normalize Eq. (14-24):

$$\frac{Q^2 + 1}{Q^2} \frac{V_o}{V_i}$$

$$= \frac{\left(\dfrac{s}{\omega_o}\right)^2 - \left(\dfrac{s}{\omega_o}\right)\dfrac{1}{Q} + 1}{\left(\dfrac{s}{\omega_o}\right)^2 + \left(\dfrac{s}{\omega_o}\right)\dfrac{1}{Q} + 1 + \dfrac{s/\omega_o}{\mathrm{GB}/\omega_o}\left[\left(\dfrac{s}{\omega_o}\right)^2 + \left(\dfrac{s}{\omega_o}\right)\left(2Q + \dfrac{1}{Q}\right) + 1\right]}.$$

$$(14\text{-}25)$$

The magnitude characteristic given by Eq. (14-25) is plotted in Fig. 14-7 for three values of Q and for $\mathrm{GB} = 200\omega_o$. Note the pre-$\omega_o$ peaking and the subsequent dipping of the magnitude response near ω_o; note also the upper 3-dB frequency (at $\omega/\omega_o = 200$). With all other values held constant, a higher GB results in less peaking and dipping near ω_o and in wider 3-dB bandwidth.

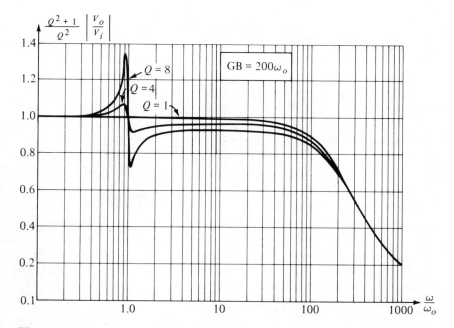

Fig. 14-7

AP *Network with Four Amplifiers*

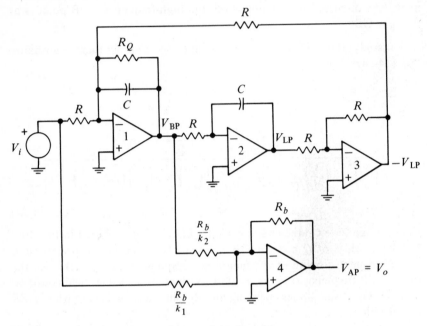

Fig. 14-8   Four-amplifier realization of AP function

$$\frac{V_o}{V_i} = -k_1 \frac{s^2 + s\frac{\omega_o}{Q}\left(1 - \frac{k_2}{k_1}Q\right) + \omega_o^2}{s^2 + s\frac{\omega_o}{Q} + \omega_o^2}, \tag{14-26}$$

$$\omega_o = \frac{1}{RC}, \qquad Q = \frac{R_Q}{R}, \qquad BW = \frac{\omega_o}{Q}. \tag{14-27}$$

To obtain the AP function, set (k_2/k_1) to

$$\frac{k_2}{k_1} = \frac{2}{Q}. \tag{14-28}$$

To realize a given ω_o and Q, use the following adjustment procedure:

1. By monitoring the BP output, set the frequency of peaking using the RC product,

$$RC = \frac{1}{\omega_o}.$$

Set the BW by using the R_Q/R ratio,

$$\frac{R_Q}{R} = \frac{\omega_o}{BW}.$$

2. By monitoring the AP output, adjust the (k_2/k_1) ratio to obtain a flat magnitude response,

$$\frac{k_2}{k_1} = 2\frac{BW}{\omega_o}.$$

Refer to Section 11-6 for a discussion of the frequency limitations concerning the band-pass portion of the network.

14-4 SUMMARY

All-pass functions have pole–zero structures that are quadrantally symmetrical. As a result, the magnitude characteristic is frequency-indepedent, while the phase characteristic varies as a function of frequency. The phase of the output of an AP network lags the phase of the input sine wave.

The first-order AP function has a pole at $s = -\alpha$ and a zero at $s = \alpha$. At a given frequency, the phase can be altered by changing α. Alternatively, for a given α, different phase values result if the frequency is changed. The second-order AP function has, in general, complex poles $(-\alpha \pm j\beta)$ and complex zeros $(\alpha \pm j\beta)$. By selecting the ω_o and Q of the poles, the phase can be made to exhibit different characteristics, which can be employed in applications involving phase corrections.

Most all-pass realizations are obtained by subtracting a band-pass output signal from the input signal. If the gain of the subtraction circuit is properly adjusted, an all-pass response is obtained. In active realizations involving high-ω_o and high-Q values, the magnitude response shows peaking and dipping near ω_o. It also falls off at higher frequencies. The higher the gain-bandwidth product of the operational amplifier, the less the peaking and dipping near ω_o and the wider is the bandwidth of the system.

PROBLEMS

14-1   When the R_1/R_2 ratio is properly adjusted, both networks of Fig. 14-9 become AP. Show this.

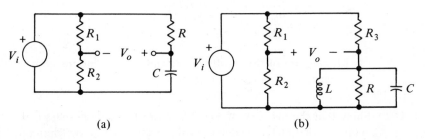

(a)   (b)

Fig. 14-9

14-2   Show that the function

$$T(s) = \frac{1 - Z(s)}{1 + Z(s)}$$

is all-pass if $Z(s)$ represents the input impedance of an LC network.

14-3   (a) Show that the network shown in Fig. 14-10 possesses all-pass characteristics.
        (b) Realize

$$T(s) = \pm \frac{1}{2} \frac{s^2 - s\dfrac{\omega_o}{Q} + \omega_o^2}{s^2 + s\dfrac{\omega_o}{Q} + \omega_o^2}$$

with this network.

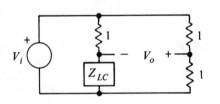

Fig. 14-10

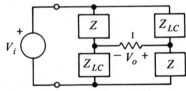

Fig. 14-11

14-4   In Fig. 14-11, $Z = (1/Z_{LC})$.
        (a) Find V_o/V_i and show that it is all-pass.
        (b) Calculate the input impedance Z_i.
        (c) With this network realize

$$\frac{V_o}{V_i} = \frac{s^2 - s\dfrac{\omega_o}{Q} + \omega_o^2}{s^2 + s\dfrac{\omega_o}{Q} + \omega_o^2} .$$

        (d) Making use of the results of (a) and (b), outline the construction of a fourth-order all-pass network.

14-5   Show that the network of Fig. 14-12 can be made all-pass if the output is taken either at V_{o1} (switch in position 1) or at V_{o2} (switch in position 2).

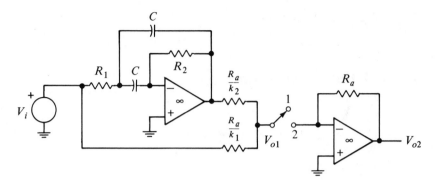

Fig. 14-12

14-6 (a) In Fig. 14-13, that should be k to make V_o/V_i all-pass?
   (b) k is adjusted to the value obtained in (a). The input is a sine wave. Show that by varying R a constant-amplitude, variable-phase sine wave is obtained at the output.
   (c) Interchange the R and the C. Repeat (a) and (b).

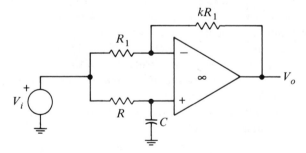

Fig. 14-13

14-7 Show that the transfer function of the network of Fig. 13-21 can be made all-pass if the R_2/R_1 ratio is properly adjusted.

14-8 Show that the transfer function of the network of Fig. 13-23 can be made all-pass if R_5 is adjusted properly.

15

RC Oscillators

Filter theory is based on the sinusoidal properties of networks. Filter networks and a great variety of other networks are designed to achieve a prescribed sinusoidal response. A great body of knowledge and experience exists on the sinusoidal testing of these networks. The testing cannot be done without the variable-frequency, constant-amplitude oscillator, which is perhaps the most useful tool of the circuit designer. In this chapter, a variety of *RC* oscillators is presented. In particular, the characteristics of the Wien-bridge oscillator are discussed in detail. Methods are given for stabilizing the amplitude of oscillation through nonlinear gain characteristics and automatic gain control. Also presented are oscillators using amplifiers with less-than-unity gain and oscillators which can be varied in frequency by varying a single element.

15-1 THE WIEN-BRIDGE OSCILLATOR

A linear sinusoidal oscillator is a network with a pair of imaginary-axis poles. The poles of many active *RC* networks can be readily placed on the imaginary axis. In particular, the poles of *all* the *KRC* networks discussed in the preceding five chapters can be moved onto the imaginary axis by adjusting the gain K of the amplifier. (The poles move in a constant-ω_o manner as a function of K.)

A very popular *KRC* oscillator, known as the Wien-bridge oscillator, is shown in Fig. 15-1a. The operational amplifier in conjunction with the R_a and R_b feedback network serves as the K amplifier. The feedback to the positive terminal is through the *RC* network. Equivalently, the Wien-bridge oscillator can be represented as in Fig. 15-1b.

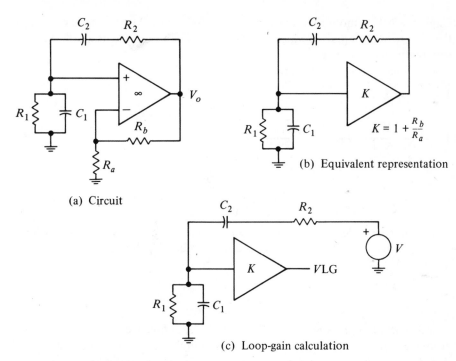

(a) Circuit

(b) Equivalent representation

(c) Loop-gain calculation

Fig. 15-1   The Wien-bridge oscillator

To determine the position of the poles, the loop gain must be found first. This can be done by inspection of Fig. 15-1c:

$$LG = K \frac{\dfrac{R_1(1/sC_1)}{R_1 + (1/sC_1)}}{\dfrac{R_1(1/sC_1)}{R_1 + (1/sC_1)} + R_2 + \dfrac{1}{sC_2}}$$

$$= \frac{K \dfrac{s}{R_2 C_1}}{s^2 + s\left(\dfrac{1}{R_1 C_1} + \dfrac{1}{R_2 C_2} + \dfrac{1}{R_2 C_1}\right) + \dfrac{1}{R_1 R_2 C_1 C_2}} = \frac{Hs}{s^2 + s\dfrac{\omega_o}{Q} + \omega_o^2}. \quad (15\text{-}1)$$

Note that the loop gain is a band-pass function. Therefore, its magnitude peaks at

$$\omega = \omega_o = \frac{1}{\sqrt{R_1 R_2 C_1 C_2}}.$$

However, the network, being RC, has poor selectivity. This is seen by calculating the 3-dB bandwidth of the loop gain:

$$BW = \frac{\omega_o}{Q} = \left(\sqrt{\frac{R_1 C_1}{R_2 C_2}} + \sqrt{\frac{R_2 C_2}{R_1 C_1}} + \sqrt{\frac{R_1 C_2}{R_2 C_1}}\right)\omega_o > 2\omega_o. \quad (15\text{-}2)$$

Thus, the BW is more than twice the center frequency. Most often the R's and the C's are made equal, in which case BW $= 3\omega_o$. This represents a rather broad selectivity characteristic.

The poles of the network are found from

$$1 - LG = 0, \tag{15-3}$$

$$1 - \frac{K \dfrac{s}{R_2 C_1}}{s^2 + s\left(\dfrac{1}{R_1 C_1} + \dfrac{1}{R_2 C_2} + \dfrac{1}{R_2 C_1}\right) + \dfrac{1}{R_1 R_2 C_1 C_2}} = 0,$$

$$s^2 + s\left[\frac{1}{R_1 C_1} + \frac{1}{R_2 C_2} + \frac{1}{R_2 C_1}(1 - K)\right] + \frac{1}{R_1 R_2 C_1 C_2} = 0. \tag{15-4}$$

The network poles can be put on the imaginary axis by making the coefficient of the s-term zero in Eq. (15-4). This is done by adjusting the gain K to the critical value of K_c:

$$K_c = 1 + \frac{R_2}{R_1} + \frac{C_1}{C_2}. \tag{15-5}$$

If $K < K_c$, the poles are in the left half-plane, and hence, once started, oscillations cannot be sustained; the sine-wave amplitude decays exponentially. If $K > K_c$, the poles are in the right half-plane, and hence oscillations grow exponentially in amplitude until the linear dynamic range of the amplifier is exceeded. If $K = K_c$, the poles are at

$$\pm j\omega_o = \pm j\frac{1}{\sqrt{R_1 R_2 C_1 C_2}},$$

and hence the network oscillates with the frequency ω_o. Note that ω_o is determined solely by the RC band-pass network employed in the feedback loop. For the equal-R, equal-C case, the output can be written as

$$v_o(t) \cong V_m e^{\frac{1}{2}\Delta K \omega_o t} \sin(\omega_o t + \theta), \tag{15-6}$$

where ΔK represents the departure of the gain from the critical value, $K_c = 3$. Because of ΔK, the poles are not at $\pm j\omega_o$ but are approximately at $\frac{1}{2}\Delta K\omega_o \pm j\omega_o$. For $\Delta K = 0$, the output is a sine wave of amplitude V_m. For $\Delta K > 0$, the output grows, whereas for $\Delta K < 0$, the output decays. The time constant associated with the growth or decay is

$$\tau = \frac{2}{\Delta K \omega_o} = \frac{T}{\pi \Delta K}, \tag{15-7}$$

where T is the period of the sine wave. For example, if K decreases by 0.01% from the critical value, then

$$\frac{\tau}{T} = \frac{100}{\pi \times 3 \times 0.01} \cong 1000.$$

Consequently, after a time interval of 1000 periods, the amplitude of oscillation decreases to 37% of its initial value.

Since the network is not driven, Eq. (15-6) represents the natural response of the network. Any disturbance, such as the application of dc sources to activate the amplifier, excites this response. The main problem in the design of oscillators is to maintain this response with $\Delta K = 0$ and $V_m = $ constant.

In order for ΔK to equal zero at all times, the gain of the amplifier must be held precisely at the critical value given by Eq. (15-5). Since this is impossible to achieve in practice, additional circuitry is used to sustain oscillations with as little distortion as possible.

The network oscillates with $\omega_o = 1/\sqrt{R_1 R_2 C_1 C_2}$ if

$$K = 1 + \frac{R_2}{R_1} + \frac{C_1}{C_2}.$$

Hence, resistor and capacitor ratios must be kept constant for variable-frequency oscillators. This is done by ganging either the R's or the C's. It should be realized, however, that it is practically impossible to maintain a constant R_2/R_1 or C_1/C_2 ratio while the R's or the C's are varied. It is therefore more difficult to design a variable-frequency oscillator than a fixed-frequency oscillator.

Two schemes are generally used to sustain oscillations at constant amplitude. One scheme introduces a nonlinearity in the amplifier's gain characteristic. The other scheme provides for an automatic adjustment of the gain characteristic. These schemes are presented in the next two sections.

EXAMPLE 15-1

How do the oscillation frequency and the critical value of the gain change if a one-pole rolloff model is used for the amplifier? Assume that the R's and the C's are equal.

SOLUTION

Obtain from Eq. (15-4) the characteristic equation when the amplifier is ideal:

$$s^2 + \frac{s}{RC}(3 - K) + \frac{1}{R^2 C^2} = 0.$$

Now let

$$K = \frac{GB}{s + \frac{GB}{K_0}} \quad \text{(see Section 7-3)},$$

and obtain the modified characteristic equation:

$$s^2 + \frac{s}{RC}\left(3 - \frac{GB}{s + \frac{GB}{K_0}}\right) + \frac{1}{R^2C^2} = 0,$$

$$s^3 + s^2\underbrace{\left(\frac{GB}{K_0} + \frac{3}{RC}\right)}_{a_2} + s\underbrace{\left[\frac{GB}{RC}\left(\frac{3}{K_0} - 1\right) + \frac{1}{R^2C^2}\right]}_{a_1} + \underbrace{\frac{GB}{K_0}\frac{1}{R^2C^2}}_{a_0} = 0.$$

In order to produce imaginary-axis poles, the a coefficients must satisfy $a_2 a_1 = a_0$. Hence,

$$\left(\frac{GB}{K_0} + \frac{3}{RC}\right)\left[\frac{GB}{RC}\left(\frac{3}{K_0} - 1\right) + \frac{1}{R^2C^2}\right] = \frac{GB}{K_0}\frac{1}{R^2C^2},$$

$$(GBRC)^2(3 - K_0) + 3K_0(GBRC)(3 - K_0) + 3K_0^2 = 0.$$

Let $K_0 = 3 + \Delta K$ and simplify:

$$(GBRC)^2\Delta K + 3(3 + \Delta K)(GBRC)\Delta K - 3(3 + \Delta K)^2 = 0.$$

Assume

$$GBRC \gg 9, \qquad |\Delta K| \ll 3,$$

and approximate:

$$(GBRC)^2\Delta K - 27 = 0,$$

$$\Delta K = \frac{27}{(GBRC)^2} = 27\left(\frac{\omega_o}{GB}\right)^2. \qquad Ans.$$

Hence, the critical value of the gain must be set higher than for the ideal case to sustain oscillations. The higher the frequency of oscillation, the higher is the required gain for oscillation. Consequently, unless additional circuitry is used the oscillator will drop out of oscillation as the frequency is changed to a higher value.

To obtain the modified frequency of oscillation, use $\omega_{om} = \sqrt{a_0/a_2}$:

$$\omega_{om} = \sqrt{\frac{\frac{GB}{K_0}\omega_o^2}{\frac{GB}{K_0} + 3\omega_o}} = \omega_o\sqrt{\frac{1}{1 + 3K_0\frac{\omega_o}{GB}}} \cong \omega_o\left(1 - \frac{9}{2}\frac{\omega_o}{GB}\right),$$

$$\frac{\Delta\omega_o}{\omega_o} \cong -\frac{9}{2}\frac{\omega_o}{GB}. \qquad Ans.$$

Hence, the higher the ω_o/GB ratio, the lower becomes the frequency of oscillation compared to the ideal value (assuming that K_0 is adjusted to the critical value).

15-2 NONLINEAR AMPLIFIER GAIN

This method employs an amplifier with a fixed but nonlinear gain characteristic. For small amplitudes of the output signal, the gain is greater than the
critical value, to assure the starting and the growing of oscillations. For larger
signal levels, the incremental gain ($\Delta v_o / \Delta v_a$) is less. As a result, the amplitude
of oscillation is eventually stabilized.

An example of an amplifier with nonlinear gain characteristic is shown in
Fig. 15-2. The back-to-back zeners are used to achieve a break in the gain

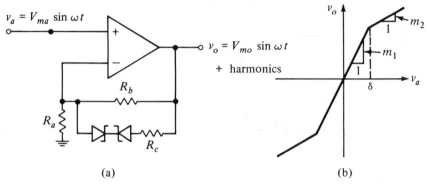

(a)   (b)

Fig. 15-2   Amplifier with nonlinear characteristic

characteristic. As long as the magnitude of the voltage across R_b is less than the
zener breakdown voltage, V_z, the amplifier gain is constant and has the value of

$$1 + \frac{R_b}{R_a} = m_1 = \frac{v_o}{v_a}.$$

When the magnitude of the voltage across R_b exceeds V_z, the zeners conduct,
thereby shunting R_b with R_c. As long as the zeners conduct, the incremental gain
is given by

$$\left(1 + \frac{\dfrac{R_b R_c}{R_b + R_c}}{R_a}\right) = m_2 = \frac{\Delta v_o}{\Delta v_a}.$$

The break point occurs when the signal at the amplifier input, v_a, reaches

$$v_a = \delta = \frac{V_z}{m_1 - 1} = V_z \frac{R_a}{R_b}.$$

In practice, because of nonideal zeners, this break point is not as sharp as indicated in the v_o-vs.-v_a curve.

If v_a is periodic, so must be v_o. If the v_a swing covers both the m_1 range as well
as m_2, v_o is no longer directly related to v_a. In this instance, to speak of the gain
of the amplifier is meaningless. Nonetheless, the concept of gain may be extended

to cover this type of operation, provided gain is interpreted as ratios of output-harmonics to input-harmonics (magnitude and phase), taken one harmonic at a time. Thus, the magnitude of the gain for the fundamental component is given by (output magnitude of fundamental)/(input magnitude of fundamental). Taken in this sense, the gain becomes a function of the harmonic number.

When this nonlinear gain amplifier is incorporated in the Wien-bridge RC network, the oscillator shown in Fig. 15-3 results. The capacitors are ganged to

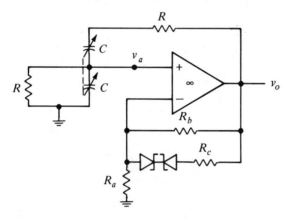

Fig. 15-3   Oscillator with nonlinear gain

vary the frequency of oscillation. Because equal resistors and equal capacitors are used, the critical value of the gain for this circuit is given by Eq. (15-5) as $K_c = 3$.

The gain $m_1 = 1 + (R_b/R_a)$ of the amplifier is chosen slightly greater than 3 to assure that the output builds up exponentially until the amplifier is driven part of the time into the m_2 range of operation (near the peak swing of the output). As a result, both v_a and v_o become nonsinusoidal, periodic signals. Since the operation is nonlinear, linear circuit theory cannot be applied to obtain the steady-state output.

As Eq. (15-1) shows, the RC-feedback network to the noninverting terminal of the amplifier is a band-pass network. The peak response occurs at the frequency of oscillation, that is, $\omega = (1/RC)$. Thus, the selective filtering action of the RC network results in a signal at v_a which is somewhat more nearly sinusoidal than the signal at v_o. (The network, being RC, presents a rather broad selectivity characteristic. See Problem 15-3 for a circuit with much narrower selectivity.) Therefore, to a first-order approximation, the signal at v_a may be considered sinusoidal. When this sinusoidal signal passes through the amplifier, its top portion is distorted symmetrically about the peak value (indicating the presence of odd harmonics only), according to the gain characteristic given in Fig. 15-2b. (In practice, because v_a is not truly sinusoidal, a slight skewing is also present in the top portion of the output waveform.)

How far is the output signal v_o driven into the m_2 range by the input signal v_a? The amplitude of v_a will determine this, but this amplitude itself is unknown even though v_a is assumed to be sinusoidal. The peak value of the fundamental component of the output signal, V_{mo}, depends upon the peak value of the amplifier input signal, V_{ma}. The larger V_{ma}, the larger V_{mo}. But the (V_{mo}/V_{ma}) ratio, i.e., the magnitude of the gain at the fundamental frequency, must decrease with increase in V_{ma} because the output looks less and less sinusoidal. That this is indeed so can be demonstrated by assuming $v_a = V_{ma} \sin \omega t$, and then finding the fundamental component of the resulting v_o. Figure 15-2b is used for calculating v_o, which is then expanded in the Fourier series to obtain the fundamental component, $v_{of} = V_{mo} \sin \omega t$; when this is done, the following expression for the gain of the fundamental component is obtained:

$$\frac{V_{mo}}{V_{ma}} = m_1 \qquad (V_{ma} < \delta)$$

$$= m_2 + \frac{2}{\pi}(m_1 - m_2)\left[\sin^{-1}\left(\frac{\delta}{V_{ma}}\right) + \left(\frac{\delta}{V_{ma}}\right)\sqrt{1 - \left(\frac{\delta}{V_{ma}}\right)^2}\right] \qquad (V_{ma} > \delta).$$

$$(15\text{-}8)$$

Equation (15-8) is plotted in Fig. 15-4 for two values of the startup gain, m_1; m_2 is used as a parameter. As long as the peak value of v_a, V_{ma}, is less than δ (the break-point voltage in the amplifier characteristic), the gain is m_1, which is made slightly greater than 3 to start the oscillation. Oscillations continue growing

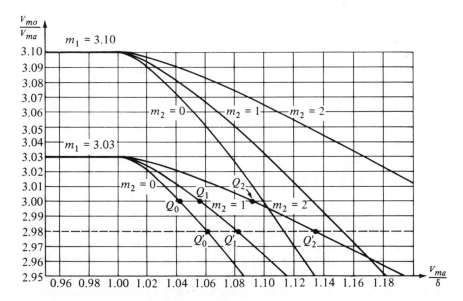

Fig. 15-4   The gain at the fundamental frequency

until the gain at the fundamental frequency is reduced to 3 by the action of the m_2 region. When steady state is reached, the fundamental component of the output is

$$v_{of} = V_{mo} \sin \omega t,$$

and the signal at the output of the RC band-pass feedback network is $(V_{mo}/3)$ $\sin (\omega t + 0°)$. (See Problem 15-1 for details.) The amplifier then amplifies this signal by 3 to produce the output. In other words, the loop gain is 1. Thus, for $m_1 = 3.03$, operation stabilizes at the point Q_0 if $m_2 = 0$; at the point Q_1 if $m_2 = 1$; or at the point Q_2 if $m_2 = 2$. The larger m_2, the less is the distortion in the output.

If care is not exercised in ganging the capacitors (and in considering stray capacitances), the attenuation provided by the RC network will not be $\frac{1}{3}$ when the phase shift around the loop is zero. In this case, the gain at the fundamental frequency must assume a value different from 3 to provide the unity loop-gain condition required for oscillation. (Nonideal amplifiers complicate matters further.) For example, if K_c drops from 3 to 2.98 due to imperfect tracking of the ganged capacitors, equilibrium will be established at the point Q_2' in Fig. 15-4, rather than at Q_2. As a result, the output amplitude will be about 4% larger than previously. This effect can be minimized by making m_2 smaller. However, the distortion then becomes higher because of the more abrupt break in the amplifier-gain characteristic.

If the startup gain m_1 is made considerably greater than 3, the output must spend a correspondingly longer time in the m_2 region in order to result in a gain of 3 for the fundamental component. As a result, the amplitude of oscillation is larger, and more distortion is present in the output.

It should be emphasized again that the quantitative analysis presented in this section is predicated on low output distortion, so that v_a can be assumed to be sinusoidal. The two requirements, constant oscillation amplitude and low distortion, are contradictory requirements. One can be bettered only at the expense of the other. For this reason, nonlinear amplifier gain alone is seldom used for variable-frequency, constant-amplitude oscillators.

15-3 AUTOMATIC GAIN CONTROL

Automatic gain control is widely used in the design of variable-frequency oscillators in order to achieve constant oscillation amplitude with low distortion. Here, the gain of the amplifier is made dependent upon the amplitude of oscillation. The gain is made high to start the oscillations. As the amplitude of the oscillation grows, the gain is automatically reduced to the value needed to sustain oscillations. Figure 15-5 illustrates the necessary amplifier characteristic to achieve this.

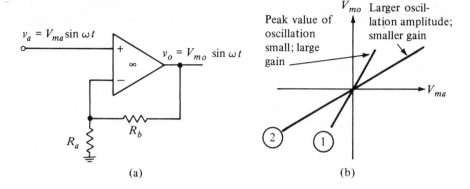

Fig. 15-5   Automatic adjustment of the gain

Since the gain of the amplifier is

$$1 + \frac{R_b}{R_a},$$

it can be varied by changing either R_a or R_b. In Fig. 15-5a, R_a is made a function of the input signal amplitude. The functional dependence is such that the larger V_{ma}, the larger R_a, and hence the smaller the gain. As a result, the gain characteristic shown in Fig. 15-5b is obtained. For low-amplitude signals, operation is along line 1 (higher gain); for high-amplitude signals, operation is along line 2 (lower gain). As V_{ma} is gradually increased, operation shifts from 1 to 2 in a continuous manner.

When this amplifier is used in conjunction with the equal-R, equal-C Wien-bridge network, R_a and R_b are chosen such that the gain is 3 for the desired amplitude of oscillation. If for any reason the amplitude of oscillation decreases slightly, R_a decreases too, thereby increasing the gain and thus counteracting the decrease in oscillation amplitude. On the other hand, an increase in amplitude from its nominal value causes an increase in R_a, which results in reduced gain and hence reduced amplitude, thereby counteracting again the original change. Thus R_a changes automatically to produce the necessary critical gain for oscillation, which may change with the frequency of operation (due to imperfect tracking of ganged capacitors or due to other reasons). How close the amplitude of oscillation remains to its nominal value while these changes occur depends upon how R_a varies with the peak value of oscillation. Up to a point, the larger (dR_a/dV_{ma}) at and near the nominal value of the oscillation amplitude, the less the variation in amplitude. However, too large a value of (dR_a/dV_{ma}) may cause an amplitude-modulated sine wave, because the gain is alternately made too large (resulting in a rapid buildup) and too small (resulting in a rapid decay).

As long as the steady-state operation is along a straight line, as shown in Fig. 15-5b, there is no distortion in the output.

Two devices, the characteristics of which are controllable, are widely used for R_a or for part of R_a. One is an incandescent lamp and the other a field-effect transistor.

Lamp-Stabilized Wien-Bridge Oscillator

An incandescent lamp operating below luminous levels is used as part of R_a in the Wien-bridge oscillator shown in Fig. 15-6. The lamp's resistance has a positive temperature coefficient. The higher the voltage across the lamp, the hotter it gets, and the higher the value of the resistance becomes.

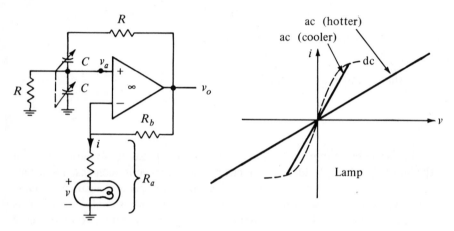

Fig. 15-6   Lamp-stabilized Wien-bridge oscillator

While the dc i–v characteristic of the lamp is nonlinear, as shown by the dashed curve, for sinusoidal signals above 1 Hz, the steady-state characteristic is a straight line, the slope of which depends on the peak value of the signal. Thus, when the lamp assumes an average temperature for a given sinusoidal signal, it can be replaced by an equivalent resistor. As the peak value of the sinusoidal voltage across the lamp is gradually increased, the i–v characteristic slowly turns clockwise about the origin and assumes a new linear relationship at a higher value of resistance. This is the type of controlling action that is desired in the Wien-bridge oscillator. The "cold" value of R_a is chosen such that the gain of the amplifier is greater than 3. This allows the starting of oscillations. As the oscillation's amplitude grows, the value of R_a increases, thereby decreasing the gain until it equals 3, which is the required value to maintain imaginary-axis poles (if the resistors and capacitors are equal and the amplifier is ideal). Suppose, however, that, at a new frequency of operation, the capacitor values differ by 5%, so that the critical value of the gain is 3.05. If the automatic controlling action of the lamp were absent, the gain would stay at 3 and oscillations would die out. But with the lamp in the circuit, the decaying amplitude causes R_a to decrease, thereby automatically increasing the gain until it is 3.05, Thus, oscillations are maintained with a slightly lower amplitude. The size of the amplitude change depends upon the sensitivity of R_a to variations of the peak value of voltage across it.

Throughout the preceding discussion, the lamp is assumed to be operating under steady-state conditions right away. However, because of the thermal time constant of the lamp, the change in R_a may lag sufficiently behind the desired controlling action, particularly at low frequencies of operation. This delay in response may cause a periodic variation in the amplitude of oscillations (envelope instability). For this reason, the lamp stabilization scheme is not used for frequencies lower than 1 Hz. A slight nonlinearity in the amplifier-gain characteristic greatly improves envelope stability, while causing very little distortion in the sine wave. The R_c-zener diode combination shown in Fig. 5-3 can be used for this purpose. The break in the amplifier-gain characteristic is placed near the peak value of the sine wave.

It should also be noted that the oscillation amplitude varies with ambient temperature because R_a is temperature-dependent.

EXAMPLE 15-2

(a) Design a 1-KHz Wien-bridge oscillator with 10-V peak amplitude.
(b) Let k represent the ohms/volt (peak) sensitivity of R_a. How much does the amplitude of oscillation change if the gain required for oscillation changes from K to $(K + \Delta K)$?
(c) Let $K = 60 \ \Omega/V$ and $R_a = 400 \ \Omega$. How much does the amplitude of oscillation change if the gain changes 1%?

SOLUTION

(a) Use the circuit of Fig. 15-6 with R_a representing the lamp's resistance, i.e., no other resistance is used as part of R_a. Since the desired output voltage has a peak value of 10 V, the voltage across R_a will have a peak value of

$$\frac{10}{3} = 3.33 \text{ V.}$$

Apply a 1-KHz, 3.33-V peak sinusoidal signal to the lamp by itself and, from the resulting i–v curve, measure R_a. Then make $R_b = 2R_a$. This will give the desired output. To obtain 1-KHz operation, select $RC = (10^{-3}/2\pi)$.

(b) Express V_m as a function of the voltage across R_a and differentiate:

$$V_m = KV_{R_a},$$

$$\frac{dV_m}{dK} = V_{R_a} + K\frac{dV_{R_a}}{dK} = V_{R_a} + K\frac{dV_{R_a}}{dR_a}\frac{dR_a}{dK}.$$

But

$$\frac{dV_{R_a}}{dR_a} = \frac{1}{k}, \qquad \frac{dK}{dR_a} = \frac{d}{dR_a}\left(1 + \frac{R_b}{R_a}\right) = -\frac{R_b}{R_a^2} = -\frac{K-1}{R_a}.$$

Hence,

$$\frac{dV_m}{dK} = V_{R_a} - \frac{K}{K-1}\frac{R_a}{k} = \left(1 - \frac{K^2}{K-1}\frac{R_a}{kV_m}\right)\frac{V_m}{K},$$

$$\frac{\Delta V_m}{V_m} \cong \left(1 - \frac{K^2}{K-1}\frac{R_a}{kV_m}\right)\frac{\Delta K}{K}. \qquad Ans. \tag{15-9}$$

(c) Substitute the given values in Eq. (15-9) and obtain

$$\left|\frac{\Delta V_m}{V_m}\right| \cong \left(1 - \frac{3^2}{3-1}\frac{400}{60 \times 10}\right)0.01 = 0.02.$$

Thus, if the gain changes by 1%, the oscillation amplitude changes by 2%.

FET-*Stabilized Wien-Bridge Oscillator*

In the Wien-bridge oscillator shown in Fig. 15-7a, automatic gain control is obtained by varying the drain-to-source resistance of a field-effect transistor (FET). As can be seen from the i–v characteristic of the FET (Fig. 15-7b), the value of this resistance depends upon the gate-to-source voltage, v_g. The more negative v_g, the higher the resistance. The automatic-gain-control circuit is designed such that it keeps changing v_g in the right direction until the oscillation amplitude reaches the desired level. Thus, if for some reason the peak value of the sine wave starts exceeding this level, v_g is made negative until the peak value is brought back to the previous level, while at the same time the amplifier gain reaches the new critical value. Because the FET can be used as a variable resistor only for small values of drain-to-source voltages, a resistor (R_s) is usually inserted in series with it, to boost the voltage across R_a and hence the voltage at the output.

The automatic-gain-control circuit shown in Fig. 15-7c provides the required dc voltage for changing the resistance of the FET. Two inputs drive this circuit, the output of the oscillator; v_o, and the reference voltage, V_R. Consequently, the current i is the sum of a half-wave-rectified sine wave and a constant. *In the steady state and under closed-loop conditions, the average value of this current must be zero.* (If this were not the case, v_g would contain a ramp component which would continue changing the gain.) Hence,

$$i_{ave} \cong \left(\frac{V_m}{\pi} - \frac{V_d}{2}\right)\frac{1}{R_1} - \frac{V_R}{R_2} = 0, \tag{15-10}$$

where V_d represents the forward voltage drop across the D1 diode. From Eq. (15-10), the peak value of the output sine wave is obtained in terms of the reference voltage:

$$V_m \cong \frac{\pi}{2}V_d + \pi\frac{R_1}{R_2}V_R. \tag{15-11}$$

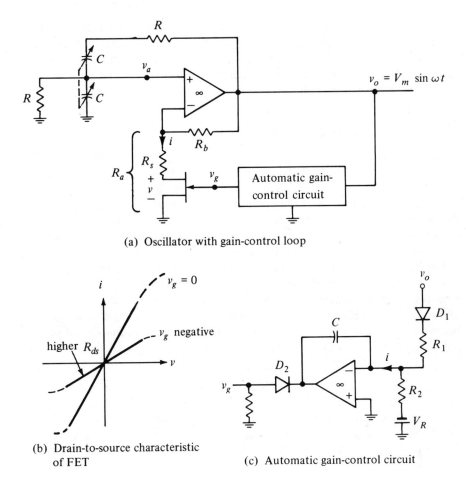

(a) Oscillator with gain-control loop

(b) Drain-to-source characteristic
of FET

(c) Automatic gain-control circuit

Fig. 15-7   FET-stabilized Wien-bridge oscillator

Note that V_m is independent of K. If for any reason K goes up, V_m increases at first. The average value of the capacitor current, i, is no longer zero but positive. As a result, v_g gradually becomes more negative as this current is integrated, and the FET's resistance increases. This reduces the gain K, which in turn reduces V_m. Equilibrium is established once again (with a more negative gate voltage) when the average value of the current is reduced to zero, which occurs when V_m is brought back to its value before the change. It is assumed that the FET's resistance has the necessary dynamic range to correct K.

If the loop gain of the gain-control circuit is too high, the amplifier gain may be overcorrected for a sudden change in V_m. As a result, V_m may at first decrease to a value lower than the equilibrium value. This in turn is sensed and quickly counteracted. This cyclic overcorrection of K may result in considerable bouncing in the output waveform before it settles down. On the other hand, if the loop gain is too low (C too large, for example), V_m may vary slowly over a long period before it reaches the value given by Eq. (15-11).

The gate voltage, v_g, is not constant in the steady state because the capacitor current is a half-wave-rectified sine wave (with an average value of zero). Consequently, v_g has some ripple. The peak-to-peak value of this ripple can be obtained readily by calculating the change in v_g when the output sine wave is negative in value. During this interval, diode $D1$ is off, and $i = -V_R/R_2$. Hence,

$$\Delta v_g = -\frac{1}{C}\int_0^{T/2} i\,dt = \frac{V_R}{R_2 C}\frac{T}{2} = \frac{\pi V_R}{\omega R_2 C} = \frac{V_m}{\omega R_1 C}. \tag{15-12}$$

In obtaining the last form of Δv_g, use is made of Eq. (15-11) with $V_d = 0$. The ripple can be made negligible by using a large C. However, if C is chosen too large, any change in V_m results in an appreciable change in the control voltage, v_g, only after several cycles of integration. Hence, corrective action is slow. On the other hand, if C is made too small, the ripple in the control voltage may cause distortion in the steady-state sine wave. Furthermore, envelope bouncing may result when V_m is disturbed from its equilibrium value.

The diode $D2$ does not conduct until the output of the integrator becomes negative. Thus, initially, the automatic-gain-control loop is broken and v_g is held at 0 which guarantees that $K > K_0$. As oscillations grow in amplitude, the average value of i becomes positive and the output of the integrator eventually becomes negative, thereby turning on $D2$. As a result, v_g becomes negative and the gain (and hence V_m) is reduced until the average value of the current, i, is zero.

EXAMPLE 15-3

The oscillator shown in Fig. 15-7 is to be designed for a 10-V peak sine-wave output. The voltage across the FET should not exceed 0.1 V. The following data is available for the source-to-drain resistance of the FET:

$v_g = 0, \qquad R_{ds} = 100 \ \Omega;$

$v_g = -1, \qquad R_{ds} = 200 \ \Omega.$

(a) Obtain the values of R_s, R_b, and V_R.
(b) To start oscillating, $K > K_c$ where

$$K_c = 1 + \frac{R_2}{R_1} + \frac{C_1}{C_2}.$$

How much a mismatch in the capacitors can be tolerated before the circuit stops oscillating? Assume the resistors are matched.

SOLUTION

(a) The required gain for oscillation is 3. Since the desired peak output voltage is 10 V, the peak voltage across R_a is

$$\frac{10}{3}V = 3.333 \ V$$

(see Fig. 15-7). Of this, 0.1 V is taken up by the FET. Hence, the resistor in series with the FET must drop the remaining voltage. Assume that the amplitude of oscillation is to be stabilized when $v_g = -1$ V which results in $R_{ds} = 200\ \Omega$:

$$R_s = \frac{3.333 - 0.100}{0.1/R_{ds}} = \frac{3.233}{0.1/200} = 6.466\ \text{K}\Omega. \qquad Ans.$$

The resistor R_b must be twice the value of R_a so as to achieve a gain of 3. Hence,

$$R_b = 2(R_s + R_{ds}) = 2(6.466 + 0.200) = 13.332\ \text{K}\Omega. \qquad Ans.$$

Let $R_2/R_1 = 2$. Use Eq. (15-11) to obtain V_R. Assume $V_d = 0.6$ V:

$$V_R = \left(\frac{V_m}{\pi} - \frac{V_d}{2}\right)\frac{R_2}{R_1} = \left(\frac{10}{\pi} - \frac{0.6}{2}\right)2 = 5.7\ \text{V}. \qquad Ans.$$

(b) The highest value of gain that the amplifier can produce occurs when $v_g = 0$, which results in $R_{ds} = 100\ \Omega$:

$$K_{max} = 1 + \frac{R_b}{R_s + (R_{ds})_{min}} = 1 + \frac{13.332}{6.466 + 0.100} = 3.030.$$

The highest value of the critical gain occurs when the C_1/C_2 ratio is at its maximum value; that is,

$$(K_c)_{max} = 2 + \left(\frac{C_1}{C_2}\right)_{max}.$$

$(K_c)_{max}$ must be smaller than K_{max} to assure the starting of oscillations. Hence,

$$2 + \left(\frac{C_1}{C_2}\right)_{max} < 3.030,$$

$$\left(\frac{C_1}{C_2}\right)_{max} < 1.030.$$

Consequently, C_1 can at most be 3% higher than C_2.

There is also a lower limit on C_1. It is established by the minimum value of gain attainable with the given FET, namely

$$K_{min} < (K_c)_{min},$$

$$1 + \frac{R_b}{R_s + (R_{ds})_{max}} < 2 + \left(\frac{C_1}{C_2}\right)_{min},$$

$$\left(\frac{C_1}{C_2}\right)_{min} > \frac{R_b}{R_s + (R_{ds})_{max}} - 1.$$

Since $(R_{ds})_{max}$ is not given, the lower limit to the C_1/C_2 ratio cannot be calculated. If $C_1/C_2 < (C_1/C_2)_{min}$, the gain of the amplifier cannot be reduced to the critical value. Hence, the poles are left in the right half-plane. The amplitude of oscillation then becomes stabilized by the non-linearities in the amplifier characteristic and hence distortion results.

15-4 OTHER *RC* OSCILLATOR CIRCUITS

A variety of *RC* oscillators can be constructed by employing a second-order, band-pass network in the feedback loop. The basic circuit arrangement is shown in Fig. 15-8a.

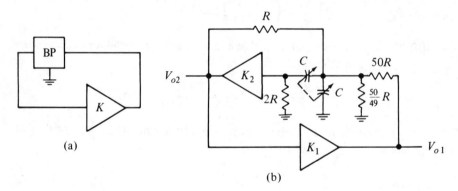

Fig. 15-8   Band-pass type oscillator circuit

The condition for oscillation is obtained from the characteristic equation which can be expressed in terms of the loop gain, LG:

$$1 - LG = 0, \tag{15-13a}$$

$$1 - KBP = 0, \tag{15-13b}$$

$$1 - K \frac{Hs}{s^2 + s\dfrac{\omega_o}{Q} + \omega_o^2} = 0; \tag{15-13c}$$

$$s^2 + s\left(\frac{\omega_o}{Q} - KH\right) + \omega_o^2 = 0. \tag{15-13d}$$

The poles can be placed on the imaginary axis by adjusting the amplifier gain to

$$K = \frac{1}{H}\frac{\omega_o}{Q} = \frac{BW}{H}, \tag{15-14}$$

where BW is the 3-dB bandwidth of the band-pass network and H is its scale factor.

The band-pass network can be *RC* or active *RC*. The Wien-bridge oscillator is a good example of an oscillator employing an *RC* band-pass network. Since the bandwidth of any *RC* band-pass network is more than twice the center frequency, not much filtering can be expected out of the BP network. On the other hand, if the BP network is an active *RC* network, the bandwidth can be made quite narrow. Consequently, any distortion introduced by the amplifier (as a result of controlling its gain, for instance) can be filtered by the BP network. An example of such a network is given in Fig. 15-8b, where the *KRC* realization discussed in Section 11-3 is used as the band-pass network. A 50:1 attenuator is included at the input of the *KRC* band-pass network, in order to obtain values for K_1 that are greater than 1. Using Eq. (11-11), the transfer function of the BP network is obtained:

$$\frac{V_{o2}}{V_{o1}} = \frac{1}{50}\frac{K_2}{RC}\frac{s}{s^2 + s\dfrac{3-K_2}{RC} + \dfrac{1}{R^2C^2}} = \frac{Hs}{s^2 + sBW + \omega_o^2}, \tag{15-15}$$

where

$$\omega_o = \frac{1}{RC}, \qquad BW = (3-K_2)\omega_o, \qquad H = \frac{1}{50}\left(3 - \frac{BW}{\omega_o}\right)\omega_o.$$

The gain at the center frequency, $s = j\omega_o$, is

$$\left.\frac{V_{o2}}{V_{o1}}\right|_{s=j\omega_o} = \frac{H}{BW} = \frac{1}{50}\left(3\frac{\omega_o}{BW} - 1\right), \tag{15-16}$$

which is less than unity for $\omega_o/BW < 17$. Hence, K_1 must have a value greater than unity so that the necessary condition for oscillation is satisfied, i.e., the loop gain becomes unity at $\omega = \omega_o$. The value of K_1 is therefore given by

$$K_1 = \frac{BW}{H} = \frac{50}{3\dfrac{\omega_a}{BW} - 1} \qquad \left(\frac{\omega_o}{BW} < 17\right). \tag{15-17}$$

For example, for $(\omega_o/BW) = 10$, which corresponds to BP pole-Q's of 10, K_1 is $(50/29)$. This is a reasonable value for the gain. If Q's higher than 17 are desired, more attenuation must be provided at the band-pass input in order to insure that $K_1 \geq 1$.

In this oscillator circuit, the gain K_2 is used to set the Q's of the BP poles,

$$K_2 = 3 - \frac{1}{Q}.$$

The frequency of oscillation is varied by varying the ganged capacitors. The output is taken at V_{o2}. Since $Q = \omega_o/BW$ is kept constant as ω_o is varied, the same amount of filtering is done regardless of ω_o. If Q is taken as 10 (which is 30 times higher than that of the BP circuit of the Wien-bridge oscillator), a sine wave of very little distortion results at V_{o2} even though the sine wave at V_{o1} may not be

stabilize the amplitude of oscillations, a combination of nonlinear
tic gain control can be applied to the gain of K_1.
... $_{1}$g. 1$_{2}$-8b, instead of the KRC band-pass circuit, the $-KRC$ or the infinite-
gain RC band-pass circuits (discussed in Sections 10-6 and 10-7) can be used.
Thus a variety of RC oscillators can be designed.

EXAMPLE 15-4

(a) In Fig. 15-9, a third-order high-pass RC network is used in the feedback
loop. Show that the circuit becomes an oscillator if K is properly chosen.
(The circuit is known as a phase-shift oscillator.)

(b) Repeat (a) if the feedback network is a third-order RC low-pass network, i.e., the R's and C's are interchanged in Fig. 15-9.

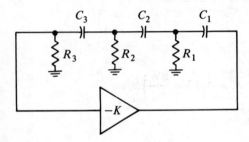

Fig. 15-9

SOLUTION

(a) Let

$$\frac{s^3}{(s + \alpha_1)(s + \alpha_2)(s + \alpha_3)} \qquad (0 < \alpha_1 < \alpha_2 < \alpha_3)$$

represent the transfer function of the RC network. The α's are functions
of the R's and the C's. Use $1 - LG = 0$ to obtain the characteristic
equation:

$$1 - \frac{(-K)s^3}{(s + \alpha_1)(s + \alpha_2)(s + \alpha_3)} = 0,$$

$$s^3\underbrace{(1 + K)}_{a_3} + s^2\underbrace{(\alpha_1 + \alpha_2 + \alpha_3)}_{a_2} + s\underbrace{(\alpha_1\alpha_2 + \alpha_1\alpha_3 + \alpha_2\alpha_3)}_{a_1} + \underbrace{\alpha_1\alpha_2\alpha_3}_{a_0} = 0.$$

To put a pair of poles on the imaginary axis, adjust K so that $a_2 a_1 = a_3 a_0$ is satisfied:

$$(\alpha_1 + \alpha_2 + \alpha_3)(\alpha_1\alpha_2 + \alpha_1\alpha_3 + \alpha_2\alpha_3) = (1 + K)\alpha_1\alpha_2\alpha_3,$$

$$K = (\alpha_1 + \alpha_2 + \alpha_3)\left(\frac{1}{\alpha_1} + \frac{1}{\alpha_2} + \frac{1}{\alpha_3}\right) - 1. \qquad Ans.$$

(b) With

$$\frac{\alpha_1 \alpha_2 \alpha_3}{(s + \alpha_1)(s + \alpha_2)(s + \alpha_3)} \qquad (0 < \alpha_1 < \alpha_2 < \alpha_3)$$

as the transfer function of the feedback network, the characteristic equation becomes

$$s^3 + s^2(\alpha_1 + \alpha_2 + \alpha_3) + s(\alpha_1 \alpha_2 + \alpha_1 \alpha_3 + \alpha_2 \alpha_3) + \alpha_1 \alpha_2 \alpha_3 (1 + K) = 0,$$

which has imaginary-axis poles for the same value of K as in (a).

15-5 OSCILLATION WITH LESS-THAN-UNITY GAIN

In order to have oscillation, the loop gain must be unity for some s on the imaginary axis, for example, $s = j\omega_o$. Generally, the RC feedback network attenuates the signal at $\omega = \omega_o$ and the amplifier makes up for the loss, so that the loop gain is unity. As is discussed in Section 5-6, it is possible to construct RC networks which have a gain greater than unity. When such an RC circuit is used in the feedback loop, oscillations can be generated with amplifiers having less-than-unity gain. An example of such an oscillator is given in Fig. 15-10a.

The loop gain is given by

$$LG = K \frac{6(sRC)^2 + 5(sRC) + 1}{(sRC)^3 + 6(sRC)^2 + 5(sRC) + 1}. \qquad (15\text{-}18)$$

Hence, the characteristic equation becomes

$$1 - LG = 0,$$

$$(sRC)^3 + 6(sRC)^2(1 - K) + 5(sRC)(1 - K) + (1 - K) = 0, \qquad (15\text{-}19)$$

which has roots on the imaginary axis when K is adjusted to

$$K = \frac{29}{30}. \qquad (15\text{-}20)$$

The oscillation frequency is

$$\omega_o = \frac{1}{\sqrt{6}RC}. \qquad (15\text{-}21)$$

When all the impedances in the network of Fig. 15-10a are multiplied by sRC and then s is replaced by $1/(sR^2C^2)$, the network of Fig. 15-10b results. Because impedance scaling does not change the loop gain, the characteristic equation of the network of Fig. 15-10b is readily obtained from Eq. (15-19) by transforming the complex-frequency variable only:

$$\left(\frac{1}{sRC}\right)^3 + 6\left(\frac{1}{sRC}\right)^2(1 - K) + 5\left(\frac{1}{sRC}\right)(1 - K) + (1 - K) = 0. \qquad (15\text{-}22)$$

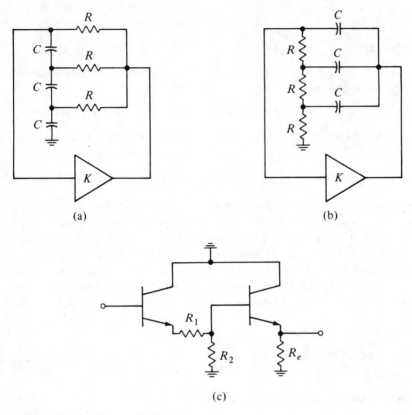

(a)

(b)

(c)

Fig. 15-10   Oscillator with $K < 1$

Hence, the condition for oscillation and the oscillation frequency are given by

$$K = \frac{29}{30}, \qquad \omega_o = \frac{\sqrt{6}}{RC}. \tag{15-23}$$

The amplifier can be constructed using the Darlington pair shown in Fig. 15-10c. The R_2/R_1 can be adjusted to provide the necessary reduction in gain from unity.

15-6 FREQUENCY CONTROL WITHOUT GANGING

In order to vary the frequency of oscillation, the poles must be moved along the imaginary axis. In all the circuits discussed so far, the imaginary part of the poles is dependent upon the product of the R's and C's of the feedback network while the real part is dependent upon their ratios and the amplifier gain K. Consequently, K is adjusted to make the real part of the poles zero. The capacitors or resistors are ganged so that the real part is maintained at zero while the imaginary part is varied. This approach requires that the ganged components track well over the entire range of variable frequencies.

By introducing an additional element which affects only the imaginary part of the poles, the frequency of oscillation can be varied without changing the critical-gain requirement. A step-by-step development of such a network is shown in Fig. 15-11.

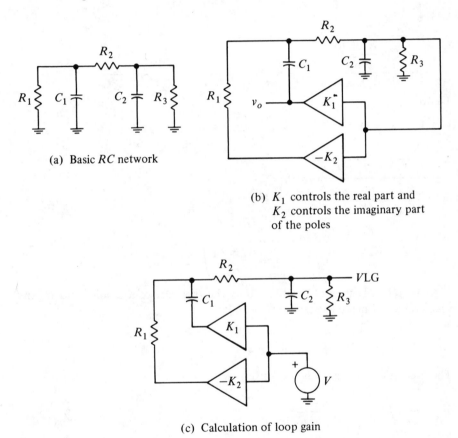

(a) Basic RC network

(b) K_1 controls the real part and K_2 controls the imaginary part of the poles

(c) Calculation of loop gain

Fig. 15-11 Variable-frequency control through K_2

The network shown in Fig. 15-11a is used as the basic two-pole RC network. To obtain complex poles, band-pass type feedback is used. As shown in Fig. 15-11b, the K_1 amplifier is used for this purpose, thus affecting only the co-efficient of the s-term in the characteristic equation (the pole loci form a circle as a function of K_1). To move the poles in a constant-α manner (with $\alpha = 0$), low-pass type feedback is used through the $-K_2$ amplifier. The gain K_2 affects only the constant coefficient in the characteristic equation. The condition for oscillation and the oscillation frequency is obtained by calculating the loop gain and then forming the characteristic equation. By inspection of Fig. 15-11c,

$$LG = K_1 BP + (-K_2)LP$$

$$= \frac{K_1}{R_2 C_2} \frac{s}{s^2 + s \dfrac{\omega_o}{Q} + \omega_o^2} - \frac{K_2}{R_1 R_2 C_1 C_2} \frac{1}{s^2 + s \dfrac{\omega_o}{Q} + \omega_o^2},$$

where

$$\omega_o = \sqrt{\frac{1 + \dfrac{R_1 + R_2}{R_3}}{R_1 R_2 C_1 C_2}}, \qquad \frac{\omega_o}{Q} = \frac{1}{R_1 C_1} + \frac{1}{R_2 C_2} + \frac{1}{R_2 C_1} + \frac{1}{R_3 C_2}.$$

Hence,

$$1 - LG = 0,$$

$$1 - \frac{\dfrac{K_1}{R_2 C_2} s - \dfrac{K_2}{R_1 R_2 C_1 C_2}}{s^2 + s \dfrac{\omega_o}{Q} + \omega_o^2} = 0,$$

$$s^2 + s\left(\frac{\omega_o}{Q} - \frac{K_1}{R_2 C_2}\right) + \omega_o^2\left(1 + \frac{K_2}{1 + \dfrac{R_1 + R_2}{R_3}}\right) = 0. \qquad (15\text{-}24)$$

As Eq. (15-24) shows, K_1 affects only the real part of the poles and K_2 the magnitude. To make the real part zero, let

$$K_1 = R_2 C_2 \frac{\omega_o}{Q} = 1 + \frac{R_2}{R_3} + \frac{C_2}{C_1}\left(1 + \frac{R_2}{R_1}\right). \qquad (15\text{-}25)$$

The oscillation frequency is then given by

$$\omega_{os} = \omega_o \sqrt{1 + \frac{K_2}{1 + \dfrac{R_1 + R_2}{R_3}}}. \qquad (15\text{-}26)$$

Thus, ω_{os} can be varied by controlling the gain of the K_2 amplifier. No ganging of components is necessary. The resistor R_3 can be made part of the gain-producing resistor in the K_2 amplifier. When equal resistors and equal capacitors are used, the critical-gain and the oscillation-frequency expressions simplify to

$$K_1 = 4, \qquad \omega_{os} = \frac{\sqrt{3}}{RC}\sqrt{1 + \frac{K_2}{3}}. \qquad (15\text{-}27)$$

The output is taken at the output of the K_1 amplifier. The amplitude at this point can be stabilized by any one, or a combination, of the schemes presented earlier in this chapter. Any distortion present in the output is filtered by the transfer function

$$\frac{-K_2 \dfrac{\omega_o}{\sqrt{3}} s}{s^2 + s\dfrac{\omega_o}{Q} + \omega_o^2\left(1 + \dfrac{K_2}{3}\right)}, \tag{15-28}$$

before it appears at the output of the K_2 amplifier. Since $Q = \frac{1}{4}\sqrt{3 + K_2}$, the K_2 amplifier output has less distortion. However, its amplitude varies with K_2.

Note that to affect a 10:1 change in the frequency of oscillation, K_2 must be changed more than 100:1, thus requiring a large dynamic range at the output of the amplifier. Furthermore, the larger K_2, the narrower the bandwidth of the K_2 amplifier, and hence the more its frequency limitations affect the operation of the oscillator. For these reasons, this scheme of frequency control works best for low-frequency oscillators with less than 10:1 frequency variation.

15-7 SUMMARY

Oscillators are networks which produce a pair of imaginary-axis poles. The condition for oscillation and the frequency of oscillation are obtained from the equation: $1 - LG = 0$. In a second-order equation, the condition for oscillation is that relationship among the elements which results in a zero coefficient for the s-term. The frequency of oscillation in rad/s is obtained by taking the square root of the constant term divided by the coefficient of the s^2-term. The most common method of making the coefficient of the s-term zero is through band-pass type feedback employing a positive-gain amplifier. The most common method of varying the frequency of oscillation is by means of ganged capacitors.

Since it is practically impossible to keep the poles on the imaginary axis over an extended period, various schemes are used to provide sustained oscillations. One scheme relies on nonlinear gain characteristics. For low signal values the gain is higher than the critical value, thus forcing growing oscillations which eventually drive the amplifier into its nonlinear region. Equilibrium is established at a signal level which makes the gain at the fundamental frequency equal to the critical gain. The more pronounced is the nonlinearity, the less is the variation in the amplitude of oscillation and the higher is the distortion in the sine wave. A much better control of poles is achieved through automatic gain control. In this scheme, the output amplitude is continuously monitored and compared to a fixed level. If for any reason a change in amplitude occurs, the gain of the amplifier is increased or reduced until the amplitude is returned to its former level. Under equilibrium conditions, operation is in the linear range of the amplifier, and hence distortion is very low. Almost all variable-frequency oscillators use automatic gain control. The most popular oscillator is the Wien-bridge oscillator. Either a lamp or a field-effect transistor is used to control the gain of the amplifier.

PROBLEMS

15-1 (a) Obtain the transfer function of the feedback network in the Wien-
          bridge oscillator. Assume equal R's and equal C's.
      (b) Sketch the magnitude and phase of this transfer function vs. ωRC.
          Designate the frequency of peaking and the magnitude and phase at
          this frequency.
      (c) How much more is the third harmonic attenuated relative to the
          fundamental?

15-2 For what value of K do the circuits shown in Fig. 15-12 oscillate? What
      is the frequency of oscillation?

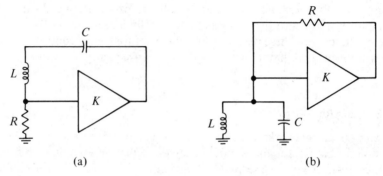

(a)   (b)

Fig. 15-12

15-3 Show that if the gain is properly adjusted, the circuit of Fig. 15-13 oscil-
      lates.

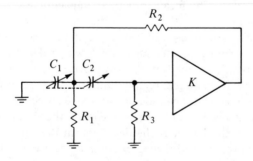

Fig. 15-13

15-4 (a) For the circuit shown in Fig. 15-14, obtain $v_o(t)$. Initial conditions on
          the capacitors are zero.
      (b) Can this circuit be used as a variable-frequency, constant-amplitude
          oscillator? Explain.

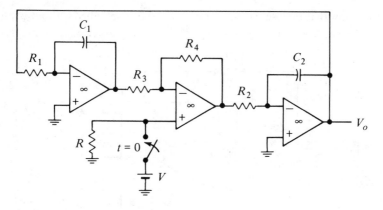

Fig. 15-14

15-5 (a) For what value of K will the circuit of Fig. 15-15 oscillate?
(b) Show that the oscillation frequency can be varied by changing R_d alone.
(c) Show that output harmonic content (due to nonlinearities in the $-K$ amplifier) diminishes as the frequency of operation is increased.
(d) Implement an amplitude-stabilizing scheme for this oscillator.
(e) How would you control the frequency with an external signal?

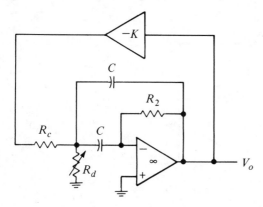

Fig. 15-15

15-6 (a) Obtain the condition for oscillation and the frequency of oscillation for the network shown in Fig. 15-16.

(b) Show that the oscillation frequency can be controlled without ganging.

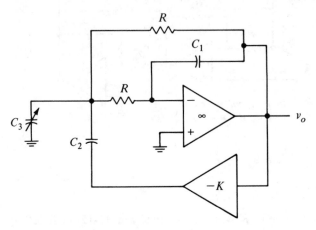

Fig. 15-16

15-7 Derive Eq. (15-28).

15-8 In the network of Fig. 15-17, K_1 can be used to put the poles on the real axis whereas $-K_2$ can be used to move the poles along the imaginary axis. Show this.

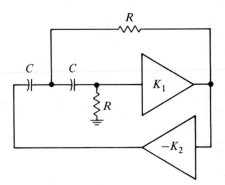

Fig. 15-17

15-9 (a) For the network shown in Fig. 15-18, find the condition for oscillation and the frequency of oscillation.

(b) Show that the network can be used as a three-phase oscillator. (The three outputs are 120° apart in phase.)

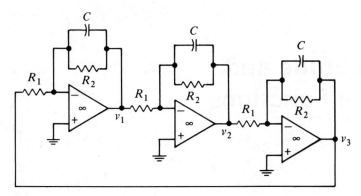

Fig. 15-18

16

Magnitude and
Phase Functions

In previous chapters, various passive and active realization techniques are developed and discussed. The presupposition is made that the transfer function is known. However, the design specifications are generally given as desired magnitude and phase characteristics. It therefore becomes necessary to translate this information into an acceptable transfer function. This can be achieved from the prescribed magnitude-squared function. The desired phase characteristics can then be approached by additional phase-correction networks. In this chapter, the properties of magnitude-squared functions and phase functions are discussed. A procedure is given for the construction of the transfer function from the magnitude characteristic.

16-1 THE DESIGN PROCEDURE

Consider a system $G(s)$ with left half-plane poles. When sinusoidally excited, this system gives a steady-state response that is sinusoidal also.

As Fig. 16-1 indicates, the input sine wave is modified in two aspects. First, its amplitude is changed from V_m to $V_m |G(j\omega)|$. Second, its phase is modified by $\theta(\omega)$ which is the angle of $G(j\omega)$. The central problem of synthesis is the following: Can a function, $G(s)$, be found such that the output has the desired magnitude and phase characteristics for $s = j\omega$? In other words, if $|G(j\omega)|$ and $\theta(\omega)$ are specified, how does one obtain $G(s)$?

486

The problem of finding the appropriate $G(s)$ from an arbitrarily specified $|G(j\omega)|$ and $\theta(\omega)$ characteristics is not easy. It is very seldom that a rational function can be found which precisely fits a given $|G(j\omega)|$-vs.-ω curve, even when the $\theta(\omega)$-vs.-ω curve is ignored. More often, the effort is concentrated on finding a $G(s)$ which for $s = j\omega$ gives a reasonable approximation to the desired $|G(j\omega)|$-vs.-ω curve. When the attention is focused solely on the magnitude characteristic, the resulting phase characteristic may depart considerably from the desired one. When this is the case, phase-equalizers can be introduced to correct the phase as much as possible without affecting the magnitude.

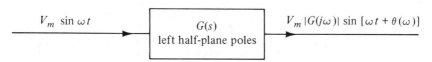

Fig. 16-1   Sinusoidal steady-state response

Realism dictates a constant give-and-take approach between the given magnitude and phase specifications and the selected rational function which is to produce the desired characteristics. While it is possible to come very close to a specified magnitude function by choosing an approximating function of sufficiently high order, practical considerations such as cost, size, ease of adjustment, etc., cannot be ignored. If a much simpler network is obtained by slightly relaxing the specifications on the magnitude and phase of $G(j\omega)$, then the trade-off between accuracy and simplicity may be worthwhile.

The rational function that approximates the desired frequency-response characteristics must be realizable. Above all, this means left half-plane poles for $G(s)$. Additional constraints are imposed on $G(s)$ if it is to be realized passively. For example, if only resistors and capacitors are to be used in the realization, then the poles of $G(s)$ must be further confined to the negative real axis.

After a suitable $G(s)$ is obtained, it can be realized passively or actively by one of the synthesis procedures outlined in the preceding chapters. In general, for frequencies less than 100 Hz, active realizations become indispensable. On the other hand, for frequencies greater than 1 MHz, passive realizations may offer the only practical solution. In between these two extremes of frequency either kind of realization may be used.

The final and most important step in the design of filters is the actual construction and testing of the network. Does it work? How close is the measured response to the desired response? Are the differences between the actual and the desired results acceptable or should a fresh start be made with a better approximating function?

The flow chart shown in Fig. 16-2 summarizes the important steps in the design procedure.

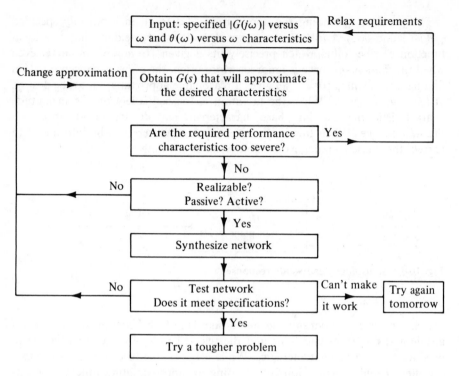

Fig. 16-2   The flow chart for filter design

16-2 THE MAGNITUDE FUNCTION

To study the behavior of the magnitude function, $G(s)$ is written in factored form first:

$$G(s) = H \frac{(s + \alpha_{z_1})(s + \alpha_{z_2}) \cdots \left(s^2 + s \dfrac{\omega_{oz_1}}{Q_{z_1}} + \omega_{oz_1}^2\right)\left(s^2 + s \dfrac{\omega_{oz_2}}{Q_{z_2}} + \omega_{oz_2}^2\right) \cdots}{(s + \alpha_{p_1})(s + \alpha_{p_2}) \cdots \left(s^2 + s \dfrac{\omega_{op_1}}{Q_{p_1}} + \omega_{op_1}^2\right)\left(s^2 + s \dfrac{\omega_{op_2}}{Q_{p_2}} + \omega_{op_2}^2\right) \cdots}.$$

$$(16\text{-}1)$$

Then, $s = j\omega$ is substituted:

$$G(j\omega) = H \frac{\begin{aligned}&(j\omega + \alpha_{z_1})(j\omega + \alpha_{z_2}) \cdots \left(-\omega^2 + j\omega \frac{\omega_{oz_1}}{Q_{z_1}} + \omega_{oz_1}^2\right)\\ &\qquad\qquad \times \left(-\omega^2 + j\omega \frac{\omega_{oz_2}}{Q_{z_2}} + \omega_{oz_2}^2\right) \cdots\end{aligned}}{\begin{aligned}&(j\omega + \alpha_{p_1})(j\omega + \alpha_{p_2}) \cdots \left(-\omega^2 + j\omega \frac{\omega_{op_1}}{Q_{p_1}} + \omega_{op_1}^2\right)\\ &\qquad\qquad \times \left(-\omega^2 + j\omega \frac{\omega_{op_2}}{Q_{p_2}} + \omega_{op_2}^2\right) \cdots\end{aligned}}.$$

$$(16\text{-}2)$$

The square of the magnitude of $G(j\omega)$ is

$$
|G(j\omega)|^2 = H^2 \frac{(\alpha_{z_1}^2 + \omega^2)(\alpha_{z_2}^2 + \omega^2) \cdots \left[(\omega_{oz_1}^2 - \omega^2)^2 + \omega^2 \frac{\omega_{oz_1}^2}{Q_{z_1}^2}\right]}{(\alpha_{p_1}^2 + \omega^2)(\alpha_{p_2}^2 + \omega^2) \cdots \left[(\omega_{op_1}^2 - \omega^2)^2 + \omega^2 \frac{\omega_{op_1}^2}{Q_{p_1}^2}\right]} \cdots
$$
$$
\times \left[(\omega_{oz_2}^2 - \omega^2)^2 + \omega^2 \frac{\omega_{oz_2}^2}{Q_{z_2}^2}\right] \cdots
$$
$$
\times \left[(\omega_{op_2}^2 - \omega^2)^2 + \omega^2 \frac{\omega_{op_2}^2}{Q_{p_2}^2}\right] \cdots
$$

(16-3)

As Eq. (16-3) shows, the magnitude-squared function is a rational function of the variable ω^2. Both the numerator and the denominator polynomial vary smoothly with ω, that is, the first derivative of either polynomial with respect to ω is continuous. Since $G(s)$ does not have any imaginary-axis poles, the denominator polynomial is never zero; hence $|G(j\omega)|^2$ is never infinite. However, $|G(j\omega)|^2$ may become zero. For example,

$$
|G(j\omega_{oz_1})|^2 = 0 \quad \text{if} \quad Q_{z_1} = \infty;
$$

that is, the magnitude-squared function becomes zero only at those frequencies which coincide with an imaginary-axis zero of $G(s)$. These properties of the magnitude-squared function are illustrated in Fig. 16-3a. The exact shape of the response depends upon the location of the poles and zeros of $G(s)$.

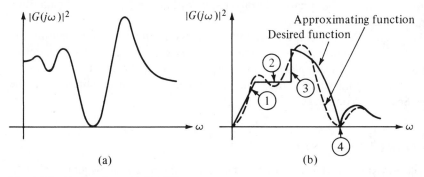

Fig. 16-3 Properties of $|G(j\omega)|^2$

The magnitude-squared function shown by the solid curve in Fig. 16-3b could not possibly have come from the rational function described by Eq. (16-3). It is impossible to obtain the piecewise-linear characteristic shown by line ①. Neither is it possible to obtain the piecewise-constant line ②. The jump at ③ cannot be produced. Neither can the discontinuity in the first derivative occurring at ④ be implemented. In other words, there is no way of placing zeros and left half-plane poles so as to generate the magnitude-squared characteristics shown in Fig. 16-3b. Such a $|G(j\omega)|^2$ specification can at best be approached by an approximation function (shown by the dashed curve) which is described by Eq. (16-3). How close can the approximation function come to the desired function? This depends upon the cost, size, complexity, etc., of the final network. Once a decision is made on the allowable number of poles and zeros for the approximating $G(s)$, then these poles and zeros can be moved around until a satisfactory magnitude response is obtained. Usually, the digital computer is used to minimize the error between the desired and the approximation function according to some chosen error criterion.

The magnitude of $G(j\omega)$ is not affected if $G(s)$ is multiplied by an all-pass function; that is,

$$\left| G(s) \underbrace{\left[\frac{-(s-\alpha)(s^2 - s\dfrac{\omega_o}{Q} + \omega_o^2)\cdots}{(s+\alpha)(s^2 + s\dfrac{\omega_o}{Q} + \omega_o^2)\cdots} \right]}_{\text{All-pass}} \right|_{s=j\omega} = |G(j\omega)|. \tag{16-4}$$

Because the poles (left half-plane) and zeros (right half-plane) of the all-pass function are symmetrically located with respect to the imaginary axis, the magnitude of the all-pass function does not depend on ω. (Graphically speaking, the length of the vector from each left half-plane pole to a point on the imaginary axis is equal to the length of the vector from the corresponding right half-plane zero to the same point on the imaginary axis.) Thus, the all-pass function can be used to alter the phase characteristic without affecting the magnitude characteristic.

The phase of $G(j\omega)$ can also be altered without affecting its magnitude by converting some or all left half-plane zeros of $G(s)$ into right half-plane zeros, that is, $(s - \alpha_z \pm j\beta_z)$ is replaced by $(s + \alpha_z \pm j\beta_z)$. Since the distance from either zero to the imaginary axis is the same, the magnitude characteristic remains the same. See also Problem 16-6.

16-3 FROM $|G(j\omega)|^2$ TO $G(s)$

Synthesis starts with a given $G(s)$. Therefore, the magnitude-squared characteristic that is generated to meet a specified response needs to be converted to $G(s)$. If $|G(j\omega)|^2$ is in the form of Eq. (16-3), the conversion is easy. Each factor of Eq. (16-3) is replaced by the corresponding factor of Eq. (16-1).

If $|G(j\omega)|^2$ is not in the factored form displayed by Eq. (16-3) but is given as the ratio of two polynomials in ω^2, then ω^2 is replaced by $-s^2$. Since

$$|G(j\omega)|^2 = G(j\omega)G(-j\omega), \tag{16-5}$$

it follows that

$$|G(j\omega)|^2_{\omega^2 = -s^2} = G(s)G(-s). \tag{16-6}$$

The product $G(s)G(-s)$ contains not only the desired poles and zeros of $G(s)$, but also the undesired poles and zeros arising from $G(-s)$. But the poles and zeros of $G(-s)$ are the negatives of the poles and zeros of $G(s)$. Therefore, the pole–zero pattern of $[G(s)G(-s)]$ is symmetrical with respect to the imaginary axis. There are no imaginary-axis poles and the imaginary-axis zeros are of even order. [This symmetrical distribution of the critical frequencies is automatically guaranteed because the approximation function is chosen to have the properties given by Eq. (16-3).] The problem now is to segregate the poles and zeros of $G(s)$ from those of $G(-s)$. To assure a stable system, all the left half-plane poles of $G(s)G(-s)$ are assigned to $G(s)$; this is a simple task. The segregation of the zeros, however, is not so clearcut; in fact, there is *no* unique segregation. As long as complex-conjugate zeros are paired together and all imaginary-axis zeros are equally divided, the zeros can be grouped together in many ways. For example, both right half-plane as well as left half-plane zeros can be assigned to $G(s)$. A simple solution would be to assign to $G(s)$ all the left half-plane zeros and one-half of the imaginary-axis zeros. A $G(s)$ constructed this way, i.e., a $G(s)$ *with no right half-plane zeros, is called minimum-phase* $G(s)$. The magnitude and phase of minimum-phase functions are not independent. When one is known, the other can be found. Additional independent control on the phase can be achieved only by cascading the minimum-phase $G(s)$ with an all-pass function.

EXAMPLE 16-1

A certain magnitude-squared characteristic is given by

$$|G(j\omega)|^2 = \frac{(1 - \tfrac{1}{4}\omega^2)^2}{1 + \omega^6}.$$

Obtain the corresponding $G(s)$.

SOLUTION

Method 1   Write $|G(j\omega)|^2$ as

$$|G(j\omega)|^2 = \frac{(1 - \tfrac{1}{4}\omega^2)^2}{1 + \omega^6} = \frac{(1 - \tfrac{1}{4}\omega^2)^2}{(1 + \omega^2)(1 - \omega^2 + \omega^4)}$$

$$= \frac{\tfrac{1}{16}(4 - \omega^2)^2}{(1 + \omega^2)[(1 - \omega^2)^2 + \omega^2]}. \tag{16-7}$$

Compare Eq. (16-7) with Eq. (16-3) and note that it is in standard form. Hence, by Eq. (16-1),

$$G(s) = \frac{\tfrac{1}{4}(s^2 + 2^2)}{(s + 1)(s^2 + s + 1)} = \frac{\tfrac{1}{4}(s^2 + 4)}{s^3 + 2s^2 + 2s + 1}. \qquad Ans.$$

Method 2   First obtain $G(s)G(-s)$:

$$G(s)G(-s) = |G(j\omega)|^2_{\omega^2 = -s^2} = \frac{(1 - \frac{1}{4}\omega^2)^2}{1 + \omega^6}\bigg|_{\omega^2 = -s^2} = \frac{(1 + \frac{1}{4}s^2)^2}{1 - s^6}.$$

Then, determine the six roots of $(s^6 - 1) = 0$:

$$s^6 = 1 = e^{jk2\pi} \qquad \text{where } k = -2, -1, 0, 1, 2, 3, \tag{16-8}$$

$$s = e^{j(k2\pi/6)};$$

$$s_1 = e^{-j(2\pi/3)}, \qquad s_2 = e^{-j(\pi/3)}, \qquad s_3 = e^{j0},$$

$$s_4 = e^{j(\pi/3)}, \qquad s_5 = e^{j(2\pi/3)}, \qquad s_6 = e^{j\pi}.$$

Write $G(s)G(-s)$ in factored form and then form $G(s)$ by putting together the left half-plane poles and half of the imaginary-axis zeros. Associate the minus sign in the numerator with $G(-s)$:

$$G(s)G(-s)$$

$$= \frac{-\frac{1}{16}(s^2 + 4)^2}{[s - e^{-j(2\pi/3)}][s - e^{-j(\pi/3)}][s - e^{j0}][s - e^{j(\pi/3)}][s - e^{j(2\pi/3)}][s - e^{j\pi}]}$$

$$= \frac{[-\frac{1}{4}(s^2 + 4)][\frac{1}{4}(s^2 + 4)]}{\left(s + \frac{1}{2} + j\frac{\sqrt{3}}{2}\right)\left(s - \frac{1}{2} + j\frac{\sqrt{3}}{2}\right)(s - 1)\left(s - \frac{1}{2} - j\frac{\sqrt{3}}{2}\right)};$$

$$\times \left(s + \frac{1}{2} - j\frac{\sqrt{3}}{2}\right)(s + 1)$$

$$G(s) = \frac{\frac{1}{4}(s^2 + 4)}{\left(s + \frac{1}{2} + j\frac{\sqrt{3}}{2}\right)\left(s + \frac{1}{2} - j\frac{\sqrt{3}}{2}\right)(s + 1)}$$

$$= \frac{\frac{1}{4}(s^2 + 4)}{(s^2 + s + 1)(s + 1)} = \frac{\frac{1}{4}(s^2 + 4)}{s^3 + 2s^2 + 2s + 1}. \qquad \textit{Ans.}$$

Since $G(s)$ contains no right half-plane zeros, $G(s)$ is minimum-phase.

16-4 THE PHASE FUNCTION

The phase associated with $G(j\omega)$ is obtained from Eq. (16-2):

$$\theta(\omega) = \tan^{-1}\left(\frac{\omega}{\alpha_{z_1}}\right) + \tan^{-1}\left(\frac{\omega}{\alpha_{z_2}}\right) + \cdots$$

$$+ \tan^{-1}\left(\frac{\frac{\omega\omega_{oz_1}}{Q_{z_1}}}{\omega_{oz_1}^2 - \omega^2}\right) + \tan^{-1}\left(\frac{\frac{\omega\omega_{oz_2}}{Q_{z_2}}}{\omega_{oz_2}^2 - \omega^2}\right) + \cdots$$

$$
-\left[\tan^{-1}\left(\frac{\omega}{\alpha_{p_1}}\right) + \tan^{-1}\left(\frac{\omega}{\alpha_{p_2}}\right) + \cdots \right.
$$

$$
\left. + \tan^{-1}\left(\frac{\dfrac{\omega\omega_{op_1}}{Q_{p_1}}}{\omega_{op_1}^2 - \omega^2}\right) + \tan^{-1}\left(\frac{\dfrac{\omega\omega_{op_2}}{Q_{p_2}}}{\omega_{op_2}^2 - \omega^2}\right) + \cdots \right]. \tag{16-9}
$$

The characteristics of the phase function are studied by examining the behavior of two typical terms that appear in Eq. (16-9). Consider first the phase associated with a pole on the negative real axis:

$$
\theta_p(\omega) = -\tan^{-1}\left(\frac{\omega}{\alpha_p}\right). \tag{16-10}
$$

In Fig. 16-4, Eq. (16-10) is plotted for $\alpha_p = 1$. (If $\alpha_p \neq 1$, interpret the real

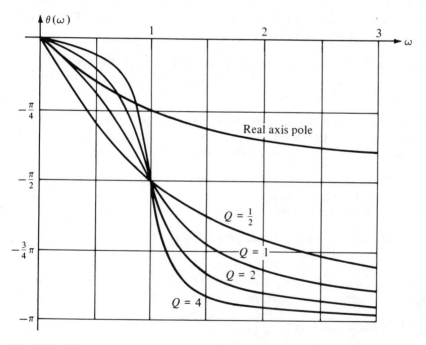

Fig. 16-4   Phase characteristics of left half-plane poles

axis as ω/α_p.) The phase varies between 0 and $-(\pi/2)$. At $\omega = \alpha_p$, the phase is $-(\pi/4)$.

Consider next the phase associated with a pair of complex-conjugate poles in the left half-plane.

$$\theta_p(\omega) = -\tan^{-1}\left(\frac{\dfrac{\omega\omega_{op}}{Q_p}}{\omega_{op}^2 - \omega^2}\right) = -\tan^{-1}\left[\frac{\dfrac{1}{Q_p}\dfrac{\omega}{\omega_{op}}}{1 - \left(\dfrac{\omega}{\omega_{op}}\right)^2}\right]. \tag{16-11}$$

Equation (16-11) is plotted in Fig. 16-4 for various values of Q and for $\omega_{op} = 1$. (If $\omega_{op} \neq 1$, interpret the real axis as ω/ω_{op}.) Regardless of the value of the pole Q, the phase varies between 0 and $-\pi$. At $\omega = \omega_{op}$, the phase is $-(\pi/2)$. The higher the Q of the poles, the more rapidly the phase varies about $\omega = \omega_{op}$. The slope of the phase at $\omega = \omega_{op}$ is $-2Q$. For $Q \geq (1/\sqrt{3}) = 0.578$, the second derivative of the phase of the complex pair of poles is zero at

$$\omega = \sqrt{\sqrt{4 - \frac{1}{Q^2}} - 1},$$

and hence the phase characteristic bends first one way and then the opposite.

The phase curves given in Fig. 16-4 are for left half-plane poles. These curves can also be used for zeros, as follows:

1. A zero on the negative real axis produces a phase that is the negative of the phase given by Eq. (16-10). Hence, its phase curve is the negative of the real-axis pole curve shown in Fig. 16-4.

2. A pair of left half-plane complex zeros produces a phase that is the negative of the phase given by Eq. (16-11). Hence, the corresponding curves are the negatives of the complex-pole curves shown in Fig. 16-4.

3. A zero on the positive real axis has a phase that equals π plus the phase given by Eq. (16-10). Hence, its phase curve is obtained by shifting the real-axis pole curve of Fig. 16-4 up π radians.

4. A pair of right half-plane complex zeros produces the same phase as a pair of left half-plane complex poles.

Finally, a zero at the origin produces $(\pi/2)$ radians of phase whereas a pair of imaginary-axis zeros at $s = \pm j\omega_o$ produce a phase that is 0 for $0 \leq \omega < \omega_o$ and π for $\omega > \omega_o$.

As a result of property #4, the functions

$$\frac{G(s)}{(s + \alpha)^2 + \beta^2} \quad \text{and} \quad G(s)[(s - \alpha)^2 + \beta^2] \tag{16-12}$$

have the same phase characteristic. The magnitude characteristics, however, are different. Thus, the magnitude response of a system can be changed without changing the phase response by converting some of the left half-plane complex poles into zeros located at mirror-image positions.

Similarly, as a result of property #3, the functions

$$\frac{G(s)}{s + \alpha} \quad \text{and} \quad -G(s)(s - \alpha) \tag{16-13}$$

have the same phase but different magnitude characteristics. (See also Problem 16-6.)

16-5 AMPLITUDE AND PHASE DISTORTION

So far attention has been focused on what happens to the magnitude and phase of a *single* sine wave as it goes through a network. Since filters are designed to process a *band* of frequencies in a predetermined way, the relationship between sine waves of different frequencies needs also to be considered. Consider, for example, an input consisting of two neighboring sine waves of frequencies ω and $(\omega + \Delta\omega)$, respectively.

Input

$$A_1 \sin \omega t + A_2 \sin[(\omega + \Delta\omega)t] \tag{16-14}$$

The system function $G(s)$ operates on this input to produce the following steady-state output.

Output

$$A_1 |G(j\omega)| \sin[\omega t + \theta(\omega)] + A_2 |G[j(\omega + \Delta\omega)]| \sin[(\omega + \Delta\omega)t + \theta(\omega + \Delta\omega)]$$

$$= A_1 |G(j\omega)| \sin\left\{\omega\left[t + \frac{\theta(\omega)}{\omega}\right]\right\} + A_2 |G[j(\omega + \Delta\omega)]|$$

$$\times \sin\left\{(\omega + \Delta\omega)\left[t + \frac{\theta(\omega + \Delta\omega)}{(\omega + \Delta\omega)}\right]\right\}. \tag{16-15}$$

Since ω and $\omega + \Delta\omega$ are very close to each other, the Taylor-series expansion of $|G[j(\omega + \Delta\omega)]|$ about ω can be approximated by

$$|G[j(\omega + \Delta\omega)]| \cong |G(j\omega)| + \frac{d|G(j\omega)|}{d\omega} \Delta\omega \qquad [G(j\omega) \neq 0]. \tag{16-16}$$

Similarly, the phase of $G(j\omega)$ can be approximated by

$$\theta(\omega + \Delta\omega) \cong \theta(\omega) + \frac{d\theta(\omega)}{d\omega} \Delta\omega \tag{16-17}$$

Also, for $|\Delta\omega| \ll \omega$,

$$\frac{1}{1 + \dfrac{\Delta\omega}{\omega}} \cong 1 - \frac{\Delta\omega}{\omega}.$$

Hence,

$$\frac{\theta(\omega + \Delta\omega)}{\omega + \Delta\omega} \cong \frac{\left[\theta(\omega) + \dfrac{d\theta(\omega)}{d\omega}\Delta\omega\right]}{\omega}\left(1 - \frac{\Delta\omega}{\omega}\right) \cong \frac{\theta(\omega)}{\omega} + \left[\frac{d\theta(\omega)}{d\omega} - \frac{\theta(\omega)}{\omega}\right]\frac{\Delta\omega}{\omega}.$$

$$(16\text{-}18)$$

The output can then be expressed as

Output

$$A_1|G(j\omega)|\sin\left[\omega\left(t + \frac{\theta(\omega)}{\omega}\right)\right] + A_2\left[|G(j\omega)| + \frac{d|G(j\omega)|}{d\omega}\Delta\omega\right]$$

$$\times \sin\left\{(\omega + \Delta\omega)\left[t + \frac{\theta(\omega)}{\omega} + \left(\frac{d\theta(\omega)}{d\omega} - \frac{\theta(\omega)}{\omega}\right)\frac{\Delta\omega}{\omega}\right]\right\}. \quad (16\text{-}19)$$

If the frequencies ω and $(\omega + \Delta\omega)$ fall in the pass-band of the filter, then the output is an undistorted replica of the input only when

$$\frac{d|G(j\omega)|}{d\omega} = 0, \tag{16-20}$$

$$\frac{d\theta(\omega)}{d\omega} - \frac{\theta(\omega)}{\omega} = 0. \tag{16-21}$$

The output then is a scaled and time-displaced version of the input. The scale factor is $|G(j\omega)|$, and the time displacement is $[\theta(\omega)/\omega]$.

Output

$$|G(j\omega)|\left\{A_1 \sin\left[\omega\left(t + \frac{\theta(\omega)}{\omega}\right)\right] + A_2 \sin\left[(\omega + \Delta\omega)\left(t + \frac{\theta(\omega)}{\omega}\right)\right]\right\}. \quad (16\text{-}22)$$

The negative of the time displacement is called phase delay and is given by

$$\tau_{ph} = -\frac{\theta(\omega)}{\omega}. \tag{16-23}$$

If the condition stated by Eq. (16-20) is not satisfied, i.e., the magnitude of $G(j\omega)$ is not constant in the pass-band, amplitude distortion occurs.

If the condition stated by Eq. (16-21) is not satisfied, phase distortion occurs. Equation (16-21) is satisfied only when $\theta(\omega) = k\omega$, where k is a constant. Thus, the phase must vary linearly with frequency throughout the pass-band if phase distortion is not to occur. (This corresponds to a constant phase delay in the pass-band.) A linear phase variation is impossible to achieve, as can be seen by referring to the phase characteristics given by Eq. (16-9) and by studying the phase curves displayed in Fig. 16-4. At best, the phase can be made *approximately linear* over a band of frequencies.

The negative of the derivative of the phase of $G(j\omega)$ with respect to ω is called the group delay (a more localized indication of delay), and it is given by

$$\tau_{gr} = -\frac{d\theta(\omega)}{d\omega}. \qquad (16\text{-}24)$$

As Eq. (16-21) shows, the group delay must be equal to the phase delay if phase distortion is not to occur; or, stated differently, the group delay must be constant. Whereas the phase delay is a function involving inverse-tangent functions, the group delay (because of the differentiation) is a rational function in ω^2. Therefore, it is mathematically easier to obtain an approximation to a constant group delay than to a constant phase delay.

It should be mentioned that not all filters are designed to approximate constant-magnitude and linear phase characteristics. For example, an integrator must produce a magnitude characteristic that varies inversely with ω, and its phase must be constant $(-\pi/2)$.

EXAMPLE 16-2

(a) Obtain the phase and group delays associated with a pair of complex-conjugate poles.
(b) For ω small, obtain approximate expressions for the phase and group delays.
(c) Plot the exact and the approximate expressions for the delay functions.

SOLUTION

(a) The phase associated with the poles is given by Eq. (16-11):

$$\theta_p(\omega) = -\tan^{-1}\left[\frac{\dfrac{1}{Q}\left(\dfrac{\omega}{\omega_o}\right)}{1 - \left(\dfrac{\omega}{\omega_o}\right)^2}\right].$$

The phase delay is

$$\tau_{ph} = -\frac{\theta_p(\omega)}{\omega} = \frac{1}{\omega}\tan^{-1}\left[\frac{\dfrac{1}{Q}\left(\dfrac{\omega}{\omega_o}\right)}{1 - \left(\dfrac{\omega}{\omega_o}\right)^2}\right]. \qquad Ans. \qquad (16\text{-}25)$$

The group delay is

$$
\tau_{gr} = -\frac{d\theta_p(\omega)}{d\omega} = \frac{d}{d\omega} \left[\frac{\dfrac{1}{Q}(\omega/\omega_o)}{1 - (\omega/\omega_o)^2} \right]
$$

$$
= \frac{1}{\omega_o Q} \left[\frac{1 + \left(\dfrac{\omega}{\omega_o}\right)^2}{1 - \left(\dfrac{\omega}{\omega_o}\right)^2 \left(2 - \dfrac{1}{Q^2}\right) + \left(\dfrac{\omega}{\omega_o}\right)^4} \right]. \qquad Ans. \qquad (16\text{-}26)
$$

(b) Obtain an approximate expression for τ_{ph} by using the first two terms of the Maclaurin-series expansion of the inverse-tangent function,

$$
\tan^{-1} x \cong x - \frac{x^3}{3} \qquad (x^2 < 1):
$$

$$
\tau_{ph} = \frac{1}{\omega} \left\{ \tan^{-1} \left[\frac{\dfrac{1}{Q}\left(\dfrac{\omega}{\omega_o}\right)}{1 - \left(\dfrac{\omega}{\omega_o}\right)^2} \right] \right\}
$$

$$
\cong \frac{1}{\omega} \left\{ \frac{\dfrac{1}{Q}\left(\dfrac{\omega}{\omega_o}\right)}{1 - \left(\dfrac{\omega}{\omega_o}\right)^2} - \frac{1}{3}\left[\frac{\dfrac{1}{Q}\left(\dfrac{\omega}{\omega_o}\right)}{1 - \left(\dfrac{\omega}{\omega_o}\right)^2} \right]^3 \right\}.
$$

Simplify further by using:

$$
\frac{1}{1 - \left(\dfrac{\omega}{\omega_o}\right)^2} \cong 1 + \left(\dfrac{\omega}{\omega_o}\right)^2 \qquad \left(\dfrac{\omega}{\omega_o}\right) \ll 1,
$$

and by keeping only first-order terms:

$$
\tau_{ph} \cong \frac{1}{Q\omega_o}\left[1 + \left(1 - \frac{1}{3Q^2}\right)\left(\frac{\omega}{\omega_o}\right)^2 \right]. \qquad Ans. \qquad (16\text{-}27)
$$

Similarly, approximate τ_{gr}:

$$\tau_{gr} = \frac{1}{Q\omega_o} \left[\frac{1 + \left(\dfrac{\omega}{\omega_o}\right)^2}{1 - \left(2 - \dfrac{1}{Q^2}\right)\left(\dfrac{\omega}{\omega_o}\right)^2 + \left(\dfrac{\omega}{\omega_o}\right)^4} \right]$$

$$\cong \frac{1}{Q\omega_o}\left[1 + \left(\frac{\omega}{\omega_o}\right)^2\right]\left[1 + \left(2 - \frac{1}{Q^2}\right)\left(\frac{\omega}{\omega_o}\right)^2\right] \qquad \left(\frac{\omega}{\omega_o}\right) \ll 1$$

$$\cong \frac{1}{Q\omega_o}\left[1 + \left(3 - \frac{1}{Q^2}\right)\left(\frac{\omega}{\omega_o}\right)^2\right]. \qquad \textit{Ans.} \tag{16-28}$$

(c) To normalize the delay at $\omega = 0$ to unity, multiply by $Q\omega_o$ the exact and the approximate expressions for the phase and group delay functions:

$$\tau_{ph}Q\omega_o = Q\left(\frac{\omega_o}{\omega}\right)\tan^{-1}\left[\frac{\dfrac{1}{Q}\left(\dfrac{\omega}{\omega_o}\right)}{1 - \left(\dfrac{\omega}{\omega_o}\right)^2}\right] \tag{16-29a}$$

$$\cong 1 + \left(1 - \frac{1}{3Q^2}\right)\left(\frac{\omega}{\omega_o}\right)^2; \tag{16-29b}$$

$$\tau_{gr}Q\omega_o = \frac{1 + \left(\dfrac{\omega}{\omega_o}\right)^2}{1 - \left(2 - \dfrac{1}{Q^2}\right)\left(\dfrac{\omega}{\omega_o}\right)^2 + \left(\dfrac{\omega}{\omega_o}\right)^4} \tag{16-30a}$$

$$\cong 1 + 3\left(1 - \frac{1}{3Q^2}\right)\left(\frac{\omega}{\omega_o}\right)^2. \tag{16-30b}$$

In Fig. 16-5, the exact and approximate expressions (normalized) for the phase and group delays are plotted against (ω/ω_o) with Q as a parameter. The dotted curves represent the approximate expressions. Also plotted are the delays associated with a first-order pole at $s = -\alpha$:

$$\alpha\tau_{ph} = \frac{\alpha}{\omega}\tan^{-1}\left(\frac{\omega}{\alpha}\right) \tag{16-31a}$$

$$\cong 1 - \frac{1}{3}\left(\frac{\omega}{\alpha}\right)^2; \tag{16-31b}$$

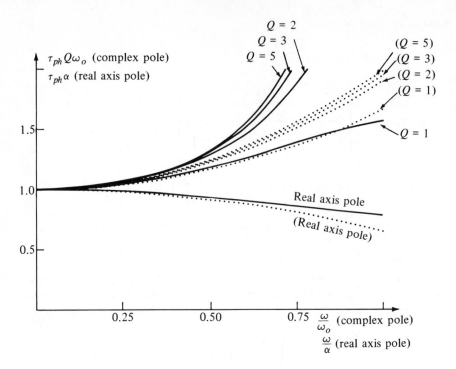

Fig. 16-5a Phase delay

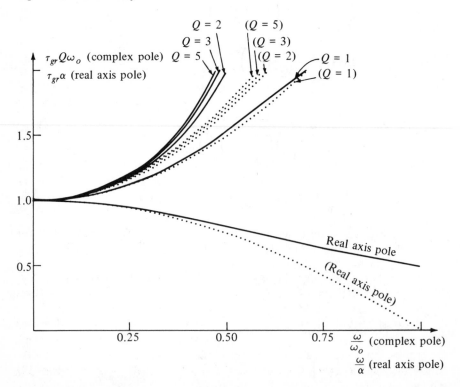

Fig. 16-5b Group delay

500

$$\alpha\tau_{gr} = \cfrac{1}{1 + \left(\cfrac{\omega}{\alpha}\right)^2} \qquad\qquad\qquad (16\text{-}32\text{a})$$

$$\cong 1 - \left(\frac{\omega}{\alpha}\right)^2. \qquad\qquad\qquad (16\text{-}32\text{b})$$

Note that all delay functions have zero slope at $\omega = 0$ and therefore start out flat. For $Q = (1/\sqrt{3})$, the second derivatives of the delays with respect to ω are also zero at $\omega = 0$, and therefore a nearly constant delay is maintained over a wider range of frequencies. By cascading second-order sections, some with Q's greater than $(1/\sqrt{3})$ and some with Q's less than $(1/\sqrt{3})$, the delays can be kept fairly constant over a wide range of frequencies. Delays associated with zeros are the negatives of the delays associated with poles. Hence, phase and group advances take place.

16-6 SUMMARY

Filter design starts with a frequency-response specification; i.e., the $j\omega$-axis behavior of the desired network is prescribed. Within given tolerances, these specifications must be met by a magnitude-squared function and by a group-delay function, both of which are rational functions of ω^2.

The next step in the filter-design procedure is the construction of $G(s)$. By substituting $-s^2$ for ω^2, $G(s)G(-s)$ is obtained from the magnitude-squared function. The poles and zeros of the resulting function are found next. $G(s)$ is then constructed out of the left half-plane poles and zeros and out of one-half of the imaginary-axis zeros. The resulting $G(s)$ is called minimum-phase because it has no right half-plane zeros. When one or more right half-plane zeros are associated with $G(s)$, a nonminimum-phase function results. Using the $G(s)$'s thus obtained, passive and active realizations can be developed by the procedures given in previous chapters.

The phase function, being the sum of inverse-tangent functions, is more difficult to deal with than the magnitude-squared function. It is generally easier to consider the group-delay function, which is the negative of the derivative of the phase function with respect to ω. The group-delay function is a rational function in ω^2. If phase distortion is to be kept to a minimum in the pass-band, the group-delay function must be as nearly constant as possible over the desired bandwidth. This corresponds to having as nearly a linear phase as possible in the frequency range of operation.

PROBLEMS

16-1 Show that

$$G(s) = H\frac{(-s)^n + a_{n-1}(-s)^{n-1} + \cdots + a_1(-s) + a_0}{s^n + a_{n-1}s^{n-1} + \cdots + a_1s + a_0}$$

is all-pass.

16-2 (a) Show that any nonminimum-phase function can be expressed as the product of a minimum-phase function with an all-pass function.

      (b) Apply the principle expressed by (a) to

$$G(s) = \frac{s[(s-1)^2 + 1]}{[(s+2)^2 + 1](s+3)}.$$

16-3 $|G(j\omega)|^2 = \dfrac{(1-\omega^2)^2}{1+\omega^8}.$

Obtain the minimum-phase $G(s)$.

16-4 $G(s)G(-s) = \dfrac{N(s)N(-s)}{D(s)D(-s)}.$

If

$$N(s)N(-s) = -s^2(s^2 + 1)^2(s + 1)(s - 1)[(s + 1)^2 + 1][(s - 1)^2 + 1],$$

what is $N(s)$?

16-5 The setup shown in Fig. 16-6 can be used to find the phase of $G(j\omega)$. Show that $\theta(\omega)$ can be calculated from the measured values of three voltages:

V_{m_1}, V_{m_2}, and V_{m_3}.

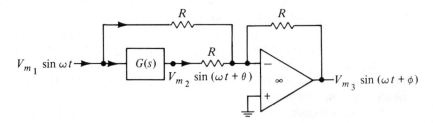

Fig. 16-6

16-6 Which characteristic is affected, magnitude or phase, if $G(s)$ is multiplied by the pole–zero structures shown in Fig. 16–7? Under what conditions, if any, is the resulting function realizable as a voltage ratio? [The poles of $G(s)$ are in the left half-plane.]

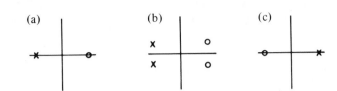

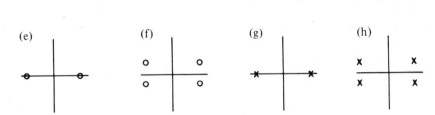

Fig. 16-7

16-7 $G(s) = \dfrac{(s-1)^2 + 1}{(s+1)[(s+2)^2 + 1]}$.

Construct a third-order function with left half-plane poles that has:
(a) the same magnitude as $G(j\omega)$ but different phase;
(b) the same phase as $G(j\omega)$ but different magnitude.

16-8 Consider as input an amplitude-modulated signal consisting of three frequencies: The carrier ω_c and the two sidebands

$$\omega_c - \omega_m \qquad \text{and} \qquad \omega_c + \omega_m.$$

(a) Show that the input can be represented by

$$(A_c + 2A_m \cos \omega_m t)\sin \omega_c t,$$

where A_c is the amplitude of the carrier and A_m is the amplitude of the sidebands.

(b) Assume that the magnitude of the system function is constant over the pass-band and that the Taylor-series expansion for the phase given by Eq. (16-17) is applicable. Obtain the output, and show that the envelope-time-delay is controlled by the group delay while the carrier-time-delay is controlled by the phase delay.

16-9 (a) Obtain the phase and group delays associated with a pole on the negative real axis.
(b) Obtain first-order approximations that show how the delay functions depart from the ideal characteristics.

16-10 Show that the group delay is a rational function of ω^2.

16-11 Obtain the phase and group delays associated with the magnitude-squared function

$$|G(j\omega)|^2 = \frac{1}{1 + \omega^6}.$$

17

Low-Pass
Approximations

The ideal low-pass function is characterized by a magnitude function that is constant in the pass-band and zero in the stop-band. The corresponding phase is linear in the pass-band. Such characteristics are unattainable with constant, lumped, and linear networks; therefore the requirements imposed by the ideal characteristics must be relaxed. This is done by allowing the magnitude and phase characteristics to stay within prescribed tolerances in the pass- and stop-bands. The functions that obey these specified constraints represent approximations to the ideal. In this chapter, a number of approximations are presented. In particular, the characteristics of the maximally flat (Butterworth) and the equal-ripple (Chebyshev) approximations are discussed in detail. The maximally-linear phase (Bessel) approximation is also presented and discussed. A simple procedure is given for obtaining a band-limited linear-amplitude approximation.

17-1 THE IDEAL AND ACTUAL LOW-PASS CHARACTERISTICS

The ideal low-pass characteristics are shown in Fig. 17-1. As the solid curves indicate, all sine waves with frequencies up to the cutoff frequency ω_c are passed with no amplitude and phase distortion. Sine waves with frequencies greater than ω_c are completely attenuated. Since $|G(j\omega)|^2 = 0$ in the stop-band, the shape of the $\theta(\omega)$-vs.-ω characteristic there is of no interest.

Since $G(s)$ is a rational function of s, the magnitude-squared characteristic for $s = j\omega$ cannot possibly be constant in the pass-band and zero in the stop-band. Neither can the phase be linear in the pass-band. Therefore, the actual $|G(j\omega)|^2$ and $\theta(\omega)$ characteristics can at best come close to the ideal characteristics. Consequently, the ideal specifications are relaxed and bounds are set for

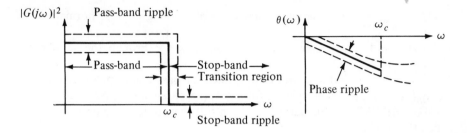

Fig. 17-1 Low-pass characteristics

acceptable performance characteristics. Thus, the pass-band magnitude-squared characteristic is allowed to vary within specified limits. The variation is usually specified in the form of an allowable ripple in the pass-band. In the stop-band, instead of infinite attenuation, a finite but acceptable value of attenuation (beyond a designated frequency) is specified. Since there can be no abrupt division between the pass-band and the stop-band, a transition region must also be allowed between the two bands. Similar constraints are specified for the phase. Any function which produces magnitude and phase curves within the specified bounds is an acceptable approximation function.

A suitable approximation to the ideal low-pass characteristics can be obtained by starting with the general expression of the magnitude-squared function,

$$|G(j\omega)|^2 = \frac{N(\omega^2)}{D(\omega^2)}.$$ (17-1)

For frequencies in the pass-band ($\omega < \omega_c$), this function should not depart appreciably from a constant, and for frequencies in the stop-band ($\omega > \omega_c$), it should be as small as possible. Even though the $N(\omega^2)$ and $D(\omega^2)$ polynomials are functions of ω^2, their ratio can be kept constant for $\omega < \omega_c$ if the polynomials track each other as a function of ω. On the other hand, for $\omega > \omega_c$, $D(\omega^2)$ should increase with ω much faster than $N(\omega^2)$ does, in order to provide attenuation. These desirable relationships between the numerator and the denominator polynomials are more easily seen if Eq. (17-1) is written as:

$$|G(j\omega)|^2 = \frac{N(\omega^2)}{N(\omega^2) + P(\omega^2)}.$$ (17-2)

The polynomial $P(\omega^2)$ should be chosen such that its value is much smaller than $N(\omega^2)$ for frequencies in the pass-band and much larger than $N(\omega^2)$ for frequencies in the stop-band. Such a functional relationship can be obtained by a variety of choices for $P(\omega^2)$ and $N(\omega^2)$. Some of these are discussed in the following sections.

In order to establish an equitable basis for comparing the various approximations, the cutoff frequency is chosen as 1. With frequency scaling, the cutoff frequency can later be set to any desired value. Furthermore, three other conditions are imposed on $|G(j\omega)|^2$.

1. $|G(j\omega)|^2_{max} = 1$; i.e., the maximum value of the magnitude is normalized to unity.

2. $|G(j1)|^2 = 1/(1 + \varepsilon^2)$, i.e., all approximations go through the same point when $\omega = \omega_c = 1$.

3. $1/(1 + \varepsilon^2) \le |G(j\omega)|^2 \le 1$ for $\omega \le \omega_c$; i.e., the pass-band ripple is confined to $[10 \log(1 + \varepsilon^2)]$ dB. In particular, for $\varepsilon = 1$, the pass-band ripple is $10 \log 2 = 3.010 \cong 3$ dB. Because the parameter ε controls the ripple in the pass-band, it is called the ripple factor.

These conditions are diagrammatically illustrated in Fig. 17-2. The two approximations that are sketched meet the pass-band specifications and therefore are acceptable approximations. The stop-band attenuation requirements can be met by choosing the order of the approximation.

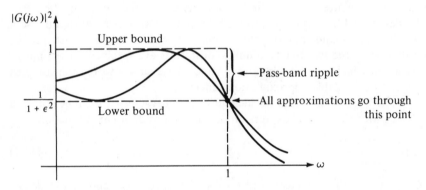

Fig. 17-2   Establishing a basis for comparing approximations

17-2 THE MAXIMALLY FLAT APPROXIMATION

The lower the degree of $N(\omega^2)$ compared to the degree of $P(\omega^2)$ in Eq. (17-2), the more attenuation is provided by $|G(j\omega)|^2$ at high frequencies. Therefore, let $N(\omega^2) = 1$. Then Eq. (17-2) becomes

$$|G(j\omega)|^2 = \frac{1}{1 + P(\omega^2)} \qquad (17\text{-}3)$$

$$= \frac{1}{1 + (a_0 + a_2\omega^2 + a_4\omega^4 + \cdots + a_{2n-2}\omega^{2n-2} + a_{2n}\omega^{2n})}. \qquad (17\text{-}4)$$

To make $|G(j0)|^2 = 1$, a_0 must be set to 0. To make $|G(j1)|^2 = 1/(1 + \varepsilon^2)$, the a coefficients must satisfy

$$a_2 + a_4 + \cdots + a_{2n-2} + a_{2n} = \varepsilon^2. \qquad (17\text{-}5)$$

Equation (17-4) then becomes

$$|G(j\omega)|^2 = \frac{1}{1 + a_2\omega^2 + a_4\omega^4 + \cdots + a_{2n-2}\omega^{2n-2} + (\varepsilon^2 - a_2 - a_4 - \cdots - a_{2n-2})\omega^{2n}}.$$

$$(17\text{-}6)$$

The remaining $(n-1)$ coefficients, a_2 through a_{2n-2}, are now used to improve the approximation. If all these coefficients are adjusted to make the magnitude-squared curve as constant as possible at $\omega = 0$, the resulting approximation is called the maximally-flat (at $\omega = 0$) approximation. This can be achieved by making the $(n-1)$ derivatives of $|G(j\omega)|^2$ with respect to ω^2 zero at $\omega = 0$ (see Problem 17-1). This requires that all the a coefficients in Eq. (17-6) be zero. The resulting approximation,

$$|G(j\omega)|^2 = \frac{1}{1 + \varepsilon^2\omega^{2n}}, \qquad (17\text{-}7)$$

is called the Butterworth approximation. It gives the best fit to the ideal low-pass magnitude-squared curve at the low end of the pass-band at the expense of increasingly greater departures from the ideal characteristic at the high end of the pass-band. At $\omega = 1$, the departure is the greatest: $[10 \log(1 + \varepsilon^2)]$ dB. The higher order n of the approximation, the closer the overall match is to unity in the pass-band, the steeper is the slope at cutoff, and the greater is the attenuation at high frequencies.

The slope of the magnitude is negative for all $\omega > 0$. The slope at cutoff is

$$\left.\frac{d|G(j\omega)|}{d\omega}\right|_{\omega=1} = -\frac{\varepsilon^2}{(1 + \varepsilon^2)^{3/2}}\,n. \qquad (17\text{-}8)$$

Hence, the larger ε (the more the ripple in the pass-band), the steeper is the slope at cutoff for $\varepsilon^2 \le 2$.

The magnitude and phase of $G(j\omega)$ are plotted in Fig. 17-3a and Fig. 17-3b, respectively, for $\varepsilon = 1$ and for even values of n from 2 to 10. Since $\varepsilon = 1$, the pass-band ripple is 3 dB. Note in particular the flat nature of the magnitude at $\omega = 0$ and fairly linear nature of the phase in the pass-band.

The dB attenuation produced by the magnitude characteristic is given by

$$\alpha = 20|\log|G(j\omega)|| = 10 \log(1 + \varepsilon^2\omega^{2n}). \qquad (17\text{-}9)$$

Equation (17-9) is plotted in Fig. 17-3c for $\varepsilon = 1$. This figure can be used to determine the order of the approximation if a specified amount of attenuation is to be produced beyond a certain frequency in the stop-band. For example, to produce 60 dB of attenuation for $\omega > 2$ calls for $n \ge 10$.

Thus, the parameter ε is selected to meet the pass-band ripple requirement, and the order n is chosen to meet the stop-band attenuation requirement.

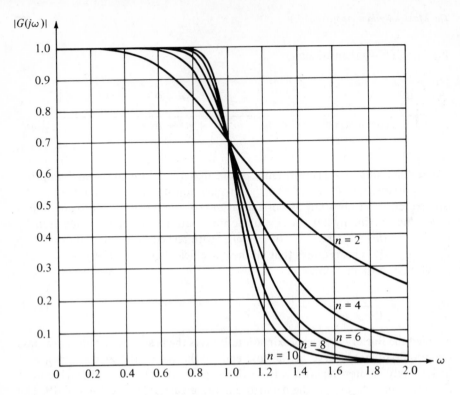

Fig. 17-3a Magnitude of the maximally flat approximation ($\varepsilon = 1$)

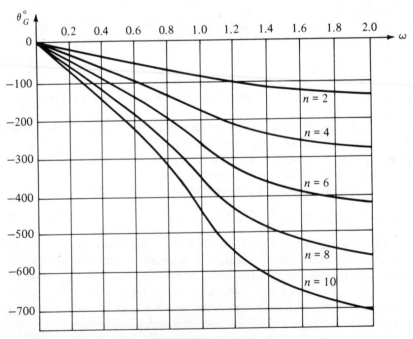

Fig. 17-3b Phase of the maximally flat approximation ($\varepsilon = 1$)

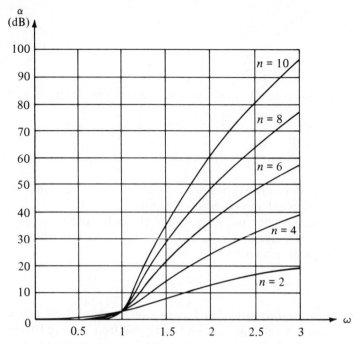

Fig. 17-3c   dB attenuation produced by the maximally flat approximation ($\varepsilon = 1$)

Determination of the Poles

From the magnitude-squared function of Eq. (17-7), the poles of $G(s)$ are obtained by the methods given in Section 16-3. The poles are given by the n left half-plane roots of:

$$1 + (-s^2)^n = 0 \qquad (\varepsilon = 1). \tag{17-10}$$

These are located at

$$s = \exp\left[j\frac{\pi}{2}\left(1 + \frac{1 + 2k}{n}\right)\right] \tag{17-11a}$$

$$= -\sin\frac{\pi}{2n}(1 + 2k) + j\cos\frac{\pi}{2n}(1 + 2k) \qquad \text{where } k = 0, 1, 2, \ldots, (n - 1). \tag{17-11b}$$

As Eq. (17-11a) shows, the poles are on a unit circle and are equally spaced at (π/n) radian intervals. Using these poles, the nth-order function, $G_n(s)$, can be constructed.

The pair of poles with the highest Q is found by using $k = 0$ and $k = (n - 1)$ in Eq. (17-11b):

$$s = \underbrace{-\sin\frac{\pi}{2n}}_{-\alpha} \pm j\cos\frac{\pi}{2n} = -\frac{1}{2Q} \pm j\sqrt{1 - \frac{1}{4Q^2}}. \tag{17-12}$$

The resulting Q is

$$Q = \frac{1}{2\alpha} = \frac{1}{2 \sin \dfrac{\pi}{2n}}. \tag{17-13}$$

Thus, the tenth-order Butterworth approximation requires realization of pole Q's no greater than

$$\frac{1}{2 \sin \dfrac{\pi}{20}} = 3.20.$$

Because of these low pole-Q values, passive and active circuits realizing the Butterworth approximation are implemented with ease.

In Table 17-1, the denominator polynomials and the poles associated with the maximally flat approximations are listed. Using this table, the nth-order Butterworth function can be constructed:

$$G_n(s) = \frac{1}{D_n(s)}.$$

The resulting magnitude characteristic has a pass-band ripple of 3 dB and a cutoff frequency of 1 rad/s. To change the pass-band ripple from 3 dB to [10 log $(1 + \varepsilon^2)$] dB, replace s by $\varepsilon^{1/n}s$. To change the cutoff frequency from 1 to ω_c, replace s by s/ω_c. Thus, to obtain the Butterworth function with [10 log$(1 + \varepsilon^2)$] dB pass-band ripple and with ω_c cutoff frequency, merely replace s by $\varepsilon^{1/n}(s/\omega_c)$, that is,

$$G_n(s)\big|_{s \text{ replaced by } \varepsilon^{1/n}(s/\omega_c)}. \tag{17-14}$$

If $\varepsilon = 0.5089$, the ripple in the pass-band is 1 dB. If $\varepsilon = 0.7648$, the ripple is 2 dB. Thus by choosing ε appropriately, the pass-band ripple can be set to any desired value. The curves given in Fig. 17-3 can still be used provided the ω-axis is interpreted as $(\varepsilon^{1/n}\omega/\omega_c)$. To find the modified pole locations, replace s in Table 17-1 by $(\varepsilon^{1/n}s/\omega_c)$.

EXAMPLE 17-1

Obtain the expression for the fourth-order maximally flat (at $\omega = 0$) low-pass approximation. The cutoff is at ω_c and the pass-band ripple is 1 dB.

SOLUTION

From Table 17-1, obtain the fourth-order, 3-dB ripple, 1 rad/s cutoff-frequency approximation and the corresponding pole locations:

$$G_4(s) = \frac{1}{[(s + 0.3827)^2 + 0.9239^2][(s + 0.9239)^2 + 0.3827^2]}, \tag{17-15}$$

$$p_{1,2} = -0.3827 \pm j0.9239, \qquad p_{3,4} = -0.9239 \pm j0.3827.$$

Table 17-1 Maximally flat (at $\omega = 0$) approximation $G_n(s) = \dfrac{1}{D_n(s)}$ (3.01-dB ripple)

$D_1 = s + 1$

$D_2 = s^2 + 1.4142s + 1 = (s + 0.7071)^2 + 0.7071^2$

$D_3 = s^3 + 2.0000s^2 + 2.0000s + 1 = (s + 1.0000)[(s + 0.5000)^2 + 0.8660^2]$

$D_4 = s^4 + 2.6131s^3 + 3.4142s^2 + 2.6131s + 1$
$\quad = [(s + 0.3827)^2 + 0.9239^2][(s + 0.9239)^2 + 0.3827^2]$

$D_5 = s^5 + 3.2361s^4 + 5.2361s^3 + 5.2361s^2 + 3.2361s + 1$
$\quad = (s + 1.0000)[(s + 0.3090)^2 + 0.9511^2][(s + 0.8090)^2 + 0.5878^2]$

$D_6 = s^6 + 3.8637s^5 + 7.4641s^4 + 9.1416s^3 + 7.4641s^2 + 3.8637s + 1$
$\quad = [(s + 0.2588)^2 + 0.9659^2][(s + 0.7071)^2 + 0.7071^2][(s + 0.9659)^2 + 0.2588^2]$

$D_7 = s^7 + 4.4940s^6 + 10.0978s^5 + 14.5918s^4 + 14.5918s^3 + 10.0978s^2 + 4.4940s + 1$
$\quad = (s + 1.0000)[(s + 0.2225)^2 + 0.9749^2][(s + 0.6235)^2 + 0.7818)^2]$
$\qquad\qquad\qquad\qquad\qquad\qquad\qquad\qquad \times [(s + 0.9010)^2 + 0.4339^2]$

$D_8 = s^8 + 5.1258s^7 + 13.1371s^6 + 21.8462s^5 + 25.6884s^4 + 21.8462s^3 + 13.1371s^2$
$\qquad\qquad\qquad\qquad\qquad\qquad\qquad\qquad\qquad\qquad\qquad + 5.1258s + 1$
$\quad = [(s + 0.1951)^2 + 0.9808^2][(s + 0.5556)^2 + 0.8315^2][(s + 0.8315)^2 + 0.5556^2]$
$\qquad\qquad\qquad\qquad\qquad\qquad\qquad\qquad \times [(s + 0.9808)^2 + 0.1951^2]$

$D_9 = s^9 + 5.7588s^8 + 16.5817s^7 + 31.1634s^6 + 41.9864s^5 + 41.9864s^4 + 31.1634s^3$
$\qquad\qquad\qquad\qquad\qquad\qquad\qquad\qquad\qquad + 16.5817s^2 + 5.7588s + 1$
$\quad = (s + 1.0000)[(s + 0.1737)^2 + 0.9848^2][(s + 0.5000)^2 + 0.8660^2]$
$\qquad\qquad\qquad\qquad \times [(s + 0.7660)^2 + 0.6428^2][(s + 0.9397)^2 + 0.3420^2]$

$D_{10} = s^{10} + 6.3925s^9 + 20.4317s^8 + 42.8021s^7 + 64.8824s^6 + 74.2334s^5 + 64.8824s^4$
$\qquad\qquad\qquad\qquad\qquad\qquad\qquad + 42.8021s^3 + 20.4317s^2 + 6.3925s + 1$
$\quad = [(s + 0.1564)^2 + 0.9877^2][(s + 0.4540)^2 + 0.8910^2][(s + 0.7071)^2 + 0.7071^2]$
$\qquad\qquad\qquad\qquad \times [(s + 0.8910)^2 + 0.4540^2][(s + 0.9877)^2 + 0.1564^2]$

To change the cutoff to ω_c and the pass-band ripple to 1 dB ($\varepsilon = 0.5089$), replace s with

$$\frac{s\varepsilon^{1/4}}{\omega_c} = \frac{s(0.5089)^{1/4}}{\omega_c} = \frac{0.8445s}{\omega_c}.$$

The desired approximation is

$$G(s) = \frac{1}{\left[\left(\dfrac{0.8445s}{\omega_c} + 0.3827\right)^2 + 0.9239^2\right]\left[\left(\dfrac{0.8445s}{\omega_c} + 0.9239\right)^2 + 0.3827^2\right]}.$$

$$(17\text{-}16)$$

The poles of $G(s)$ are at

$$(-0.3827 \pm j0.9239)\frac{\omega_c}{0.8445}, \qquad (-0.9239 \pm j0.3827)\frac{\omega_c}{0.8445}.$$

The magnitude, phase, and attenuation characteristics of the desired function can be studied by referring to Fig. 17-3 and using the $n = 4$ curves. The ω-axis labelling must be changed to $0.8445(\omega/\omega_c)$.

17-3 MAXIMALLY FLAT APPROXIMATION WITH IMAGINARY-AXIS ZEROS

A system function with a pair of imaginary-axis zeros at $\pm j\omega_o$ is needed to stop a sine wave of frequency ω_o in the steady state. The numerator of $G(s)$ must therefore contain $(s^2 + \omega_o^2)$ as a factor. Using this constraint and Eq. (17-2), a low-pass approximation with a stop-band zero at $\omega = \omega_o$ can be constructed:

$$|G(j\omega)|^2 = \frac{(\omega_o^2 - \omega^2)^2}{(\omega^2 - \omega_o^2)^2 + P(\omega^2)} \tag{17-17}$$

$$= \frac{(\omega_o^2 - \omega^2)^2}{(\omega_o^2 - \omega^2)^2 + a_0 + a_2\omega^2 + a_4\omega^4 + \cdots + a_{2n-2}\omega^{2n-2} + a_{2n}\omega^{2n}}. \tag{17-18}$$

To make $|G(j0)|^2 = 1$, a_0 must be set to zero. To make $|G(j1)|^2 = \frac{1}{2}$ (which corresponds to 3-dB pass-band ripple), the a coefficients must satisfy

$$a_2 + a_4 + \cdots + a_{2n-2} + a_{2n} = (\omega_o^2 - 1)^2. \tag{17-19}$$

Equation (17-18) then becomes

$$|G(j\omega)|^2 = \frac{(\omega_o^2 - \omega^2)^2}{\begin{array}{l}(\omega_o^2 - \omega^2)^2 + a_2\omega^2 + a_4\omega^4 + \cdots + a_{2n-2}\omega^{2n-2} \\ + [(\omega_o^2 - 1)^2 - a_2 - a_4 - \cdots - a_{2n-2}]\omega^{2n}\end{array}}. \tag{17-20}$$

The remaining $(n - 1)$ a coefficients are chosen to make $|G(j\omega)|^2$ maximally flat at $\omega = 0$. This calls for $a_{2i} = 0$ $(i = 1, \ldots, n - 1)$. See Problem 17-5. The result is

$$|G(j\omega)|^2 = \frac{(\omega_o^2 - \omega^2)^2}{(\omega_o^2 - \omega^2)^2 + (\omega_o^2 - 1)^2\omega^{2n}} \qquad (\omega_o > 1). \tag{17-21}$$

Since the magnitude-squared function given by Eq. (17-21) is maximally flat at $\omega = 0$, is 3 dB down at $\omega = 1$, and is forced to zero at $\omega = \omega_o$, it does have a better pass-band magnitude characteristic than the Butterworth function. In particular, the slope at cutoff is steeper, namely,

$$\left.\frac{d|G(j\omega)|}{d\omega}\right|_{\omega=1} = -\frac{n}{2\sqrt{2}} - \frac{1}{\sqrt{2}(\omega_o^2 - 1)}. \tag{17-22}$$

The first term in Eq. (17-22) is the slope of the Butterworth approximation. The second term shows the benefit derived from the added imaginary-axis zeros.

The closer ω_o is to 1, the steeper is the slope at cutoff. However, this improvement in the cutoff region is offset by the degradation of attenuation at higher frequencies. This is seen by studying Fig. 17-4, which shows the effect of the zero placement on the magnitude characteristic for $n = 5$. As ω_o is decreased from 1.5 to 1.1 (in steps of 0.1), the cutoff becomes sharper but the peaking in the stop-band becomes more pronounced.

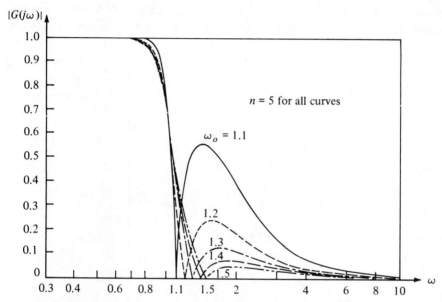

Fig. 17-4   Effect of imaginary-axis zeros on the magnitude characteristic

For the nth-order approximation, the peak occurs at

$$\omega_p = \omega_o \sqrt{\frac{n}{n-2}}, \tag{17-23}$$

and has the value

$$|G(j\omega_p)| = \frac{1}{\sqrt{1 + \left(\dfrac{n-2}{2}\right)^2 \left(\dfrac{n}{n-2}\right)^n \omega_o^{2n} \left(\dfrac{\omega_o^2 - 1}{\omega_o^2}\right)^2}}. \tag{17-24}$$

Equations (17-22) and (17-24) can be used to effect a proper tradeoff between the slope at cutoff and the attenuation beyond ω_o. Note that as $\omega_o \to \infty$, the magnitude function, given by Eq. (17-21), approaches the Butterworth function.

The Butterworth-function poles are equally spaced on a unit circle. The introduction of a pair of imaginary-axis zeros forces these poles to change location in order to preserve the maximally flat nature of the approximation at $\omega = 0$. With $-s^2$ substituted for ω^2, Eq. (17-21) becomes

$$G(s)G(-s) = \frac{(s^2 + \omega_o^2)^2}{(s^2 + \omega_o^2)^2 + (-1)^n(\omega_o^2 - 1)^2 s^{2n}}. \tag{17-25}$$

For a given n and ω_o, the n left half-plane zeros of the denominator polynomial can be found using the digital computer. Then $G(s)$ can be constructed.

It is interesting to plot the poles of $G(s)$ as a function of ω_o, to see how they move to preserve maximal flatness at $\omega = 0$. The resulting pole loci for $n = 5$ are shown in Fig. 17-5. As ω_o varies from 1.1 to infinity, the poles move inward at

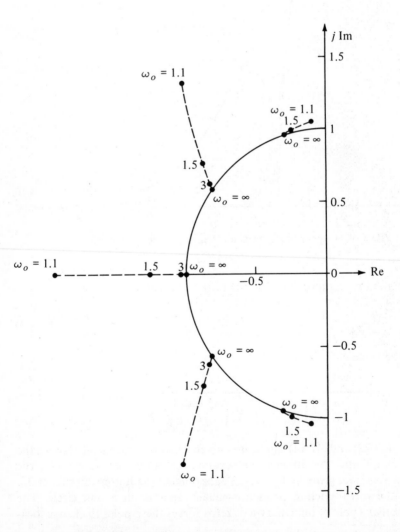

Fig. 17-5   Pole loci for $n = 5$

different rates. For $\omega_o = \infty$, the poles are on the unit circle, at the regular Butterworth-pole locations. For any other value of ω_o, the poles do not fall on any specific curve in the s-plane.

17-4 THE EQUAL-RIPPLE APPROXIMATION

In Eq. (17-2), let

$$N(\omega^2) = 1 \quad \text{and} \quad P(\omega^2) = \varepsilon^2 \, C_n^2(\omega).$$

Then

$$|G(j\omega)|^2 = \frac{1}{1 + \varepsilon^2 C_n^2(\omega)}. \tag{17-26}$$

The polynomial $C_n^2(\omega)$ must be a function of ω^2, should stay within prescribed bounds for $\omega \le \omega_c$, and should increase rapidly with ω for $\omega > \omega_c$. Such a characteristic is exhibited by the function

$$C_n^2(\omega) = \cos^2(n \cos^{-1}\omega) \tag{17-27a}$$

$$= \cosh^2(n \cosh^{-1}\omega). \tag{17-27b}$$

Although either form of $C_n^2(\omega)$ can be employed, it is convenient to use Eq. (17-27a) for $\omega \le 1$ and Eq. (17-27b) for $\omega \ge 1$. It will presently be shown that this particular function also produces a pass-band magnitude characteristic that is equal-ripple in nature.

Since

$$\cos^2 (n \cos^{-1} \omega) \le 1 \quad \text{for all } \omega \le 1,$$

$C_n^2 (\omega)$ is bounded in the pass-band.

Since

$$\cosh^2(n \cosh^{-1}\omega) \ge 1 \quad \text{for all } \omega \ge 1,$$

and

$$\cosh^2(n \cosh^{-1}\omega) \cong 2^{2n-2}\omega^{2n} \quad \text{for } \omega \gg 1,$$

$C_n^2(\omega)$ increases rapidly with ω in the stop-band.

It remains now to be shown that $C_n^2(\omega)$ is a function of ω^2. $C_0 (\omega)$ and $C_1(\omega)$ are readily evaluated from either form of Eq. (17-27):

$$C_0 (\omega) = 1, \quad C_1(\omega) = \omega. \tag{17-28}$$

The following recursion formula can also be easily obtained:

$$C_{n+1}(\omega) = 2\omega C_n (\omega) - C_{n-1}(\omega). \tag{17-29}$$

Starting with $C_0 (\omega)$ and $C_1(\omega)$, and using the recursion formula repeatedly, the higher-order functions can be constructed:

$$C_2(\omega) = 2\omega^2 - 1, \qquad C_3(\omega) = \omega(4\omega^2 - 3), \qquad (17\text{-}30a)$$

$$C_4(\omega) = 8\omega^4 - 8\omega^2 + 1, \qquad C_5(\omega) = \omega(16\omega^4 - 20\omega^2 + 5). \qquad (17\text{-}30b)$$

For n even, $C_n(\omega)$ is an even polynomial; for n odd, it is an odd polynomial. Hence, $C_n^2(\omega)$ is a function of ω^2.

The polynomials $C_n(\omega)$ are called the Chebyshev polynomials. The resulting low-pass approximation, Eq. (17-26), is called the Chebyshev approximation. It is also called the equal-ripple approximation because in the pass-band the $|G(j\omega)|^2$ curve swings between 1 and $1/(1 + \varepsilon^2)$ exactly n times. Note that

$$|G(j\omega)|^2 = 1 \qquad \text{when } C_n^2(\omega) = 0,$$

which occurs when $\omega = \cos(\tfrac{1}{2}k\pi/n)$, $k = 1, 3, 5, \ldots$, and

$$|G(j\omega)|^2 = \frac{1}{1 + \varepsilon^2} \qquad \text{when } C_n^2(\omega) = 1,$$

which occurs when $\omega = \cos(l\pi/n)$, $l = 0, 1, 2, \ldots$.

The slope of the magnitude characteristic at cutoff is

$$\frac{d|G(j\omega)|}{d\omega}\bigg|_{\omega=1} = -\frac{\varepsilon^2}{(1 + \varepsilon^2)^{3/2}} n^2. \qquad (17\text{-}31)$$

For 3-dB pass-band ripple ($\varepsilon = 0.9976 \cong 1$),

$$\frac{d|G(j\omega)|}{d\omega}\bigg|_{\omega=1} = \underbrace{\left(-\frac{n}{2\sqrt{2}}\right)}_{\substack{\text{Butterworth} \\ \text{slope}}} n. \qquad (17\text{-}32)$$

When compared to the Butterworth approximation, the Chebyshev magnitude approximation is n times steeper at cutoff, a feature that is extremely desirable. This and other aspects of the approximations are displayed in Fig. 17-6 for $n = 4$ and $\varepsilon = 1$. Note that in the pass-band the Chebyshev approximation bounces between $(1/\sqrt{2})$ and 1 four times, but in the stop-band it is below the Butterworth approximation. A numerical measure of this improvement in the stop-band attenuation is obtained by calculating the dB attenuation produced by the Chebyshev magnitude characteristic. From Eq. (17-26),

$$\alpha \cong 20[\log|G(j\omega)|] = 10 \log[1 + \varepsilon^2 C_n^2(\omega)]. \qquad (17\text{-}33)$$

For $\omega \gg 1$,

$$C_n(\omega) \cong 2^{n-1}\omega^n \qquad (n \neq 0),$$

and hence

$$\alpha \cong 20 \log(2^{n-1}\varepsilon\omega^n) \cong 6(n - 1) + \underbrace{20 \log \varepsilon + 20n \log \omega}_{\substack{\text{Butterworth} \\ \text{attenuation}}} \qquad (\omega \gg 1). \qquad (17\text{-}34)$$

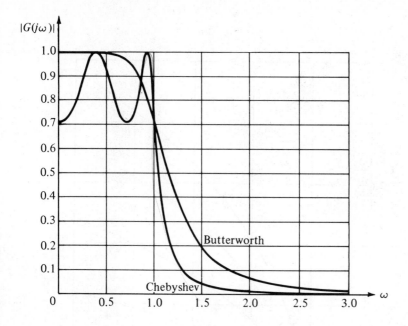

Fig. 17-6a   Fourth-order Chebyshev and Butterworth magnitude characteristics

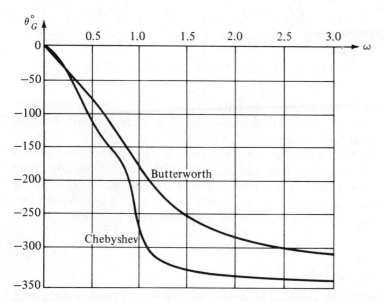

Fig. 17-6b   Fourth-order Chebyshev and Butterworth phase characteristics

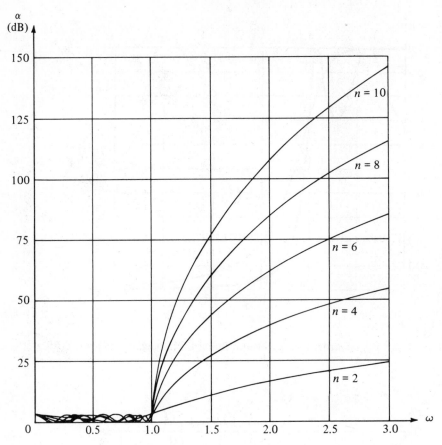

Fig. 17-6c   dB attenuation produced by the equal-ripple approximation

Therefore, for the same ripple in the pass-band, the Chebyshev approximation provides an additional $6(n - 1)$-dB attenuation over the Butterworth function [see Eq. (17-9)] at high frequencies.

Equation (17-33) is plotted in Fig. 17-6c for $\varepsilon = 1$ and for even values of n from 2 to 10. These curves can be used to determine the order of the 3-dB ripple approximation to achieve a prescribed stop-band attenuation. For example, to produce 60 dB of attenuation for $\omega > 2$ calls for $n \geq 6$, which is considerably less than 10, the order of the Butterworth approximation which meets the same specification.

Determination of the Poles

The poles of $G(s)$ are obtained by replacing ω with $-js$ in the denominator of Eq. (17-26) and then solving for the n left half-plane roots of the resulting equation (see Problem 17-9):

$$s = -\sin\left[\frac{\pi}{2n}(1+2k)\right]\sinh\left(\frac{1}{n}\sinh^{-1}\frac{1}{\varepsilon}\right) + j\cos\left[\frac{\pi}{2n}(1+2k)\right]\cosh\left(\frac{1}{n}\sinh^{-1}\frac{1}{\varepsilon}\right)$$

$$= -\alpha_k + j\beta_k \qquad k = 0, 1, 2, \ldots, (n-1).$$ (17-35)

Since

$$\left[\frac{\alpha_k}{\sinh\left(\frac{1}{n}\sinh^{-1}\frac{1}{\varepsilon}\right)}\right]^2 + \left[\frac{\beta_k}{\cosh\left(\frac{1}{n}\sinh^{-1}\frac{1}{\varepsilon}\right)}\right]^2 = 1,$$ (17-36)

the poles of the nth-order Chebyshev approximation fall on an ellipse with real- and imaginary-axis intercepts of

$$\pm\sinh\left(\frac{1}{n}\sinh^{-1}\frac{1}{\varepsilon}\right) \qquad \text{and} \qquad \pm\cosh\left(\frac{1}{n}\sinh^{-1}\frac{1}{\varepsilon}\right),$$

respectively. The higher the order n or the larger the ripple factor ε, the smaller the values of the major- and the minor-axis intercepts.

Using $k = 0$ and $k = n - 1$, the complex-conjugate pair of poles with the highest Q is obtained from Eq. (17-35):

$$s = -\sin\frac{\pi}{2n}\sinh\left(\frac{1}{n}\sinh^{-1}\frac{1}{\varepsilon}\right) \pm j\cos\frac{\pi}{2n}\cosh\left(\frac{1}{n}\sinh^{-1}\frac{1}{\varepsilon}\right)$$ (17-37)

$$= -\frac{\omega_o}{2Q} \pm j\omega_o\sqrt{1 - \frac{1}{4Q^2}}.$$

The Q of this pair of poles is

$$Q = \underbrace{\left(\frac{1}{2\sin\frac{\pi}{2n}}\right)}_{\substack{\text{Butterworth} \\ Q}}\sqrt{1 + \left[\frac{\cos\frac{\pi}{2n}}{\sinh\left(\frac{1}{n}\sinh^{-1}\frac{1}{\varepsilon}\right)}\right]^2}.$$ (17-38)

When compared with the corresponding Q of the Butterworth approximation, Eq. (17-13), the Chebyshev pole-Q is much higher because of the multiplying factor involving the square root. For $n = 10$ and $\varepsilon = 1$, the highest pole-Q value of the Chebyshev approximation is 35.91, which is 11.24 times more than the Q of the corresponding pole of the tenth-order Butterworth approximation. This is why care must be exercised in the design of high-order passive and active Chebyshev filters.

Table 17-2 lists the denominator polynomials and the poles associated with the nth-order Chebyshev function, $G_n(s)$. The table is constructed for 3-dB pass-band ripple and for 1 rad/s cutoff frequency. To change the cutoff frequency to ω_c rad/s, replace s by s/ω_c. Unlike the Butterworth approximation, the ripple in the pass-band cannot be changed by merely frequency-scaling. A new table must be constructed for each value of pass-band ripple.

Table 17-2   Equal-ripple (exactly 3-dB) approximation $G_n(s) = \dfrac{H}{D_n(s)}$.

Choose H to make $G_n(0) = 1$ for n odd and $(1/\sqrt{1.9953})$ for n even.

$D_1 = s + 1.0024$

$D_2 = s^2 + 0.6449s + 0.7080 = [(s + 0.3225)^2 + 0.7772^2]$

$D_3 = s^3 + 0.5972s^2 + 0.9284s + 0.2506 = (s + 0.2986)[(s + 0.1493)^2 + 0.9038^2]$

$D_4 = s^4 + 0.5816s^3 + 1.1691s^2 + 0.4048s + 0.1770$
$\quad = [(s + 0.0852)^2 + 0.9465^2][(s + 0.2056)^2 + 0.3921^2]$

$D_5 = s^5 + 0.5744s^4 + 1.4150s^3 + 0.5489s^2 + 0.4079s + 0.0626$
$\quad = (s + 0.1775)[(s + 0.0548)^2 + 0.9659^2][(s + 0.1436)^2 + 0.5970^2]$

$D_6 = s^6 + 0.5707s^5 + 1.6629s^4 + 0.6906s^3 + 0.6991s^2 + 0.1634s + 0.0443$
$\quad = [(s + 0.0382)^2 + 0.9764^2][(s + 0.1045)^2 + 0.7148^2][(s + 0.1427)^2 + 0.2616^2]$

$D_7 = s^7 + 0.5684s^6 + 1.9116s^5 + 0.8314s^4 + 1.0519s^3 + 0.3000s^2 + 0.1462s + 0.0157$
$\quad = (s + 0.1265)[(s + 0.0282)^2 + 0.9827^2][(s + 0.0789)^2 + 0.7881^2]$
$\qquad\qquad\qquad\qquad\qquad\qquad\qquad\qquad\quad \times\, [(s + 0.1140)^2 + 0.4373^2]$

$D_8 = s^8 + 0.5670s^7 + 2.1607s^6 + 0.9720s^5 + 1.4667s^4 + 0.4719s^3 + 0.3208s^2 + 0.0565s$
$\qquad\qquad\qquad\qquad\qquad\qquad\qquad\qquad\qquad\qquad\qquad\qquad\quad + 0.0111$

$\quad = [(s + 0.0216)^2 + 0.9868^2][(s + 0.0615)^2 + 0.8365^2][(s + 0.0920)^2 + 0.5590^2]$
$\qquad\qquad\qquad\qquad\qquad\qquad\qquad\qquad\qquad \times\, [(s + 0.1085)^2 + 0.1963^2]$

$D_9 = s^9 + 0.5659s^8 + 2.4101s^7 + 1.1123s^6 + 1.9438s^5 + 0.6789s^4 + 0.5835s^3$
$\qquad\qquad\qquad\qquad\qquad\qquad\qquad\qquad\qquad + 0.1314s^2 + 0.0476s + 0.0039$

$\quad = (s + 0.0983)[(s + 0.0171)^2 + 0.9896^2][(s + 0.0491)^2 + 0.8702^2]$
$\qquad\qquad\qquad\qquad \times\, [(s + 0.0753)^2 + 0.6459^2][(s + 0.0924)^2 + 0.3437^2]$

$D_{10} = s^{10} + 0.5652s^9 + 2.6597s^8 + 1.2527s^7 + 2.4834s^6 + 0.9211s^5 + 0.9499s^4$
$\qquad\qquad\qquad\qquad\qquad\qquad + 0.2492s^3 + 0.1278s^2 + 0.0180s + 0.0028$

$\quad = [(s + 0.0138)^2 + 0.9915^2][(s + 0.0401)^2 + 0.8945^2][(s + 0.0625)^2 + 0.7099^2]$
$\qquad\qquad\qquad\qquad \times\, [(s + 0.0788)^2 + 0.4558^2][(s + 0.0873)^2 + 0.1571^2]$

17-5 TRANSITIONAL BUTTERWORTH–CHEBYSHEV APPROXIMATIONS

The low-pass Butterworth magnitude approximation is excellent at and near $\omega = 0$. The low-pass Chebyshev magnitude approximation is superior at and near $\omega = \omega_c$. The desirable attributes of these two approximations can be combined in a single approximation that is given by

$$|G(j\omega)|^2 = \frac{1}{1 + \varepsilon^2 \omega^{2k} C_{n-k}^2(\omega)}. \qquad (17\text{-}39)$$

When $k = n$, $|G(j\omega)|^2$ is identical with the Butterworth function. On the other hand, when $k = 0$, $|G(j\omega)|^2$ is identical with the Chebyshev function. For any other value of k, Eq. (17-39) possesses both Butterworth-like and Chebyshev-like characteristics, behaving more like the former as $k \to n$ and like the latter as $k \to 0$. Thus, the parameter k is used to obtain, for any order n, a set of new approximations called transitional Butterworth–Chebyshev approximations.

The power-series expansion of Eq. (17-39) for $|\varepsilon^2 \omega^{2k} C_{n-k}^2(\omega)| < 1$ is

$$|G(j\omega)|^2 = 1 - \varepsilon^2 \omega^{2k} C_{n-k}^2(\omega) + \text{higher-order terms}, \tag{17-40}$$

where

$$C_{n-k}^2(\omega) = \begin{cases} 1 + a_2 \omega^2 + a_4 \omega^4 + \cdots & (n-k) \text{ even}, \\ b_2 \omega^2 + b_4 \omega^4 + \cdots & (n-k) \text{ odd}. \end{cases}$$

Hence, for $(n - k)$ even, $(2k - 1)$ derivatives of $|G(j\omega)|^2$ with respect to ω are zero at $\omega = 0$; for $(n - k)$ odd, $(2k + 1)$ derivatives are zero. Thus, the larger the k, the flatter is the magnitude-squared characteristic of the transitional approximation at $\omega = 0$.

The slope of the magnitude characteristic at cutoff is

$$\frac{d|G(j\omega)|}{d\omega}\bigg|_{\omega=1} = -\frac{2}{(1 + \varepsilon^2)^{3/2}} [(n - k)^2 + k]. \tag{17-41}$$

Hence, the smaller k, the steeper the cutoff.

The dB attenuation produced in the stop-band is

$$20[\log|G(j\omega)|] = 10 \log[1 + \varepsilon^2 \omega^{2k} C_{n-k}^2(\omega)] \tag{17-42}$$

$$\cong 20 \log(2^{n-k-1} \varepsilon \omega^n) \qquad (\omega \gg 1), \qquad (k \neq n), \tag{17-43}$$

$$\cong 6(n - k - 1) + 20 \log \varepsilon + 20n \log \omega \qquad (\omega \gg 1). \tag{17-44}$$

In Fig. 17-7 the magnitude and phase characteristics and the upper half-plane pole locations of the eighth-order transitional approximation are displayed for $k = 0, 2, 4, 6, 8$. The curves are drawn for 3-dB ripple in the pass-band. Note in particular the decrease of the pole-Q values as k is increased from 0 to 8.

To obtain the poles of the transitional approximation, Eq. (17-39) is written as

$$|G(j\omega)|^2 = \frac{1}{1 + \dfrac{\varepsilon^2}{2} \omega^{2k} \left[1 + C_{2(n-k)}(\omega) \right]}, \tag{17-45}$$

where use is made of the result presented in Problem 17-7d. Thus, the necessity of squaring Chebyshev polynomials is avoided. It should be emphasized that the poles of $G(s)$ are not a mixture of the regular Butterworth and Chebyshev poles. The poles of the transitional function must be calculated from Eq. (17-45) using the methods presented in Section (16-3). The scale factor of the resulting $G(s)$ is chosen such that $G(0) = 1$.

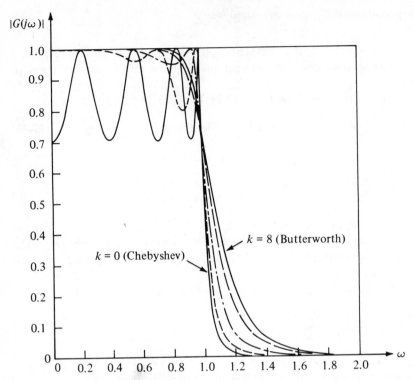

Fig. 17-7a Magnitude characteristics of eighth-order transitional Butterworth–Chebyshev functions for $k = 0, 2, 4, 6, 8$

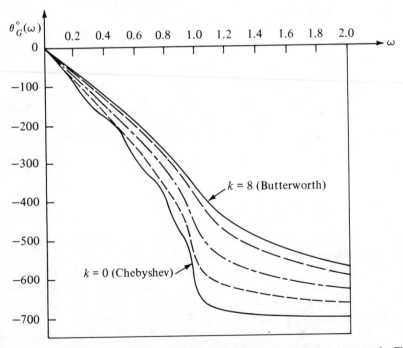

Fig. 17-7b Phase characteristics of eighth-order transitional Butterworth–Chebyshev functions for $k = 0, 2, 4, 6, 8$

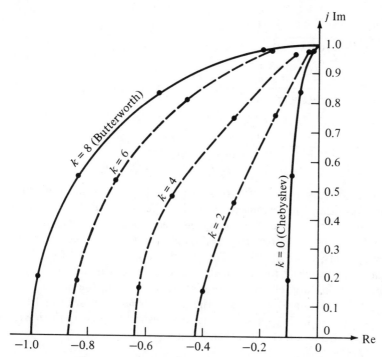

Fig. 17-7c Upper-plane poles of eighth-order transitional Butterworth-Chebyshev functions for $k = 0, 2, 4, 6, 8$

17-6 THE MAXIMALLY LINEAR PHASE APPROXIMATION

The methods presented so far have concentrated on approximating the ideal low-pass magnitude-squared characteristic. The ensuing phase characteristic has been accepted without question. In this section, attention is focused on making the phase as linear as possible while maintaining the low-pass nature of the approximation. Therefore, as a starting point, an all-pole function, one with all zeros at infinity, is chosen:

$$G(s) = \frac{1}{1 + a_1 s + a_2 s^2 + \cdots + a_n s^n}. \tag{17-46a}$$

$G(j\omega)$, the square of its magnitude, its phase, and the associated group delay are given by

$$G(j\omega)$$

$$= \frac{1}{(1 - a_2 \omega^2 + a_4 \omega^4 - a_6 \omega^6 + \cdots) + j\omega(a_1 - a_3 \omega^2 + a_5 \omega^4 - a_7 \omega^6 + \cdots)}, \tag{17-46b}$$

$$|G(j\omega)|^2 =$$

$$\frac{1}{(1 - a_2\omega^2 + a_4\omega^4 - a_6\omega^6 + \cdots)^2 + \omega^2(a_1 - a_3\omega^2 + a_5\omega^4 - a_7\omega^6 + \cdots)^2},$$
(17-46c)

$$\theta_G(\omega) = -\tan^{-1}\frac{\omega(a_1 - a_3\omega^2 + a_5\omega^4 - a_7\omega^6 + \cdots)}{1 - a_2\omega^2 + a_4\omega^4 - a_6\omega^6 + \cdots},$$
(17-46d)

$$\tau_{gr}(\omega) = \frac{a_1 + \omega^2(a_1a_2 - 3a_3) + \omega^4(5a_5 - 3a_1a_4 + a_2a_3)}{1 + \omega^2(a_1^2 - 2a_2) + \omega^4(a_2^2 - 2a_1a_3 + 2a_4)}.$$
(17-46e)

If all the a coefficients are adjusted to make the phase as linear as possible at $\omega = 0$, the resulting approximation is called the maximally linear phase (at $\omega = 0$) approximation. Since linearizing the phase is equivalent to making the group delay constant, the resulting approximation is also called the maximally flat group-delay approximation. The group delay, being a rational function, is mathematically easier to work with than the phase which is an inverse-tangent function. Therefore, effort is now directed at making the group delay flat at $\omega = 0$. Without loss of generality, the delay for $\omega = 0$ is set equal to 1 second by making $a_1 = 1$. (If τ seconds of delay is desired for $\omega = 0$, s can later be replaced by $s\tau$.) Equation (17-46e) then becomes

$$\tau_{gr}(\omega) = \frac{1 + \omega^2(a_2 - 3a_3) + \omega^4(5a_5 - 3a_4 + a_2a_3)}{1 + \omega^2(1 - 2a_2) + \omega^4(a_2^2 - 2a_3 + 2a_4)}.$$
(17-47)

A rational function, such as $\tau_{gr}(\omega)$, can be made maximally flat at the origin by equating the numerator and denominator coefficients of equal powers of ω. (See Problem 17-4b.) Thus,

$$a_2 - 3a_3 = 1 - 2a_2$$
$$5a_5 - 3a_4 + a_2a_3 = a_2^2 - 2a_3 + 2a_4$$
$$5a_6 + a_3a_4 - 3a_2a_5 - 7a_7 = a_3^2 + 2a_5 - 2a_2a_4 - 2a_6$$
(17-48)

$$\cdot$$
$$\cdot$$
$$\cdot$$

The constraints expressed by Eq. (17-48) can be simplified to

$$a_2 - a_3 = \tfrac{1}{3}$$
$$3(a_5 - a_4) + a_3 = \tfrac{1}{15}$$
$$15(a_6 - a_7) + a_4 - 6a_5 = \tfrac{1}{105}$$
(17-49)

$$\cdot$$
$$\cdot$$
$$\cdot$$

This set of equations can then be solved for the desired a coefficients. Consider, for example, a fourth-order low-pass approximation:

$$G(s) = \frac{1}{a_4 s^4 + a_3 s^3 + a_2 s^2 + s + 1} .$$

If its group delay is to be maximally flat at $\omega = 0$, the a coefficients must satisfy

$$a_2 - a_3 = \tfrac{1}{3},$$
$$-3a_4 + a_3 = \tfrac{1}{15},$$
$$a_4 = \tfrac{1}{105}.$$

Solving for the a's gives

$$a_2 = \tfrac{3}{7}, \qquad a_3 = \tfrac{2}{21}, \qquad a_4 = \tfrac{1}{105}.$$

The resulting approximation is:

$$G(s) = \frac{105}{s^4 + 10s^3 + 45s^2 + 105s + 105} . \tag{17-50}$$

Whereas $G(j\omega)$ obtained from Eq. (17-50) possesses excellent phase linearity at $\omega = 0$, it should be realized that this is achieved by ignoring altogether any improvement that can be made on the magnitude characteristic. By contrast, the Butterworth approximation is derived solely on the basis of making the magnitude characteristic flat at $\omega = 0$ and ignoring any improvement that can be made for linearizing the phase. To see the differences in the two approaches, a graphical comparison of the magnitude and phase curves is presented in Fig. 17-8. Since $|G(j1)|$ obtained from Eq. (17-50) is equal to 0.9300, frequency

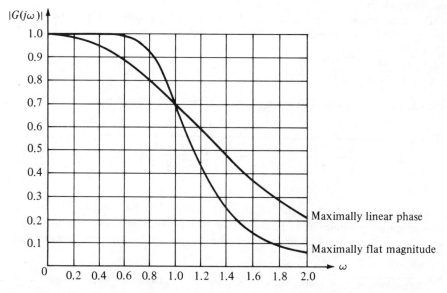

Fig. 17-8a   Fourth-order Butterworth and Bessel magnitude approximations

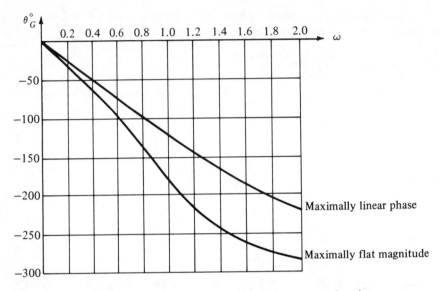

Fig. 17-8b   Fourth-order Butterworth and Bessel phase approximations

scaling must be used to make the magnitude curve go through the 3-dB point at cutoff. To achieve this, s is replaced by $s/0.4731$ in Eq. (17-50). The curves of Fig. 17-8a show the rather poor cutoff characteristic (relative to the Butterworth) associated with the maximally linear phase approximation. On the other hand, the curves of Fig. 17-8b show its practically linear phase characteristic throughout the pass-band.

It can be shown* that maximally flat group delay results if the a coefficients in Eq. (17-46a) are chosen according to the relation:

$$\frac{a_{i+1}}{a_i} = \frac{2(n-i)}{(2n-i)(i+1)} \qquad (i = 1, 2, \ldots, n-1), \tag{17-51}$$

with a_1 taken as unity. Equation (17-46a) then can be written as

$$G(s) = \frac{1}{1 + s + \dfrac{(n-1)}{2n-1}\left\{ s^2 + \dfrac{2(n-2)}{3(2n-2)}\left[s^3 + \dfrac{2(n-3)}{4(2n-3)}\left(s^4 + \cdots + \dfrac{2s^n}{n(n+1)} \right)\right]\right\}}. \tag{17-52}$$

For any order n, the poles of $G(s)$ are in the left half-plane. Because the denominator polynomial is related to Bessel polynomials, the corresponding magnitude-squared function is called the Bessel approximation. Its excellent

---

* Louis Weinberg, *Network Analysis and Synthesis* (McGraw-Hill Book Company, Inc., New York, 1962), pp. 499–506.

phase characteristic in the pass-band is offset by its poor characteristic at cutoff. For this reason, the Bessel approximation is used when the stop-band requirements are not very stringent.

EXAMPLE 17-2

Show that Eq. (17-52), which results in the Bessel approximation, approaches e^{-s} as $n \to \infty$.

SOLUTION

From Eq. (17-51),

$$\frac{a_{i+1}}{a_i} = \frac{2(n-i)}{(2n-i)(i+1)}\bigg|_{n \to \infty} \to \frac{1}{i+1},$$

which is precisely the ratio of succeeding coefficients in the Maclaurin-series expansion of e^s. Hence

$$G(s)\bigg|_{n \to \infty} = \frac{1}{e^s} = e^{-s}. \qquad Ans.$$

Note that the resulting phase, $\theta(\omega) = -\omega$, varies indeed linearly with frequency. However, the magnitude characteristic is no longer low-pass. It is all-pass.

There are many other low-pass approximations, each a little different in some sense from those mentioned in this chapter. Using the basic principles involved in the low-pass approximation, the reader can construct his own approximations. Knowledge of different kinds of polynomials, which abound in the mathematical literature, is helpful in this respect.

17-7 LOW-PASS, LINEAR-AMPLITUDE APPROXIMATION

The system function $G(s) = s$ has interesting applications. It operates on the excitation, $E(s)$, to produce a response of $sE(s)$. Since multiplication by s in the complex-frequency domain corresponds to differentiation with respect to t in the time domain, the response is the derivative of the excitation. If the input is sinusoidal, $V_m \sin \omega t$, then the output is $\omega V_m \sin (\omega t + \pi/2)$. Thus, the peak value of the output sine wave becomes linearly dependent on the frequency of operation. If the peak value of this sine wave is detected, a dc signal proportional to the frequency of the sine wave is obtained.

Because of the differentiation, fast changes in the input produce large output signals. This may be bothersome if the changes are due to unwanted signals, e.g., noise. In that case, it is desirable to band-limit the differentiation by imposing a low-pass constraint on the system function. This can be achieved by multiplying the desired magnitude characteristic, ω, with the low-pass Butterworth approximation; that is,

$$|G(j\omega)| = \omega \times \frac{1}{\sqrt{1 + \varepsilon^2 \left(\dfrac{\omega}{\omega_c}\right)^{2n}}}, \tag{17-53}$$

where ω_c represents the cutoff frequency. For $\omega \ll \omega_c$, $|G(j\omega)| = \omega$; thus, the desired linear amplitude characteristic is produced. For $\omega \gg \omega_c$,

$$|G(j\omega)| = \omega / \left[\varepsilon\left(\frac{\omega}{\omega_c}\right)^{n}\right];$$

thus, the high frequencies are attenuated. At $\omega = \omega_c$,

$$|G(j\omega)| = \omega_c / \sqrt{1 + \varepsilon^2};$$

hence, the magnitude is down by $[10 \log (1 + \varepsilon^2)]$ dB from the value it would have assumed if it were not band-limited. The parameter ε controls the amount by which the magnitude curve departs from the linear curve at cutoff. For $\varepsilon = 1$, the departure is 3.01 dB. The higher the n, the sharper the cutoff characteristic.

The Maclaurin-series expansion of Eq. (17-53) is

$$|G(j\omega)| = \omega\left[1 - \frac{1}{2}\varepsilon^2\left(\frac{\omega}{\omega_c}\right)^{2n} + \cdots\right] \qquad \left(\frac{\omega}{\omega_c} < 1\right). \tag{17-54}$$

Hence, the first $2n$ derivatives of the error between $|G(j\omega)|$ and ω are zero at $\omega = 0$, thereby assuring a maximally linear magnitude characteristic at $\omega = 0$.

Figure 17-9 illustrates the magnitude and the phase of the fourth-order, maximally linear magnitude approximation, which departs 3 dB from linearity at the cutoff frequency, $\omega = 1$. Whereas true differentiation requires a constant $(\pi/2)$-radian phase shift, the band-limited differentiator produces an almost linear phase shift.

The system function is constructed by dividing into s the appropriate order Butterworth polynomial obtained from Table 17-1. Thus, the fourth-order, band-limited, linear magnitude function is

$$T(s) = \frac{s}{B_4(s)} = \frac{s}{s^4 + 2.6131s^3 + 3.4142s^2 + 2.6131s + 1}.$$

The magnitude of this function departs from linearity by 3.01 dB at $\omega = 1$. To change this to $[10 \log (1 + \varepsilon^2)]$ dB at $\omega = \omega_c$, replace s in the denominator polynomial with $(s\varepsilon^{1/n}/\omega_c)$. To change the slope of the $|T(j\omega)|$-vs.-ω curve from 1 to k, multiply $T(s)$ by k.

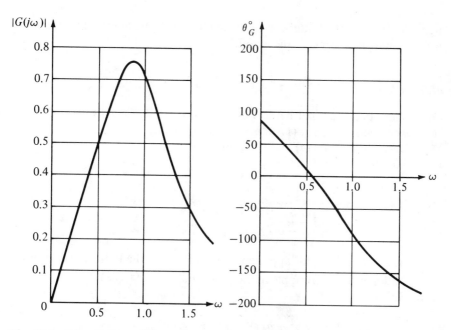

Fig. 17-9   Magnitude and phase characteristics of linear-amplitude filter

17-8 SUMMARY

The maximally flat (Butterworth) low-pass approximation is an all-pole approximation. The poles are spaced equally on a unit circle. The resulting magnitude characteristic has $(2n - 1)$ zero-derivatives at $\omega = 0$. The equal-ripple (Chebyshev) low-pass approximation is also an all-pole approximation. The poles fall on an ellipse. The resulting magnitude characteristic is equal-ripple in nature in the pass-band. Compared to the Butterworth approximation, the Chebyshev approximation has a steeper slope at cutoff and provides more attenuation in the stop-band. However, the Chebyshev-approximation phase characteristic departs from the linear curve more than the Butterworth does. Moreover, Chebyshev approximations require higher pole-Q realizations, thus demanding tighter component tolerances. Transitional Butterworth–Chebyshev approximations combine the best features of the Butterworth and Chebyshev approximations. The resulting magnitude characteristics are flatter than the Chebyshev at $\omega = 0$ and steeper than the Butterworth at cutoff.

The maximally linear phase (or maximally flat group-delay) approximation is an all-pole approximation that uses the Bessel polynomials. While excellent phase linearity is obtained with this approximation, the magnitude characteristic exhibits a rather slow transition between pass-band and stop-band.

A low-pass, linear-amplitude characteristic can be obtained by multiplying the maximally flat magnitude characteristic with ω. The corresponding functions of s are obtained by dividing the Butterworth polynomials into s. The higher the order of the Butterworth polynomials, the more linear is the magnitude at $\omega = 0$ and the sharper is the cutoff.

PROBLEMS

17-1   Show that maximal flatness about $\omega = 0$ is achieved by setting all the a coefficients in Eq. (17-6) equal to zero.

17-2   Derive Eqs. (17-11a) and (17-11b).

17-3   Sketch

$$|G(j\omega)|^2 = \cfrac{1}{1 + \left(\cfrac{\omega^2 - \omega_f^2}{1 - \omega_f^2}\right)^n} \qquad (n \text{ even})$$

vs. ω. How does this approximation differ from the Butterworth approximation? For 3-dB ripple in the pass-band, $\omega_f < (1/\sqrt{2})$. Why?

17-4   Given

$$|G(j\omega)|^2 = \frac{1 + a_2\omega^2 + a_4\omega^4 + \cdots + a_{2m}\omega^{2m}}{1 + b_2\omega^2 + b_4\omega^4 + \cdots + b_{2n}\omega^{2n}} \qquad (m < n).$$

(a) Using long division, expand $|G(j\omega)|^2$ in the Maclaurin series.

(b) If $|G(j\omega)|^2$ is to be maximally flat at the origin, $a_{2i} = b_{2i}$, $(i = 1, \ldots, n-1)$. Show this.

(c) Apply the results of (b) to obtain Eq. (17-21).

17-5   Show that the Maclaurin-series expansion of Eq. (17-21) is

$$|G(j\omega)|^2 = 1 - \left(\frac{\omega_o^2 - 1}{\omega_o^2}\right)^2 \omega^{2n} + \text{higher-order terms.}$$

17-6   Obtain the nth-order low-pass, magnitude-squared function that is maximally flat at $\omega = 0$ and has two arbitrarily placed zeros in the stop-band. The pass-band ripple is 3 dB and cutoff is at 1 rad/s.

17-7   Show that
(a) $\cos(n \cos^{-1}\omega) = \cosh(n \cosh^{-1}\omega)$,
(b) $\cosh(n \cosh^{-1}\omega) = 2^{n-1}\omega^n$ for $\omega \gg 1$,
(c) $C_{n+1}(\omega) = 2\omega C_n(\omega) - C_{n-1}(\omega)$,
(d) $C_n^2(\omega) = \frac{1}{2}[1 + C_{2n}(\omega)]$.

17-8   Derive Eq. (17-31).

17-9   (a) Derive Eq. (17-35).
(b) Show that the Chebyshev-approximation poles can be obtained directly from the Butterworth-approximation poles.

17-10 An oscilloscope is fed the following inputs:
Horizontal amplifier: $\cos \omega t$;
Vertical amplifier: $\cos n\omega t$.

(a) Sketch the resulting CRT pattern for $n = 0, 1, 2, 3, 4$.

(b) Obtain the expression for the vertical deflection as a function of the horizontal deflection.

17-11 The inverse Chebyshev low-pass approximation is given by

$$|G(j\omega)|^2 = \frac{\varepsilon^2 C_n^2\left(\dfrac{1}{\omega}\right)}{1 + \varepsilon^2 C_n^2\left(\dfrac{1}{\omega}\right)},$$

where $C_n(\omega)$ represents the Chebyshev polynomial of order n. Sketch $|G(j\omega)|^2$ vs. ω.

17-12 Figure 17-10 gives the low-pass specifications in terms of ε, ω_c, ω_a, and A. The pass-band ripple is determined by ε. As shown, the cutoff frequency is ω_c, and the dB attenuation for $\omega > \omega_a$ is greater than $20 \log A$.

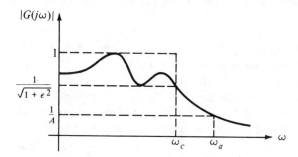

Fig. 17-10

(a) Obtain the expression for $|G(j\omega)|^2$ if it is to be maximally flat at $\omega = 0$.

(b) Obtain the expression for $|G(j\omega)|^2$ if it is to be equal ripple in the pass-band.

(c) For (a) and (b), obtain the expression for the required order n of the approximation in terms of the given constraints.

(d) Obtain the order n for each case if the allowable variation in the pass-band is 10% of the maximum value and 40 dB $(A = 100)$ or more attenuation is to be produced for $\omega \geq 1.5\omega_c$.

(e) Repeat (d) if 20% variation is acceptable in the pass-band. The attenuation requirement is the same as before.

17-13 Discuss the approximation

$$|G(j\omega)|^2 = \frac{1}{1 + \dfrac{\omega^n}{2}\left[1 + C_n(\omega)\right]} \qquad (n \text{ even}).$$

17-14 Evaluate the poles of Eq. (17-39) for $k = 1$, $n = 2$, and $\varepsilon = 1$.

17-15 Derive Eq. (17-46e).

17-16 Show that for the indicated responses, the two circuits given in Fig. 17-11 are true differentiators.

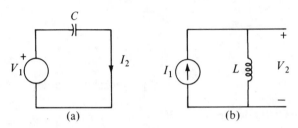

(a)   (b)

Fig. 17-11

17-17 Obtain the system function which has the following linear magnitude characteristic:

Slope: 1;

Pass-band: $0 - 1$ Krad/s;

Departure from linearity in pass-band ≤ 3 dB;

Attenuation for ω large ≥ 18 dB/octave.

17-18 Explain qualitatively the operation of the system shown in Fig. 17-12. The output of the system under test is applied to the vertical amplifier of the oscilloscope and the output of the peak detector to the horizontal.

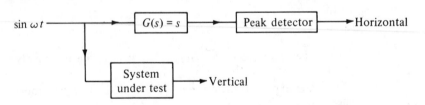

Fig. 17-12

18

Low-Pass
Realizations

Resistively terminated *LC* ladder networks are used to obtain passive realiza-tions of low-pass functions. The realization procedure is simple when ideal elements are considered. However, actual networks constructed on the basis of ideal elements have frequency-response characteristics that may depart from the ideal. In order to obtain the desired characteristics with lossy inductors, the transfer functions must be predistorted first and then realized. Two such pre-distortion techniques are discussed in this chapter. Examples are given to show the various steps that are undertaken in the design of low-pass networks with lossy elements. Active realizations are given in terms of cascaded second-order stages. The frequency response of a fourth-order, low-pass filter is given in terms of the responses of the two second-order stages.

18-1 PASSIVE REALIZATION (LOSSLESS INDUCTORS)

The all-pole, low-pass function

$$\frac{V_o}{V_i} = \frac{1}{s^n + a_{n-1}s^{n-1} + \cdots + a_1 s + a_0} \tag{18-1}$$

can be realized passively by a ladder network with resistive terminations at either or both ends. The inductors are placed in the series branches of the ladder and the capacitors in the shunt branches. Each inductor and each capacitor produce a zero at infinity. Because there are n zeros at infinity, the realization requires a total of n reactive elements. If n is even, the number of inductors is the same as the number of capacitors. If n is odd, the number of inductors is either one more

533

or one less than the number of capacitors, depending upon the kind of source and load terminations that are used.

The first step in the synthesis procedure is to partition the denominator polynomial of Eq. (18-1) into its even and odd parts (see Section 5-3):

$$\frac{V_o}{V_i} = \frac{1}{\text{Ev} + \text{Od}}. \tag{18-2}$$

If a resistive-source termination is desired, $Z_i = \text{Ev}/\text{Od}$ is formed. Then, depending upon the position of the first element, either Z_i (first element in series) or Y_i (first element in shunt) is developed as the input impedance or admittance of an LC network, as shown in Fig. 18-1a. On the other hand, if a resistive-load termination is desired, $Z_o = \text{Od}/\text{Ev}$ is formed. Then, either Z_o (first element in series) or Y_o (first element in shunt) is developed as the output impedance or admittance of an LC network, as shown in Fig. 18-1b. In either case, the impedance or the admittance is developed in Cauer I form so that the inductors appear in the series arms and the capacitors in the shunt arms of the ladder.

Note that the network can be drawn without resorting to Fig. 18-1 if the order n (even or odd) and the kind of termination (resistive source or load) is known. One simply starts the ladder with either an inductor in series or a capacitor in shunt, and proceeds to the right until n elements are used. Only the correct start produces the correct end element, which is either a shunt capacitor or the resistive load. The impedance (first element in series) or admittance (first element in shunt) facing the resistor is determined by inspection of the network. If the impedance or admittance is zero for $s = 0$, then the Od part must be on top. If the impedance or admittance is infinite for $s = 0$, then the Od part must be at the bottom.

Almost always, the low-pass approximations are listed for 1 rad/s cutoff frequency. Hence, V_o/V_i, which is formed from tables of poles or polynomials (Tables 17-1 and 17-2), produces a low-pass realization with $\omega_c = 1$ rad/s. To move the cutoff frequency from 1 to ω_c requires the substitution of ω/ω_c for ω and hence s/ω_c for s. It follows that sL becomes

$$\left(\frac{s}{\omega_c}\right) L = s\left(\frac{L}{\omega_c}\right)$$

and sC becomes

$$\left(\frac{s}{\omega_c}\right) C = s\left(\frac{C}{\omega_c}\right).$$

Thus, the networks in Fig. 18-1 can be made to have a cutoff frequency of ω_c by merely dividing all inductors and capacitors by ω_c. The resistor values remain the same. Since multiplication of all *impedances* by the same constant k does not change the voltage ratio, the impedance level can be changed by multiplying all resistors and inductors by k and dividing all capacitors by k. In this way a source or load termination other than $1\,\Omega$ is obtained.

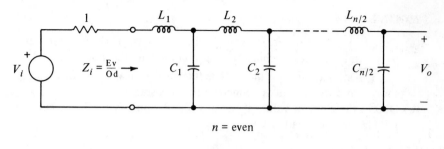

$$n = \text{even}$$

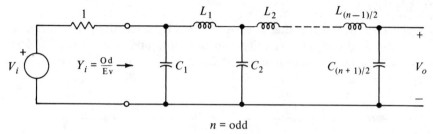

$$n = \text{odd}$$

(a) Resistive source termination

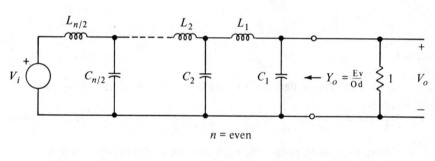

$$n = \text{even}$$

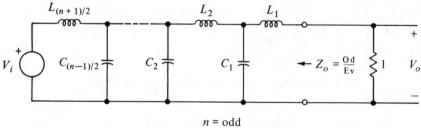

$$n = \text{odd}$$

(b) Resistive load termination

Fig. 18-1   Development of low-pass function $V_o/V_i = 1/(\text{Ev} + \text{Od})$

EXAMPLE 18-1

(a) Design a fourth-order, maximally flat, low-pass network that has a pass-band magnitude characteristic confined to the bounds of 1 and $1/\sqrt{1+\varepsilon^2}$. The cutoff is at ω_c. The filter is driven with an ideal voltage source, and the output is connected to a resistance of value R.

(b) Obtain the element values if the pass-band ripple is 1 dB, the cutoff is 1 KHz, and the load is 1 KΩ.

SOLUTION

(a) The order of the filter is even ($n = 4$). It is resistively terminated at the load end. Hence, using two inductors and two capacitors, draw the network as shown in Fig. 18-2. From Table 17-1, obtain the fourth-

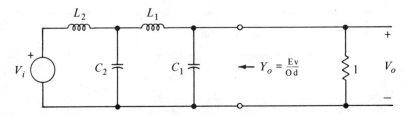

Fig. 18-2

order, maximally flat, 3-dB ripple ($\varepsilon = 1$, $\omega_c = 1$) approximation:

$$\frac{V_o}{V_i} = \frac{1}{s^4 + 2.6131s^3 + 3.4142s^2 + 2.6131s + 1}. \tag{18-3}$$

From Eq. (18-3), obtain the even and the odd parts and form Y_o:

$$Ev = s^4 + 3.4142s^2 + 1, \qquad Od = 2.6131s^3 + 2.6131s;$$

$$Y_o = \frac{Ev}{Od} = \frac{s^4 + 3.4142s^2 + 1}{2.6131s^3 + 2.6131s}. \tag{18-4}$$

Expand Y_o in Cauer I form (see Section 4-2 for details):

$$Y_o = 0.383s + \cfrac{1}{1.082s + \cfrac{1}{1.577s + \cfrac{1}{1.531s}}}$$

		1	3.4142	1
0.383	—			
	2.6131	2.6131		
1.082				
	2.4142	1		
1.577	—			
	1.5307			
1.531	—			
	1			

Hence, the element values are

$$C_1 = 0.383, \qquad L_1 = 1.082, \qquad C_2 = 1.577, \qquad L_2 = 1.531.$$

To confine the pass-band magnitude characteristic between 1 and $1/\sqrt{1 + \varepsilon^2}$, replace s with $(\varepsilon^{1/4}s)$. To move the cutoff to ω_c, replace s with s/ω_c. To have an R-ohm resistive-load termination, multiply all impedances by R.

All three changes are accomplished in one step by multiplying the two capacitor values by $\varepsilon^{1/4}/(\omega_c R)$ and the two inductor values by $\varepsilon^{1/4}R/\omega_c$. The result is

$$C_1 = \frac{0.383\varepsilon^{1/4}}{\omega_c R}, \qquad L_1 = \frac{1.082\varepsilon^{1/4}R}{\omega_c}, \qquad C_2 = 1.577\frac{\varepsilon^{1/4}}{\omega_c R},$$

$$L_2 = \frac{1.531\varepsilon^{1/4}R}{\omega_c}. \qquad Ans.$$

(b) For $\varepsilon = 0.5089$ (1-dB ripple), $\omega_c = 2\pi \times 10^3$, and $R = 10^3$, the element values become

$$C_1 = 515 \text{ pF}, \qquad L_1 = 1.455 \text{ mH}, \qquad C_2 = 2120 \text{ pF}, \qquad L_2 = 2.058 \text{ mH}.$$
$$Ans.$$

EXAMPLE 18-2

Show that for n even or n odd, element values of the networks in Fig. 18-1a are related to the element values of the networks in Fig. 18-1b.

SOLUTION

Consider n even, and refer to Fig. 18-3. In the resistive-source termination, the first element is a series element. Hence, the Cauer I expansion is performed on $Z_i = \text{Ev}/\text{Od}$ [Od appears at the bottom because $Z_i(0) = \infty$]. On the other hand, in the resistive-load termination, the first element is a shunt element. Hence, the Cauer I expansion is performed on $Y_o = \text{Ev}/\text{Od}$ [Od appears at the bottom because $Y_o(0) = \infty$].

Since both expansions are for the same Ev/Od, the resulting numerical values are the same. Thus, the value of L_1 in Fig. 18-3a is the same as the value of C_a in Fig. 18-3b. Similarly, C_1 of Fig. 18-3a and L_a of Fig. 18-3b have the same numerical value, etc. Thus, if the element values of one network are known, the other network can be constructed by transferring the known values in the proper sequence.

The same one-to-one correspondence is applicable for n odd.

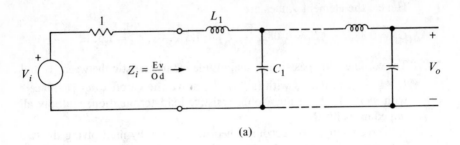

(a)

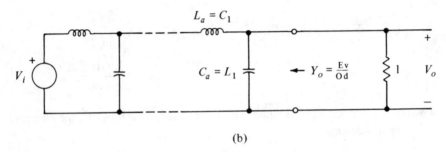

(b)

Fig. 18-3

18-2 LOW-PASS NETWORK WITH IMAGINARY-AXIS ZEROS

A low-pass function with a pair of imaginary-axis zeros can be readily implemented by modifying the terminal reactive elements of all-pole, low-pass realizations. Consider, for example,

$$\frac{V_o}{V_i} = \frac{H(s^2 + \omega_z^2)}{s^n + a_{n-1}s^{n-1} + \cdots + a_1s + a_0},$$ (18-5)

where ω_z represents the frequency of infinite attenuation. To produce the desired transfer-function zeros at $\pm j\omega_z$, either a parallel LC network must be introduced in a series arm of the ladder, or a series LC network must appear in a shunt arm of the ladder. This can be achieved by first developing Eq. (18-5) as if all the transfer-function zeros are at infinity (thereby realizing the poles of V_o/V_i). After all the element values are obtained, the terminal elements, L_p and C_p, are split, as shown in Fig. 18-4, to produce the imaginary-axis zeros. If the desired ω_z is less than $1/\sqrt{L_p C_p}$ and the resistive-source termination is used, then C_p is split as shown in Fig. 18-4a. Part of it (C_q) is put in the series arm and the rest (C_r), together with L_p, is connected in the shunt arm. Thus, the two zeros at infinity produced by L_p and C_p are taken out and, instead, the desired imaginary-axis zeros are produced by the $L_p C_r$ network. (C_q does not introduce a zero at the origin because of the shunt C_r.) If the desired ω_z is greater than $1/\sqrt{L_p C_p}$, then

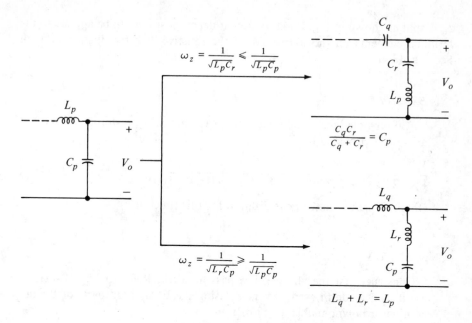

(a) Resistive-source termination

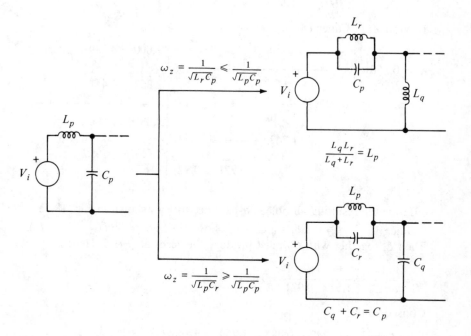

(b) Resistive-load termination

Fig. 18-4 Imaginary-axis zero-producing sections

L_p is split as shown in Fig. 18-4a. A similar decomposition can be applied to the terminal elements on the source side if the resistive-load termination is used. See Fig. 18-4b.

EXAMPLE 18-3

A four-pole, maximally flat approximation with a stop-band zero at $\omega = 1.5$ is provided by

$$\frac{V_o}{V_i} = \frac{1.8}{2.25} \cdot \frac{s^2 + 1.5^2}{s^4 + 2.860s^3 + 4.089s^2 + 3.441s + 1.800}. \qquad (18\text{-}6)$$

Obtain a passive network realization for this function.

SOLUTION

Construct first the all-pole, low-pass network. Refer to Fig. 18-1a and use it as a guide (n even). Form Z_i using the Ev and Od parts of the denominator polynomial [Eq. (18-6)]:

$$Z_i = \frac{\text{Ev}}{\text{Od}} = \frac{s^4 + 4.089s^2 + 1.800}{2.860s^3 + 3.441s}.$$

Expand Z_i in Cauer I form:

$$Z_i = 0.350s + \cfrac{1}{0.991s + \cfrac{1}{1.742s + \cfrac{1}{0.921s}}}$$

	1.000	4.089	1.800
0.350	—		
	2.860	3.441	
0.991			
	2.886	1.800	
1.742	—		
	1.657		
0.921	—		
	1.8		

Draw the resulting all-pole low-pass network as shown in Fig. 18-5a.

Since $1/\sqrt{L_2 C_2} < 1.5$, split L_2 as shown in Fig. 18-5b ($L_2 = L_q + L_r$). Then L_r together with C_2 must produce the zeros at $\pm j1.5$. Hence,

$$L_r = \frac{1}{\omega_z^2 C_2} = \frac{1}{1.5^2 \times 0.921} = 0.483. \qquad (18\text{-}7)$$

Consequently,

$$L_q = L_2 - L_r = 1.742 - 0.483 = 1.259. \qquad \textit{Ans.}$$

The resulting network with the desired transfer-function zeros is given in Fig. 18-5b.

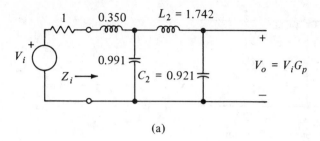

(a)

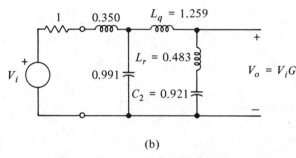

(b)

Fig. 18-5

18-3 NETWORK WITH LINEAR-AMPLITUDE CHARACTERISTIC

The band-limited, linear-amplitude function is discussed in Section 17-7. It is described by

$$G(s) = \frac{s}{s^n + a_{n-1}s^{n-1} + \cdots + a_1 s + a_0},$$ (18-8)

which can be readily implemented as a voltage ratio by arranging the input and output connections properly. For resistive-source termination,

$$Z_i = \text{Od}/\text{Ev} \quad (n \text{ odd}) \qquad \text{or} \qquad Y_i = \text{Ev}/\text{Od} \quad (n \text{ even})$$

is developed from the source end in Cauer I form, and the output is taken across the last inductor, L_l, as shown in Fig. 18-6a. For resistive-load termination,

$$Z_o = \text{Ev}/\text{Od} \quad (n \text{ even}) \qquad \text{or} \qquad Y_o = \text{Od}/\text{Ev} \quad (n \text{ odd})$$

is developed from the load end in Cauer I form, and the voltage source is introduced in series with the last capacitor, C_l, as shown in Fig. 18-6b. Thus, either L_l or C_l is properly positioned to obtain the desired zero at the origin.

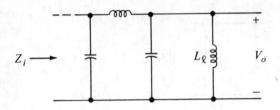

(a) Resistive-source termination

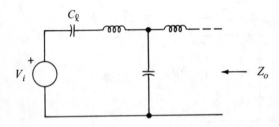

(b) Resistive-load termination

Fig. 18-6   Load and source terminations for producing a zero at the origin

EXAMPLE 18-4

The six-pole, maximally linear, band-limited magnitude approximation is characterized by the transfer function

$$\frac{V_o}{V_i} = \frac{Hs}{s^6 + 3.8637s^5 + 7.4641s^4 + 9.1416s^3 + 7.4641s^2 + 3.8637s + 1}.$$

(18-9)

Obtain a passive realization for this function.

SOLUTION

To obtain the network with resistive-source termination, develop $Y_i =$ Ev/Od in Cauer I form and take the output across the last inductor, as shown in Fig. 18-7:

$$Y_i = \frac{s^6 + 7.4641s^4 + 7.4641s^2 + 1}{3.8637s^5 + 9.1416s^3 + 3.8637s}$$

$$= 0.259s + \cfrac{1}{0.758s + \cfrac{1}{1.202s + \cfrac{1}{1.553s + \cfrac{1}{1.759s + \cfrac{1}{1.553s}}}}} \quad . \qquad Ans.$$

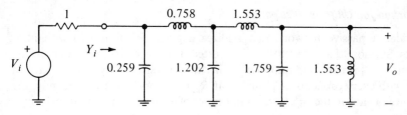

Fig. 18-7

18-4 PASSIVE REALIZATION (LOSSY INDUCTORS)

The passive realizations presented so far are difficult to implement in practice because of the nonideal nature of the actual elements. Whereas resistors and capacitors may come close to being ideal, such is not the case with inductors. As a result, the actual characteristics of the networks may depart appreciably from the expected characteristics. For example, the equal-ripple pass-band characteristic of a Chebyshev filter may no longer be equal-ripple and its cutoff may be less sharp.

The solution to this problem lies in the development of a synthesis procedure which takes into account the nonideal nature of the inductors. Two such procedures are presented in this section. Both use a more realistic model for the inductors by incorporating a series resistor as part of the inductors, as shown in Fig. 18-8b.

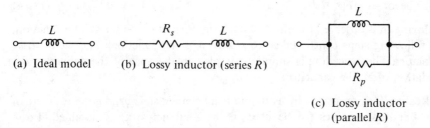

(a) Ideal model   (b) Lossy inductor (series R)

(c) Lossy inductor
(parallel R)

Fig. 18-8   Models for inductors

Equivalently, the inductor losses can be represented by a parallel resistance, as shown in Fig. 18-8c. It is possible to make the series representation equivalent to the parallel representation *at one frequency* of operation if R_s and R_p are related by

$$R_s = \frac{\omega^2 L^2}{R_p}.$$

$$(18\text{-}10)$$

Equation (18-10) is based on the approximation that $\omega L/R_s \gg 1$. (See Problem 18-7).

Predistortion for Uniform Dissipation

Consider a passive network that realizes a given function $T(s)$ with ideal elements. Suppose that a resistor R_{si} is inserted in series with every inductor, L_i, such that L_i/R_{si} is constant. Suppose further that a resistor, R_{pi}, is connected in parallel with every capacitor, C_i, such that $R_{pi} C_i$ is constant. Consequently, the impedance of each inductor and the admittance of each capacitor are transformed as follows:

$$sL_i \quad \text{becomes} \quad sL_i + R_{si} = L_i\left(s + \frac{R_{si}}{L_i}\right); \tag{18-11a}$$

$$sC_i \quad \text{becomes} \quad sC_i + \frac{1}{R_{pi}} = C_i\left(s + \frac{1}{R_{pi} C_i}\right). \tag{18-11b}$$

By appropriate selection of the added resistors it is possible to make all inductor time constants (L_i/R_{si}) and capacitor time constants $(R_{pi} C_i)$ equal; that is,

$$\frac{R_{s1}}{L_1} = \frac{R_{s2}}{L_2} = \cdots = \frac{1}{R_{p_1}C_1} = \frac{1}{R_{p_2}C_2} = \cdots = d, \tag{18-12}$$

where d represents the common dissipation factor. It is the inverse of the common time constant. When the dissipation is thus made uniform throughout the network, Eq. (18-11) simplifies to the following:

$$sL_i \quad \text{becomes} \quad L_i(s + d), \tag{18-13a}$$

$$sC_i \quad \text{becomes} \quad C_i(s + d). \tag{18-13b}$$

As a result,

$$T(s) \quad \text{becomes} \quad T(s + d). \tag{18-14}$$

Having thus observed the effect of uniform dissipation on the system function, the following steps can be taken to realize a given function, $T(s)$, with lossy inductors. The procedure is simple to implement because of the deliberate degradation of all the capacitors.

1. Realize $T(s)$, on paper, by assuming ideal elements. Thus, a good estimate of actual inductor values (to be used in the final network) is obtained. From measurements done on the inductors, determine the greatest R_s/L ratio. Let d represent this value.

2. Predistort the function to be realized by substituting $(s - d)$ for s, that is, construct $T(s - d)$. Thus, in anticipation of the forthcoming losses, the poles and zeros of the desired function are moved *to the right* by d units. However, the poles of the predistorted function must not be in the right half-plane if a realizable passive network is to result. Thus, if $-\alpha$ represents the real part of the pole of $T(s)$ that is nearest to the imaginary axis, then $d < \alpha$. (It may be necessary to use inductors with smaller dissipation to meet this condition.)

3. Realize the predistorted function with ideal elements.

4. Undo the predistortion by connecting a resistor of value $L_i d$ in series with each inductor and a resistor of value $1/(C_i d)$ in parallel with each capacitor. (Note that $L_i d$ represents the total series resistance including the actual resistance of the coil.) The introduction of the resistors moves the poles and zeros to the left d units, thereby positioning them in the desired locations [as determined by $T(s)$].

The dc gain of the resulting network is less than unity because of the series and shunt resistors in the ladder sections.

Most often realizations are obtained first on the basis of unity cutoff frequency. For example, all the approximations presented in Tables 17-1 and 17-2 are frequency-normalized. Hence, when the frequency-normalized functions are predistorted, d/ω_c, rather than d, must be used for the dissipation factor. (When the frequency normalization is later undone by replacing s with s/ω_c, the correct displacement of s then takes place.)

EXAMPLE 18-5

(a) Obtain a fourth-order, maximally flat, low-pass realization using the uniform dissipation method. Assume $d/\omega_c = 0.1$.
(b) What are the element values if $\omega_c = 10^6$ rad/s and the load is 50 Ω? What is the resulting d?

SOLUTION

(a) Using Table 17-1, construct the desired transfer function (3-dB ripple, 1-rad/s cutoff):

$$G(s) = \frac{1}{s^4 + 2.6131s^3 + 3.4142s^2 + 2.6131s + 1}.$$

Using $d = 0.1$, calculate the predistorted function by replacing s with $(s - 0.1)$. (Note that the poles of the desired function have real parts of -0.3827, and -0.9239, which are all greater than 0.1 in magnitude.) Then

$$G_{pd} = \frac{H}{(s - 0.1)^4 + 2.6131(s - 0.1)^3 + 3.4142(s - 0.1)^2}$$
$$+ 2.6131(s - 0.1) + 1$$

$$(18\text{-}15)$$

$$= \frac{H}{s^4 + 2.2131s^3 + 2.6903s^2 + 2.0047s + 0.7703}$$

$$= \frac{H}{Ev + Od}.$$

To obtain the low-pass network (with resistive-load termination) which realizes the predistorted function, form $Y_o = \text{Ev}/\text{Od}$ and develop it in Cauer I form:

$$Y_o = \frac{s^4 + 2.6903s^2 + 0.7703}{2.2131s^3 + 2.0047}.$$

	1	2.6903	0.7703
0.452	—		
	2.2131	2.0047	
1.240			
	1.7845	0.7703	
1.701	—		
	1.0494		
1.362	—		
	0.7703		

The resulting network is shown in Fig. 18-9a. To get the element values

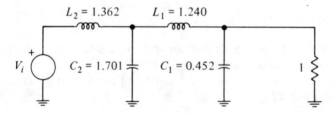

$L_2 = 1.362$ $L_1 = 1.240$

V_i $C_2 = 1.701$ $C_1 = 0.452$ 1

(a) Network corresponding to
predistorted function

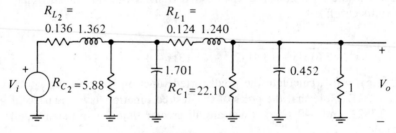

$R_{L_2} =$
0.136 1.362

$R_{L_1} =$
0.124 1.240

V_i $R_{C_2} = 5.88$ 1.701 $R_{C_1} = 22.10$ 0.452 1 V_o

(b) Network realizing the actual
transfer function

Fig. 18-9

of the desired network, resistors are added in series with each inductor and in shunt with each capacitor, such that

$$d = \frac{R_{L_1}}{L_1} = \frac{R_{L_2}}{L_2} = \frac{1}{R_{C_1}C_1} = \frac{1}{R_{C_2}C_2}.$$

Thus, the resistor values are obtained:

$$R_{L_1} = 0.1 \times 1.240 = 0.124, \qquad R_{L_2} = 0.1 \times 1.362 = 0.136,$$

$$R_{C_1} = \frac{10}{0.452} = 22.10, \qquad R_{C_2} = \frac{10}{1.701} = 5.88.$$

The final network is shown in Fig. 18-9b. Except for the scale factor, this network produces the same transfer function as the two networks shown in Fig. 18-2.

(b) To move the cutoff to 10^6 rad/s and to make the load 50 Ω, multiply all resistors by 50, all inductors by 50×10^{-6}, and all capacitors by 0.02×10^{-6}. The resulting d is 0.1×10^6.

Predistortion for Lossy Inductors

Consider the low-pass realizations shown in Fig. 18-10 where a resistor is

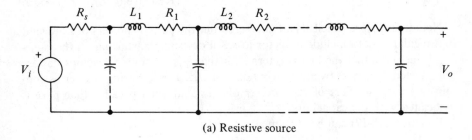

(a) Resistive source

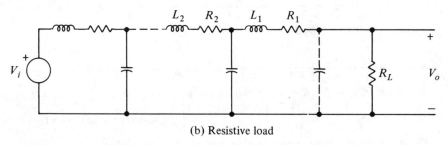

(b) Resistive load

Fig. 18-10

included in series with every inductor. The resistor values are chosen such that

$$\frac{R_1}{L_1} = \frac{R_2}{L_2} = \cdots = d.$$

However, unlike the uniform-dissipation case, resistors are not connected in parallel with the capacitors.

First assume that the inductors are ideal, that is, $R_1 = R_2 = \cdots = 0$. Express the resulting transfer function as

$$\frac{V_o}{V_i} = \frac{b_0}{b_n s^n + \cdots + b_5 s^5 + b_4 s^4 + b_3 s^3 + b_2 s^2 + b_1 s + b_0} = \frac{b_0}{\sum\limits_{i=0}^{n} b_i s^i}. \qquad (18\text{-}16)$$

When inductor losses are introduced (equal-time-constant type), Eq. (18-16) is modified as follows (see Problem 18-8):

$$\frac{V_o}{V_i} = \frac{b_0}{\cdots b_5 s^3 (s+d)^2 + b_4 s^2 (s+d)^2 + b_3 s^2 (s+d) + b_2 s (s+d) + b_1 s + b_0}$$

$$\text{(Resistive source)}, \qquad (18\text{-}17a)$$

$$\frac{V_o}{V_i} = \frac{b_0}{\cdots b_5 s^2 (s+d)^3 + b_4 s^2 (s+d)^2 + b_3 s (s+d)^2 + b_2 s (s+d) + b_1 (s+d) + b_0}$$

$$\text{(Resistive load)}. \qquad (18\text{-}17b)$$

Note that the *b*-coefficients are unaltered but the various powers of *s* are split in a regular way to include inductor losses. For example, starting on the right-hand side, note that the first two terms for the resistive-source development are not affected. The next two terms contain $(s + d)^1$. The following two terms contain $(s + d)^2$, etc. Part of each power of *s* is unaffected because that part is associated with the capacitors.

Equation (18-17) can be written as

$$\frac{V_o}{V_i} = \frac{b_0}{\cdots + s^3 (b_3 + 2b_4 d + b_5 d^2 + b_6 d^3)}$$
$$+ s^2 (b_2 + b_3 d + b_4 d^2) + s(b_1 + b_2 d) + b_0$$

$$\text{(Resistive source)}, \qquad (18\text{-}18a)$$

$$\frac{V_o}{V_i} = \frac{b_0}{\cdots + s^3 (b_3 + 2b_4 d + 3b_5 d^2 + b_6 d^3 + b_7 d^4)}$$
$$+ s^2 (b_2 + 2b_3 d + b_4 d^2 + b_5 d^3) + s(b_1 + b_2 d + b_3 d^2) + (b_0 + b_1 d)$$

$$\text{(Resistive load)}. \qquad (18\text{-}18b)$$

These equations must agree (within a scale factor) with the given transfer function which is specified in terms of the *a*-coefficients, namely

$$\frac{V_o}{V_i} = \frac{H}{a_n s^n + \cdots + a_3 s^3 + a_2 s^2 + a_1 s + a_0}. \qquad (18\text{-}19)$$

Hence, in Eqs. (18-18) and (18-19), the coefficients of the various powers of *s* must be identical. Equating the corresponding coefficients, *n* equations are obtained which are then solved simultaneously for the *b*-coefficients (in terms of the known *a*-coefficients and *d*).

Once the *b*-coefficients are determined, the predistorted function can be constructed:

$$\frac{V_o}{V_i} = \frac{b_0}{b_n s^n + b_{n-1}s^{n-1} + \cdots + b_1 s + b_0}.\tag{18-20}$$

Equation (18-20) is then realized with ideal elements as a low-pass network. Finally, the predistortion is undone by inserting a resistor of value $L_i d$ in series with every inductor in the network. (Note that $L_i d$ represents the total series resistance including the actual resistance of the coil.) This step converts Eq. (18-20) to Eq. (18-17) which, by construction, agrees with the desired transfer function [Eq. (18-19)]. Thus, a realization with lossy inductors is achieved.

If the desired function is based on a cutoff frequency of 1 rad/s, and the final network is to have a cutoff of ω_c rad/s, then d/ω_c (instead of d) must be used for predistortion. As in the case of uniform dissipation, d is obtained from the most lossy inductor that is used in the actual network. Since d is not known until the actual inductor values are obtained, an estimate of d can be made on the basis of inductor values obtained from a lossless-*L* realization. Unless losses are very great, the actual *L*'s will not differ appreciably from the ideal *L*'s. Hence, the *d* values of the inductors can be measured and the highest value of *d* can be determined. It should be realized that if actual inductor losses are too large and the desired pole locations are too close to the imaginary axis, this predistortion technique may result in negative values for the elements. It simply is not possible to obtain high-Q poles with poor inductors.

The various steps involved in this predistortion technique are best illustrated by means of an example.

EXAMPLE 18-6

(a) Obtain a fourth-order, equal-ripple (3-dB), low-pass realization using lossy inductors. Assume $d/\omega_c = 0.05$. The resistive termination is on the source side.
(b) Plot the magnitude characteristics corresponding to the desired and predistorted transfer functions. Also plot the magnitude characteristic produced by the network that is developed on the basis of ideal inductors but is then constructed with lossy inductors.

SOLUTION

(a) Using Table 17-2, obtain the fourth-order, 3-dB ripple, low-pass function ($\omega_c = 1$):

$$\frac{V_o}{V_i} = \frac{0.1770}{\underbrace{1s^4}_{a_4} + \underbrace{0.5816s^3}_{a_3} + \underbrace{1.1691s^2}_{a_2} + \underbrace{0.4048s}_{a_1} + \underbrace{0.1770}_{a_0}}.\tag{18-21}$$

Using Eq. (18-17a) and lossy inductors ($d = 0.05$), construct the transfer function that is produced by the actual fourth-order, low-pass network (resistive source):

$$\frac{V_o}{V_i} = \frac{b_0}{b_4 s^2 (s + 0.05)^2 + b_3 s^2 (s + 0.05) + b_2 s(s + 0.05) + b_1 s + b_0}$$

(18-22)

$$= \frac{b_0}{b_4 s^4 + (b_3 + 0.1 b_4) s^3 + (b_2 + 0.05 b_3 + 0.0025 b_4) s^2 + (b_1 + 0.05 b_2) s + b_0}.$$

(18-23)

Since the transfer function produced by the network must agree with the desired transfer function, equate the coefficients of equal powers of s:

$$
\begin{aligned}
b_4 &= a_4 = 1, \\
0.1 b_4 + b_3 &= a_3 = 0.5816, \\
0.0025 b_4 + 0.05 b_3 + b_2 &= a_2 = 1.1691, \\
0.05 b_2 + b_1 &= a_1 = 0.4048, \\
b_0 &= a_0 = 0.1770.
\end{aligned}
$$

Solve for the b-coefficients and obtain

$$b_4 = 1, \quad b_3 = 0.4816, \quad b_2 = 1.1425, \quad b_1 = 0.3477,$$
$$b_0 = 0.1770. \quad (18\text{-}24)$$

Using these coefficients, construct the predistorted function:

$$\left(\frac{V_o}{V_i}\right)_{pd} = \frac{b_0}{b_4 s^4 + b_3 s^3 + b_2 s^2 + b_1 s + b_0}$$

$$= \frac{0.1770}{s^4 + 0.4816 s^3 + 1.1425 s^2 + 0.3477 s + 0.1770}.$$

(18-25)

Realize the predistorted function using ideal inductors [refer to Fig. 18-1a ($n =$ even) for details]:

$$Z_i = \frac{Ev}{Od} = \frac{s^4 + 1.1425 s^2 + 0.1770}{0.4816 s^3 + 0.3477 s}.$$

(18-26)

Expand Z_i in Cauer I form:

$$
\begin{array}{l|lll}
 & 1 & 1.1425 & 0.1770 \\
Z_1 = 2.076s & \text{—} & & \\
 & 0.4816 & 0.3477 & \\
\hline
Y_2 = 1.145s & & & \\
 & 0.4205 & 0.1770 & \\
Z_3 = 2.906s & \text{—} & & \\
 & 0.1440 & & \\
Y_4 = 0.817s & \text{—} & & \\
 & 0.1770 & &
\end{array}
$$

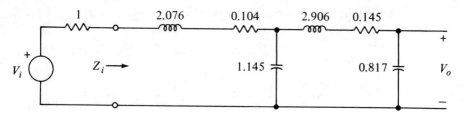

Fig. 18-11

Draw the resulting network and add a resistor of value $0.05 \times L$ in series with each inductor. The desired realization (employing lossy inductors) is shown in Fig. 18-11.

(b) Obtain the desired function from Eq. (18-21):

$$\left(\frac{V_o}{V_i}\right)_{desired} = \frac{0.1770}{s^4 + 0.5816s^3 + 1.1691s^2 + 0.4048s + 0.1770} \tag{18-27a}$$

$$= \frac{0.1770}{[(s + 0.0852)^2 + 0.9465^2][(s + 0.2056)^2 + 0.3921^2]}. \tag{18-27b}$$

The magnitude of the desired function [Eq. (18-27a)] is plotted in Fig. 18-12 (solid curve). Its upper half-plane poles are also shown. Obtain

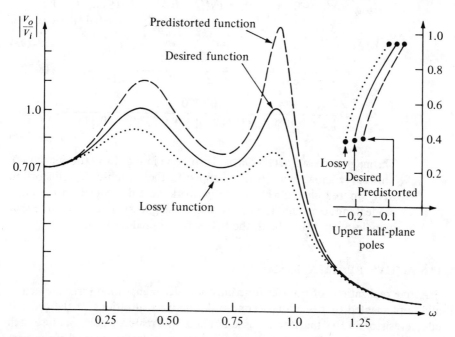

Fig. 18-12

the predistorted function from Eq. (18-25):

$$\left(\frac{V_o}{V_i}\right)_{\text{predistorted}} = \frac{0.1770}{s^4 + 0.4816s^3 + 1.1425s^2 + 0.3477s + 0.1770}$$

(18-28a)

$$= \frac{0.1770}{[(s + 0.0615)^2 + 0.9478^2][(s + 0.1793)^2 + 0.4050^2]}.$$

(18-28b)

The magnitude of the predistorted function [Eq. (18-28a)] is plotted in Fig. 18-12 (dashed curve). This is the response that the network of Fig. 18-11 produces if the loss resistors are taken out. Note that the poles of the predistorted function are closer to the imaginary axis than the poles of the desired function. The difference in the locations is mainly in the real parts of the poles. When the loss resistors are introduced, the predistorted poles move to the left and coincide with the desired pole locations. Because of the higher pole-Q's, the predistorted function has more than 3-dB ripple in the pass-band.

The transfer function of the network that is developed on the basis of ideal inductors is given by Eq. (18-17a). If loss resistors ($d = 0.05$) are added in series with the ideal inductors, the transfer function becomes

$$\left(\frac{V_o}{V_i}\right)_{\text{lossy}} = \frac{0.1770}{\begin{aligned}s^2(s + 0.05)^2 &+ 0.5816s^2(s + 0.05) + 1.1691s(s + 0.05) \\ &+ 0.4048s + 0.1770\end{aligned}}$$

(18-29a)

$$= \frac{0.1770}{s^4 + 0.6816s^3 + 1.2007s^2 + 0.4633s + 0.1770}$$

(18-29b)

$$= \frac{0.1770}{[(s + 0.1080)^2 + 0.9448^2][(s + 0.2328)^2 + 0.3762^2]}.$$

(18-29c)

The magnitude of Eq. (18-29a) is also plotted in Fig. 18-12 (dotted curve). This curve shows what would happen to the response if predistortion were ignored altogether and the network were developed on the basis of ideal inductors and then constructed using lossy inductors. Note that its poles would then be to the left of the desired locations.

18-5 ACTIVE REALIZATIONS

Active realizations of transfer functions of order higher than three are generally implemented by cascading second-order active realizations. If the order is odd, a resistor and a capacitor are used to realize the real-axis pole. Because each second-order section has a low output impedance (ideally zero), there is very little interaction between succeeding sections. Thus, each pair of complex poles can be individually tuned, a feature that is very desirable.

For an example, consider the realization of the fourth-order, maximally-flat, low-pass function as a voltage ratio:

$$\frac{V_o}{V_i} = G(s) = \left[\frac{1}{(s + 0.9239)^2 + 0.3827^2}\right]\left[\frac{1}{(s + 0.3827)^2 + 0.9239^2}\right]. \tag{18-30}$$

The two second-order transfer functions are

$$G_1(s) = \frac{H_1}{(s + 0.9239)^2 + 0.3827^2} = \frac{H_1}{s^2 + 1.8478s + 1} \tag{18-31a}$$

and

$$G_2(s) = \frac{H_2}{(s + 0.3827)^2 + 0.9239^2} = \frac{H_2}{s^2 + 0.7654s + 1}. \tag{18-31b}$$

A realization of

$$G(s) = HG_1(s)G_2(s)$$

is given in Fig. 18-13. Two second-order *KRC* networks, the properties of which are discussed in detail in Sections 10-3 and 10-4, are connected in cascade to obtain the desired response.

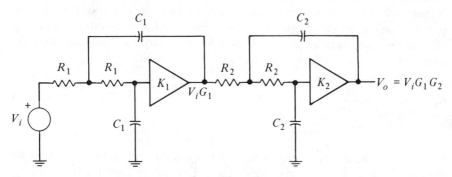

Fig. 18-13   Fourth-order low-pass realization

From Eqs. (10-11) and (10-12), the transfer functions are obtained:

$$G_1 = \frac{K_1\omega_{o1}^2}{s^2 + s\dfrac{\omega_{o1}}{Q_1} + \omega_{o1}^2}, \qquad \text{where } \omega_{o1} = \frac{1}{R_1C_1}, \qquad \text{and} \qquad Q_1 = \frac{1}{3 - K_1};$$

$$\tag{18-32a}$$

$$G_2 = \frac{K_2\omega_{o2}^2}{s^2 + s\dfrac{\omega_{o2}}{Q_2} + \omega_{o2}^2}, \qquad \text{where } \omega_{o2} = \frac{1}{R_2C_2}, \qquad \text{and} \qquad Q_2 = \frac{1}{3 - K_2}.$$

$$\tag{18-32b}$$

Comparison of Eq. (18-32) with Eq. (18-31) gives

$$\omega_{o1} = 1, \quad Q_1 = \frac{1}{1.8478} = 0.5412, \quad \omega_{o2} = 1, \quad Q_2 = \frac{1}{0.7654} = 1.3065.$$

Hence,

$$R_1C_1 = 1, \quad K_1 = 1.152, \quad R_2C_2 = 1, \quad K_2 = 2.235. \tag{18-33}$$

In each stage, the RC product sets the ω_o of the poles, while K adjusts the Q. Because pole-ω_o's are unity, both stages can be designed with identical values for the resistors and capacitors; for example,

$$R_1 = R_2 = R \quad \text{and} \quad C_1 = C_2 = C.$$

Then either R or C is arbitrarily chosen, and the remaining element is calculated from $RC = 1$. The resulting network has a cutoff frequency of 1 rad/s. To move the cutoff to ω_c rad/s, divide all capacitors by ω_c. If desired, the impedances of the elements in each stage can also be scaled up or down. Thus, in the first stage, the resistors can be multiplied by k and the capacitors divided by k without affecting the voltage ratio.

A very desirable feature of realizations that use cascaded second-order sections is that each second-order stage can be individually checked and, if necessary, tuned to the correct ω_o and Q. As shown in Fig. 18-14, the magnitude and phase characteristics of each stage can be used for this purpose. Note also that the transfer function of each stage evaluated at $s = j\omega_o$ yields:

$$G(j\omega_o) = \left. \frac{K\omega_o^2}{s^2 + s\dfrac{\omega_o}{Q} + \omega_o^2} \right|_{s = j\omega_o} = KQe^{-j(\pi/2)}. \tag{18-34}$$

Hence, when the sine-wave frequency is set at the value of ω_o (which represents the magnitude of the complex poles), the output phase must lag the input by exactly $\pi/2$ rad. If measurements indicate a lower phase shift, then the actual ω_o is too high and the RC product must be increased to lower the ω_o (since $\omega_o = 1/RC$). Furthermore, as Eq. (18-34) shows, the magnitude of the gain at $\omega = \omega_o$ is Q times the magnitude of the gain at dc. If it is lower, the gain K is increased until the correct value of the Q [$Q = 1/(3 - K)$] is obtained. When the two second-order stages (operating alone) produce the correct magnitude or phase responses, then the correct overall response is obtained when the stages are cascaded.

Because the poles of $G_2(s)$ have a higher Q, $G_2(s)$ is realized as the second stage in the cascaded structure. This way, neither amplifier is overdriven. (If the first and second stages are interchanged in Fig. 18-13, the first may be overdriven for frequencies near $\omega = 0.84/RC$.)

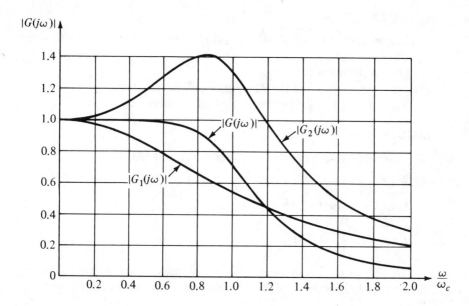

Fig. 18-14a   Magnitude of $G_1(j\omega)$, $G_2(j\omega)$, and $G(j\omega)$

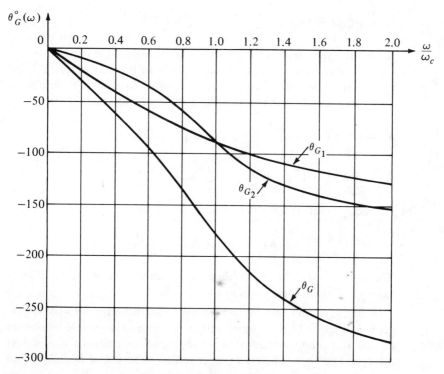

Fig. 18-14b   Phase of $G_1(j\omega)$, $G_2(j\omega)$, and $G(j\omega)$

EXAMPLE 18-7

The low-pass network of Fig. 18-13 is to be designed for a cutoff frequency of 100 Hz.
(a) Obtain the element values.
(b) Find the dc gain and the value of the maximum gain.
(c) Suggest schemes for varying the cutoff frequency continuously.

SOLUTION

(a) From Eq. (18-33) obtain the values for $\omega_c = 1$ rad/s:

$$R_1 C_1 = 1, \qquad K_1 = 1.152, \qquad R_2 C_2 = 1, \qquad K_2 = 2.235.$$

Let $C_1 = C_2 = 1$ μf; then,

$$R_1 = R_2 = \frac{1}{C_1} = 1 \ \text{M}\Omega.$$

To change the cutoff to $\omega_c = 2\pi \times 100$ rad/s, divide all capacitors by $2\pi \times 100$. Thus, the first- and second-stage element values become

$$R_1 = 1 \ \text{M}\Omega, \qquad C_1 = \frac{1}{200\pi} \ \mu\text{f}, \qquad K_1 = 1.152,$$

$$R_2 = 1 \ \text{M}\Omega, \qquad C_2 = \frac{1}{200\pi} \ \mu\text{f}, \qquad K_2 = 2.235.$$

An equally valid set of element values is obtained if the impedances are divided by 10 in the first stage and by 100 in the second stage. (This operation is indicated here for this example to show the flexibility, rather than the desirability, of such a scaling.) The result is

$$R_1 = 100 \ \text{K}\Omega, \qquad C_1 = \frac{1}{20\pi} \ \mu\text{f}, \qquad K_1 = 1.152,$$

$$R_2 = 10 \ \text{K}\Omega, \qquad C_2 = \frac{1}{2\pi} \ \mu\text{f}, \qquad K_2 = 2.235.$$

(b) Use Eq. (18-32) with $s = 0$ to obtain the dc gain:

$$\frac{V_o}{V_i} = G_1(0)G_2(0) = K_1 K_2 = 1.152 \times 2.235 = 2.575.$$

The dc gain is also the maximum gain, as can be seen from the $|G(j\omega)|$-vs.-ω/ω_c characteristic given in Fig. 18-14a.

If the individual and overall responses are to agree with the curves given in Fig. 18-14, the source resistance must be negligible relative to R_1 and R_2. A common error in the testing of such circuits is to forget this fact. For high-Q poles, a source resistance of 1% of R_1 (or R_2) can appreciably alter the selectivity characteristic.

(c) In order to vary the cutoff frequency, the poles of $G(s)$ [and hence $G_1(s)$

and $G_2(s)$] must be varied along radial lines, i.e., along constant-Q lines. Stated differently, the complex-frequency variable, s, must be scaled by multiplying it with k. For $k > 1$, the frequency scale is contracted, and for $k < 1$ the frequency scale is expanded.

From Eq. (10-12), it is seen that Q is a function of resistor ratios, capacitor ratios, and the amplifier gain K. Hence, Q can be kept constant by keeping constant the resistor and capacitor ratios and K. The cutoff frequency can then be changed by varying the RC product. The RC product can be continuously changed while Q is kept constant by ganging all four resistors or all four capacitors. The cutoff frequency can also be controlled smoothly without altering the pass-band and stop-band characteristics by using matched thermistors for the four resistors. These resistors can then be placed in an oven. The temperature of the oven sets the cutoff frequency.

Light sensitive resistors, FET's, or other devices, the characteristic of which can be controlled by a common variable, can be used to vary the cutoff frequency. The main problem is the matching of component values over the desired range of operation.

18-6 SUMMARY

Resistively terminated LC ladder networks and cascaded second-order RC-active networks are used to implement low-pass functions.

If all the zeros of the low-pass function are at infinity, all the inductors and all the capacitors are placed in the series and shunt arms of the ladder, respectively. Either Z or Y facing the resistive termination is formed from the even and odd parts of the denominator polynomial and developed in Cauer I form. If the first element is a series element, Z is constructed and developed. If the first element is a shunt element, Y is constructed and developed. Z or Y is formed as Od/Ev if $Z(0)$ or $Y(0)$ is zero (as determined by inspection of the network). The ratio is Ev/Od if $Z(0)$ or $Y(0)$ is infinite (as determined by inspection of the network).

A pair of imaginary-axis zeros can be readily implemented by splitting the terminal LC elements and putting either a parallel LC circuit in the series branch or a series LC circuit in the shunt branch. These circuits are tuned to provide infinite or zero impedance, respectively, at the desired rejection frequency. Similarly, a zero at the origin can be implemented by a series capacitor or a shunt inductor.

Low-pass filters can be implemented with lossy inductors if the function to be realized is predistorted first. In the uniform-dissipation development, a series resistor is included with every inductor and a shunt resistor is provided with every capacitor. Predistortion is achieved by substituting $(s - d)$ for s. The factor d represents the R/L ratio of all the inductors and the $1/RC$ ratio of all the capacitors. In the lossy-inductor development, a series resistor is included with every inductor. The R/L ratios are made the same for all the inductors.

PROBLEMS

18-1 Draw the schematic diagram of two first-order, low-pass networks. At $\omega = \omega_c$ the magnitude of the transfer function, V_o/V_i, is $[10 \log (1 + \varepsilon^2)]$ dB lower than at $\omega = 0$. The resistor value is R ohms. What are the element values?

18-2 Obtain a passive, fourth-order, maximally flat, low-pass realization which has a pass-band ripple of 3 dB and a cutoff frequency of ω_c. The source termination is R and the load is infinite.

18-3 Obtain two passive realizations that have a magnitude characteristic given by $V_o/V_i = 1/\sqrt{1 + \omega^6}$.

18-4 (a) Develop a synthesis procedure for resistively terminated LC networks which realize a given I_o/V_i.
(b) Repeat (a) for V_o/I_i.
(c) Repeat (a) for I_o/I_i.

18-5 For the transfer function given in Example 18-3, obtain a different realization.

18-6 For the transfer function given in Example 18-4, obtain a different realization.

18-7 Consider the terminal behavior of two circuits. One circuit is composed of an inductor, L_s, in series with a resistor, R_s. The other circuit is composed of an inductor, L_p, in parallel with a resistor, R_p. Show that the two circuits are equivalent at a given frequency if

$$L_p = L_s \quad \text{and} \quad R_p = \frac{\omega^2 L_s^2}{R_s}.$$

Assume losses are small, that is, $\omega L_s/R_s \gg 1$.

18-8 Consider resistively terminated LC networks where the inductors appear in the series arms and the capacitors in the shunt arms of the ladder. Show that when losses with equal R/L ratios are added to the inductors, the denominator polynomial of the transfer function changes from

$$b_0 + b_1 s + b_2 s^2 + b_3 s^3 + b_4 s^4 + b_5 s^5 + b_6 s^6 + \cdots$$

to

$$b_0 + b_1(s + d) + b_2 s(s + d) + b_3 s(s + d)^2 + b_4 s^2(s + d)^2 + b_5 s^2(s + d)^3$$
$$+ b_6 s^3(s + d)^3 + \cdots \qquad \text{(Resistive load)},$$

or to

$$b_0 + b_1 s + b_2 s(s + d) + b_3 s^2(s + d) + b_4 s^2(s + d)^2 + b_5 s^3(s + d)^2$$
$$+ b_6 s^3(s + d)^3 + \cdots \qquad \text{(Resistive source)}.$$

18-9 Obtain a fifth-order, maximally flat, low-pass network using inductors with equal R/L ratios. The network is driven from a resistive source, $d = 0.1$, $\omega_c = 1$.

18-10 Consider the fourth-order, maximally flat, low-pass function with $\omega_c = 1$. Using the resistive-source termination, obtain a passive realization employing:
   (a) ideal reactive elements,
   (b) inductors with equal R/L ratios $(d = 0.1)$,
   (c) inductors and capacitors with uniform dissipation $(d = 0.1)$.

18-11 The fourth-order KRC low-pass realization given in Fig. 18-13 is to be implemented with unity-gain amplifiers.
   (a) Give the design equations.
   (b) Obtain the element values if the magnitude characteristic is to be maximally flat and the cutoff frequency is to be at 1 KHZ.

18-12 Obtain a $-KRC$ realization that produces a third-order, maximally-flat, low-pass characteristic $(\omega_c = 1)$.

18-13 Obtain a KRC (unity-gain) realization that produces the 3-dB ripple, fourth-order Chebyshev low-pass characteristic. Cutoff is at 1 KHZ.

18-14 Obtain an active network with the following low-pass magnitude specifications: Allowable pass-band ripple ≤ 3 dB;
$$\text{Cutoff: } \omega_c;$$
$$\text{Attenuation at } \omega = 1.3\omega_c: \quad \text{greater than 20 dB.}$$

18-15 Give two active realizations for $V_o/V_i = Hs$.

18-16 Draw the schematic diagram (no values) of an active network that realizes

$$\left|\frac{V_o}{V_i}\right| = \frac{\omega}{\sqrt{1 + \omega^{10}}}.$$

High-Pass Functions and Realizations

The low-pass-to-high-pass transformation allows the construction of high-pass functions and passive networks from known low-pass functions and passive networks. In this chapter the properties of this transformation are discussed in detail. Several examples are given to show the usefulness of the transformation and the ease with which it can be implemented. The design of crossover networks is also presented.

19-1 THE LOW-PASS-TO-HIGH-PASS TRANSFORMATION

By means of a simple transformation, the vast knowledge available for low-pass (LP) functions and networks can be readily converted to high-pass (HP) functions and networks. The transformation is achieved by replacing s with $1/s$ in the low-pass functions. Thus, the HP function, $G_{HP}(s)$, is obtained from the LP function, $G_{LP}(s)$, by

$$G_{HP}(s) = G_{LP}(s)\Big|_{s \text{ replaced by } 1/s} . \qquad (19\text{-}1)$$

To distinguish between the two variables, let the LP complex-frequency variable be designated by s_{LP}. Then Eq. (19-1) can be written as

$$G_{HP}(s) = G_{LP}(s_{LP})\Big|_{s_{LP} = 1/s} . \qquad (19\text{-}2)$$

Since $s = 1/s_{LP}$, the poles and zeros of the HP function are the reciprocals of the poles and zeros of the LP function. Thus, an LP pole at $re^{j\theta}$ becomes, after transformation, an HP pole at

$$\frac{1}{re^{j\theta}} = \frac{1}{r}e^{-j\theta},$$

i.e., the magnitude of the HP pole is the inverse of the LP pole's magnitude and the angle of the HP pole is the negative of the angle of the LP pole. Consequently, a pair of complex-conjugate HP poles have the same Q as the corresponding LP poles.

What about the magnitude and phase of $G_{HP}(j\omega)$? From Eq. (19-2),

$$G_{HP}(j\omega) = G_{LP}(j\omega_{LP})\Big|_{j\omega_{LP}=1/j\omega}.$$

In terms of magnitudes and angles,

$$|G_{HP}(j\omega)|e^{j\theta_{HP}(\omega)} = \left\{ |G_{LP}(j\omega_{LP})|\Big|_{\omega_{LP}=1/\omega} \right\} e^{-j\theta_{LP}(\omega_{LP})}\Big|_{\omega_{LP}=1/\omega}.$$

Hence,

$$|G_{HP}(j\omega)| = |G_{LP}(j\omega_{LP})|\Big|_{\omega_{LP}=1/\omega}, \tag{19-3a}$$

$$\theta_{HP}(\omega) = -\theta_{LP}(\omega_{LP})\Big|_{\omega_{LP}=1/\omega}. \tag{19-3b}$$

Thus, the magnitude of the HP function at ω is equal to the magnitude of the LP function at $\omega_{LP} = 1/\omega$. The phase of the HP function at ω is the negative of the phase of the LP function at $\omega_{LP} = 1/\omega$. In other words, the magnitudes of the LP and HP functions are identical at respective frequencies which have a geometric mean of unity (or which are inverses of each other). At these same two frequencies, the phase of one is the negative of the phase of the other. As a result, the magnitude and phase of the LP function are converted to the magnitude and phase of the HP function, as shown in Fig. 19-1. Note that the

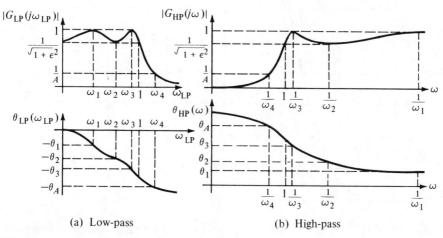

(a) Low-pass   (b) High-pass

Fig. 19-1   Low-pass-to-high-pass transformation

LP peak occurring at $\omega_{LP} = \omega_1$ is converted to the HP peak of the same value occurring at $\omega = 1/\omega_1$. The LP minimum occurring at $\omega = \omega_2$ occurs as an HP minimum of the same value at $\omega = 1/\omega_2$, etc. Observe also the phase relationships at these frequencies. The phases are equal but opposite in sign.

The cutoff frequency, $\omega_c = 1$, stays the same under the transformation. The LP frequencies from 0 to 1 become stretched to HP frequencies from ∞ to 1; on the other hand, the LP frequencies from 1 to ∞ become compressed to HP frequencies from 1 to 0.

The cutoff frequency, the pass-band ripple, and the stop-band attenuation requirement completely specifiy the HP function. These specifications are identical with the LP function provided the pass- and stop-band frequencies are interchanged. Therefore, any suitable LP function can serve as an HP function if $1/s$ is substituted for s. (If LP cutoff is at $\omega = \omega_c$ and the HP cutoff is also to be at $\omega = \omega_c$, then substitute ω_c^2/s for s).

In Fig. 19-2, the magnitude and phase of two fourth-order HP approximations are compared. The Butterworth magnitude characteristic is maximally flat about

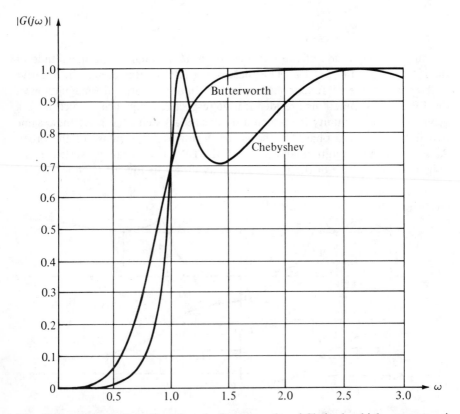

Fig. 19-2a   Magnitudes of fourth-order Butterworth and Chebyshev high-pass approximations

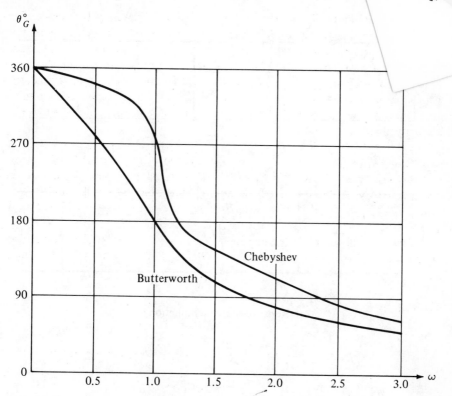

Fig. 19-2b Phases of fourth-order Butterworth and Chebyshev high-pass approximations

infinity. On the other hand, the Chebyshev magnitude characteristic has a narrower transition region between the pass- and stop-bands. Note also its ripply pass-band characteristic.

19-2 PASSIVE REALIZATIONS

The HP function can be realized as a voltage ratio of a passive network in one of two ways. In the first procedure, the function to be realized is split into its even and odd parts:

$$\frac{V_o}{V_i} = \frac{s^n}{s^n + a_{n-1}s^{n-1} + \cdots + a_1 s + a_0} \tag{19-4}$$

$$= \frac{s^n}{\text{Ev} + \text{Od}}.$$

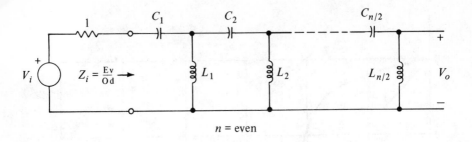

$n = \text{even}$

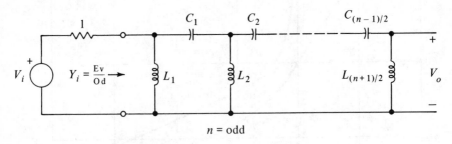

$n = \text{odd}$

(a) Resistive source termination

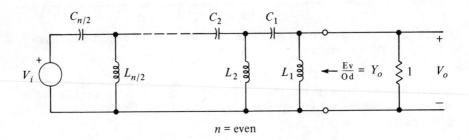

$n = \text{even}$

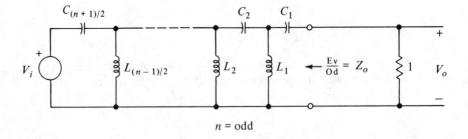

$n = \text{odd}$

(b) Resistive load termination

Fig. 19-3   High-pass networks $\dfrac{V_o}{V_i} = \dfrac{s^n}{(\text{Ev} + \text{Od})}$

Then, depending on the order of the function (even or odd) and the kind of resistive termination (source or load), the appropriate driving-point function is formed and expanded in Cauer II form. This expansion assures that all capacitors are placed in the series branches and all inductors in the shunt branches of the ladder network. In this way the n transfer-function zeros at the origin are realized. Figure 19-3 illustrates the resulting networks. It also shows which driving function is expanded and how it is formed from the odd and even parts of the denominator polynomial.

The second procedure of realization is more direct, and involves very little work provided the corresponding low-pass network, called the prototype, is known. The transformation is directly applied to the impedance of the inductors and the admittance of the capacitors of the LP network, as shown in Fig. 19-4.

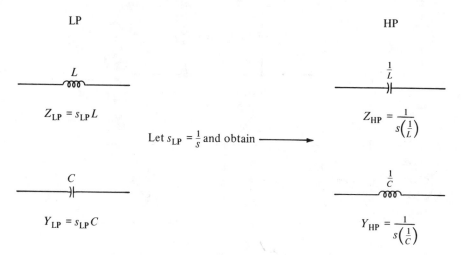

Fig. 19-4 Transformation of network elements

Thus, an inductor of value L in the LP network becomes transformed into a capacitor of value $1/L$ in the HP network. Similarly, a capacitor of value C becomes an inductor of value $1/C$. Resistor values remain the same. When the transformations are applied to all the elements of the LP network, the resulting network is the desired HP network.

EXAMPLE 19-1

(a) Obtain a passive, fourth-order maximally flat (about infinity), high-pass network that has a 3-dB pass-band ripple. The cutoff is at ω_c. The filter is driven with an ideal voltage source and the output is taken across a resistor of value R.

(b) Obtain the network if inductors with equal R/L ratios are employed. Let $d = R/L = 0.1$.

SOLUTION

(a) The desired specifications, when considered in terms of normalized LP specifications, are identical with those of Example 18-1 ($\varepsilon = 1$). Hence the LP prototype with 1-rad/s cutoff frequency is known. For convenience, it is reproduced in Fig. 19-5a. Construct the HP network (with 1-rad/s cutoff and 1-Ω load) from the LP prototype by

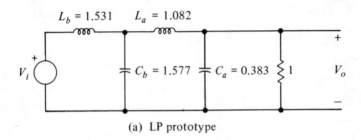

(a) LP prototype

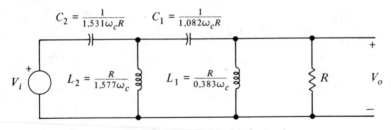

(b) HP network (ideal inductors)

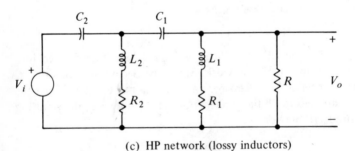

(c) HP network (lossy inductors)

Fig. 19-5   From LP network to HP network

1. replacing all inductors with capacitors and all capacitors with inductors; and

2. inverting all element values.

To move the cutoff frequency to ω_c, divide all inductors and all capacitors by ω_c. Finally, to get the desired load termination, multiply all impedances by R. The resulting network is shown in Fig. 19-5b.

(b) The high-pass network with lossy inductors cannot be constructed from the low-pass, lossy prototype. (The loss-producing resistors would appear in series with the capacitors of the HP network.) Instead, the high-pass function is obtained first and then predistorted. The predistorted function is realized with ideal elements. Then losses are introduced to undo the predistortion.

To obtain the HP function, construct the corresponding LP function, using Table 17-1, and then substitute $1/s$ for s:

$$G_{\text{LP}}(s) = \frac{1}{s^4 + 2.6131s^3 + 3.4142s^2 + 2.6131s + 1}, \qquad (19\text{-}5)$$

$$G_{\text{HP}}(s) = \frac{s^4}{s^4 + 2.6131s^3 + 3.4142s^2 + 2.6131s + 1}. \qquad (19\text{-}6)$$

This is the function to be realized with lossy inductors. To handle the effect of the losses explicitly, use the procedure discussed in Section 18-4 and (referring to Fig. 19-5b) obtain

$$\frac{V_o}{V_i} = \frac{s^2(s+d)^2}{b_4 s^2(s+d)^2 + b_3 s(s+d)^2 + b_2 s(s+d) + b_1(s+d) + b_0}, \qquad (19\text{-}7)$$

where $1/d$ represents the loss time constant of the inductors. Equate the denominator polynomials of Eqs. (19-6) and (19-7), and calculate the b-coefficients:

$$b_4 = 1, \qquad b_3 = 2.4131, \qquad b_2 = 2.9216, \qquad b_1 = 2.2968, \qquad b_0 = 0.7703.$$

Next, construct the predistorted transfer function using these b-coefficients:

$$G_{pd} = \frac{Hs^4}{s^4 + 2.4131s^3 + 2.9216s^2 + 2.2968s + 0.7703} = \frac{Hs^4}{\text{Ev} + \text{Od}}. \qquad (19\text{-}8)$$

Realize this transfer function as in Fig. 19-3b (n even), using ideal inductors:

$$Y_o = \frac{\text{Ev}}{\text{Od}} = \frac{0.7703 + 2.9216s^2 + s^4}{2.2968s + 2.4131s^3}.$$

Obtain the Cauer II continued-fraction expansion for Y_o:

$$
\begin{array}{c|ll}
 & 0.7703 \quad 2.9216 \quad 1 \\
\dfrac{1}{L_a} = 0.335 & - \\
 & 2.2968 \quad 2.4131 \\
\dfrac{1}{C_a} = 1.087 & \rule{4cm}{0.4pt} \\
 & 2.1123 \quad 1 \\
\dfrac{1}{L_b} = 1.593 & - \\
 & 1.3257 \\
\dfrac{1}{C_b} = 1.326 & - \\
 & 1
\end{array}
$$

Undo the predistortion by inserting $R_a = 0.1L_a$ in series with L_a, and $R_b = 0.1L_b$ in series with L_b. Finally, apply frequency and impedance scaling to obtain the network of Fig. 19-5c, where

$$L_1 = \frac{R}{0.335\omega_c}, \qquad R_1 = \frac{0.1R}{0.335\omega_c},$$

$$L_2 = \frac{R}{1.593\omega_c}, \qquad R_2 = \frac{0.1R}{1.593\omega_c},$$

$$C_1 = \frac{1}{1.087\omega_c R}, \qquad C_2 = \frac{1}{1.326\omega_c R}.$$

It should be realized that the network of Fig. 19-5c produces only two transfer-function zeros at the origin. These are due to the two series capacitors. Because of the lossy inductors, the other two zeros are not at the origin but at $s = -d\omega_c$. [See Eq. (19-7).] For d small, these two zeros can be considered, for all practical purposes, to be at the origin.

19-3 ACTIVE REALIZATIONS

Active realizations are implemented by cascading second-order, high-pass active networks. The properties of these networks are discussed in detail in Chapter 12. An example of a fourth-order high-pass realization is given below.

EXAMPLE 19-2

Obtain an active realization that meets the specifications given in Example 19-1.

SOLUTION

First, construct the prototype LP function by using Table 17-1:

$$G(s) = \left[\frac{1}{(s+0.9239)^2 + 0.3827^2}\right]\left[\frac{1}{(s+0.3827)^2 + 0.9239^2}\right]. \tag{19-9}$$

Next, obtain the 1-rad/s cutoff HP function by substituting $1/s$ for s:

$$G(s) = \left[\frac{1}{\left(\frac{1}{s}+0.9239\right)^2 + 0.3827^2}\right]\left[\frac{1}{\left(\frac{1}{s}+0.3827\right)^2 + 0.9239^2}\right]$$

$$= \left[\frac{s^2}{(1+0.9239s)^2 + (0.3827s)^2}\right]\left[\frac{s^2}{(1+0.3827s)^2 + (0.9239s)^2}\right]$$

$$= \left[\frac{s^2}{(s+0.9239)^2 + 0.3827^2}\right]\left[\frac{s^2}{(s+0.3827)^2 + 0.9239^2}\right]$$

$$= \left(\frac{s^2}{s^2 + 1.8478s + 1}\right)\left(\frac{s^2}{s^2 + 0.7654s + 1}\right). \tag{19-10}$$

Because the poles of the maximally flat LP function have unity magnitude, the poles of the maximally flat HP function are identical with the LP poles. (This is not the case with the Chebyshev function.)

An active realization of Eq. (19-10) is given in Fig. 19-6. The second-order *KRC* networks are discussed in detail in Section 12-3.

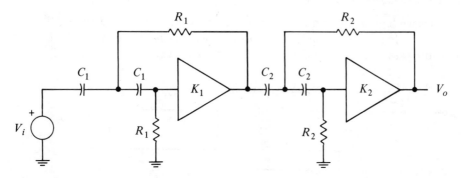

Fig. 19-6 Fourth-order high-pass realization

Obtain the transfer function of each second-order stage by using Eq. (12-11):

$$G_1 = \frac{K_1 s^2}{s^2 + s\dfrac{\omega_{o1}}{Q_1} + \omega_{o1}^2} = \frac{K_1 s^2}{s^2 + 1.8478s + 1},$$

where

$$\omega_{o1} = 1 = \frac{1}{R_1 C_1}, \qquad Q_1 = \frac{1}{1.8478} = \frac{1}{3 - K_1},$$

(19-11a)

$$G_2 = \frac{K_2 s^2}{s^2 + s \dfrac{\omega_{o2}}{Q_2} + \omega_{o2}^2} = \frac{K_2 s^2}{s^2 + 0.7654s + 1},$$

where

$$\omega_{o2} = 1 = \frac{1}{R_2 C_2}, \qquad Q_2 = \frac{1}{0.7654} = \frac{1}{3 - K_2}.$$

(19-11b)

Because the poles of each stage have the same magnitude, that is, $\omega_{o1} = \omega_{o2} = 1$, identical values can be used for all four resistors and all four capacitors. Equation (19-11) then gives

$$RC = 1, \qquad K_1 = 1.152, \qquad K_2 = 2.235.$$

(19-12)

Either R or C can be arbitrarily chosen and the remaining element calculated from $RC = 1$. The resulting network has a high-frequency gain of $K_1 K_2$ and a cutoff of 1 rad/s. To move the cutoff to ω_c rad/s, divide all capacitors by ω_c. If desired, the impedances of each stage can be independently scaled up or down by multiplication with the same constant.

19-4 CROSSOVER NETWORKS

Depending upon frequency, crossover networks direct the power flow to different loads. An example of a crossover network is shown in Fig. 19-7 where a low-pass network terminated in 1 Ω is connected in series with a high-pass

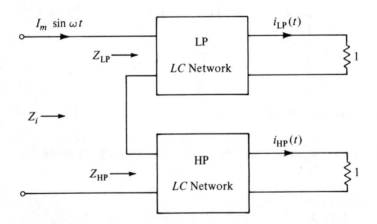

Fig. 19-7 Crossover network

network also terminated in 1 Ω. At low frequencies, power is absorbed by the LP network and rejected by the HP network. Thus power is delivered to the upper resistor. At high frequencies, the converse is true, and power is delivered to the lower resistor.

Let $G_{LP}(s)$ and $G_{HP}(s)$ represent respectively the transfer functions (I_2/I_1) of the two networks. The steady-state values of the load currents are given by

$$i_{LPss}(t) = I_m |G_{LP}(j\omega)| \sin[\omega t + \theta_{LP}(\omega)], \tag{19-13a}$$

$$i_{HPss}(t) = I_m |G_{HP}(j\omega)| \sin[\omega t + \theta_{HP}(\omega)]. \tag{19-13b}$$

Consequently, the average power delivered to each load is

$$P_{LP}(\omega) = \tfrac{1}{2} I_m^2 |G_{LP}(j\omega)|^2, \tag{19-14a}$$

$$P_{HP}(\omega) = \tfrac{1}{2} I_m^2 |G_{HP}(j\omega)|^2. \tag{19-14b}$$

Because the networks are lossless (*LC*), the output power must equal the input power, which is absorbed by the real part of the input impedance:

$$\tfrac{1}{2} I_m^2 |G_{LP}(j\omega)|^2 = \tfrac{1}{2} I_m^2 \, \text{Re}\{Z_{LP}(j\omega)\}, \tag{19-15a}$$

$$\tfrac{1}{2} I_m^2 |G_{HP}(j\omega)|^2 = \tfrac{1}{2} I_m^2 \, \text{Re}\{Z_{HP}(j\omega)\}. \tag{19-15b}$$

Hence,

$$\text{Re}\{Z_{LP}(j\omega)\} = |G_{LP}(j\omega)|^2, \tag{19-16a}$$

$$\text{Re}\{Z_{HP}(j\omega)\} = |G_{HP}(j\omega)|^2. \tag{19-16b}$$

Now suppose that the LP network is designed to possess a maximally-flat magnitude characteristic with 3-dB ripple in the pass-band and the HP network is obtained from the LP network by substituting ω_c^2/s for s. This means that

$$|G_{LP}(j\omega)|^2 = \frac{1}{1 + \left(\dfrac{\omega}{\omega_c}\right)^{2n}}, \tag{19-17a}$$

$$|G_{HP}(j\omega)|^2 = |G_{LP}(j\omega)|^2_{\omega \text{ replaced by } \omega_c^2/\omega} = \frac{\left(\dfrac{\omega}{\omega_c}\right)^{2n}}{1 + \left(\dfrac{\omega}{\omega_c}\right)^{2n}}, \tag{19-17b}$$

where ω_c is the cutoff frequency at which the LP magnitude is down by 3 dB from the dc value. Using Eq. (19-14), the power delivered to each load and the combined power output are obtained:

$$P_{LP}(\omega) = \frac{1}{2} I_m^2 \frac{1}{1 + \left(\dfrac{\omega}{\omega_c}\right)^{2n}}, \tag{19-18a}$$

$$P_{\text{HP}}(\omega) = \frac{1}{2} I_m^2 \frac{\left(\dfrac{\omega}{\omega_c}\right)^{2n}}{1 + \left(\dfrac{\omega}{\omega_c}\right)^{2n}},$$ (19-18b)

$$P_{\text{total}}(\omega) = P_{\text{LP}}(\omega) + P_{\text{HP}}(\omega) = \tfrac{1}{2} I_m^2.$$ (19-18c)

Note that the sum of the powers delivered to the two loads is constant; i.e., it is independent of frequency. Consequently, the total power supplied by the input is constant, which implies that the real part of the input impedance, $Z_i(j\omega)$, is constant. As seen from Fig. 19-7,

$$\text{Re}\{Z_i(j\omega)\} = \text{Re}\{Z_{\text{LP}}(j\omega)\} + \text{Re}\{Z_{\text{HP}}(j\omega)\}$$

$$= |G_{\text{LP}}(j\omega)|^2 + |G_{\text{HP}}(j\omega)|^2 = \frac{1}{1 + \left(\dfrac{\omega}{\omega_c}\right)^{2n}} + \frac{\left(\dfrac{\omega}{\omega_c}\right)^{2n}}{1 + \left(\dfrac{\omega}{\omega_c}\right)^{2n}} = 1.$$

(19-19)

Furthermore, because of maximal flatness,

$$\text{Im}\{Z_{\text{LP}}(j\omega)\} = -\text{Im}\{Z_{\text{HP}}(j\omega)\}.$$ (19-20)

Therefore, it follows that

$$Z_i(j\omega) = 1 \qquad \text{and} \qquad Z_i(s) = 1.$$ (19-21)

Thus, the input impedance of the crossover network is constant.

EXAMPLE 19-3

As shown in Fig. 19-8, a third-order LP network is connected in parallel with a third-order HP network.
(a) Design the LP network so that V_{oLP}/V_i has a maximally flat magnitude characteristic. The pass-band ripple is 3 dB and cutoff is at ω_c.
(b) Obtain the HP network by using the LP-to-HP transformation.
(c) Show that the resulting network can be used as a crossover network in a speaker system. The upper and lower resistors represent the resistance of a woofer and tweeter, respectively.
(d) What impedance does the voltage source, V_i, see?
(e) Obtain the system functions

$$\frac{V_{oLP} + V_{oHP}}{V_i} \qquad \text{and} \qquad \frac{V_{oLP} - V_{oHP}}{V_i}.$$

What do these system functions represent?

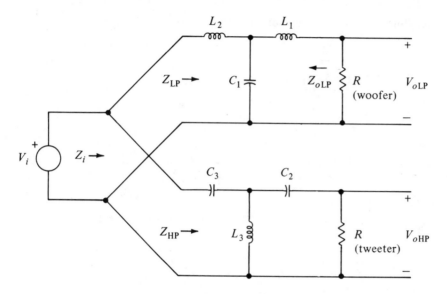

Fig. 19-8

SOLUTION

(a) Using Table 17-1, construct the 3-dB ripple, low-pass Butterworth function:

$$\frac{V'_{oLP}}{V_i} = \frac{1}{s^3 + 2s^2 + 2s + 1} = \frac{1}{\text{Ev} + \text{Od}}.$$

Form $Z'_{oLP} = \text{Od}/\text{Ev}$ and obtain the Cauer I (or Foster I) network based on 1-Ω termination and 1-rad/s cutoff frequency:

$$Z'_{oLP} = \frac{s^3 + 2s}{2s^2 + 1} = \frac{s}{2} + \frac{1}{\dfrac{4}{3}s + \dfrac{2}{s3}} = sL'_1 + \frac{1}{sC'_1 + \dfrac{1}{sL'_2}},$$

$$L'_1 = \tfrac{1}{2}, \qquad C'_1 = \tfrac{4}{3}, \qquad L'_2 = \tfrac{3}{2}.$$

To obtain R ohms of termination, multiply L'_1 and L'_2 by R and divide C'_1 by R. To obtain ω_c-rad/s cutoff frequency, divide the inductors and the capacitors by ω_c. The result is

$$L_1 = \frac{R}{2\omega_c}, \qquad C_1 = \frac{4}{3R\omega_c}, \qquad L_2 = \frac{3R}{2\omega_c}. \qquad Ans.$$

(b) In the LP network replace s with ω_c^2/s and obtain directly the element values of the HP network:

$$C_2 = \frac{1}{\omega_c^2 L_1} = \frac{2}{R\omega_c}, \qquad L_3 = \frac{1}{\omega_c^2 C_1} = \frac{3R}{4\omega_c}, \qquad C_3 = \frac{1}{\omega_c^2 L_2} = \frac{2}{3R\omega_c}. \qquad Ans.$$

(c) Since V_{oLP}/V_i is a third-order 3-dB Butterworth function,

$$\left|\frac{V_{oLP}}{V_i}\right|^2 = \frac{1}{1 + \left(\dfrac{\omega}{\omega_c}\right)^6}.$$

Consequently,

$$\left|\frac{V_{oHP}}{V_i}\right|^2 = \left|\frac{V_{oLP}}{V_i}\right|^2\bigg|_{\omega \text{ replaced by } \omega_c{}^2/\omega} = \frac{\left(\dfrac{\omega}{\omega_c}\right)^6}{1 + \left(\dfrac{\omega}{\omega_c}\right)^6}.$$

Hence, the total power supplied to the woofer and tweeter is

$$P_o = \frac{1}{2}\frac{V_m^2}{R}\frac{1}{1 + \left(\dfrac{\omega}{\omega_c}\right)^6} + \frac{1}{2}\frac{V_m^2}{R}\frac{\left(\dfrac{\omega}{\omega_c}\right)^6}{1 + \left(\dfrac{\omega}{\omega_c}\right)^6} = \frac{1}{2}\frac{V_m^2}{R},$$

where V_m represents the maximum value of the input sine wave. The total power output is constant. Hence, the network can be used as a crossover network which directs power to the woofer or tweeter according to frequency.

(d) Use the element values obtained in (a) and (b) to calculate Y_{LP} and Y_{HP}:

$$Y_{LP} = \frac{1}{3R}\frac{2\left(\dfrac{s}{\omega_c}\right)^2 + 4\left(\dfrac{s}{\omega_c}\right) + 3}{\left(\dfrac{s}{\omega_c}\right)^3 + 2\left(\dfrac{s}{\omega_c}\right)^2 + 2\left(\dfrac{s}{\omega_c}\right) + 1},$$

$$Y_{HP} = \frac{1}{3R}\frac{3\left(\dfrac{s}{\omega_c}\right)^2 + 4\left(\dfrac{s}{\omega_c}\right)^2 + 2\left(\dfrac{s}{\omega_c}\right)}{\left(\dfrac{s}{\omega_c}\right)^3 + 2\left(\dfrac{s}{\omega_c}\right)^2 + 2\left(\dfrac{s}{\omega_c}\right) + 1}.$$

Hence, the input impedance is

$$Z_i = \frac{1}{Y_{LP} + Y_{HP}} = R. \qquad Ans.$$

(e) From (a),

$$\frac{V_{oLP}}{V_i} = \frac{1}{\left(\dfrac{s}{\omega_c}\right)^3 + 2\left(\dfrac{s}{\omega_c}\right)^2 + 2\left(\dfrac{s}{\omega_c}\right) + 1}.$$

Replace s with ω_c^2/s and obtain

$$\frac{V_{oHP}}{V_i} = \frac{\left(\dfrac{s}{\omega_c}\right)^3}{\left(\dfrac{s}{\omega_c}\right)^3 + 2\left(\dfrac{s}{\omega_c}\right)^2 + 2\left(\dfrac{s}{\omega_c}\right) + 1},$$

$$\frac{V_{oLP} + V_{oHP}}{V_i} = \frac{1 + \left(\dfrac{s}{\omega_c}\right)^3}{\left(\dfrac{s}{\omega_c}\right)^3 + 2\left(\dfrac{s}{\omega_c}\right)^2 + 2\left(\dfrac{s}{\omega_c}\right) + 1}$$

$$= \frac{\left(\dfrac{s}{\omega_c} + 1\right)\left[\left(\dfrac{s}{\omega_c}\right)^2 - \left(\dfrac{s}{\omega_c}\right) + 1\right]}{\left(\dfrac{s}{\omega_c} + 1\right)\left[\left(\dfrac{s}{\omega_c}\right)^2 + \left(\dfrac{s}{\omega_c}\right) + 1\right]}$$

$$= \frac{\left(\dfrac{s}{\omega_c}\right)^2 - \left(\dfrac{s}{\omega_c}\right) + 1}{\left(\dfrac{s}{\omega_c}\right)^2 + \left(\dfrac{s}{\omega_c}\right) + 1}, \qquad Ans.$$

which represents an all-pass function. Similarly,

$$\frac{V_{oLP} - V_{oHP}}{V_i} = \frac{1 - \left(\dfrac{s}{\omega_c}\right)^3}{\left(\dfrac{s}{\omega_c}\right)^3 + 2\left(\dfrac{s}{\omega_c}\right)^2 + 2\left(\dfrac{s}{\omega_c}\right) + 1}$$

$$= -\frac{\left(\dfrac{s}{\omega_c} - 1\right)\left[\left(\dfrac{s}{\omega_c}\right)^2 + \left(\dfrac{s}{\omega_c}\right) + 1\right]}{\left(\dfrac{s}{\omega_c} + 1\right)\left[\left(\dfrac{s}{\omega_c}\right)^2 + \left(\dfrac{s}{\omega_c}\right) + 1\right]}$$

$$= -\frac{\left(\dfrac{s}{\omega_c} - 1\right)}{\left(\dfrac{s}{\omega_c} + 1\right)}, \qquad Ans.$$

which also represents an all-pass function.

19-5 SUMMARY

By replacing s with $1/s$, low-pass functions and networks are converted to high-pass functions and networks. Thus, the transformation interchanges pass- and stop-bands about $\omega = 1$. If the low-pass cutoff is at $\omega = \omega_c$, then s is replaced by ω_c^2/s to put the high-pass cutoff at $\omega = \omega_c$. A complex-conjugate pair of LP poles designated by (ω_{oLP}, Q_{LP}) becomes, upon transformation, a pair of HP poles that are designated by $(1/\omega_{oLP}, Q_{LP})$. Thus the magnitude of the poles changes, but not the Q. The desired pass-band characteristic determines the kind of approximation (Butterworth, Chebyshev) that is used. The allowable pass-band ripple determines the ε of the approximation, and the required stop-band attenuation fixes the order n.

The passive HP network is obtained from the low-pass prototype (1-rad/s cutoff and 1-Ω termination) by replacing all inductors with capacitors and all capacitors with inductors, and by inverting, at the same time, all element values. Frequency scaling can then be applied to put the cutoff at ω_c by dividing all inductors and capacitors by ω_c. Furthermore, magnitude scaling can be used on all impedances to provide the desired resistive termination. The active HP network is best obtained directly from the HP function by cascading second-order HP sections.

Crossover networks use LP and HP networks connected either in series or in parallel. The low-pass prototype is based on the maximally flat (3-dB ripple) approximation. When the crossover network is driven with an ideal voltage or current source, the input power, and hence the total output power, is constant, i.e., independent of frequency. However, the power to each load is a function of frequency.

PROBLEMS

19-1 Using the specifications of Example 19-1, construct the output admittance of the LC network. Expand Y_o in Cauer II form and obtain the HP network. Check the result with Fig. 19-5b.

19-2 Obtain the HP network which meets the specifications of Example 19-1. The network is driven by a resistive source.

19-3 The 1-dB ripple, fourth-order Chebyshev LP function is given by

$$G_{LP} = \frac{0.2457}{s^4 + 0.9528s^3 + 1.4539s^2 + 0.7426s + 0.2756}$$

$$= \frac{0.2457}{[(s + 0.1395)^2 + 0.9834^2][(s + 0.3369)^2 + 0.4073^2]} \, .$$

(a) Obtain the corresponding HP function.
(b) Sketch in the same s-plane the location of the LP and HP poles.
(c) Obtain two passive realizations for the HP function. What is the scale factor H for each realization?

19-4 (a) Apply the $s \to 1/s$ transformation to the LP active network shown in Fig. 19-9. Draw the resulting network.
(b) Multiply by $1/s$ all the impedances of the network obtained in (a) and compare the resulting network with Fig. 19-6.

(c) The transfer function of the LP network is

$$\frac{V_o}{V_i} = \frac{K}{(sRC)^2 + sRC(3 - K) + 1}.$$

What is the transfer function of the network of (b)?

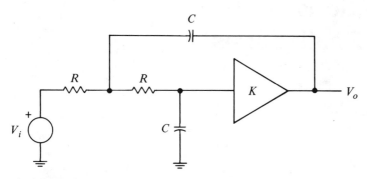

Fig. 19-9

19-5 Obtain two first-order high-pass networks with cutoff at ω_c. At $\omega = \omega_c$ the magnitude of the transfer function, $|V_o/V_i|$, is $[10 \log(1 + \varepsilon^2)]$-dB lower than at $\omega = \infty$. The resistor value is R ohms. What is the resulting transfer function?

19-6 The LP prototype function is

$$\frac{V_o}{V_i} = \frac{0.4913}{s^3 + 0.9883s^2 + 1.2384s + 0.4913} \qquad \text{(1 dB, Chebyshev)}.$$

   (a) Obtain the passive realization (with element values) for a 1-Ω load termination.
   (b) Draw the schematic diagram (no element values) of the active realization employing the infinite-gain realization.

19-7 (a) A crossover network is to be designed such that the combined average power output (woofer + tweeter) is independent of frequency. Both speakers are represented by an impedance of R ohms. The low-pass network is of third order and is connected in series with the high-pass network. The crossover frequency is ω_c. Obtain the element values of the crossover network, which is driven by a current source.
   (b) Obtain the input impedance of the crossover network.

19-8 (a) Design a crossover network employing second-order LP and HP sections. The speaker impedances are represented by R ohms and the combined average power output is independent of frequency. The cutoff frequency is ω_c.
   (b) What is the input impedance of the crossover network?

Band-Pass
Functions
and Realizations

The band-pass network is used more widely than any other kind of filter network. In Chapter 11, the second-order band-pass function and its passive and active realizations are discussed in detail. In this chapter, higher-order functions and networks are considered. The low-pass-to-band-pass transformation is used to generate the band-pass functions and the passive band-pass networks. The properties of the transformation are examined in detail and graphs are given for the pole locations. Butterworth- and Chebyshev-type magnitude and phase characteristics are compared with the characteristics obtained from mutually coupled coils. Expressions for the pole locations and for the bandwidth are obtained in terms of the coefficient of coupling and the parameters of the tuned circuits. Positions of the poles are displayed graphically for flat- and equal-ripple-type responses. The RC active equivalent of the mutually coupled coils is developed and discussed. Finally, a frequency discriminator is given which possesses a linear magnitude characteristic about the center frequency. Throughout the chapter, examples are introduced to demonstrate the various principles that are discussed in the text.

20-1 THE LOW-PASS-TO-BAND-PASS TRANSFORMATION

Band-pass (BP) functions and passive BP networks can be readily obtained from low-pass (LP) functions and passive LP networks by means of a simple transformation. The transformation requires the replacement of the variable s in LP functions with the variable $(s^2 + \omega_m^2)/s$BW. The constants ω_m and BW represent respectively the center frequency (rad/s) and the bandwidth (rad/s) of the BP filter. Thus, the BP function, $G_{BP}(s)$, is obtained from the LP function, $G_{LP}(s)$, by

$$G_{BP}(s) = G_{LP}(s)\Big|_{s \text{ replaced by } (s^2 + \omega_m^2)/sBW}. \tag{20-1}$$

To distinguish between the two variables, let the LP complex-frequency variable be designated by s_{LP}. Then Eq. (20-1) can be written as

$$G_{BP}(s) = G_{LP}(s_{LP})\Big|_{s_{LP} = (s^2 + \omega_m^2)/sBW}. \tag{20-2}$$

Pole Locations

To see how poles (or zeros) are transformed, consider a pair of complex-conjugate LP poles described by

$$s_{LP}^2 + s_{LP}\frac{\omega_{oLP}}{Q_{LP}} + \omega_{oLP}^2 = 0. \tag{20-3}$$

Replacement of s_{LP} by $(s^2 + \omega_m^2)/sBW$ results in the following expression for the poles of the BP function:

$$\left(\frac{s^2 + \omega_m^2}{sBW}\right)^2 + \left(\frac{s^2 + \omega_m^2}{sBW}\right)\frac{\omega_{oLP}}{Q_{LP}} + \omega_{oLP}^2 = 0. \tag{20-4}$$

Note that under this transformation each LP pole maps into two BP poles. Equation (20-4) can be written as

$$\left(\frac{s}{\omega_m}\right)^4 + \left(\frac{BW\omega_{oLP}}{\omega_m Q_{LP}}\right)\left(\frac{s}{\omega_m}\right)^3 + \left(2 + \frac{BW^2\omega_{oLP}^2}{\omega_m^2}\right)\left(\frac{s}{\omega_m}\right)^2$$
$$+ \left(\frac{BW\omega_{oLP}}{\omega_m Q_{LP}}\right)\left(\frac{s}{\omega_m}\right) + 1 = 0. \tag{20-5}$$

The roots of this equation depend upon four variables. Two of the variables (ω_{oLP}, Q_{LP}) come from the LP poles, and the remaining two (ω_m, BW) come from the transformation. With the substitutions

$$s_n = \frac{s}{\omega_m}, \qquad \delta = \left(\frac{BW}{\omega_m}\right)\omega_{oLP}, \tag{20-6a}$$

the roots of Eq. (20-5) are made dependent upon two variables (Q_{LP}, δ):

$$s_n^4 + \frac{\delta}{Q_{LP}}s_n^3 + (2 + \delta^2)s_n^2 + \frac{\delta}{Q_{LP}}s_n + 1 = 0. \tag{20-6b}$$

Let (ω_{oBP1}, Q_{BP1}) and (ω_{oBP2}, Q_{BP2}) designate the BP pole locations, which in factored form, produce

$$\left[s_n^2 + \left(\frac{\omega_{oBP1}}{\omega_m}\right)\frac{1}{Q_{BP1}}s_n + \left(\frac{\omega_{oBP1}}{\omega_m}\right)^2\right]\left[s_n^2 + \left(\frac{\omega_{oBP2}}{\omega_m}\right)\frac{1}{Q_{BP2}}s_n + \left(\frac{\omega_{oBP2}}{\omega_m}\right)^2\right] = 0.$$
$$\tag{20-7}$$

By equating the coefficients of equal powers of s_n in Eqs. (20-6b) and (20-7), the BP poles can be expressed in terms of Q_{LP}, δ, and ω_m (see Problem 20-1):

$$Q_{BP1} = Q_{BP2} = Q_{BP} = \frac{Q_{LP}}{\sqrt{2}} \sqrt{1 + \frac{4}{\delta^2} + \sqrt{\left(1 + \frac{4}{\delta^2}\right)^2 - \frac{4}{\delta^2 Q_{LP}^2}}}, \qquad (20\text{-}8a)$$

$$\omega_{oBP2,1} = \frac{\omega_m}{2}\left[\delta\frac{Q_{BP}}{Q_{LP}} \pm \sqrt{\left(\delta\frac{Q_{BP}}{Q_{LP}}\right)^2 - 4}\right]. \qquad (20\text{-}8b)$$

Both complex-conjugate BP poles have the same Q. The geometric mean of the magnitudes of the poles is ω_m, that is,

$$\sqrt{\omega_{oBP1}\,\omega_{oBP2}} = \omega_m.$$

These relationships are illustrated in Fig. 20-1. The transformation also yields a pair of zeros at the origin, as shown.

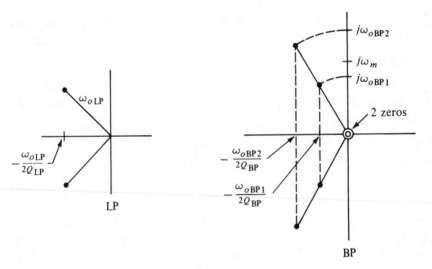

Fig. 20-1   LP to BP transformation of poles

The Q and the two ω_o's of the BP poles are plotted in Fig. 20-2 as a function of the low-pass Q and the parameter δ. These curves can be used to determine the location of the BP poles if the LP poles and the constants of the transformations are given. Consider, for example,

$$\omega_{oLP} = 1, \qquad Q_{LP} = 3, \qquad BW = 0.1, \qquad \omega_m = 1.$$

Then

$$\delta = \frac{BW\omega_{oLP}}{\omega_m} = 0.1.$$

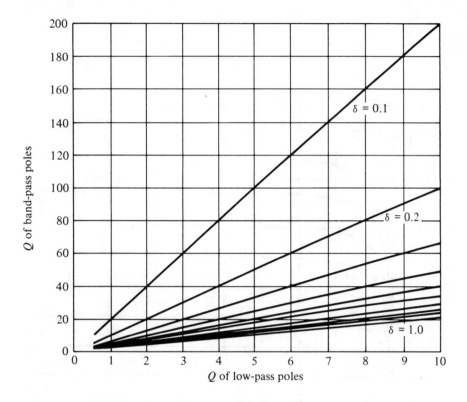

Fig. 20-2a   Band-pass Q as a function of low-pass Q and δ; $\delta = (BW/\omega_m)\, \omega_{oLP}$

With $\delta = 0.1$ and $Q_{LP} = 3$, Fig. 20-2a gives $Q_{BP} = 60$, and Fig. 20-2b gives

$$\omega_{oBP2} = 1.05, \qquad \omega_{oBP1} = 0.95.$$

As long as

$$Q_{LP} \gg \frac{1}{\sqrt{2}}, \tag{20-9}$$

Eq. (20-8) can be approximated by

$$Q_{BP} = Q_{LP}\sqrt{1 + \frac{4}{\delta^2}}, \tag{20-10a}$$

$$\omega_{oBP2,1} = \omega_m\left(\sqrt{1 + \frac{\delta^2}{4}} \pm \frac{\delta}{2}\right). \tag{20-10b}$$

Further insight into the behavior of the BP poles is gained by studying Fig. 20-3, which is a plot of the upper-half-plane roots of Eq. (20-6b) as a function of Q_{LP} and δ. The solid lines represent constant-Q_{LP} loci; the dashed lines

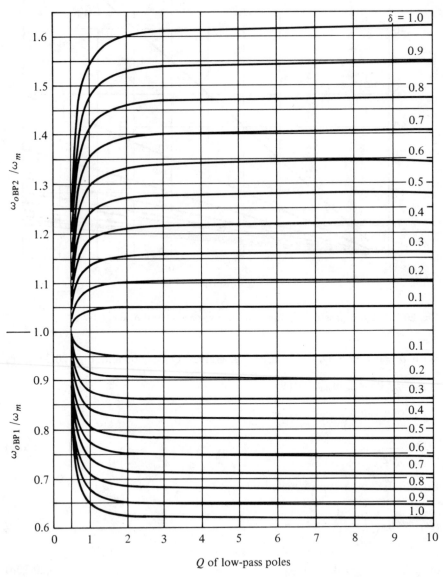

Fig. 20-2b   Band-pass ω_o's as a function of low-pass Q and δ; $\delta = (BW/\omega_m)\,\omega_{oLP}$

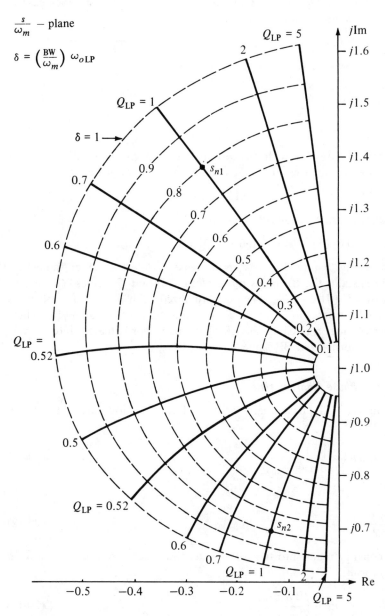

Fig. 20-3 Upper-half-plane band-pass poles as a function of Q_{LP} and δ

represent constant-δ loci. The BP poles are at the intersection of the constant-Q_{LP} lines with the constant-δ lines. Because there are two solid lines for each given Q_{LP} (except for $Q_{LP} = 0.5$), two upper-half-plane poles are located for every LP pole. As an example, consider:

LP pole: $\omega_{oLP} = 0.5$, $Q_{LP} = 1$;

Transformation constants: $BW = 0.8$, $\omega_m = 0.5$.

The resulting δ is

$$\frac{0.8 \times 0.5}{0.5} = 0.8.$$

The intersection of $\delta = 0.8$ with $Q_{LP} = 1$ in Fig. 20-3 gives the points s_{n1} and s_{n2} (a line drawn through s_{n1} and s_{n2} passes through the origin):

$s_{n1} = -0.265 + j1.385$, $s_{n2} = -0.134 + j0.694$.

Since $s_n = s/\omega_m = 2s$, the four BP poles are at

$s = -0.1325 \pm j0.6925$, $s = -0.670 \pm j0.347$.

Note that as the bandwidth, BW, is decreased while the center frequency, ω_m, is held constant, the BP poles move along constant-Q_{LP} lines ($Q_{LP} = 1$ in this case), toward $\pm j1.0$ ($\pm j\omega_m$ when denormalized). For $\delta \leq 0.1$, the BP poles for all practical purposes are on a circle of radius $\delta/2$ centered at $\pm j1.0$. Thus, only for δ small does a circle in the LP plane map approximately into a circle about $j1$ in the normalized BP plane.

Magnitude and Phase Characteristics

The magnitude and phase characteristics of BP functions are obtained by using Eq. (20-2):

$$G_{BP}(j\omega) = G_{LP}(j\omega_{LP})\Big|_{j\omega_{LP} = \frac{\omega_m^2 - \omega^2}{j\omega BW}}. \tag{20-11}$$

Since

$$\omega_{LP} = \frac{\omega^2 - \omega_m^2}{\omega BW}, \tag{20-12}$$

each LP frequency maps into two BP frequencies. These are found by solving Eq. (20-12) for ω:

$$\omega = \frac{BW\omega_{LP}}{2} \pm \sqrt{\omega_m^2 + \left(\frac{BW\omega_{LP}}{2}\right)^2}. \tag{20-13}$$

One of these frequencies is positive, the other negative. Two other BP frequencies, one positive, the other negative, are obtained when $-\omega_{LP}$ is used. This is because the $-\omega_{LP}$-to-ω_{LP} range maps into two BP ranges, one along the $+\omega$ axis and other along the $-\omega$ axis. (See Problem 20-2). Since the magnitude and phase for $s = +j\omega$ is of interest, the two positive BP frequencies are obtained from

$$\omega = \sqrt{\omega_m^2 + \left(\frac{\mathrm{BW}\omega_{\mathrm{LP}}}{2}\right)^2} \pm \frac{\mathrm{BW}\omega_{\mathrm{LP}}}{2}. \tag{20-14}$$

Let $|G_{\mathrm{LP}}(j\omega_{\mathrm{LP}})|$ and $\theta_{\mathrm{LP}}(\omega_{\mathrm{LP}})$ represent the magnitude and phase of the LP function at the frequency ω_{LP}. Then, the magnitude and phase of the BP function are given by

$$|G_{\mathrm{BP}}(j\omega)| = |G_{\mathrm{LP}}(j\omega_{\mathrm{LP}})|, \tag{20-15a}$$

$$\theta_{\mathrm{BP}}(\omega) = \pm\theta_{\mathrm{LP}}(\omega_{\mathrm{LP}}), \tag{20-15b}$$

where ω is determined by Eq. (20-14) and the proper association of the $\pm$ signs is observed. Thus, at

$$\omega = \sqrt{\omega_m^2 + \left(\frac{\mathrm{BW}\omega_{\mathrm{LP}}^2}{2}\right)} + \frac{\mathrm{BW}\omega_{\mathrm{LP}}}{2}$$

and at

$$\omega = \sqrt{\omega_m^2 + \left(\frac{\mathrm{BW}\omega_{\mathrm{LP}}}{2}\right)^2} - \frac{\mathrm{BW}\omega_{\mathrm{LP}}}{2},$$

the magnitude of the BP function is identical with the magnitude of the LP function evaluated at ω_{LP}. At the higher frequency, the phase is identical with the phase of the LP function evaluated at ω_{LP}, whereas at the lower frequency it is the negative of the LP phase.

Note that the geometric mean of the two frequencies given by Eq. (20-14) is ω_m. Hence, the BP magnitude is the same at the two BP frequencies, which have a geometric mean of ω_m. At these two frequencies the BP phases are of equal value but are of opposite sign. Equation (20-14) also indicates that the difference between these two BP frequencies is $\mathrm{BW}\omega_{\mathrm{LP}}$. In particular, *if ω_{LP} is taken as unity, the difference becomes* BW, *which can be taken as a measure for the bandwidth of the* BP *pass-band* (assuming that $\omega_{\mathrm{LP}} = 1$ represents the LP bandwidth).

Figure 20-4 shows the correspondence between the LP and BP magnitude and phase characteristics. Note in particular the relationship between the two magnitude and phase characteristics at the points o, p, and c. If the curve opc represents the pass-band in the LP plane, then the curve $cpopc$ represents the pass-band in the BP plane. The upper and lower BP edge frequencies (cutoff frequencies) and the bandwidth are given by

$$\omega_{2,1} = \sqrt{\omega_m^2 + \left(\frac{\mathrm{BW}}{2}\right)^2} \pm \frac{\mathrm{BW}}{2}, \tag{20-16a}$$

$$\omega_2 - \omega_1 = \mathrm{BW}. \tag{20-16b}$$

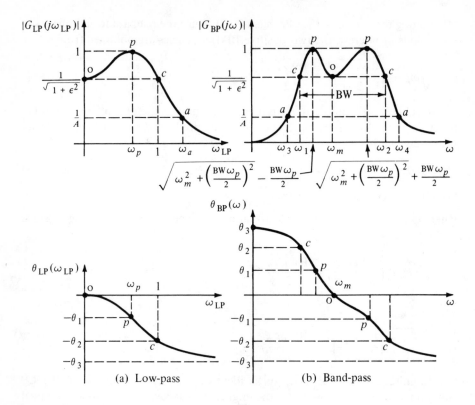

Fig. 20-4   Low-pass-to-band-pass transformation

As Fig. 20-4 shows, the given BP specifications can be transformed into corresponding LP specifications, with which a suitable LP function can be constructed. The allowable pass-band ripple fixes the ε of the LP approximation. The amount of required attenuation, (20 log A) dB, at or beyond a specified stop-band frequency, ω_a, fixes the order n of the LP function. The resulting LP function is then transformed into the BP function by substituting $(s^2 + \omega_m^2)/s\mathrm{BW}$ for s. The transformation constant ω_m is found from the geometric mean of the desired band edge frequencies and the constant BW is obtained from the difference; that is,

$$\omega_m = \sqrt{\omega_2 \omega_1}, \qquad \mathrm{BW} = \omega_2 - \omega_1. \tag{20-17}$$

EXAMPLE 20-1

A BP function is to be obtained with the following specifications:
Pass-band ripple: 3 dB or less; Pass-band: 41.4 rad/s to 241.4 rad/s;
Attenuation: at least 20 dB for $\omega > 300$ rad/s and for $\omega < 30$ rad/s.

SOLUTION

To confine the pass-band ripple to 3 dB or less requires that $\varepsilon \le 1$. Let $\varepsilon = 1$. From the pass-band edge frequencies, obtain the ω_m and the BW of the transformation:

$$\omega_m = \sqrt{\omega_2 \omega_1} = \sqrt{241.4 \times 41.4} = 100 \text{ rad/s},$$

$$BW = \omega_2 - \omega_1 = 241.4 - 41.4 = 200 \text{ rad/s}.$$

Transform the BP attenuation requirements to constraints on the LP function by using Eq. (20-12). Follow the designations introduced in Fig. 20-4 and obtain:

$$\omega_{a1} = -\frac{\omega_3^2 - \omega_m^2}{\omega_3 \, BW} = -\frac{30^2 - 100^2}{30 \times 200} = 1.51 \text{ rad/s},$$

$$\omega_{a2} = \frac{\omega_4^2 - \omega_m^2}{\omega_4 \, BW} = \frac{300^2 - 100^2}{300 \times 200} = 1.33 \text{ rad/s}.$$

Since $\omega_{a2} < \omega_{a1}$, choose the order of the LP function such that more than 20-dB attenuation is secured for $\omega_{LP} > 1.33$. This can be achieved with the eighth-order Butterworth approximation (see Fig. 17-3c) or with the fourth-order Chebyshev approximation (see Fig. 1-76c). Using the latter approximation, refer to Table 17-2 and construct the fourth-order, 3-dB ripple Chebyshev low-pass function:

$$G_{LP}(s) = \frac{0.1770/\sqrt{1.9953}}{s^4 + 0.5816s^3 + 1.1691s^2 + 0.4048s + 0.1770}. \tag{20-18}$$

Replace s with $(s^2 + 10^4)/200s$ and obtain the desired BP approximation:

$$G_{BP}(s) = \frac{0.1770/\sqrt{1.9953}}{\left(\dfrac{s^2 + 10^4}{200s}\right)^4 + 0.5816\left(\dfrac{s^2 + 10^4}{200s}\right)^3 + 1.1691\left(\dfrac{s^2 + 10^4}{200s}\right)^2} \tag{20-19}$$

$$+ 0.4048\left(\dfrac{s^2 + 10^4}{200s}\right) + 0.1770$$

$$= \frac{2.0049 \times 10^8 s^4}{s^8 + 116.32s^7 + 8.6764 \times 10^4 s^6 + 672.8 \times 10^4 s^5}$$

$$+ 1818.48 \times 10^6 s^4 + 672.8 \times 10^8 s^3 + 867.64 \times 10^{10} s^2$$

$$+ 116.32 \times 10^{12} s + 10^{16} \tag{20-20}$$

The desired BP function is of eighth order with four zeros at the origin. The magnitude and phase of this BP function are plotted in Fig. 20-5. For comparison purposes, the magnitude and phase of the eighth-order Butterworth approximation are also given. The frequency axis is normalized to ω/ω_m. Note that the geometric symmetry about $\omega_m = 1$ causes crowding of the magnitude characteristic at the lower end of the pass-band and expansion at the higher end. Note further the sharper skirts produced by the Chebyshev BP.

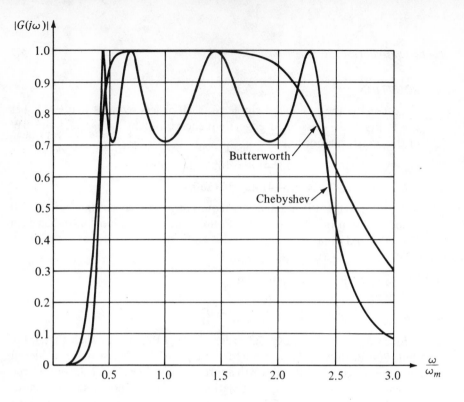

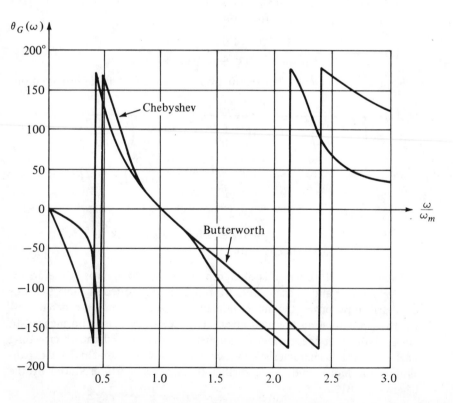

Fig. 20-5 (a) Eighth-order Butterworth and Chebyshev magnitude characteristics (b) Eighth-order Butterworth and Chebyshev phase characteristics

20-2 PASSIVE REALIZATIONS

A passive BP network can be developed either from the given BP function or obtained directly from the prototype LP network. In the latter approach, the LP-to-BP transformation is applied to the impedance of inductors and the admittance of capacitors of the LP network, as shown in Fig. 20-6.

LP $\qquad s_{LP} = \dfrac{s^2 + \omega_m^2}{BW} \qquad$ BP

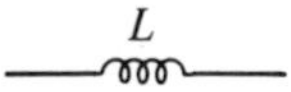

$$Z = s_{LP} L \quad \text{becomes} \quad Z = s\left(\frac{L}{BW}\right) + \frac{1}{s\left(\dfrac{BW}{\omega_m^2 L}\right)}$$

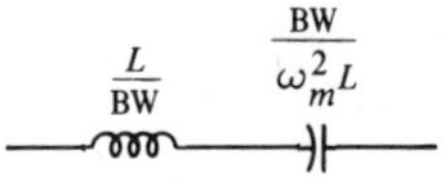

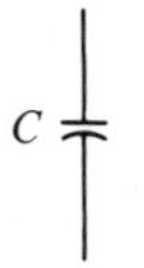

$$Y = s_{LP} C \quad \text{becomes} \quad Y = s\left(\frac{C}{BW}\right) + \frac{1}{s\left(\dfrac{BW}{\omega_m^2 C}\right)}$$

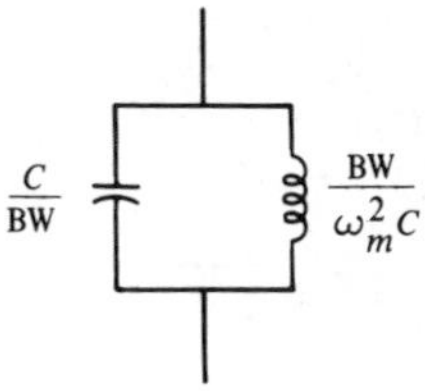

Fig. 20-6 Transformation of network elements

Thus, an inductor of value L in the LP network becomes transformed into the series combination of an inductor of value L/BW and a capacitor of value $BW/(\omega_m^2 L)$. Similarly, a capacitor of value C becomes the parallel combination of a capacitor of value C/BW with an inductor of value $BW/(\omega_m^2 C)$. Consequently, the BP center frequency can be changed without affecting the bandwidth by dividing all series capacitors and shunt inductors by ω_m^2. On the other hand, division of all series inductors and shunt capacitors by BW and multiplication of all series capacitors and shunt inductors by BW changes the bandwidth of the BP network without affecting the center frequency.

EXAMPLE 20-2

(a) Obtain a passive, eighth-order, maximally flat (about ω_m) BP network that has 3-dB pass-band ripple. The center frequency is ω_m and the bandwidth BW. The filter is driven from an ideal voltage source and the output is taken across a resistor of value R.

(b) Obtain low- and high-frequency equivalent circuits from the BP network.

SOLUTION

(a) The desired specifications, when considered in terms of the LP specifications, are identical with those of Example 18-1 ($\varepsilon = 1$). Hence the LP

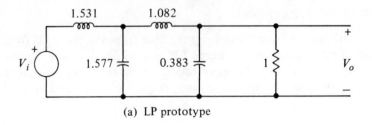

(a) LP prototype

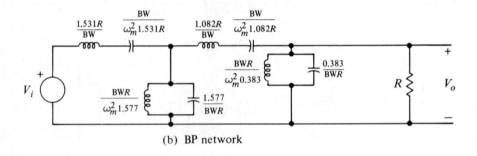

(b) BP network

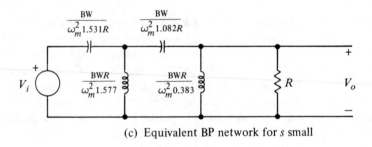

(c) Equivalent BP network for s small

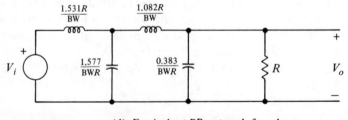

(d) Equivalent BP network for s large

Fig. 20-7   From LP to BP network

prototype with 1 rad/s cutoff is known. For convenience, this fourth-order LP network is reproduced in Fig. 20-7a.

The eighth-order BP network with center frequency ω_m and 3-dB bandwidth BW is constructed from the LP prototype network by using the transformation of elements, as shown in Fig. 20-6. The result is Fig. 20-7b where, in addition, all impedances are multiplied by R to produce the desired load termination.

(b) For s small, the impedance of all the series inductors is negligible in comparison with the impedance of the series capacitors; therefore, short circuits are substituted for all the series inductors. The shunt capacitors are open-circuited since their impedance is much higher than the impedance of the shunt inductors. The result is Fig. 20-7c, which is an HP network. Thus, at low frequencies, the BP network acts like an HP network. For s large, the impedance of all the series capacitors is negligible in comparison with the impedance of the series inductors, and the impedance of all the shunt inductors is much higher than the impedance of the shunt capacitors. Consequently, at high frequencies, the BP network acts like the LP network shown in Fig. 20-7d.

For $s = j\omega_m$, the impedance of all the series LC circuits is 0 and the impedance of all the shunt LC circuits is infinite. Hence, perfect transmission occurs at this frequency; that is,

$$\left. \frac{V_o}{V_i} \right|_{s=j\omega_m} = 1.$$

EXAMPLE 20-3

Explain how inductor losses can be incorporated in BP networks that realize a given transfer function.

SOLUTION

Consider the LP section shown in Fig. 20-8a. Assume that it is obtained by applying uniform-dissipation-type predistortion to the LP function, so that $R_L/L = 1/(R_c C)$.

Transform the LP section and obtain the BP section shown in Fig. 20-8b. Associate R_L with the series inductor and R_c with the shunt inductor. As shown in Fig. 20-8c, R_c can be converted into an equivalent resistance, $(\omega_m^2 L_{sh}^2)/R_c$, appearing in series with L_{sh}. (The equivalence is approximate and valid only for the frequency ω_m. See Problem 18-7.) Measure the losses associated with inductors L_s and L_{sh}. Based on the most lossy inductor, insert enough resistance in the series and shunt arms to obtain the same R/L ratio for all the BP inductors. For the BP section shown in Fig. 20-8b,

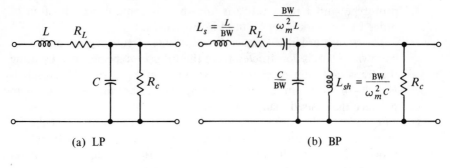

(a) LP   (b) BP

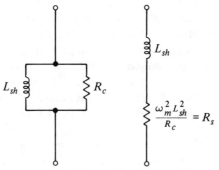

(c) R_c to R_s transformation

Fig. 20-8   LP-to-BP transformation

$$\frac{R}{L} \text{ for the series inductor} = \frac{R_L}{L_s} = \frac{R_L\,\text{BW}}{L}, \tag{20-21a}$$

$$\frac{R}{L} \text{ for the shunt inductor} = \frac{\dfrac{\omega_m^2 L_{sh}^2}{R_c}}{L_{sh}} = \frac{\text{BW}}{R_c C}. \tag{20-21b}$$

Equate the two ratios and obtain d:

$$\frac{R_L\,\text{BW}}{L} = \frac{\text{BW}}{R_c C} = d. \tag{20-22}$$

Divide through by BW to obtain the dissipation factor for the LP function:

$$\frac{R_L}{L} = \frac{1}{R_c C} = \frac{d}{\text{BW}}. \tag{20-23}$$

To obtain the predistorted LP function, replace s with $[s + (d/\text{BW})]$. The details for obtaining the LP network from the predistorted LP function are presented in Section 18-4.

20-3 COUPLED TUNED CIRCUITS

Coupled tuned circuits are widely used as band-pass networks. The basic building block is the second-order tuned circuit shown in Fig. 20-9a. The ω_o and Q of the poles of this circuit are given by

$$\omega_o = \frac{1}{\sqrt{LC}}, \qquad Q = \omega_o RC = R\sqrt{\frac{C}{L}}. \tag{20-24}$$

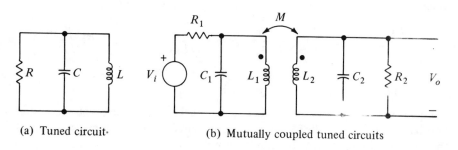

(a) Tuned circuit· (b) Mutually coupled tuned circuits

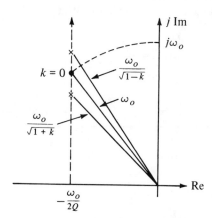

(c) k dependence of the upper half-plane poles

$$(\omega_{o1} = \omega_{o2} = \omega_o, \; Q_1 = Q_2 = Q)$$

Fig. 20-9 Band-pass network

Two such circuits, one tuned to (ω_{o1}, Q_1) and the other tuned to (ω_{o2}, Q_2) can be mutually coupled through the magnetic field, as shown in Fig. 20-9b. Here M represents the mutual inductance and k the coupling coefficient; M is given by:

$$M = k\sqrt{L_1 L_2} \qquad (0 \le k \le 1). \tag{20-25}$$

The resulting fourth-order transfer function is given by

$$\frac{V_o}{V_i} = k \,\frac{\omega_{o1}\omega_{o2}}{R_1\sqrt{C_1 C_2}}$$

$$\times \frac{s}{\left(s^2 + s\dfrac{\omega_{o1}}{Q_1} + \omega_{o1}^2\right)\left(s^2 + s\dfrac{\omega_{o2}}{Q_2} + \omega_{o2}^2\right) - k^2 s^2\left(s + \dfrac{\omega_{o1}}{Q_1}\right)\left(s + \dfrac{\omega_{o2}}{Q_2}\right)},$$

$$(20\text{-}26)$$

where

$$\omega_{o1} = \frac{1}{\sqrt{L_1 C_1}}, \qquad Q_1 = R_1\sqrt{\frac{C_1}{L_1}}, \qquad \omega_{o2} = \frac{1}{\sqrt{L_2 C_2}}, \qquad Q_2 = R_2\sqrt{\frac{C_2}{L_2}}.$$

Pole Locations

Equation (20-26) shows that when k approaches 0, the poles of the system approach the poles of the individually tuned circuits. As k is increased from 0, the two tuned circuits interact, and the pole positions change. Therefore, the frequency response changes. To simplify the study of this network, consider tuned circuits that are identical:

$$\omega_{o1} = \omega_{o2} = \omega_o \qquad \text{and} \qquad Q_1 = Q_2 = Q.$$

Then Eq. (20-26) can be written as

$$\frac{V_o}{V_i} = \frac{k}{Q}\,\frac{\left(\dfrac{s}{\omega_o}\right)}{\left[\left(\dfrac{s}{\omega_o}\right)^2 + \left(\dfrac{s}{\omega_o}\right)\dfrac{1}{Q} + 1\right]^2 - k^2\left(\dfrac{s}{\omega_o}\right)^2\left(\dfrac{s}{\omega_o} + \dfrac{1}{Q}\right)^2}$$

$$= \frac{k}{1 - k^2}\,\frac{1}{Q}$$

$$\times \frac{\left(\dfrac{s}{\omega_o}\right)}{\left(\dfrac{s}{\omega_o}\right)^4 + \left(\dfrac{s}{\omega_o}\right)^3\dfrac{2}{Q} + \left(\dfrac{s}{\omega_o}\right)^2\left(\dfrac{2}{1 - k^2} + \dfrac{1}{Q^2}\right) + \left(\dfrac{s}{\omega_o}\right)\dfrac{2}{Q(1 - k^2)} + \dfrac{1}{1 - k^2}}.$$

$$(20\text{-}27)$$

The coefficient of the $(s/\omega_o)^3$ term, $2/Q$, is independent of k. Since this coefficient represents the normalized negative sum of the four poles, the center of gravity of the four poles is independent of k. In fact, all four poles have the same real part, namely

$$\frac{1}{4}\left(\frac{-2\omega_o}{Q}\right) = -\frac{\omega_o}{2Q}.$$

Therefore, the denominator polynomial in Eq. (20-27) can be written in factored form as

$$
\frac{V_o}{V_i} = \frac{k}{1-k^2}\frac{1}{Q}\frac{\left(\dfrac{s}{\omega_o}\right)}{\left[\left(\dfrac{s}{\omega_o}+\dfrac{1}{2Q}\right)^2+\left(\dfrac{1}{1+k}-\dfrac{1}{4Q^2}\right)\right]\left[\left(\dfrac{s}{\omega_o}+\dfrac{1}{2Q}\right)^2+\left(\dfrac{1}{1-k}-\dfrac{1}{4Q^2}\right)\right]}
$$

(20-28)

$$
= \frac{k}{1-k^2}\frac{1}{Q}\frac{\left(\dfrac{s}{\omega_o}\right)}{\left[\left(\dfrac{s}{\omega_o}\right)^2+\left(\dfrac{s}{\omega_o}\right)\dfrac{1}{Q}+\dfrac{1}{1+k}\right]\left[\left(\dfrac{s}{\omega_o}\right)^2+\left(\dfrac{s}{\omega_o}\right)\dfrac{1}{Q}+\dfrac{1}{1-k}\right]}.
$$

(20-29)

The validity of the factorization of the denominator polynomial can be checked by multiplication and comparison.

Equation (20-29) shows that the poles are functions of k. The ω_{oBP} and Q_{BP} of the coupled circuits can be obtained in terms of the ω_o and Q of the uncoupled circuits:

$$
\omega_{oBP1} = \frac{\omega_o}{\sqrt{1+k}}, \qquad Q_{BP1} = \frac{Q}{\sqrt{1+k}}, \qquad \text{and} \qquad \omega_{oBP2} = \frac{\omega_o}{\sqrt{1-k}},
$$

$$
Q_{BP2} = \frac{Q}{\sqrt{1-k}}. \qquad (20\text{-}30)
$$

The upper-half-plane pole positions are drawn in Fig. 20-9c. The tighter the coupling (the larger the k), the more the poles are separated along the constant-α line.

Magnitude and Phase Characteristics

The magnitude and phase characteristics associated with Eq. (20-29) are plotted in Fig. 20-10 for $Q = 20$ and $k = 0.025, 0.05, 0.1, 0.2$. As k is increased from 0 (the poles separated farther from each other), the magnitude characteristic broadens, and hence the bandwidth increases. For $k > 0.05$, a dip and two humps appear about $\omega = \omega_o$. As k gets larger, the dip becomes larger. If $0.05 < k \le 0.1$, the maxima are of equal amplitude and occur at frequencies located very nearly equidistantly from $\omega = \omega_o$. The phase varies between $\pi/2$ and $-3\pi/2$. At $\omega = \omega_o$, the phase is $-\pi/2$ for all k.

At what frequency do the maxima and the minima occur? What is the bandwidth? Approximate analytical answers to these questions can be obtained if the coupling is loose and the Q of the tuned circuits is high. When these constraints are met, the imaginary part of the poles may be approximated by

$$
\beta = \sqrt{\frac{1}{1\pm k}-\frac{1}{4Q^2}} \cong \sqrt{\frac{1}{1\pm k}} \cong \sqrt{1\mp k} \cong 1\mp\frac{k}{2} \qquad \left(k \ll 1, \ Q \gg \frac{1}{2}\right),
$$

(20-31)

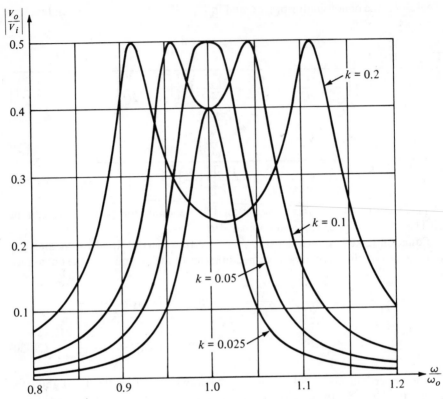

Fig. 20-10a   Magnitude characteristic of coupled tuned circuits ($Q = 20$)

and Eq. (20-28) may then be written as:

$$\frac{V_o}{V_i} \cong \frac{k}{Q} \frac{\dfrac{s}{\omega_o}}{\left[\dfrac{s}{\omega_o} + \dfrac{1}{2Q} - j\left(1 - \dfrac{k}{2}\right)\right]\left[\dfrac{s}{\omega_o} + \dfrac{1}{2Q} + j\left(1 - \dfrac{k}{2}\right)\right]}$$

$$\times \left[\dfrac{s}{\omega_o} + \dfrac{1}{2Q} - j\left(1 + \dfrac{k}{2}\right)\right]\left[\dfrac{s}{\omega_o} + \dfrac{1}{2Q} + j\left(1 + \dfrac{k}{2}\right)\right]$$

$$= \frac{k}{Q} \frac{\dfrac{s}{\omega_o}}{\left[\left(\dfrac{s}{\omega_o} + \dfrac{1}{2Q} - j\right)^2 + \dfrac{k^2}{4}\right]\left[\left(\dfrac{s}{\omega_o} + \dfrac{1}{2Q} + j\right)^2 + \dfrac{k^2}{4}\right]} . \qquad (20\text{-}32)$$

This approximation assumes that the poles are very close to the imaginary axis and near $s = \pm j\omega_o$. See Fig. 20-9c. Since the frequency region of interest is in the neighborhood of $\omega = \omega_o$, the frequency response is almost solely determined by the two upper-half-plane poles. Being quite far away, the two lower-half-plane poles and the zero at the origin play insignificant roles in

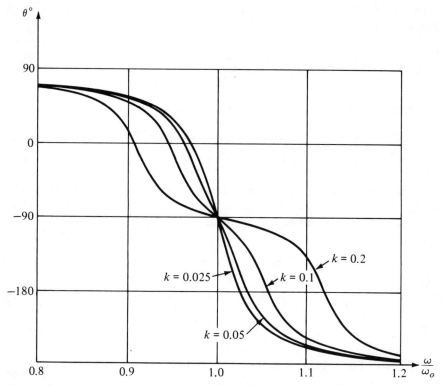

Fig. 20-10b   Phase characteristic of coupled tuned circuits ($Q = 20$)

the shaping of the frequency response. (Think in terms of the variations in the pole-to-$j\omega$ and zero-to-$j\omega$ distance vectors.) Consequently, all vectors except those vectors emanating from the upper-half-plane poles can be treated as constant vectors. Therefore, for $s = j\omega$ and ω near ω_o, Eq. (20-32) can be written as

$$
\frac{V_o}{V_i} = \frac{k}{Q} \frac{j\dfrac{\omega}{\omega_o}}{\left[\left(\dfrac{j\omega}{\omega_o} + \dfrac{1}{2Q} - j\right)^2 + \dfrac{k^2}{4}\right]\left[\left(\dfrac{j\omega}{\omega_o} + \dfrac{1}{2Q} + j\right)^2 + \dfrac{k^2}{4}\right]}
$$

$$
\cong \frac{k}{Q} \frac{j}{\left[\left(j\dfrac{\omega}{\omega_o} + \dfrac{1}{2Q} - j\right)^2 + \dfrac{k^2}{4}\right](2j)^2}
$$

$$
= \frac{k}{4Q} \frac{-j}{\left[\dfrac{1}{4Q^2} + \dfrac{k^2}{4} - \left(\dfrac{\omega}{\omega_o} - 1\right)^2\right] + j\dfrac{1}{Q}\left(\dfrac{\omega}{\omega_o} - 1\right)}. \qquad (20\text{-}33)
$$

To concentrate on the frequencies near ω_o, let $\omega = \omega_o + \Delta\omega$. Equation (20-33) then becomes

$$\frac{V_o}{V_i} = \frac{k}{4Q} \frac{-j}{\dfrac{1}{4Q^2} + \dfrac{k^2}{4} - \left(\dfrac{\Delta\omega}{\omega_o}\right)^2 + j\dfrac{1}{Q}\left(\dfrac{\Delta\omega}{\omega_o}\right)}. \tag{20-34}$$

The magnitude associated with Eq. (20-34) is

$$\left|\frac{V_o}{V_i}\right| = \frac{k}{4Q} \frac{1}{\sqrt{\left[\dfrac{1}{4Q^2} + \dfrac{k^2}{4} - \left(\dfrac{\Delta\omega}{\omega_o}\right)^2\right]^2 + \dfrac{1}{Q^2}\left(\dfrac{\Delta\omega}{\omega_o}\right)^2}}. \tag{20-35}$$

The maxima and the minimum of the magnitude characteristic can be obtained by differentiating the square of the denominator polynomial with respect to $(\Delta\omega/\omega_o)$, setting the result equal to zero, and solving for $\Delta\omega/\omega_o$:

$$2\left[\frac{1}{4Q^2} + \frac{k^2}{4} - \left(\frac{\Delta\omega}{\omega_o}\right)^2\right]\left[-2\left(\frac{\Delta\omega}{\omega_o}\right)\right] + \frac{2}{Q^2}\left(\frac{\Delta\omega}{\omega_o}\right) = 0,$$

$$\left(\frac{\Delta\omega}{\omega_o}\right)\left[\left(\frac{\Delta\omega}{\omega_o}\right)^2 - \frac{1}{4}\left(k^2 - \frac{1}{Q^2}\right)\right] = 0,$$

$$\frac{\Delta\omega}{\omega_o} = 0, \qquad \frac{\Delta\omega}{\omega_o} = \pm\frac{1}{2}\sqrt{k^2 - \frac{1}{Q^2}} \qquad \left(k > \frac{1}{Q}\right). \tag{20-36}$$

A study of Eqs. (20-35) and (20-36) shows that

1. For $k \leq 1/Q$, a maximum occurs at $\omega = \omega_o$. The maximum value of the magnitude is $kQ/(1 + k^2Q^2)$, which reduces to $\frac{1}{2}$ for $k = 1/Q$.

2. For $k > 1/Q$, a minimum occurs at $\omega = \omega_o$ and maxima at

$$\omega_{p2,1} = \omega_o \pm \frac{\omega_o}{2}\sqrt{k^2 - \frac{1}{Q^2}}.$$

The minimum value of the magnitude is $kQ/(1 + k^2Q^2)$. The maximum value is $\frac{1}{2}$.

These results, obtained with the assumption that $k \ll 1$ and $Q \gg \frac{1}{2}$, are in close agreement with the values presented by the actual magnitude curves shown in Fig. 20-10a for $Q = 20$.

For $k > 1/Q$, the magnitude characteristic is double-humped. For $k = 1/Q$, it is the flattest possible at $\omega = \omega_o$. However, this flatness should not be confused with the maximal flatness achieved by transforming the second-order Butterworth low-pass function to the fourth-order band-pass function, because the transformed function has two zeros at the origin (not one) and its poles are on constant-Q lines (not constant-α).

Now consider $k \geq 1/Q$, in which case the maxima have the value of $\frac{1}{2}$. To calculate the frequencies at which the magnitude is down by $[10 \log (1 + \varepsilon^2)]$ dB from the maximum value, set Eq. (20-35) equal to

$$\frac{1}{2}\frac{1}{\sqrt{1 + \varepsilon^2}},$$

and solve for the two ω's.

$$\left(\frac{k}{4Q}\right)^2 \frac{1}{\left[\frac{1}{4Q^2} + \frac{k^2}{4} - \left(\frac{\Delta\omega}{\omega_o}\right)^2\right]^2 + \frac{1}{Q^2}\left(\frac{\Delta\omega}{\omega_o}\right)^2} = \frac{1}{4}\frac{1}{1+\varepsilon^2}, \tag{20-37}$$

$$\left(\frac{\Delta\omega}{\omega_o}\right)^2 = \frac{1}{4}\left(k^2 - \frac{1}{Q^2}\right) \pm \frac{\varepsilon}{2}\frac{k}{Q}; \tag{20-38}$$

$$\left(\frac{\Delta\omega}{\omega_o}\right)_{inner} = \pm\frac{1}{2}\sqrt{k^2 - \frac{1}{Q^2} - \frac{2\varepsilon k}{Q}} \qquad k \geq \frac{1}{Q}\left(\varepsilon + \sqrt{1+\varepsilon^2}\right), \tag{20-39a}$$

$$\left(\frac{\Delta\omega}{\omega_o}\right)_{outer} = \pm\frac{1}{2}\sqrt{k^2 - \frac{1}{Q^2} + \frac{2\varepsilon k}{Q}} \qquad k \geq \frac{1}{Q}. \tag{20-39b}$$

For

$$k < \frac{1}{Q}\left(\varepsilon + \sqrt{1+\varepsilon^2}\right),$$

there are only two frequencies, given by Eq. (20-39b), at which the magnitude is $1/\sqrt{1+\varepsilon^2}$ times the maximum value. For

$$k > \frac{1}{Q}\left(\varepsilon + \sqrt{1+\varepsilon^2}\right),$$

the magnitude assumes the value of

$$\frac{1}{2}\frac{1}{\sqrt{1+\varepsilon^2}}$$

at two additional frequencies which are given by Eq. (20-39a). These frequencies are designated as inner frequencies and are present only when the magnitude dips below the value of $\frac{1}{2}/\sqrt{1+\varepsilon^2}$ at $\omega = \omega_o$. Otherwise, these two inner frequencies are imaginary and therefore nonexistent.

The bandwidth is defined as the difference of the two outer frequencies. From Eq. (20-39b), the upper and lower band-edge frequencies and the bandwidth are obtained:

$$\omega_2 = \omega_o + \frac{\omega_o}{2}\sqrt{k^2 - \frac{1}{Q^2} + \frac{2\varepsilon k}{Q}}, \tag{20-40a}$$

$$\omega_1 = \omega_o - \frac{\omega_o}{2}\sqrt{k^2 - \frac{1}{Q^2} + \frac{2\varepsilon k}{Q}}; \tag{20-40b}$$

$$BW = \omega_2 - \omega_1 = \omega_o\sqrt{k^2 - \frac{1}{Q^2} + \frac{2\varepsilon k}{Q}}. \tag{20-40c}$$

To obtain the 3-dB bandwidth, let $\varepsilon = 1$ in Eq. (20-40c).

A summary of all the important results that have been derived for the mutually-coupled tuned circuits appears in Fig. 20-11. In particular, note that for

$$k = \frac{1}{Q}, \qquad (20\text{-}41)$$

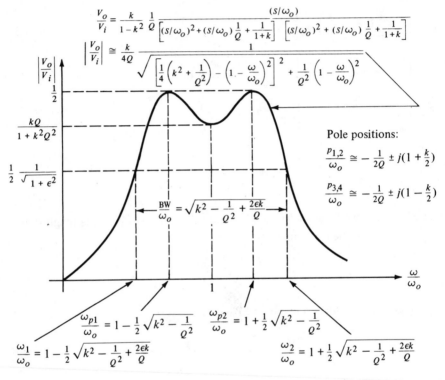

Pole positions:

$$\frac{p_{1,2}}{\omega_o} \cong -\frac{1}{2Q} \pm j(1 + \frac{k}{2})$$

$$\frac{p_{3,4}}{\omega_o} \cong -\frac{1}{2Q} \pm j(1 - \frac{k}{2})$$

Fig. 20-11   Summary of results on coupled tuned circuits $\left(\dfrac{1}{Q} \leq k \ll 1 \right)$

the frequencies of maxima coalesce at $\omega = \omega_o$. As a result, the flattest possible top is obtained. On the other hand, for

$$k = \frac{1}{Q}(\varepsilon + \sqrt{1 + \varepsilon^2}), \qquad (20\text{-}42)$$

the minimum at $\omega = \omega_o$ has the value of $\frac{1}{2}/\sqrt{1 + \varepsilon^2}$. Hence, the magnitude has the same value at the center frequency and at the two band-edge frequencies. As a result, the pass-band magnitude becomes equal-ripple.

EXAMPLE 20-4

Refer to the summary of results on coupled tuned circuits $(1/Q \leq k \ll 1)$ given in Fig. 20-11. These results are also valid for fourth-order transfer functions which are characterized by a zero at the origin and by poles that lie along a constant-α line.

(a) The bandwidth is to be varied while maintaining the flattest possible magnitude characteristic at the center frequency. How should the poles be varied to achieve this? Plot the resulting magnitude characteristic.

(b) The bandwidth is to be varied while maintaining an equal-ripple magnitude characteristic in the pass-band. How should the poles be varied to achieve this? Plot the resulting magnitude characteristic.

(c) How should the poles be varied to achieve the transition from a flat-top-type characteristic to an equal-ripple-type characteristic while at the same time the bandwidth is broadened? Plot the resulting magnitude characteristic.

(d) How should the poles be varied to achieve transition from a flat-top characteristic to an equal-ripple-type characteristic while maintaining a constant bandwidth? Plot the resulting magnitude characteristic.

SOLUTION

(a) Obtain the expression for the bandwidth from Fig. 20-11 and impose the condition for maximum flatness at the center frequency:

$$\frac{BW}{\omega_o} = \sqrt{k^2 - \frac{1}{Q^2} + \frac{2\varepsilon k}{Q}}\Bigg|_{k=1/Q} = \frac{\sqrt{2\varepsilon}}{Q}. \tag{20-43}$$

The corresponding pole positions are

$$\frac{p}{\omega_o} = -\frac{1}{2Q} \pm j\left(1 \pm \frac{k}{2}\right)\Bigg|_{k=1/Q} = -\frac{1}{2Q} \pm j\left(1 \pm \frac{1}{2Q}\right) = \pm j - \frac{1}{2Q}(1 \pm j). \tag{20-44}$$

As Eq. (20-44) shows, the vector drawn from j to either upper-half-plane pole is at an angle of 45° with respect to the horizontal. Hence, variable bandwidth with maximum flatness at $\omega/\omega_o = 1$ can be achieved by positioning the poles along lines drawn at $\pm 45°$, as shown in Fig. 20-12. The farther out the poles are along lines 1-2, the wider becomes the bandwidth. From Eq. (20-44), note also that the separation between the two upper-half-plane poles is twice the real part of the poles.

The design procedure is as follows.

1. From the desired center frequency, determine ω_o:

ω_o = desired center frequency.

For coupled tuned circuits, $\omega_o = \dfrac{1}{\sqrt{LC}}$.

2. From the desired bandwidth, determine Q:

$$Q = \sqrt{2\varepsilon}\,\frac{\omega_o}{BW}.$$

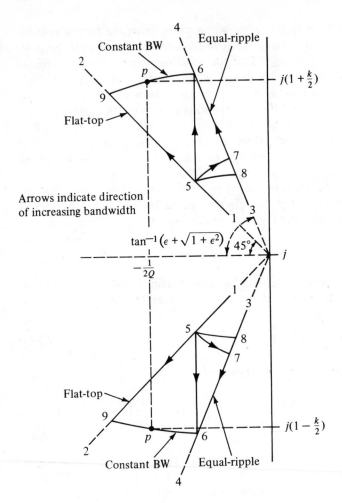

Fig. 20-12   Upper-half-plane pole locations for various kinds of responses $(Q \gg 1)$

Bandwidth is taken as the difference between the upper and lower band-edge frequencies. The band-edge frequencies are the frequencies at which the magnitude is down by $[10 \log (1 + \varepsilon^2)]$ dB. For 3-dB ripple in the pass-band, $\varepsilon = 1$.

For coupled tuned circuits, $Q = R\sqrt{C/L}$.

3. Position the upper-half-plane poles at

$$p_{1,2} = -\frac{\omega_o}{2Q} + j\omega_o\left(1 \pm \frac{1}{2Q}\right).$$

For coupled tuned circuits, choose $k = 1/Q$.

4. To maintain the flat-top characteristic while varying the bandwidth, vary k and Q such that their product is constant; that is,

$$kQ = 1.$$

The resulting magnitude characteristic is given by

$$\left|\frac{V_o}{V_i}\right| \cong \frac{1}{4Q^2} \frac{1}{\sqrt{\left[\frac{1}{2}\frac{1}{Q^2} - \left(1 - \frac{\omega}{\omega_o}\right)^2\right]^2 + \frac{1}{Q^2}\left(1 - \frac{\omega}{\omega_o}\right)^2}}$$

$$= \frac{1}{2} \frac{1}{\sqrt{1 + \left[\sqrt{2}Q\left(1 - \frac{\omega}{\omega_o}\right)\right]^4}}. \tag{20-45}$$

Figure 20-13a shows the variation in the BW as a function of Q. These curves are only slightly different from the curves obtained by transforming the second-order LP Butterworth function into the fourth-order BP function. (See Problem 20-3).

(b) Impose the equal-ripple condition on the expression for the bandwidth, and obtain

$$\frac{\text{BW}}{\omega_o} = \sqrt{k^2 - \frac{1}{Q^2} + \frac{2\varepsilon k}{Q}}\bigg|_{k=(1/Q)(\varepsilon + \sqrt{1+\varepsilon^2})}$$

$$= \frac{2}{Q}\sqrt{\varepsilon^2 + \varepsilon\sqrt{1 + \varepsilon^2}}. \tag{20-46}$$

The corresponding pole positions are

$$\frac{p}{\omega_o} = -\frac{1}{2Q} \pm j\left(1 \pm \frac{k}{2}\right)\bigg|_{k=(1/Q)(\varepsilon + \sqrt{1+\varepsilon^2})}$$

$$= -\frac{1}{2Q} \pm j\left[1 \pm \frac{1}{2Q}(\varepsilon + \sqrt{1 + \varepsilon^2})\right]$$

$$= \pm j - \frac{1}{2Q}[1 \pm j(\varepsilon + \sqrt{1 + \varepsilon^2})]. \tag{20-47}$$

As Eq. (20-47) shows, the vector drawn from j to either upper-half-plane pole is at an angle of

$$\tan^{-1}(\varepsilon + \sqrt{1 + \varepsilon^2}) \tag{20-48}$$

with respect to the horizontal. For 3-dB bandwidth ($\varepsilon = 1$), this angle is

$$\tan^{-1}(1 + \sqrt{2}) = 67.5°.$$

Hence, variable bandwidths with equal-ripple characteristics can be achieved by positioning the poles along lines drawn at $\pm\tan^{-1}(\varepsilon + \sqrt{1 + \varepsilon^2})$ degrees as shown in Fig. 20-12. The farther out the poles are along lines 3-4, the wider becomes the bandwidth.

The design procedure is as follows.

1. From the desired center frequency determine ω_o:

ω_o = desired center frequency in rad/s.

For coupled circuits, $\omega_o = \dfrac{1}{\sqrt{LC}}$.

2. From the desired bandwidth determine Q:

$$Q = 2\sqrt{\varepsilon^2 + \varepsilon\sqrt{1 + \varepsilon^2}} \; \frac{\omega_o}{\text{BW}}.$$

For coupled circuits, $Q = R\sqrt{\dfrac{C}{L}}$.

3. Position the upper-half-plane poles at

$$p_{1,2} = -\frac{\omega_o}{2Q} + j\omega_o\left[1 \pm \frac{1}{2Q}(\varepsilon + \sqrt{1 + \varepsilon^2})\right].$$

For coupled tuned circuits, $k = \dfrac{1}{Q}(\varepsilon + \sqrt{1 + \varepsilon^2})$.

4. To maintain the equal-ripple characteristic while varying the bandwidth, vary k and Q such that kQ is constant, that is,

$$kQ = \varepsilon + \sqrt{1 + \varepsilon^2}.$$

The resulting magnitude characteristic is given by

$$\left|\frac{V_o}{V_i}\right| \cong \frac{\varepsilon + \sqrt{1 + \varepsilon^2}}{4Q^2}$$

$$\times \frac{1}{\sqrt{\left[\frac{1}{2}\left(\frac{1 + \varepsilon^2 + \varepsilon\sqrt{1 + \varepsilon^2}}{Q^2}\right) - \left(1 - \frac{\omega}{\omega_o}\right)^2\right]^2 + \frac{1}{Q^2}\left(1 - \frac{\omega}{\omega_o}\right)^2}}$$

$$= \frac{1}{2}\frac{1}{\sqrt{1 + \varepsilon^2}} \frac{1}{\sqrt{1 - \left[\dfrac{2Q\left(1 - \dfrac{\omega}{\omega_o}\right)}{1 + \varepsilon^2 + \varepsilon\sqrt{1 + \varepsilon^2}}\right]^2}} \qquad (20\text{-}49)$$

$$\times \left[\varepsilon(\varepsilon + \sqrt{1 + \varepsilon^2}) - Q^2\left(1 - \frac{\omega}{\omega_o}\right)^2\right].$$

When $\varepsilon = 1$ (3-dB pass-band ripple), this expression simplifies to

$$\left|\frac{V_o}{V_i}\right| \approx \frac{1}{2\sqrt{2}} \frac{1}{\sqrt{1 - \left[\dfrac{2Q\left(1 - \dfrac{\omega}{\omega_o}\right)}{2 + \sqrt{2}}\right]^2 \left[1 + \sqrt{2} - Q^2\left(1 - \dfrac{\omega}{\omega_o}\right)^2\right]}} .$$

(20-50)

Figure 20-13b shows the variation in the BW as a function of Q. These curves are only slightly different from the curves obtained by transforming the second-order 3-dB ripple LP Chebyshev function into the fourth-order BP function.

(c) Refer to Fig. 20-12. When poles are positioned along lines 1-2, a flat-top characteristic is obtained. Along lines 3-4 an equal-ripple characteristic is obtained. Therefore, when the two upper-half-plane poles are moved along any vertical line, such as 5-6, transitional characteristics are obtained.

When the poles are placed at the two points marked 5, the magnitude characteristic is flat and the bandwidth is given by

$$\left(\frac{BW}{\omega_o}\right)_{min} = \frac{\sqrt{2\varepsilon}}{Q}, \qquad \left(k_{min} = \frac{1}{Q}\right).$$

(20-51)

When the poles are moved to the two points marked 6 by increasing k, the magnitude characteristic becomes equal-ripple and the bandwidth is given by

$$\left(\frac{BW}{\omega_o}\right)_{max} = \frac{2}{Q}\sqrt{\varepsilon^2 + \varepsilon\sqrt{1 + \varepsilon^2}} = \left(\frac{BW}{\omega_o}\right)_{min}\left[\sqrt{2}\sqrt{\varepsilon + \sqrt{1 + \varepsilon^2}}\right],$$

$$\left[k_{max} = \frac{1}{Q}\left(\varepsilon + \sqrt{1 + \varepsilon^2}\right)\right]. \qquad (20\text{-}52)$$

Thus, a $[\sqrt{2}\sqrt{\varepsilon + \sqrt{1 + \varepsilon^2}}]$-to-1 change in bandwidth is accomplished by moving the poles along lines 5-6. For 3-dB pass-band ripple ($\varepsilon = 1$), this corresponds to a change of 2.2-to-1. The resulting magnitude curves are displayed in Fig. 20-13c for $Q = 20$. Had a higher Q been used, a narrower bandwidth (covering a 2.2-to-1 range) would have resulted.

As in the other two cases, the allowable pass-band ripple establishes the value of ε. The center frequency determines the ω_o, and the minimum value of BW fixes the Q. By reference to Fig. 20-12 it should be clear that the poles can be moved from one boundary line (corresponding to flat-top characteristic) to the other boundary line (corresponding to equal-ripple characteristic) along paths other than the vertical one described here. For example, a much smaller change in BW results if the poles are moved along circular paths from 5 to 7.

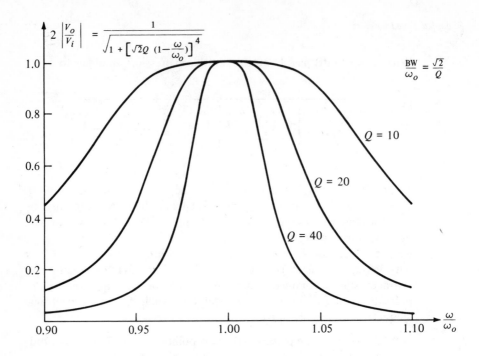

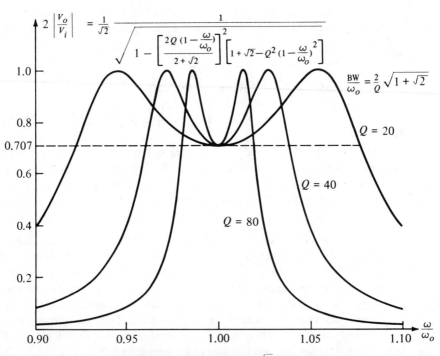

Fig. 20-13 (a) $kQ = 1$ $(Q \gg 1)$ (b) $kQ = 1 + \sqrt{2}$ $(Q \gg 1)$

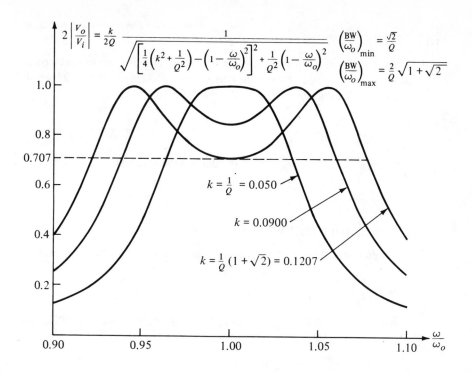

$$2\left|\frac{V_o}{V_i}\right| = \frac{k}{2Q} \frac{1}{\sqrt{\left[\frac{1}{4}\left(k^2+\frac{1}{Q^2}\right)-\left(1-\frac{\omega}{\omega_o}\right)^2\right]^2+\frac{1}{Q^2}\left(1-\frac{\omega}{\omega_o}\right)^2}}$$

$$\left(\frac{BW}{\omega_o}\right)_{min} = \frac{\sqrt{2}}{Q}$$

$$\left(\frac{BW}{\omega_o}\right)_{max} = \frac{2}{Q}\sqrt{1+\sqrt{2}}$$

$$k = \frac{1}{Q} = 0.050$$

$$k = 0.0900$$

$$k = \frac{1}{Q}(1+\sqrt{2}) = 0.1207$$

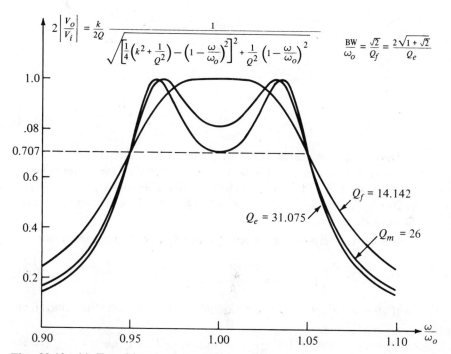

$$2\left|\frac{V_o}{V_i}\right| = \frac{k}{2Q} \frac{1}{\sqrt{\left[\frac{1}{4}\left(k^2+\frac{1}{Q^2}\right)-\left(1-\frac{\omega}{\omega_o}\right)^2\right]^2+\frac{1}{Q^2}\left(1-\frac{\omega}{\omega_o}\right)^2}}$$

$$\frac{BW}{\omega_o} = \frac{\sqrt{2}}{Q_f} = \frac{2\sqrt{1+\sqrt{2}}}{Q_e}$$

$$Q_f = 14.142$$

$$Q_e = 31.075$$

$$Q_m = 26$$

Fig. 20-13 (c) Transitional characteristics for $Q = 20$ (flat to equal ripple) (d) Constant bandwidth (BW $= 0.1\omega_o$)

(d) Transition from a flat-top characteristic to an equal-ripple characteristic can also be achieved at constant bandwidth. To see how this is achieved, consider the expression for the bandwidth:

$$\frac{BW}{\omega_o} = \sqrt{k^2 - \frac{1}{Q^2} + \frac{2\varepsilon k}{Q}}. \tag{20-53}$$

As long as k and Q are varied in such a manner that

$$k^2 - \frac{1}{Q^2} + \frac{2\varepsilon k}{Q}$$

is kept constant, the BW remains constant.

When k is at its minimum value, the flat-top curve is obtained. Hence,

$$k_{min} = \frac{1}{Q_f}.$$

When k is at its maximum value, the equal-ripple curve is obtained. Hence,

$$k_{max} = \frac{1}{Q_e}(\varepsilon + \sqrt{1 + \varepsilon^2}).$$

In either case, the BW is the same. Therefore,

$$\sqrt{\left(k^2 - \frac{1}{Q^2} + \frac{2\varepsilon k}{Q}\right)}\bigg|_{k = k_{min}} = \sqrt{\left(k^2 - \frac{1}{Q^2} + \frac{2\varepsilon k}{Q}\right)}\bigg|_{k = k_{max}},$$

$$\frac{\sqrt{2\varepsilon}}{Q_f} = \frac{2}{Q_e}\sqrt{\varepsilon^2 + \varepsilon\sqrt{1 + \varepsilon^2}},$$

$$\frac{Q_e}{Q_f} = \sqrt{2}\sqrt{\varepsilon + \sqrt{1 + \varepsilon^2}}, \qquad \frac{k_{max}}{k_{min}} = \sqrt{\frac{\varepsilon + \sqrt{1 + \varepsilon^2}}{2}}. \tag{20-54}$$

Thus, ε determines the ratio of the extreme values of the Q's. For $\varepsilon = 1$, this ratio is 2.2. The corresponding ratio of the extreme values of the k's is 1.1. Two sets of constant BW pole loci are shown in Fig. 20-12: the set 5-8 and the set 6-9. The set 6-9 corresponds to wider bandwidth.

If a flat-top response is desired, Q_f and k_{min} are determined from:

$$Q_f = \sqrt{2\varepsilon}\,\frac{\omega_o}{BW}, \qquad k_{min} = \frac{1}{Q_f}. \tag{20-55}$$

If an equal-ripple response is desired, Q_e and k_{max} are determined from

$$Q_e = 2\sqrt{\varepsilon^2 + \varepsilon\sqrt{1 + \varepsilon^2}}\,\frac{\omega_o}{BW}, \qquad k_{max} = \frac{1}{Q_e}(\varepsilon + \sqrt{1 + \varepsilon^2}). \tag{20-56}$$

The in-between responses are obtained by selecting Q and k such that Eq. (20-53) is satisfied for the desired bandwidth.

All three kinds of responses are shown in Fig. 20-13d for BW $= 0.1\omega_o$. The sharpest cutoff characteristics are produced by the equal-ripple response.

EXAMPLE 20-5

Instead of using mutual inductance, a mutual impedance can be used to couple two tuned circuits and thus obtain a fourth-order band-pass network. This is shown in Fig. 20-14a, where Z_c is used as the coupling impedance.

(a) Obtain the transfer function V_o/V_i.

(b) One pair of complex poles is to be moved in a constant-α manner by adjusting Z_c. Choose Z_c to achieve this.

(c) Assume that the bandwidth is much less than the center frequency. What must be the value of Z_c to achieve the flattest possible magnitude characteristic in the pass-band? What is the resulting center frequency and the 3-dB bandwidth?

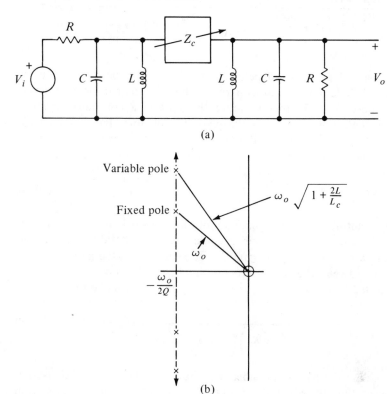

(a)

(b)

Fig. 20-14

SOLUTION

(a) Let

$$Y = sC + \frac{1}{R} + \frac{1}{sL}.$$

Express the transfer function in terms of Y and Z_c.

$$\frac{V_o}{V_i} = \frac{1}{R} \frac{1}{2Y + Z_c Y^2} = \frac{1}{RYZ_c} \frac{1}{Y + \dfrac{2}{Z_c}}.$$

With

$$\omega_o = \frac{1}{\sqrt{LC}} \text{ and } Q = R\sqrt{\frac{C}{L}},$$

the transfer function becomes

$$\frac{V_o}{V_i} = \frac{1}{RC^2} \frac{s}{Z_c} \frac{s}{\left(s^2 + \dfrac{s\omega_o}{Q} + \omega_o^2\right)\left[\left(s^2 + \dfrac{s\omega_o}{Q} + \omega_o^2\right) + \dfrac{2}{C}\dfrac{s}{Z_c}\right]}. \qquad Ans.$$

$$(20\text{-}57)$$

(b) One pair of poles is independent of Z_c, and this pair represents the poles of the uncoupled tuned circuit. The other pair of poles is dependent upon Z_c. In order to move this pair in a constant-α manner, the coefficient of the s-term in the quadratic polynomial must not change. Therefore, let $Z_c = sL_c$ and thereby change the constant term only. Thus, the second factor in the denominator polynomial becomes

$$\left(s^2 + s\frac{\omega_o}{Q} + \omega_o^2 + \frac{2}{L_c C}\right). \qquad (20\text{-}58)$$

The resulting pole–zero diagram of this network is shown in Fig. 20-14b. Note that one pair of poles remains stationary while the other pair is varied by changing L_c. The smaller L_c, the more the interaction between the tuned circuits and hence the more the poles are separated from each other. The resulting transfer function is

$$\frac{V_o}{V_i} = \frac{1}{RL_c C^2} \frac{s}{\left(s^2 + s\dfrac{\omega_o}{Q} + \omega_o^2\right)\left(s^2 + s\dfrac{\omega_o}{Q} + \omega_o^2 + \dfrac{2}{L_c C}\right)}. \qquad (20\text{-}59)$$

(c) To achieve the flattest possible magnitude characteristic, the distance between the upper-half-plane poles should be twice the distance from the imaginary axis. (See Example 20-4a for details.)

The real part of the poles is $-(\omega_o/2Q)$. The imaginary parts of the upper-half-plane poles are

$$j\omega_o\sqrt{1 - \frac{1}{4Q^2}} \cong j\omega_o, \qquad j\omega_o\sqrt{1 + \frac{2}{\omega_o^2 L_c C} - \frac{1}{4Q^2}} \cong j\omega_o\sqrt{1 + \frac{2}{\omega_o^2 L_c C}}.$$

Hence, choose L_c according to

$$\omega_o\left[\sqrt{1 + \frac{2}{\omega_o^2 L_c C}} - 1\right] = 2\frac{\omega_o}{2Q}. \qquad (20\text{-}60)$$

Assume $Q \gg \frac{1}{2}$ and solve Eq. (20-60) for L_c. Use $\omega_o = 1/\sqrt{LC}$ and obtain

$$\frac{L_c}{L} \cong Q. \qquad Ans.$$

The center frequency is approximately the arithmetic average of the imaginary parts of the two upper-half-plane poles; that is,

$$\omega_m \cong \frac{\omega_o}{2}\left[1 + \sqrt{1 + \frac{2}{\omega_o^2 L_c C}}\right] \cong \frac{\omega_o}{2}\left[1 + \sqrt{1 + \frac{2}{Q}}\right]. \qquad Ans.$$

Use $\varepsilon = 1$ in Eq. (20-43) and obtain the 3-dB bandwidth:

$$\omega_{3dB} = \sqrt{2}\frac{\omega_o}{Q}. \qquad Ans.$$

20-4 ACTIVE REALIZATIONS

The LP-to-BP transformation results in BP functions that are of even order. The transfer-function zeros are equally divided between the origin and infinity. Therefore, a BP function of order n can be realized as the cascaded connection of $n/2$ second-order, active, BP networks, which are discussed in detail in Chapter 11. These second-order BP networks are easy to tune because the response peaks exactly at ω_o and the 3-dB bandwidth is exactly ω_o/Q. It is also possible to obtain active BP networks by cascading equal numbers of second-order LP and HP networks.

EXAMPLE 20-6

Design a fourth-order maximally flat (Butterworth) active BP network with a 1-dB bandwidth that equals 10% of the center frequency.

SOLUTION

Obtain the 3-dB bandwidth, maximally flat second-order LP function from Table 17-1:

$$G_{LP1} = \frac{1}{s^2 + \sqrt{2}s + 1}.$$

To produce 1-dB ($\varepsilon = 0.5089$) loss at $\omega = 1$, divide s by $1/\sqrt{\varepsilon} = 1\,4019$:

$$G_{LP} = \frac{1}{\left(\dfrac{s}{1.4019}\right)^2 + \sqrt{2}\left(\dfrac{s}{1.4019}\right) + 1}.$$

The transformation to BP is achieved by replacing s with

$$\frac{s^2 + \omega_m^2}{sBW} = \frac{\left(\dfrac{s}{\omega_m}\right)^2 + 1}{\left(\dfrac{s}{\omega_m}\right)\left(\dfrac{BW}{\omega_m}\right)} = \frac{\left(\dfrac{s}{\omega_m}\right)^2 + 1}{0.1\left(\dfrac{s}{\omega_m}\right)}; \qquad (20\text{-}61)$$

$$G_{BP} = \frac{\left(\dfrac{s}{\omega_m}\right)^2}{50.8847\left(\dfrac{s}{\omega_m}\right)^4 + 10.0881\left(\dfrac{s}{\omega_m}\right)^3 + 102.7694\left(\dfrac{s}{\omega_m}\right)^2 + 10.0881\left(\dfrac{s}{\omega_m}\right) + 50.8847} \qquad (20\text{-}62\text{a})$$

$$= \frac{1}{50.8847} \frac{\left(\dfrac{s}{\omega_m}\right)^2}{\left[\left(\dfrac{s}{\omega_m} + 0.0520\right)^2 + 1.0496^2\right]\left[\left(\dfrac{s}{\omega_m} + 0.0471\right)^2 + 0.9504^2\right]}. \qquad (20\text{-}62\text{b})$$

The ω_o and Q of the BP poles are

$$\frac{\omega_{o1}}{\omega_m} = \sqrt{0.0520^2 + 1.0496^2} = 1.0509, \qquad Q_1 = \frac{1.0509}{2 \times 0.0520} = 10.10, \qquad (20\text{-}63\text{a})$$

$$\frac{\omega_{o2}}{\omega_m} = \sqrt{0.0471^2 + 0.9504^2} = 0.9516, \qquad Q_2 = \frac{0.9516}{2 \times 0.0471} = 10.10. \qquad (20\text{-}63\text{b})$$

A realization of Eq. (20-62b) is given in Fig. 20-15. (See Section 11-6 for details.)

Using Eq. (11-34), the transfer function of each stage is obtained:

$$-\frac{1}{10} \frac{\omega_o s}{s^2 + s\dfrac{\omega_o}{Q} + \omega_o^2}, \qquad (20\text{-}64)$$

where

$$\omega_o = \frac{1}{RC}, \qquad Q = \frac{R_Q}{R}.$$

The gain of each stage at the frequency of peaking ($s = j\omega_o$) is $Q/10$.

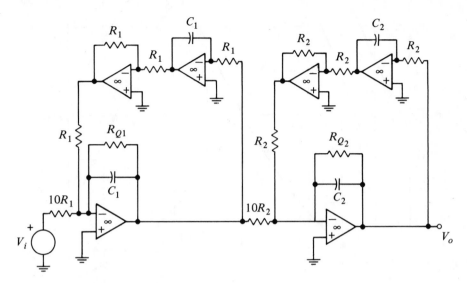

Fig. 20-15   Active BP realization

To put the BP poles in the desired locations, the following constraints must be applied to each stage:

$$1/(R_1 C_1) = \omega_{o1} = 1.0509\omega_m, \qquad R_{Q1} = R_1 Q_1 = 10.10 R_1, \qquad (20\text{-}65a)$$

$$1/(R_2 C_2) = \omega_{o2} = 0.9516\omega_m, \qquad R_{Q2} = R_2 Q_2 = 10.10 R_2. \qquad (20\text{-}65b)$$

For a given ω_m, either the resistors or the capacitors are arbitrarily chosen. Then, Eq. (20-65) is used to obtain the values of the remaining elements. The band-edge frequencies (1-dB down points) are obtained by using Eq. (20-14) with $\omega_{LP} = 1$ and $BW = 0.1\omega_m$:

$$\frac{\omega_{2,1}}{\omega_m} = \sqrt{1 + \left(\frac{BW}{2\omega_m}\right)^2} \pm \left(\frac{BW}{2\omega_m}\right) = \sqrt{1 + \left(\frac{0.1}{2}\right)^2} \pm \frac{0.1}{2};$$

$$\omega_2 = 1.051\omega_m, \qquad \omega_1 = 0.952\omega_m. \qquad (20\text{-}66)$$

Comparison of Eqs. (20-65) and (20-66) shows that the band-edge frequencies coincide with the frequency of peaking of the individual stages.

The magnitude-normalized response of each second-order stage and the overall response are given in Fig. 20-16. The first stage peaks at

$$\frac{\omega_{o1}}{\omega_m} = 1.0509$$

and has a 3-dB bandwidth of

$$\frac{BW_1}{\omega_m} = 0.0520 \times 2 = 0.1040.$$

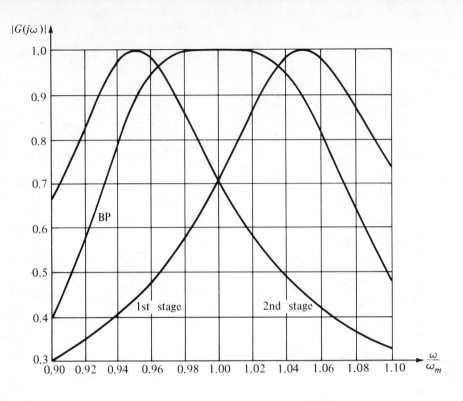

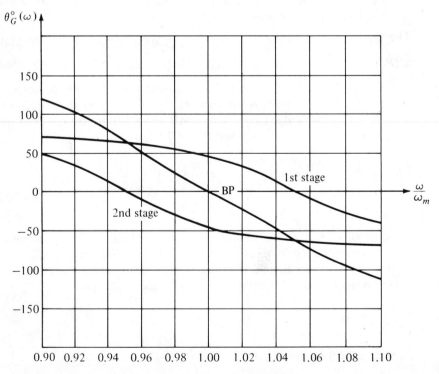

Fig. 20-16 (a) Four-pole maximally flat (1 dB) BP magnitude and its decomposition
(b) Four-pole maximally flat (1 dB) BP phase and its decomposition

The second stage peaks at

$$\frac{\omega_{o2}}{\omega_m} = 0.9516$$

and has a 3-dB bandwidth of

$$\frac{BW_2}{\omega_m} = 0.0471 \times 2 = 0.0942.$$

The product of the two magnitude curves is the desired maximally flat (at $\omega/\omega_m = 1$) magnitude curve.

For comparison purposes, the fourth-order, equal-ripple (1-dB Chebyshev) frequency-response curves are given in Fig. 20-17. Compared to the Butterworth response, each Chebyshev stage peaks at a frequency slightly closer to the center frequency and has a narrower bandwidth. Note the similarity of these changes with those presented in Fig. 20-13 for the mutually coupled tuned circuits.

The Chebyshev curves are obtained by transforming the 1-dB, second-order LP Chebyshev function:

$$G_{LP} = \frac{1}{\sqrt{0.5089}} \frac{1}{s^2 + 1.0977s + 1.1025}. \tag{20-67}$$

The resulting band-pass function is

$$G_{BP} = \frac{0.9826\left(\dfrac{s}{\omega_m}\right)^2}{100\left(\dfrac{s}{\omega_m}\right)^4 + 10.9773\left(\dfrac{s}{\omega_m}\right)^3 + 201.1025\left(\dfrac{s}{\omega_m}\right)^2 + 10.9773\left(\dfrac{s}{\omega_m}\right) + 100} \tag{20-68a}$$

$$= \frac{0.009826\left(\dfrac{s}{\omega_m}\right)^2}{\left[\left(\dfrac{s}{\omega_m} + 0.0287\right)^2 + 1.0454^2\right]\left[\left(\dfrac{s}{\omega_m} + 0.0262\right)^2 + 0.9558^2\right]}. \tag{20-68b}$$

20-5 SIMULATION OF COUPLED TUNED CIRCUITS WITH AN ACTIVE RC NETWORK

The transfer function of coupled tuned circuits has a zero at the origin. The four poles have the same real part. [See Eq. (20-28) and Fig. 20-9.] By adjusting the coefficient of coupling k, the vertical separation between the upper-half-plane poles is changed and hence the bandwidth is changed. Any active realization of this transfer function must therefore contain provision for moving the poles in a constant-α manner. Such a circuit is presented in Fig. 20-18.

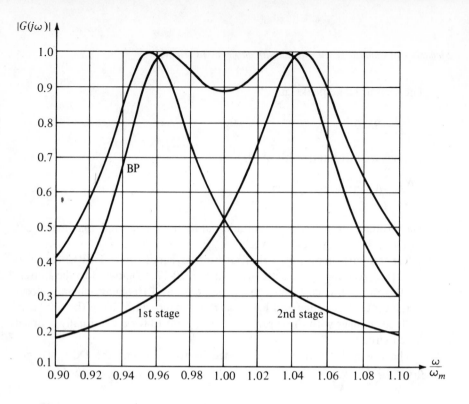

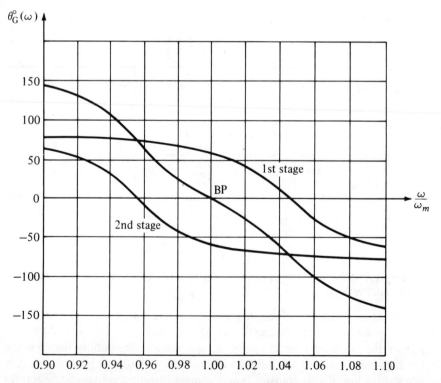

Fig. 20-17 (a) Four-pole equal-ripple (1 dB) BP magnitude and its decomposition
(b) Four-pole equal-ripple (1 dB) BP phase and its decomposition

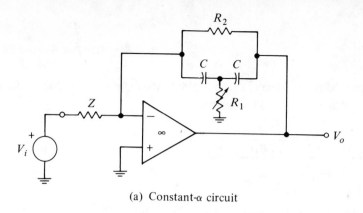

(a) Constant-α circuit

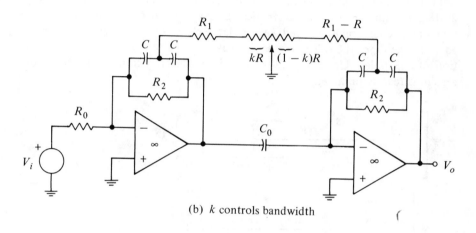

(b) k controls bandwidth

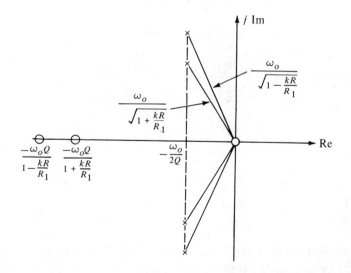

(c) Pole-zero pattern

Fig. 20-18 Active RC realization of coupled tuned circuits

To see how this circuit works, consider first the circuit of Fig. 20-18a. Its transfer function is given by

$$\frac{V_o}{V_i} = -\frac{2}{ZC} \frac{s + \dfrac{1}{2R_1 C}}{s^2 + s\dfrac{2}{R_2 C} + \dfrac{1}{R_1 R_2 C^2}}. \tag{20-69}$$

The real part of the poles, $-1/(R_2 C)$, is independent of R_1. Hence, R_1 can be used to vary only the imaginary part of the poles.

Two constant-α, second-order circuits can be cascaded, as shown in Fig. 20-18b, to simulate the characteristics of the coupled tuned circuits. As the potentiometer setting is changed (k varied), the ω_o of one pair of poles is increased while the ω_o of the other pair is decreased, thereby controlling the separation of the poles. Using Eq. (20-69), the transfer function of the four-pole system is readily obtained:

$$\frac{V_o}{V_i} = \left[-\frac{2}{R_0 C} \frac{s + \dfrac{1}{2(R_1 + kR)C}}{s^2 + s\dfrac{2}{R_2 C} + \dfrac{1}{(R_1 + kR)R_2 C^2}} \right]$$

$$\times \left[-\frac{2C_0 s}{C} \frac{s + \dfrac{1}{2(R_1 - kR)C}}{s^2 + s\dfrac{2}{R_2 C} + \dfrac{1}{(R_1 - kR)R_2 C^2}} \right]$$

$$= \frac{4C_0}{R_0 C^2} \frac{s\left(s + \dfrac{\omega_o Q}{1 + k\dfrac{R}{R_1}}\right)\left(s + \dfrac{\omega_o Q}{1 - k\dfrac{R}{R_1}}\right)}{\left(s^2 + s\dfrac{\omega_o}{Q} + \dfrac{\omega_o^2}{1 + k\dfrac{R}{R_1}}\right)\left(s^2 + s\dfrac{\omega_o}{Q} + \dfrac{\omega_o^2}{1 - k\dfrac{R}{R_1}}\right)}$$

$$(0 \le k \le 1, \quad R_1 > R), \tag{20-70}$$

where

$$\omega_o = \frac{1}{C\sqrt{R_1 R_2}}, \qquad Q = \frac{1}{2}\sqrt{\frac{R_2}{R_1}}.$$

The resulting pole–zero pattern is shown in Fig. 20-18c. Note that ω_o and Q as defined represent the pole ω_o and Q only for $k = 0$. Unlike its passive counterpart, this circuit has two additional zeros on the negative real axis. However, for narrow-band circuits, these zeros, having magnitudes much larger than the magnitude of the poles, have negligible effect on the magnitude characteristics. Therefore, Eq. (20-70) can be approximated by

$$\frac{V_o}{V_i} \cong \frac{4C_0}{R_0 C^2} \frac{\omega_o^2 Q^2}{1 - \left(k\dfrac{R}{R_1}\right)^2} \frac{s}{\left(s^2 + s\dfrac{\omega_o}{Q} + \dfrac{\omega_o^2}{1 + k\dfrac{R}{R_1}}\right)\left(s^2 + s\dfrac{\omega_o}{Q} + \dfrac{\omega_o^2}{1 - k\dfrac{R}{R_1}}\right)}.$$

$$(20\text{-}71)$$

With the exception of the scale factor, this equation is identical with the equation of coupled tuned circuits, Eq. (20-29), provided $k(R/R_1)$ is interpreted as the coefficient of coupling. The magnitude of the transfer function given by Eq. (20-70) is plotted in Fig. 20-19 for

$$\frac{4C_0}{R_0 C^2} = 25 \times 10^{-5} \quad \text{and} \quad Q = 20.$$

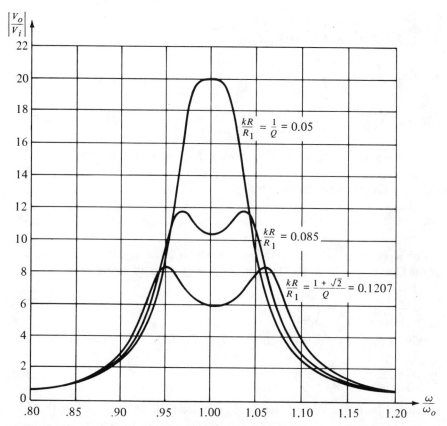

Fig. 20-19 Control of bandwidth through potentiometer setting ($Q = 20$)

The flattest possible response at $\omega_o = 1$ is obtained by adjusting kR/R_1 to

$$\frac{1}{Q} = 0.05.$$

The widest bandwidth is obtained while maintaining 3-dB ripple in the pass-band by adjusting kR/R_1 to

$$\frac{1 + \sqrt{2}}{Q} = 0.1207.$$

Unlike the passive circuit, a lower overall amplification results as kR/R_1 is increased, because kR/R_1 does not appear as a multiplying factor in the transfer function. Had $Q = 10$ been used instead of $Q = 20$, wider bandwidths and less steep cutoff characteristics would have been obtained. The ability to control the bandwidth about the center frequency with a single potentiometer is a very desirable feature of this network.

Using Eq. (20-40), the upper and lower 3-dB frequencies and the 3-dB band-width for narrow band-pass active filters can be obtained for $kR/R_1 > 1/Q$:

$$\omega_2 = \omega_o + \frac{\omega_o}{2}\sqrt{\left(\frac{kR}{R_1}\right)^2 - \frac{1}{Q^2} + \frac{kR}{R_1}\frac{2}{Q}}, \qquad (20\text{-}72\text{a})$$

$$\omega_1 = \omega_o - \frac{\omega_o}{2}\sqrt{\left(\frac{kR}{R_1}\right)^2 - \frac{1}{Q^2} + \frac{kR}{R_1}\frac{2}{Q}}; \qquad (20\text{-}72\text{b})$$

$$\text{BW} = \omega_o\sqrt{\left(\frac{kR}{R_1}\right)^2 - \frac{1}{Q^2} + \frac{kR}{R_1}\frac{2}{Q}}. \qquad (20\text{-}72\text{c})$$

When $kR/R_1 = (1/Q)(1 + \sqrt{2})$, the widest 3-dB bandwidth is achieved (for fixed Q) while maintaining 3-dB ripple in the pass-band. [See Eq. (20-42).] Equation (20-72c) then reduces to

$$\text{BW} = \frac{2\omega_o}{Q}\sqrt{1 + \sqrt{2}}. \qquad (20\text{-}73)$$

For a given normalized bandwidth, BW/ω_o, the rejection characteristics of the filter can be improved by using higher Q's and adjusting kR/R_1 to

$$\frac{kR}{R_1} = -\frac{1}{Q} + \sqrt{\frac{2}{Q^2} + \left(\frac{\text{BW}}{\omega_o}\right)^2}. \qquad (20\text{-}74)$$

When kR/R_1 is adjusted to equal to $1/Q$, Eq. (20-74) simplifies to

$$Q = \frac{\sqrt{2}}{\text{BW}/\omega_o} \qquad \left(\frac{kR}{R_1} = \frac{1}{Q}\right), \qquad (20\text{-}75)$$

which results in the flattest band-pass characteristic at $\omega = \omega_o$. On the other hand, when kR/R_1 is adjusted to equal $(1/Q)(1 + \sqrt{2})$, Eq. (20-74) simplifies to

$$Q = \frac{2\sqrt{1+\sqrt{2}}}{BW/\omega_o}, \quad \left[\frac{kR}{R_1} = \frac{1}{Q}(1+\sqrt{2})\right], \tag{20-76}$$

and the steepest cutoff characteristic is obtained without exceeding 3-dB ripple in the pass-band.

For example, consider a band-pass function with a 3-dB bandwidth of $0.1\omega_o$. The Q for the flattest and the Q for the steepest band-pass characteristic are given by

$$Q = \frac{\sqrt{2}}{0.1} = 14.142, \quad \frac{kR}{R_1} = \frac{1}{14.142} = 0.0707 \quad \text{(flattest)},$$

$$Q = \frac{2\sqrt{1+\sqrt{2}}}{0.1} = 31.075, \quad \frac{kR}{R_1} = \frac{1+\sqrt{2}}{31.08} = 0.0777 \quad \text{(steepest)}.$$

In Fig. 20-13d, these two extreme characteristics are displayed. To facilitate comparison, the peak value of the magnitude curves is normalized to unity. As Q is increased from 14.142, a progressively increasing dip occurs in the magnitude curve at $\omega = \omega_o$, and the response at the band-edge frequencies becomes steeper. For $Q = 31.075$, the dip is 3 dB, and the cutoff is sharpest.

20-6 A FREQUENCY DISCRIMINATOR

In Section 17-7, the characteristics of a band-limited, linear-amplitude function are discussed. The desired function is formed by dividing into s the Butterworth polynomial of order n; that is,

$$T_{LP} = \frac{s}{B_n(s)}. \tag{20-77}$$

The s in the numerator gives the desired linear-amplitude characteristic while $B_n(s)$ in the denominator limits the bandwidth of operation. Application of the low-pass-to-band-pass transformation to Eq. (20-77) gives

$$T_{BP} = \frac{s}{B_n(s)}\bigg|_{s \text{ replaced by } (s^2+\omega_m^2)/sBW} \tag{20-78}$$

For an example, consider the fourth-order function obtained by transforming the second-order Butterworth polynomial given in Table 17-1:

$$T_{LP} = \frac{s}{B_2(s)} = \frac{s}{s^2 + \sqrt{2}s + 1},$$

$$T_{BP} = \frac{\dfrac{s^2+1}{0.1s}}{\left(\dfrac{s^2+1}{0.1s}\right)^2 + \sqrt{2}\left(\dfrac{s^2+1}{0.1s}\right) + 1} = \frac{0.1s(s^2+1)}{s^4 + 0.1\sqrt{2}s^3 + 2.01s^2 + 0.1\sqrt{2}s + 1},$$

$$\tag{20-79}$$

where the band-center frequency is taken as 1 rad/s and the bandwidth 0.1 rad/s. The magnitude and phase of this function are plotted in Fig. 20-20. Note the maximally linear characteristic about $\omega = 1$ and the 3-dB departure from linearity at $\omega \cong 0.95$ and $\omega \cong 1.05$. Had a third-order function been transformed,

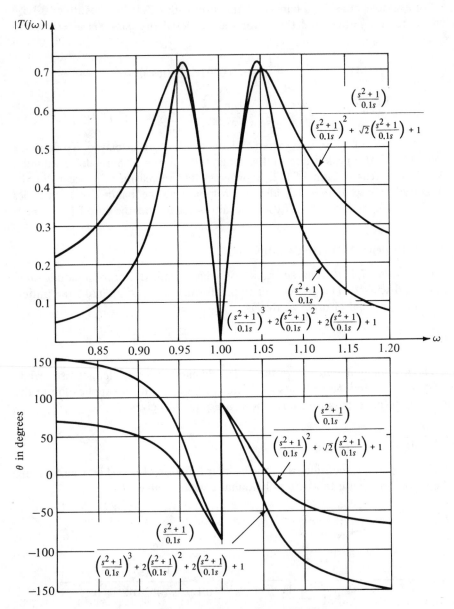

Fig. 20-20   Plot of the magnitude and phase of Eq. (20-79)

the sixth-order function, also shown in Fig. 20-20, would have resulted. Then the cutoff characteristics would have been steeper and the pass-band characteristic would have been more linear.

Although the functions depicted in Fig. 20-20 provide an amplitude discrimination that rises linearly with frequency, a distinction in the magnitude cannot be made between the frequencies that are on one or the other side of ω_m. An increase in magnitude is obtained with an increase as well as a decrease in frequency from $\omega_m = 1$. This undesirable feature is remedied by the circuit shown in block diagram form in Fig. 20-21. The system is of order four with

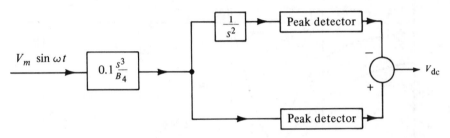

Fig. 20-21   A frequency discriminator

$\omega_m = 1$ and BW $= 0.1$. The individual blocks can be readily realized by active networks. B_4 represents the denominator polynomial of Eq. (20-79).

The resulting dc voltage at the output is given by

$$V_{dc} = V_m \frac{\dfrac{\omega^2 - 1}{0.1\omega}}{\sqrt{1 + \left(\dfrac{\omega^2 - 1}{0.1\omega}\right)^4}}. \qquad (20\text{-}80)$$

Equation (20-80) is plotted in Fig. 20-22. For comparison purposes, the output from the sixth-order system is also plotted. These curves can be obtained directly from those of Fig. 20-20 by inverting the magnitude characteristic for $\omega < 1$ and by changing the vertical scale from $|T_{BP}(j\omega)|$ to V_{dc}/V_m. The resulting curves display a maximally linear characteristic about $\omega = 1$ and may be used as a frequency discriminator. The band-center frequency can be changed from 1 to ω_m by replacing s with s/ω_m. The bandwidth is controlled by BW in the LP-to-BP transformation. At the upper and lower end of the band-pass region, the dc voltage departs 3 dB from the ideal value determined by the perfectly linear characteristics. The difference of these two frequencies is BW.

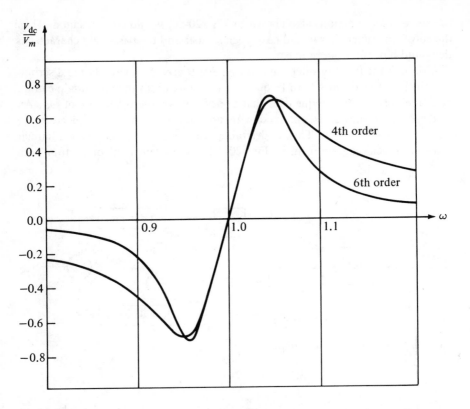

Fig. 20-22   Frequency discriminator characteristics

20-7 SUMMARY

Band-pass functions may be obtained from low-pass functions by replacing s with $(s^2 + \omega_m^2)/(s\text{BW})$, where ω_m is the center frequency and BW is the bandwidth. (The prototype low-pass function is assumed to have a cutoff frequency of unity.) If ω_1 and ω_2 are taken as the band-edge frequencies, then

$$\omega_m = \sqrt{\omega_1\omega_2} \quad \text{and} \quad \text{BW} = \omega_2 - \omega_1.$$

A complex-conjugate pair of LP poles is transformed into two complex-conjugate pairs of BP poles. The Q's of both pairs of poles are the same. The geometric mean of the ω_o's of the BP poles is ω_m.

The magnitude of the BP function has the same value at two frequencies which have a geometric mean of ω_m. At these two frequencies the BP phases are of equal value but are opposite in sign.

Passive band-pass networks can be obtained directly from the low-pass prototypes by applying the transformation to each element of the LP network.

This results in the replacement of series LP inductors with series-connected inductor–capacitor pairs. Shunt LP capacitors are replaced by shunt-connected inductor–capacitor pairs.

The poles of mutually coupled tuned circuits are on constant-α lines. The coefficient of coupling, k, determines the separation of the poles. The larger α and the more widely the upper-half-plane poles are separated from each other, the wider the bandwidth. A flat-top magnitude characteristic is obtained if the upper-half-plane poles are placed along the $\pm 45°$ lines converging at $s = j\omega_m$. Along steeper lines, the magnitude curve is double-humped. If the poles are placed along the $\pm 67.5°$ lines, the magnitude value at $\omega = \omega_m$ is 3 dB lower than the magnitude values at the hump frequencies. Anywhere between the 45° and 67.5° lines, the dip at the center is less than 3 dB.

The narrow-band, mutually coupled tuned circuits can be simulated by active RC networks. It is possible to control the separation between the poles by a single potentiometer. Active RC realizations are implemented as stagger-tuned second-order BP networks. The ω_o and Q of each second-order stage can be readily tuned.

PROBLEMS

20-1 Equate the coefficients of equal powers of s in Eqs. (20-6b) and (20-7) and show that $Q_{BP1} = Q_{BP2}$. Solve for Q_{BP1} and then obtain the expressions for ω_{oBP1} and ω_{oBP2}.

20-2 (a) Consider points along the imaginary axis in the LP plane. Find the corresponding points in the BP plane. In particular, indicate the mapping of the points represented by $\omega_{LP} = -1, -\omega, 0, \omega, 1$.

(b) Show that both positive and negative LP frequencies must be considered to obtain the positive BP frequencies. As a result, certain relationships exist between the LP and the BP magnitude and phase characteristics. Discuss these.

20-3 Consider as prototype the second-order LP Butterworth function.

(a) Obtain the ω_o and Q of the poles of the corresponding fourth-order BP function.

(b) Assume BW $\ll \omega_m$. Obtain the expression for the real and imaginary parts of the upper-half-plane poles. Show that the poles are located on $\pm 45°$ lines converging at $s = \pm j\omega_m$.

20-4 Examine the mapping of a real-axis pole at $-\alpha$ under the transformation

$$s_{LP} = \frac{s^2 + \omega_m^2}{s\text{BW}}.$$

Sketch the BP poles as a function of the LP pole location.

20-5 Show that the geometric mean of two unequal numbers is less than the arithmetic mean.

20-6 A band-pass network has been designed on the basis of $\omega_m = 1$ and $BW = 1$. How would you modify the element values so as to put the center frequency at ω_m and make the bandwidth BW?

20-7 (a) Construct a first-order *RC* and a first-order *RL* low-pass network. At $\omega = 1$, the magnitude is to be down by $[(10 \log (1 + \varepsilon^2)]$-dB from the dc value.

(b) Use the LP-to-BP transformation on the networks constructed in (a) and obtain the corresponding BP networks. Designate the pole–zero locations.

20-8 The 3-dB bandwidth of a second-order BP network is BW. Using buffer amplifiers between stages, *n* second-order BP stages are connected in cascade. What is the overall 3-dB bandwidth?

20-9 A fourth-order BP function is given by

$$\frac{s^2}{\left(s^2 + s\dfrac{\omega_{o1}}{Q} + \omega_{o1}^2\right)\left(s^2 + s\dfrac{\omega_{o2}}{Q} + \omega_{o2}^2\right)}.$$

This function is realized as a cascade connection of two second-order systems. A buffer stage between the two stages prevents one stage interacting with the other.

(a) Show that the BP function can be partitioned in three different ways.

(b) For each case, sketch the magnitude response of the individual second-order stages.

20-10 The poles of the following three BP functions are adjusted so that all three functions produce the same bandwidth:

$$\left[\frac{s}{(s + \alpha_1)^2 + \beta^2}\right]^2, \quad \left[\frac{s}{(s + \alpha)^2 + \beta_1^2}\right]\left[\frac{s}{(s + \alpha)^2 + \beta_2^2}\right],$$

$$\left(\frac{s}{s^2 + s\dfrac{\omega_{o1}}{Q} + \omega_{o1}^2}\right)\left(\frac{s}{s^2 + s\dfrac{\omega_{o2}}{Q} + \omega_{o2}^2}\right).$$

Compare qualitatively the magnitude characteristics. Assume the bandwidth is narrow compared to the center frequency.

20-11 Draw the schematic diagram (no values) of three passive, fourth-order BP networks terminated in a resistive load. Explain how each network is obtained.

20-12 Obtain a passive, eighth-order, maximally flat BP network. The passband ripple is 3 dB. The center frequency is ω_m, and the bandwidth is BW. The filter is driven from a resistive source (*R*). Give the element values in terms of ω_m, BW, and *R*.

20-13 Given a fourth-order BP network with a narrow pass-band ($\delta < 0.1$), show that the difference of the pole magnitudes, $\omega_{o2} - \omega_{o1}$, can be measured from the step response.

20-14 In Fig. 20-23, $RC = 1$.

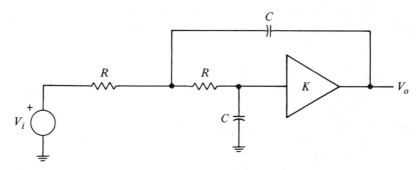

Fig. 20-23

(a) What is the value of K that will produce a pole-Q of 25?
(b) K is adjusted for $Q = 25$. How much do the ω_o and Q of the poles change if K changes by 1%?
(c) To what tolerance must K be kept if Q is to be kept within $\pm 10\%$ of 25?

20-15 Apply the transformation

$$s_{\text{LP}} = \frac{s^2 + \omega_m^2}{s\text{BW}}$$

to the network shown in Fig. 20-23 and draw the resulting network.

20-16 Using operational amplifiers in the infinite-gain mode, design a fourth-order, equal-ripple (3-dB Chebyshev) BP filter. The bandwidth is 10% of the center frequency. Use a 1000:1 resistive attenuator at the input of each second-order stage.

20-17 The fundamental and third harmonic components of a signal are to be completely rejected and the second harmonic passed. Low and high frequencies must be attenuated. Indicate approximately where the poles and zeros of the system must be placed to achieve the desired result.

20-18 Two identical tuned circuits (ω_o, Q) are coupled with the capacitor C_c as shown in Fig. 20-24.

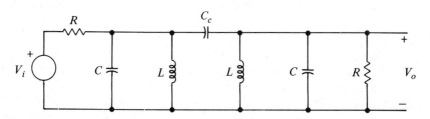

Fig. 20-24

(a) Obtain the transfer function.

(b) Describe the motion of the poles as a function of C_c.

20-19 Derive Eq. (20-74).

20-20 In Fig. 20-25, let

$$R_0 \gg (R_1 + kR), \qquad \omega_o = \frac{1}{C\sqrt{R_1 R_2}}, \qquad Q = \frac{1}{2}\sqrt{\frac{R_2}{R_1}}.$$

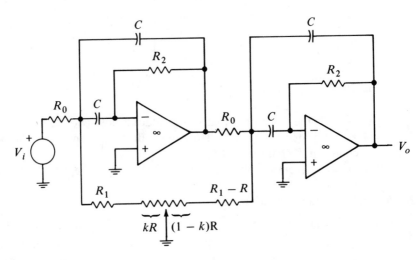

Fig. 20-25

(a) Obtain the transfer function V_o/V_i.

(b) Sketch the upper-half-plane poles as a function of k.

20-21 Obtain a passive realization (with values) for the function given by Eq. (20-79). The scale factor may be selected arbitrarily.

20-22 Derive Eq. (20-80).

21

Band-Stop
Functions
and Realizations

With a simple transformation of the complex-frequency variable, band-stop functions can be obtained from low-pass functions. Hence, all well known low-pass approximations can be readily converted to band-stop approximations. The transformation can also be applied directly to the low-pass passive network itself, thus avoiding complicated synthesis procedures. Active realizations are implemented by the cascaded connection of second-order realizations. These topics are discussed in detail in this chapter.

21-1 THE LOW-PASS-TO-BAND-STOP TRANSFORMATION

By means of the low-pass-to-band-stop transformation, band-stop (BS) functions can be obtained directly from low-pass functions. This is achieved by replacing the variable s in the LP function with the variable $s\mathrm{BW}/(s^2 + \omega_m^2)$. The constants ω_m and BW represent respectively the center frequency and the bandwidth of the BS filter. Thus, the BS function, $G_{\mathrm{BS}}(s)$, is obtained from the LP function, $G_{\mathrm{LP}}(s)$, by

$$G_{\mathrm{BS}}(s) = G_{\mathrm{LP}}(s)\Big|_{s \text{ replaced by } s\mathrm{BW}/(s^2 + \omega_m^2)} \qquad (21\text{-}1)$$

To distinguish between the two variables, let the LP complex-frequency variable be designated by s_{LP}. Then Eq. (21-1) can be written as

$$G_{\mathrm{BS}}(s) = G_{\mathrm{LP}}(s_{\mathrm{LP}})\Big|_{s_{\mathrm{LP}} = s\mathrm{BW}/(s^2 + \omega_m^2)} \qquad (21\text{-}2)$$

Pole Locations

To see how poles (or zeros) are transformed, consider a pair of complex-conjugate LP poles described by

$$s_{LP}^2 + s_{LP} \frac{\omega_{oLP}}{Q_{LP}} + \omega_{oLP}^2 = 0. \tag{21-3}$$

The poles of the corresponding BS function are obtained by substituting $sBW/(s^2 + \omega_m^2)$ for s_{LP}:

$$\left(\frac{sBW}{s^2 + \omega_m^2}\right)^2 + \left(\frac{sBW}{s^2 + \omega_m^2}\right) \frac{\omega_{oLP}}{Q_{LP}} + \omega_{oLP}^2 = 0; \tag{21-4}$$

$$\left(\frac{s}{\omega_m}\right)^4 + \left(\frac{BW}{\omega_m \omega_{oLP}} \frac{1}{Q_{LP}}\right)\left(\frac{s}{\omega_m}\right)^3 + \left(2 + \frac{BW^2}{\omega_m^2 \omega_{oLP}^2}\right)\left(\frac{s}{\omega_m}\right)^2$$
$$+ \left(\frac{BW}{\omega_m \omega_{oLP}} \frac{1}{Q_{LP}}\right)\left(\frac{s}{\omega_m}\right) + 1 = 0. \tag{21-5}$$

With

$$s_n = \frac{s}{\omega_m} \quad \text{and} \quad \gamma = \frac{BW}{\omega_m \omega_{oLP}}, \tag{21-6a}$$

Eq. (21-5) becomes

$$s_n^4 + \frac{\gamma}{Q_{LP}} s_n^3 + (2 + \gamma^2)s_n^2 + \frac{\gamma}{Q_{LP}} s_n + 1 = 0. \tag{21-6b}$$

This equation is identical with Eq. (20-6b) given for the BP function. Hence, using Eq. (20-8), the Q and ω_o of the two pairs of BS poles can be obtained:

$$Q_{BS1} = Q_{BS2} = Q_{BS} = \frac{Q_{LP}}{\sqrt{2}} \sqrt{1 + \frac{4}{\gamma^2} + \sqrt{\left(1 + \frac{4}{\gamma^2}\right)^2 - \frac{4}{\gamma^2 Q_{LP}^2}}}, \tag{21-7a}$$

$$\omega_{oBS2,1} = \frac{\omega_m}{2}\left[\gamma \frac{Q_{BS}}{Q_{LP}} \pm \sqrt{\left(\gamma \frac{Q_{BS}}{Q_{LP}}\right)^2 - 4}\right]. \tag{21-7b}$$

Both complex-conjugate BS poles have the same Q. The geometric mean of the pole magnitudes is ω_m, that is,

$$\sqrt{\omega_{oBS2} \omega_{oBS1}} = \omega_m.$$

These relationships are illustrated in Fig. 21-1. Note also the four zeros (at $\pm j\omega_m$) produced by the transformation.

When the pass is taken as stop and δ as γ, Figs. 20-2 and 20-3 can be used to determine the BS pole positions graphically.

For an example, consider

$$\omega_{oLP} = 1, \quad Q_{LP} = 3, \quad BW = 0.1, \quad \omega_m = 1.$$

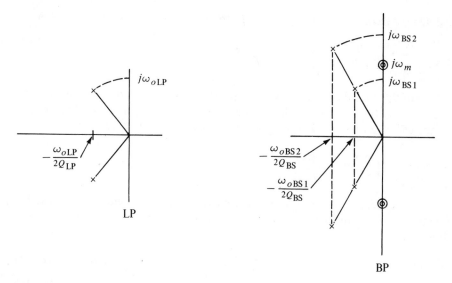

Fig. 21-1   LP-to-BS transformation

Then $\gamma = \text{BW}/(\omega_m \omega_{oLP}) = 0.1$. With $\delta = 0.1$ and $Q_{LP} = 3$, Fig. 20-2a gives $Q_{BS} = 60$ and Fig. 20-2b gives

$$\omega_{oBS2} = 1.05, \qquad \omega_{oBS1} = 0.95.$$

As long as

$$Q_{LP} \gg \frac{1}{\sqrt{2}}, \tag{21-8}$$

Eq. (21-7) may be approximated by

$$Q_{BS} = Q_{LP}\sqrt{1 + \frac{4}{\gamma^2}}, \tag{21-9a}$$

$$\omega_{oBS2,1} = \omega_m\left(\sqrt{1 + \frac{\gamma^2}{4}} \pm \frac{\gamma}{2}\right). \tag{21-9b}$$

Again taking pass as stop and δ as γ, Fig. 20-3 can be used to locate the upper-half-plane BS poles as a function of Q_{LP}. As in the BP case, the two upper-half-plane poles move along the two constant Q_{LP} lines toward $j\omega_m$ as the stop-bandwidth is decreased.

Magnitude and Phase Characteristics

The magnitude and phase characteristics of BS functions can be obtained by substituting $s = j\omega$ in Eq. (21-2):

$$G_{BS}(j\omega) = G_{LP}(j\omega_{LP})\Big|_{\omega_{LP}=\omega\text{BW}/(\omega_m^2-\omega^2)}. \tag{21-10}$$

Each LP frequency maps into two BS frequencies which are obtained from

$$\omega = -\frac{\text{BW}}{2\omega_{\text{LP}}} \pm \sqrt{\omega_m^2 + \left(\frac{\text{BW}}{2\omega_{\text{LP}}}\right)^2}.$$

One of these frequencies is positive, the other negative. Two other frequencies, one positive, the other negative, are obtained when $-\omega_{\text{LP}}$ is used. The desired frequencies are the two positive frequencies which are given by

$$\omega = \sqrt{\omega_m^2 + \left(\frac{\text{BW}}{2\omega_{\text{LP}}}\right)^2} \pm \frac{\text{BW}}{2\omega_{\text{LP}}}. \tag{21-11}$$

Let $|G_{\text{LP}}(j\omega_{\text{LP}})|$ and $\theta_{\text{LP}}(\omega_{\text{LP}})$ represent the magnitude and phase of the LP function at the frequency ω_{LP}. Then, the magnitude and phase of the BS function is given by

$$|G_{\text{BS}}(j\omega)| = |G_{\text{LP}}(j\omega_{\text{LP}})|, \tag{21-12a}$$

$$\theta_{\text{BS}}(\omega) = \mp\theta_{\text{LP}}(\omega_{\text{LP}}), \tag{21-12b}$$

where ω is determined by Eq. (21-11), and the proper association of the $\pm$ signs is observed. Thus, at

$$\omega = \sqrt{\omega_m^2 + \left(\frac{\text{BW}}{2\omega_{\text{LP}}}\right)^2} + \frac{\text{BW}}{2\omega_{\text{LP}}}$$

and at

$$\omega = \sqrt{\omega_m^2 + \left(\frac{\text{BW}}{2\omega_{\text{LP}}}\right)^2} - \frac{\text{BW}}{2\omega_{\text{LP}}},$$

the magnitude of the BS function is identical with the magnitude of the LP function evaluated at ω_{LP}. At the lower frequency, the phase is identical with the phase of the LP function (evaluated at ω_{LP}) whereas at the higher frequency it is the negative of the LP phase.

Note that the geometric mean of the two frequencies given by Eq. (21-11) is ω_m. Hence, the BS magnitude is the same at the two BS frequencies which have a geometric mean of ω_m. At these two frequencies the BS phases are of equal magnitude but opposite sign. Equation (21-11) also indicates that the difference between the two frequencies is $\text{BW}/\omega_{\text{LP}}$. In particular, *for $\omega_{\text{LP}} = 1$, the difference frequency becomes* BW, *which may be taken as a measure for the bandwidth of the stop-band* (assuming that $\omega_{\text{LP}} = 1$ is the LP bandwidth).

Figure 21-2 shows the correspondence between the LP and BS magnitude and phase characteristics. Note in particular the relationship between the two magnitude and phase characteristics at points o, p, and c. If the curve opc represents the pass-band in the LP plane, the curve opc and cp- represents the pass-band in the BS plane. Similarly, the curve cdc in the BS plane corresponds to curve

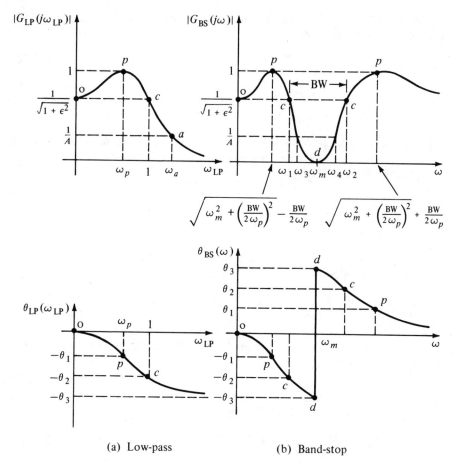

Fig. 21-2 Low-pass-to-band-stop transformation

ca- in the LP plane. The upper and lower BS edge-frequencies (cutoff frequencies) and the bandwidth of rejection are given by

$$\omega_{2,1} = \sqrt{\omega_m^2 + \left(\frac{\text{BW}}{2}\right)^2} \pm \frac{\text{BW}}{2}, \tag{21-13a}$$

$$\text{BW} = \omega_2 - \omega_1. \tag{21-13b}$$

As Fig. 21-2 shows, a given BS specification can be transformed into the corresponding LP specification, which can then be used to construct a suitable LP function. The allowable pass-band ripple fixes the ε of the LP approximation. The amount of required attenuation, $(20 \log A)$ dB, at the stop-band frequency, ω_a, fixes the order of the LP function. The resulting LP function is then transformed into the BS function by substituting $s\text{BW}/(s^2 + \omega_m^2)$ for s. The constant,

ω_m, is obtained from the geometric mean of the desired band-edge frequencies and the constant BW is obtained from the difference, that is,

$$\omega_m = \sqrt{\omega_2 \omega_1}, \qquad \text{BW} = \omega_2 - \omega_1. \tag{21-14}$$

EXAMPLE 21-1

Obtain the four-pole BS function that has maximally flat characteristics at $\omega = 0$ and $\omega = \infty$. The bandwidth taken between the 1-dB down frequencies is to be $0.1\omega_m$.

SOLUTION

Using Table 17-1, construct the 3-dB ripple, low-pass Butterworth transfer function:

$$T_{LP3}(s) = \frac{1}{s^2 + \sqrt{2}s + 1}. \tag{21-15a}$$

To convert the pass-band ripple from 3 dB to 1 dB, replace s with $s\sqrt{\varepsilon} = s\sqrt{0.5089} = 0.7133s$:

$$T_{LP1} = \frac{1}{(0.7133s)^2 + \sqrt{2}(0.7133s) + 1}. \tag{21-15b}$$

To obtain the frequency-normalized (with respect to ω_m) BS function, replace s with $s0.1/(s^2 + 1)$:

$$T_{BSn} = \frac{1}{\left(0.7133 \dfrac{0.1s}{s^2 + 1}\right)^2 + \sqrt{2}\left(0.7133 \dfrac{0.1s}{s^2 + 1}\right) + 1}$$

$$= \frac{(s^2 + 1)^2}{s^4 + 0.1009s^3 + 2.0051s^2 + 0.1009s + 1}$$

$$= \frac{(s^2 + 1)^2}{[(s + 0.0246)^2 + 0.9748^2][(s + 0.0259)^2 + 1.0252^2]}. \tag{21-16}$$

The resulting equation has the desired BS characteristics at $\omega_m = 1$. To move the rejection to $\omega = \omega_m$, replace s by s/ω_m:

$$T_{BS} = \frac{[(s/\omega_m)^2 + 1]^2}{[(s/\omega_m + 0.0246)^2 + 0.9748^2][(s/\omega_m + 0.0259)^2 + 1.0252^2]}. \qquad \textit{Ans.}$$

$$\tag{21-17}$$

The magnitude and phase of Eq. (21-17) are plotted in Fig. 21-3. Note the flat nature of the magnitude at $\omega = 0$ and $\omega = \infty$. Because of the two

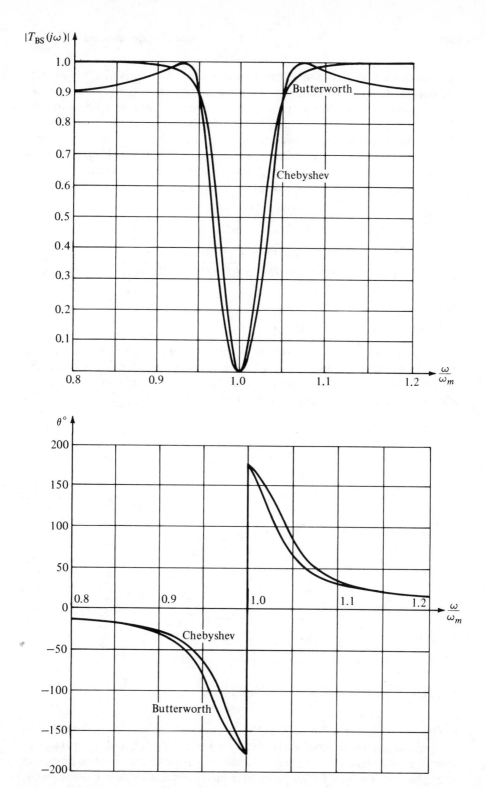

Fig. 21-3 (a) Four-pole BS magnitude characteristics (b) Four-pole BS phase characteristics

635

zeros at $s = j\omega_m$, the magnitude as well as its first derivative are zero at $\omega/\omega_m = 1$. At this frequency, the phase jumps 2π radians (π radians per zero). For comparison purposes, the fourth-order Chebyshev BS characteristics (with 1-dB pass-band ripple) are also given. Note the sharper rejection produced by the Chebyshev magnitude curve. Also, the Chebyshev curve is flatter at $\omega = \omega_m$, a desirable feature which allows good rejection even when the network is not precisely tuned or when components change slightly. The Chebyshev BS transfer function is given by

$$T_{BS1} = \frac{0.9826(s^2 + 1)^2}{1.1025s^4 + 0.1098s^3 + 2.2150s^2 + 0.1098s + 1.1025}$$

$$= \frac{0.8913(s^2 + 1)^2}{[(s + 0.0239)^2 + 0.9599^2][(s + 0.0259)^2 + 1.0411^2]}.$$

21-2 PASSIVE REALIZATIONS

A passive BS network can either be developed from the given BS function or obtained directly from the prototype LP network. In the latter, easier approach, the LP-to-BS transformation is applied to the impedance of inductors and the admittance of capacitors of the LP network, as shown in Fig. 21-4.

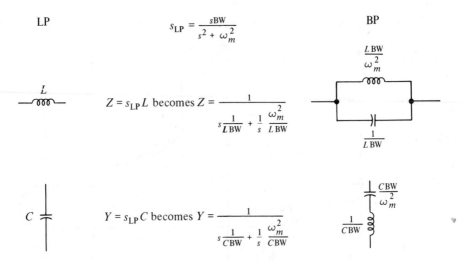

Fig. 21-4 Transformation of network elements

Thus, an inductor of value L in the LP network becomes transformed into the parallel combination of an inductor of value LBW/ω_m^2 and a capacitor of value $1/LBW$. Similarly a capacitor of value C becomes the series combination of a capacitor of value CBW/ω_m^2 with an inductor of value $1/CBW$. Consequently, the BS center frequency can be changed without affecting the bandwidth of rejection by dividing all inductors in the series branches and all capacitors in the shunt branches by ω_m^2. On the other hand, multiplication of all inductors in the series branches and all capacitors in the shunt branches by BW and division of all capacitors in the series branches and all inductors in the shunt branches by BW changes the bandwidth of rejection without affecting the rejection frequency.

EXAMPLE 21-2

Obtain a passive realization for the fourth-order BS transfer (Butterworth) function given in Eq. (21-16). Also obtain the low- and high-frequency equivalent networks.

SOLUTION

From Eq. (21-15b) obtain the LP function

$$T_{LP1} = \frac{1}{0.5089} \frac{1}{s^2 + 1.9824s + 1.9652}.$$

To construct the LP network with resistive load termination, divide the numerator and denominator of T_{LP1} by the even part of the denominator polynomial:

$$T_{LP1} = \frac{1.9652\left(\dfrac{1}{s^2 + 1.9652}\right)}{\dfrac{1.9824s}{s^2 + 1.9652} + 1} = \frac{1.9652}{s^2 + 1.9652} \frac{1}{Z_{OLC} + 1},$$

where

$$Z_{OLC} = \frac{1.9824s}{s^2 + 1.9652} = \frac{1}{s0.5044 + \dfrac{1}{s1.0088}}.$$

The resulting LP network with 1-dB pass-band ripple is shown in Fig. 21-5a. The fourth-order BS network with $0.1\omega_m$ rejection bandwidth is constructed from the LP prototype network by referring to Fig. 21-4 and using the transformation

$$s_{LP} = \frac{s0.1}{s^2 + \omega_m^2}.$$

The result is Fig. 21-5b where, in addition, all impedances are multiplied by R to provide an arbitrary load termination.

For s small, the impedance of the series branch in Fig. 21-5b is determined primarily by the inductor; on the other hand, the impedance of the shunt

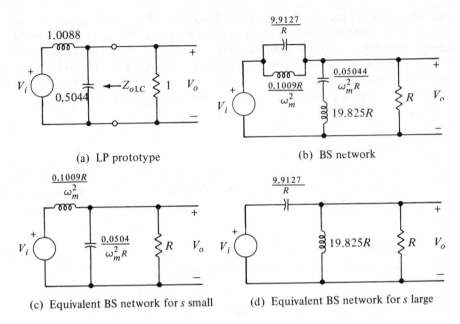

(a) LP prototype   (b) BS network

(c) Equivalent BS network for s small   (d) Equivalent BS network for s large

Fig. 21-5 From LP to BS network

branch is determined primarily by the capacitor. Thus for s small, the network of Fig. 21-5b can be redrawn as in Fig. 21-5c, which shows the LP nature of the resulting network. For s large, the other elements determine the behavior of the BS network as shown by Fig. 21-5d; note the HP nature of the resulting network. At $s = j\omega_m$, the impedance of the series branch becomes infinite and the impedance of the shunt branch becomes zero. Hence, the signal is rejected twice.

21-3 ACTIVE BS REALIZATION

Band-stop functions may be realized by active RC networks by cascading second-order sections. The transfer function of a second-order stage is of the form

$$T(s) = \frac{s^2 + \omega_m^2}{s^2 + s\dfrac{\omega_o}{Q} + \omega_o^2}, \tag{21-18}$$

where ω_o may be equal to, less than, or greater than ω_m. Consequently, the magnitude characteristic exhibits one of the forms shown in Fig. 21-6. The amount of peaking that occurs before or after ω_m depends upon the Q of the poles.

For example, if the BS function is of order four, the magnitude of one pair of complex poles is less than ω_m and the magnitude of the other pair greater

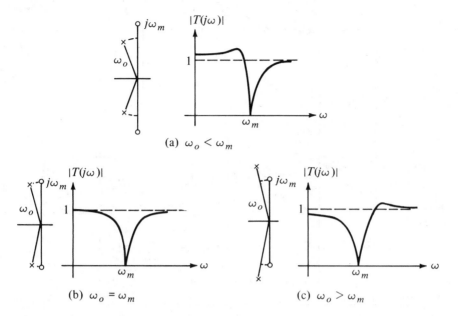

Fig. 21-6   Position of poles relative to zeros

than ω_m. Thus, one stage produces the characteristics shown in Fig. 21-6a and
the other stage produces the characteristic shown in Fig. 21-6c. The product of
these magnitude characteristics represents the magnitude of the fourth-order
system; the overall magnitude has geometric symmetry about ω_m.

One method of realizing the second-order transfer function given by Eq.
(21-18) is through the use of appropriate combinations of the input signal with
band-pass and low-pass output signals; that is,

$$V_o = V_i - V_i\,\mathrm{BP} \pm V_i\,\mathrm{LP},$$

$$\frac{V_o}{V_i} = 1 - \frac{H_1 s}{s^2 + s\dfrac{\omega_o}{Q} + \omega_o^2} \pm \frac{H_2}{s^2 + s\dfrac{\omega_o}{Q} + \omega_o^2}$$

$$= \frac{s^2 + s\left(\dfrac{\omega_o}{Q} - H_1\right) + (\omega_o^2 \pm H_2)}{s^2 + s\dfrac{\omega_o}{Q} + \omega_o^2}. \tag{21-19}$$

By adjusting H_1 to equal ω_o/Q, the zeros are placed on the imaginary axis.
The magnitude of the zeros can be made greater or less than ω_o by using the
$+$ or the $-$ sign with H_2.

To assure that the BP and the LP functions have the same poles, the three-
amplifier realization discussed in Section 11-6 is used. An inverting summer

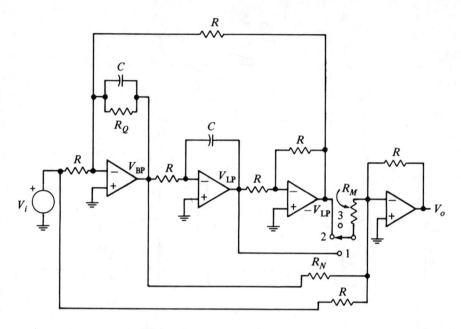

Fig. 21-7 Second-order BS building block

provides the proper scaling and summing of the three signals. Figure 21-7 illustrates the complete network. To obtain $\omega_o < \omega_m$, the switch must be at position 1; to obtain $\omega_o > \omega_m$, position 2 must be used. Since for $\omega_o = \omega_m$ no low-pass signal is required, the switch must be at position 3.

The various transfer functions involved with Fig. 21-7 are the following:

$$\frac{V_{BP}}{V_i} = -\frac{\dfrac{s}{RC}}{s^2 + s\dfrac{1}{R_Q C} + \dfrac{1}{R^2 C^2}},$$

$$\frac{V_{LP}}{V_i} = \frac{\dfrac{1}{R^2 C^2}}{s^2 + s\dfrac{1}{R_Q C} + \dfrac{1}{R^2 C^2}};$$

$$\frac{V_o}{V_i} = -\frac{s^2 + s\dfrac{1}{R_Q C}\left(1 - \dfrac{R_Q}{R_N}\right) + \dfrac{1}{R^2 C^2}\left(1 \pm \dfrac{R}{R_M}\right)}{s^2 + s\dfrac{1}{R_Q C} + \dfrac{1}{R^2 C^2}}$$

$$= -\frac{s^2 + s\dfrac{\omega_o}{Q}\left(1 - \dfrac{R_Q}{R_N}\right) + \omega_o^2\left(1 \pm \dfrac{R}{R_M}\right)}{s^2 + s\dfrac{\omega_o}{Q} + \omega_o^2} \quad \begin{cases} +\text{with switch at 1} \\ -\text{with switch at 2} \end{cases}, \quad (21\text{-}20a)$$

where

$$\omega_o = \frac{1}{RC}, \qquad Q = \frac{R_Q}{R}, \qquad \frac{\omega_o}{Q} = \frac{1}{R_Q C}. \qquad (21\text{-}20b)$$

To put the zeros on the imaginary axis, R_N is made equal to R_Q. The upper-half-plane zero is at

$$j\omega_m = j\omega_o \sqrt{1 \pm \frac{R}{R_M}};$$

the switch position determines the sign, being $+$ at position 1 and $-$ at position 2. If $\omega_m = \omega_o$, use position 3.

EXAMPLE 21-3

Obtain an active realization for the four-pole BS function given by Eq. (21-16).

SOLUTION

From Eq. (21-16), obtain the desired function and partition it as follows:

$$\frac{V_o}{V_i} = \left[\frac{s^2 + 1}{s^2 + 0.0492s + 0.9751^2} \right]\left[\frac{s^2 + 1}{s^2 + 0.0518s + 1.0255^2} \right]$$

$$= \left[\frac{s^2 + \omega_m^2}{s^2 + s\dfrac{\omega_{o1}}{Q_1} + \omega_{o1}^2} \right]\left[\frac{s^2 + \omega_m^2}{s^2 + s\dfrac{\omega_{o2}}{Q_2} + \omega_{o2}^2} \right],$$

where

$\omega_m = 1;$

$\omega_{o1} = 0.9751, \qquad Q_1 = 19.832,$

$\omega_{o2} = 1.0255, \qquad Q_2 = 19.832.$

For the design of the individual stages, refer to Fig. 21-7 and Eq. (21-20).

Stage 1 (switch at 1)
Let $C = 100\ \mu\text{F}$.
Then

$$R = \frac{1}{\omega_{o1} C} = \frac{10^6}{0.9751 \times 100} = 10.255\ \text{K}\Omega;$$

$$R_Q = RQ_1 = 10.255 \times 19.832 = 203.38\ \text{K}\Omega,$$

$$R_N = R_Q = 203.38\ \text{K}\Omega,$$

$$R_M = \frac{R}{\left(\dfrac{\omega_m}{\omega_{o1}}\right)^2 - 1} = \frac{10.255}{\left(\dfrac{1}{0.9751}\right)^2 - 1} = 198.19\ \text{K}\Omega.$$

Stage 2 (switch at 2)
Let $C = 100 \ \mu$F.
Then

$$R = \frac{1}{\omega_{o2} C} = \frac{10^6}{1.0255 \times 100} = 9.751 \ \text{K}\Omega;$$

$$R_Q = RQ_2 = 9.751 \times 19.832 = 193.379 \ \text{K}\Omega,$$

$$R_N = R_Q = 193.379 \ \text{K}\Omega,$$

$$R_M = \frac{R}{1 - \left(\dfrac{\omega_m}{\omega_{o2}}\right)^2} = \frac{9.751}{1 - \left(\dfrac{1}{1.0255}\right)^2} = 198.19 \ \text{K}\Omega.$$

To move the frequency of rejection from $\omega = 1$ to ω_m, divide all C's by ω_m.

21-4 SUMMARY

By replacing s with $s\text{BW}/(s^2 + \omega_m^2)$, the low-pass function is converted to the band-stop function. The constant ω_m represents the rejection frequency and the constant BW represents the rejection bandwidth which is taken as the difference of the band-edge frequencies. When the transformation is applied to a second-order LP function, a four-pole BS function with double zeros at $\pm j\omega_m$ is obtained. The poles are on constant-Q lines.

The magnitude of the resulting BS function possesses geometric symmetry about $\omega = \omega_m$, that is, the magnitude has the same value at two frequencies which have a geometric average of ω_m. At these two frequencies, the phases are equal in value but opposite in sign.

The passive BS network can be obtained from the passive LP network by replacing all series inductors with parallel LC combinations and all shunt capacitors with series LC combinations. Fourth- and higher-order active-RC BS networks are realized by cascading second-order BS networks. In the second-order stages, the ω_o of the poles does not coincide with ω_m. For this reason, the three-amplifier realization is used in conjunction with a summing amplifier to generate the poles and zeros of the second-order stages.

PROBLEMS

21-1   Show that a two-step procedure can be used to obtain the band-stop functions. First, the LP-to-HP transformation is used. Then, the BP transformation is used on the result of the first transformation.

21-2   Apply the LP-to-BS transformation to points on the real axis in the LP plane. What is the resulting curve in the BS plane?

21-3 $V_o/V_i = 1/(s + 1)$. Replace s with $sBW/(s^2 + \omega_m^2)$.
  (a) Sketch the poles and zeros of the resulting function. How do the poles move as a function of BW and ω_m?
  (b) Obtain a resistive load and a resistive source realization for the resulting BS function.

21-4 The magnitude characteristic shown in Fig. 21-8 is transformed by re-

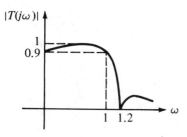

Fig. 21-8

placing ω with $0.1\omega/(1 - \omega^2)$. Sketch the magnitude characteristic of the resulting function.

21-5 Sketch the magnitude characteristic of an eighth-order BS function which is obtained by transforming the fourth-order Chebyshev LP function.

21-6 The LP-to-BS transformation results in functions which have zeros all at $\pm j\omega_m$. Consider a fourth-order BS function and discuss qualitatively the resulting change in the magnitude characteristic if the zeros are pulled apart slightly, i.e., the four zeros are placed at

$$s = \pm j(\omega_m + \Delta\omega_m), \qquad \pm j(\omega_m - \Delta\omega_m).$$

21-7 Obtain a passive realization (with resistive source termination) that meets the specifications given in Example 21-2.

22

Phase Equalization

Filters are used to pass certain bands of frequencies and attenuate others. Ideally the pass-band magnitude characteristic should be flat and the phase characteristics linear (or the group delay constant). In practice, it is impossible to build linear circuits that produce such characteristics exactly. At best, the magnitude in the pass-band is approximately flat and the phase approximately linear. The desired approximating function is usually obtained in two steps. First, the desired magnitude characteristic is approximated with little or no regard to the phase. The ensuing phase is noted. Then, the attention is directed to linearizing (equalizing) the phase as much as possible without affecting the magnitude. This is done by choosing an appropriate all-pass function.

22-1 PHASE EQUALIZATION WITH SECOND-ORDER ALL-PASS FUNCTIONS

The overall approach to phase equalization is shown in block diagram form in Fig. 22-1.

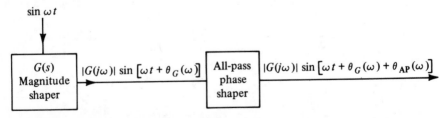

Fig. 22-1   Steady-state signals at the various stages of the filter

The aim is to make the overall phase as nearly linear as possible according to some criterion of approximation. The AP function introduces additional phase shift to achieve this. As a result, there is a greater amount of overall phase shift, and hence a greater time delay, between input and output. If the phase of the magnitude-shaper, $\theta_G(\omega)$, is quite linear to begin with, then a low-order AP function can be used to provide the necessary phase correction. On the other hand, if $\theta_G(\omega)$ is far from being linear, a higher-order AP function may be necessary to provide the desired degree of phase equalization. For example, the phase of Butterworth functions is more nearly linear than the phase of Chebyshev functions. Therefore, the phase of Butterworth functions can be brought close to the ideal more readily (with a lower-order AP function) than can the phase of Chebyshev functions.

Linearizing the Phase at $\omega = 0$

Where should the poles, and hence the zeros, of the AP function be placed in order that a prescribed amount of phase equalization is obtained? Suppose, for example, that only a *second-order* AP *function* is to be used to make the phase as linear as possible at $\omega = 0$. (This corresponds to making the time delay at $\omega = 0$ as constant as possible.) In other words, the ω_o and Q of the poles of the AP function are to be chosen such that

$$\theta(\omega) = \theta_G(\omega) + \theta_{AP}(\omega) \tag{22-1a}$$

is maximally linear at $\omega = 0$. In Eq. (22-1a), $\theta(\omega)$ represents the overall phase, $\theta_G(\omega)$ the phase of the magnitude shaper, and $\theta_{AP}(\omega)$ the phase of the AP function (see Fig. 22-1). To take a closer look at the characteristics of $\theta(\omega)$ at $\omega = 0$, obtain its power-series expansion about $\omega = 0$:

$$\theta(\omega) = k_1 \omega + k_3 \omega^3 + k_5 \omega^5 + k_7 \omega^7 + \cdots. \tag{22-1b}$$

Ideally, a linear phase characteristic is produced if all k coefficients except k_1 are zero. Because a second-order AP function has two adjustable variables, ω_o and Q, only two of the k-coefficients, k_3 and k_5, can be made zero. To see how this is achieved, consider first the AP function and its phase:

$$G_{AP}(s) = \left. \frac{s^2 - s\dfrac{\omega_{oAP}}{Q_{AP}} + \omega_{oAP}^2}{s^2 + s\dfrac{\omega_{oAP}}{Q_{AP}} + \omega_{oAP}^2} \right|_{s=j\omega} = e^{j\theta_{AP}(\omega)}; \tag{22-2a}$$

$$\theta_{AP}(\omega) = -2\tan^{-1}\left[\frac{\dfrac{1}{Q_{AP}}\left(\dfrac{\omega}{\omega_{oAP}}\right)}{1 - \left(\dfrac{\omega}{\omega_{oAP}}\right)^2} \right]. \tag{22-2b}$$

When $\theta_{AP}(\omega)$ is expanded in the Maclaurin's series (see Problem 22-2), Eq. (22-2b) becomes

$$\theta_{AP}(\omega) = -2\left[\frac{1}{Q_{AP}}\left(\frac{\omega}{\omega_{oAP}}\right) + \frac{1}{Q_{AP}}\left(1 - \frac{1}{3Q_{AP}^2}\right)\left(\frac{\omega}{\omega_{oAP}}\right)^3\right.$$

$$\left. + \frac{1}{Q_{AP}}\left(1 - \frac{1}{Q_{AP}^2} + \frac{1}{5Q_{AP}^4}\right)\left(\frac{\omega}{\omega_{oAP}}\right)^5 + \cdots\right]. \quad (22\text{-}3)$$

Similarly, the phase of the magnitude-shaper, $G(j\omega)$, is expanded in the Maclaurin's series:

$$\theta_G(\omega) = g_1\omega + g_3\omega^3 + g_5\omega^5 + g_7\omega^7 + \cdots. \quad (22\text{-}4)$$

The constants $g_1, g_3, g_5, g_7, \ldots$ are all known because the phase of $G(j\omega)$ is known. Then the overall phase is obtained by combining Eqs. (22-3) and (22-4):

$$\theta(\omega) = \theta_G(\omega) + \theta_{AP}(\omega)$$

$$= \left(g_1 - \frac{2}{Q_{AP}\omega_{oAP}}\right)\omega + \left[g_3 - \frac{2}{Q_{AP}\omega_{oAP}^3}\left(1 - \frac{1}{3Q_{AP}^2}\right)\right]\omega^3$$

$$+ \left[g_5 - \frac{2}{Q_{AP}\omega_{oAP}^5}\left(1 - \frac{1}{Q_{AP}^2} + \frac{1}{5Q_{AP}^4}\right)\right]\omega^5 + \cdots. \quad (22\text{-}5)$$

To obtain as linear a phase as possible at $\omega = 0$, ω_{oAP} and Q_{AP} are adjusted to make the coefficients of the ω^3 and ω^5 terms zero:

$$g_3 - \frac{2}{Q_{AP}\omega_{oAP}^3}\left(1 - \frac{1}{3Q_{AP}^2}\right) = 0, \quad (22\text{-}6a)$$

$$g_5 - \frac{2}{Q_{AP}\omega_{oAP}^5}\left(1 - \frac{1}{Q_{AP}^2} + \frac{1}{5Q_{AP}^4}\right) = 0. \quad (22\text{-}6b)$$

The resulting phase is then given by

$$\theta(\omega) = \left(g_1 - \frac{2}{Q_{AP}\omega_{oAP}}\right)\omega + g_7\omega^7 + \cdots. \quad (22\text{-}7)$$

Equation (22-6) can be manipulated to

$$\omega_{oAP}^{15} = \left[\frac{2}{Q_{AP}g_3}\left(1 - \frac{1}{3Q_{AP}^2}\right)\right]^5, \quad (22\text{-}8a)$$

$$\omega_{oAP}^{15} = \left[\frac{2}{Q_{AP}g_5}\left(1 - \frac{1}{Q_{AP}^2} + \frac{1}{5Q_{AP}^4}\right)\right]^3 \quad (22\text{-}8b)$$

After some algebra, Eq. (22-8) can be rearranged to

$$(Q_{AP}^2)^6 - \left(3 + 4\frac{g_5^3}{g_3^5}\right)(Q_{AP}^2)^5 + \left(\frac{18}{5} + \frac{20}{3}\frac{g_5^3}{g_3^5}\right)(Q_{AP}^2)^4 - \left(\frac{11}{5} + \frac{40}{9}\frac{g_5^3}{g_3^5}\right)(Q_{AP}^2)^3$$

$$+ \left(\frac{18}{15} + \frac{40}{27}\frac{g_5^3}{g_3^5}\right)(Q_{AP}^2)^2 - \left(\frac{3}{25} + \frac{20}{81}\frac{g_5^3}{g_3^5}\right)(Q_{AP}^2) + \left(\frac{1}{125} + \frac{4}{243}\frac{g_5^3}{g_3^5}\right) = 0. \quad (22\text{-}9)$$

The roots of this sixth-degree equation in Q_{AP}^2 can be obtained by using a digital computer. Although there are 12 values of Q_{AP} which satisfy Eq. (22-9), only the positive and real values of Q_{AP} represent the solution to the all-pass problem. Once Q_{AP} is known, ω_{oAP} can be found from:

$$\omega_{oAP} = \sqrt{\frac{g_3}{g_5} \frac{\left(1 - \frac{1}{Q_{AP}^2} + \frac{1}{5Q_{AP}^4}\right)}{\left(1 - \frac{1}{3Q_{AP}^2}\right)}}. \tag{22-10}$$

The problem of linearizing the phase of an arbitrary $G(j\omega)$ at $\omega = 0$ with a second-order AP function is thus solved. Because all the effort is directed toward making the phase maximally linear at $\omega = 0$, the phase may depart considerably from linearity at frequencies much greater than $\omega = 0$.

Refer to Chapter 14 for passive and active realizations for AP functions.

EXAMPLE 22-1

(a) Evaluate g_3 and g_5 for all-pole fourth-order systems.
(b) Obtain the second-order AP function that linearizes the phase of the 3-dB ripple, fourth-order Butterworth low-pass function at $\omega = 0$.
(c) Discuss the results.

SOLUTION

(a) The all-pole fourth-order system is described by

$$\frac{H}{\left(s^2 + s\frac{\omega_{o1}}{Q_1} + \omega_{o1}^2\right)\left(s^2 + s\frac{\omega_{o2}}{Q_2} + \omega_{o2}^2\right)}. \tag{22-11}$$

The associated phase is

$$\theta(\omega) = -\tan^{-1}\left[\frac{\frac{1}{Q_1}\frac{\omega}{\omega_{o1}}}{1 - \left(\frac{\omega}{\omega_{o1}}\right)^2}\right] - \tan^{-1}\left[\frac{\frac{1}{Q_2}\frac{\omega}{\omega_{o2}}}{1 - \left(\frac{\omega}{\omega_{o2}}\right)^2}\right]. \tag{22-12}$$

Making use of Eq. (22-3), expand $\theta(\omega)$ in the Maclaurin's series, and rearrange it as

$$\theta(\omega) = -\left(\frac{1}{Q_1\omega_{o1}} + \frac{1}{Q_2\omega_{o2}}\right)\omega$$

$$- \left[\frac{1}{Q_1\omega_{o1}^3}\left(1 - \frac{1}{3Q_1^2}\right) + \frac{1}{Q_2\omega_{o2}^3}\left(1 - \frac{1}{3Q_2^2}\right)\right]\omega^3$$

$$- \left[\frac{1}{Q_1\omega_{o1}^5}\left(1 - \frac{1}{Q_1^2} + \frac{1}{5Q_1^4}\right) + \frac{1}{Q_2\omega_{o2}^5}\left(1 - \frac{1}{Q_2^2} + \frac{1}{5Q_2^4}\right)\right]\omega^5 - \cdots. \tag{22-13}$$

Obtain g_3 and g_5 by comparing Eqs. (22-13) and (22-4):

$$g_3 = -\left[\frac{1}{Q_1 \omega_{o1}^3}\left(1 - \frac{1}{3Q_1^2}\right) + \frac{1}{Q_2 \omega_{o2}^3}\left(1 - \frac{1}{3Q_2^2}\right)\right], \qquad Ans. \qquad (22\text{-}14a)$$

$$g_5 = -\left[\frac{1}{Q_1 \omega_{o1}^5}\left(1 - \frac{1}{Q_1^2} + \frac{1}{5Q_1^4}\right) + \frac{1}{Q_2 \omega_{o2}^5}\left(1 - \frac{1}{Q_2^2} + \frac{1}{5Q_2^4}\right)\right]. \qquad Ans.$$
$$(22\text{-}14b)$$

(b) Using Table 17-1, construct the fourth-order (3-dB ripple) Butterworth low-pass function:

$$B_4 = \frac{1}{(s^2 + 1.8478s + 1)(s^2 + 0.7654s + 1)}. \qquad (22\text{-}15)$$

The ω_o and Q of each complex pair of poles are

$$\omega_{o1} = 1, \qquad Q_1 = \frac{1}{1.8478} = 0.5412, \qquad (22\text{-}16a)$$

$$\omega_{o2} = 1, \qquad Q_2 = \frac{1}{0.7654} = 1.3089. \qquad (22\text{-}16b)$$

Using Eqs. (22-14), obtain g_3, g_5, and the g_5^3/g_3^5 ratio:

$$g_3 = -\left[1.8478\left(1 - \frac{1.8478^2}{3}\right) + 0.7654\left(1 - \frac{0.7654^2}{3}\right)\right] = -0.3608, \qquad (22\text{-}17a)$$

$$g_5 = -\left[1.8478\left(1 - 1.8478^2 + \frac{1.8478^4}{5}\right)\right.$$
$$\left. + 0.7654\left(1 - 0.7654^2 + \frac{0.7654^4}{5}\right)\right] = -0.2165, \quad (22\text{-}17b)$$

$$\frac{g_5^3}{g_3^5} = 1.6595. \qquad (22\text{-}17c)$$

Substitute the g_5^3/g_3^5 ratio in Eq. (22-9), and solve for the roots of the resulting equation. Out of 12 possible roots, only one is positive and real and satisfies Eq. (22-17). This root therefore represents the Q of the poles of the AP function. [There is another positive and real root which gives the correct g_5^3/g_3^5 ratio but does not satisfy Eqs. (22-17a) and (22-17b).] The desired value of Q and the value of the corresponding ω_o, obtained by using Eq. (22-10), are

$$\omega_{oAP} = 1.100, \qquad Q_{AP} = 0.543.$$

Hence, the second-order AP function that linearizes the phase of the fourth-order low-pass Butterworth function is

$$\frac{s^2 - s\dfrac{1.100}{0.543} + 1.100^2}{s^2 + s\dfrac{1.100}{0.543} + 1.100^2}. \qquad Ans. \qquad (22\text{-}18)$$

(c) The phases of the uncompensated function (LP), the compensating function (AP), and the compensated function (combined) are plotted vs. ω in Fig. 22-2a. Note the linearity of the overall phase at $\omega = 0$ and the

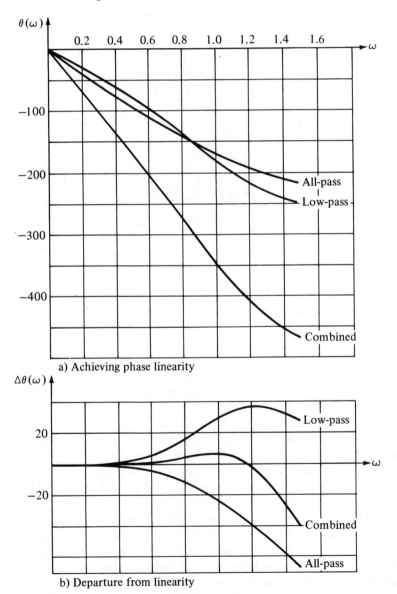

a) Achieving phase linearity

b) Departure from linearity

Fig. 22-2 How phase compensation is achieved

increased amount of overall phase shift. To see how much the phase of each function departs from being linear (as established by its initial slope), a phase-error function is constructed by subtracting the actual phase from the phase that would have resulted had the function maintained its rate of change at $\omega = 0$; that is,

$$\Delta\theta(\omega) = \left[\frac{d\theta(\omega)}{d\omega}\bigg|_{\omega=0}\right]\omega - \theta(\omega). \tag{22-19}$$

The three phase-error functions are

$$\Delta\theta_{AP}(\omega) = -\frac{2}{Q_{AP}\,\omega_{oAP}} - 2\tan^{-1}\left[\frac{\dfrac{1}{Q_{AP}}\left(\dfrac{\omega}{\omega_{oAP}}\right)}{1 - \left(\dfrac{\omega}{\omega_{oAP}}\right)^2}\right],$$

$$\Delta\theta_{LP}(\omega) = -\left(\frac{1}{Q_1\omega_{o1}} + \frac{1}{Q_2\omega_{o2}}\right)\omega$$

$$-\left[\tan^{-1}\frac{\dfrac{1}{Q_1}\left(\dfrac{\omega}{\omega_{o1}}\right)}{1 - \left(\dfrac{\omega}{\omega_{o1}}\right)^2} + \tan^{-1}\frac{\dfrac{1}{Q_2}\left(\dfrac{\omega}{\omega_{o2}}\right)}{1 - \left(\dfrac{\omega}{\omega_{o2}}\right)^2}\right],$$

$$\Delta\theta(\omega) = \Delta\theta_{AP}(\omega) + \Delta\theta_{LP}(\omega).$$

The three phase-error curves are plotted in Fig. 22-2b. The phase of the LP function curves downward as ω is increased from 0; therefore the error is positive. In order to compensate for this downward trend, the phase of the AP function must curve upward and thus produce a negative error. When the ω_o and Q of the AP function are correctly chosen, the downward and upward trends cancel each other at $\omega = 0$ and a linear phase results. The farther away the frequency is from $\omega = 0$, the larger becomes the phase error. If $\omega \leq 1$ is considered to be the pass-band, the maximum departure from linearity is about $7°$, and it occurs at $\omega = 1$.

A study of Fig. 22-2 shows that a much smaller departure of phase from linearity could be achieved in the pass-band by using an AP function with a slightly lower ω_o. This function will have a steeper initial slope and a larger phase shift at a given frequency. Consequently, the combined error curve will be lower than the one shown in Fig. 22-2b and the hump in the error curve near the cutoff frequency will be reduced. However, the phase is no longer maximally linear at $\omega = 0$. Thus, at the expense of less linearity at $\omega = 0$, a phase characteristic can be obtained that is ripply but departs less from linearity throughout the pass-band. It should also be noted that the exact positioning of the poles and zeros of the AP function in this problem is not critical because of the low value of the pole-Q.

22-2 SUMMARY

The phase of a system can be brought close to the linear characteristic by an appropriately selected all-pass function. The degree of linearity that is achieved depends upon the phase of the system and the order of the all-pass function. If the phase of the system is nearly linear to begin with, a second-order all-pass function alone may provide sufficient correction. In this chapter, a method is presented for obtaining a maximally-linear phase characteristic at $\omega = 0$ by using second-order all-pass functions. Equations are given for calculating the ω_o and Q of the all-pass function.

PROBLEMS

22-1 Let $\theta(\omega)$ represent the angle resulting from a pair of complex-conjugate roots. For what value of Q is $d^3\theta/d\omega^3 = 0$ at $\omega = 0$?

22-2 Derive Eq. (22-3).

22-3 Let $\tau(\omega)$ represent the group delay associated with a pair of complex-conjugate poles. Obtain the Maclaurin-series representation for $\tau(\omega)$.

22-4 An nth-order system is described by the transfer function given below:

$$T(s) = \frac{(s + \alpha_{z1})(s + \alpha_{z2}) \cdots \left(s^2 + s\,\dfrac{\omega_{oz1}}{Q_{z1}} + \omega_{oz1}^2\right)\left(s^2 + s\,\dfrac{\omega_{oz2}}{Q_{z2}} + \omega_{oz2}^2\right) \cdots}{(s + \alpha_{p1})(s + \alpha_{p2}) \cdots \left(s^2 + s\,\dfrac{\omega_{op1}}{Q_{p1}} + \omega_{op1}^2\right)\left(s^2 + s\,\dfrac{\omega_{op2}}{Q_{p2}} + \omega_{op2}^2\right) \cdots}.$$

The Maclaurin-series expansion of the phase of $T(j\omega)$ can be written as

$$\theta_G(\omega) = g_1\omega + g_3\,\omega^3 + g_5\,\omega^5 + \cdots.$$

Evaluate g_1, g_3, and g_5 in terms of the parameters (α, ω_o, Q) of $T(s)$.

Appendix A

A SUMMARY OF TRANSFORM PAIRS

$F(s)$	$f(t) \quad (t \geq 0)$
1	unit impulse at $t = 0$
$\dfrac{1}{s}$	1, unit step at $t = 0$
$\dfrac{1}{s + \alpha}$	$e^{-\alpha t}$
$\dfrac{\omega}{s^2 + \omega^2}$	$\sin \omega t$
$\dfrac{s}{s^2 + \omega^2}$	$\cos \omega t$
$\dfrac{\omega}{(s + \alpha)^2 + \omega^2}$	$e^{-\alpha t} \sin \omega t$
$\dfrac{s + \alpha}{(s + \alpha)^2 + \omega^2}$	$e^{-\alpha t} \cos \omega t$
$\dfrac{1}{s^2}$	t
$\dfrac{2}{s^3}$	t^2

$$\frac{1}{(s + \alpha)^2} \qquad\qquad te^{-\alpha t}$$

$$\frac{2\omega^3}{(s^2 + \omega^2)^2} \qquad\qquad \sin \omega t - \omega t \cos \omega t$$

$$\frac{2\omega^3}{[(s + \alpha)^2 + \omega^2]^2} \qquad\qquad e^{-\alpha t}(\sin \omega t - \omega t \cos \omega t)$$

Appendix B

LAPLACE TRANSFORMATION OF EQUATIONS

Let $f(t)$ represent a function of time for $t \geq 0$. The Laplace transform of $f(t)$, $F(s)$, is given by

$$F(s) = \mathscr{L}[f(t)] = \int_0^\infty f(t)\bar{e}^{st}dt.$$

It can readily be shown that*

$$\mathscr{L}\left[\frac{df(t)}{dt}\right] = sF(s) - f(0),$$

$$\mathscr{L}\left[\frac{d^2f(t)}{dt^2}\right] = s^2F(s) - sf(0) - \frac{df(t)}{dt}\bigg|_{t=0},$$

$$\mathscr{L}\left[\int_0^t f(t)dt\right] = \frac{F(s)}{s},$$

$$\mathscr{L}[af(t)] = aF(s) \quad (a \text{ not a function of time}),$$

$$\mathscr{L}[f_1(t) \pm f_2(t)] = F_1(s) \pm F_2(s).$$

Using these relations, the element-defining equations can be transformed:

$$v(t) = Ri(t) \qquad\qquad V(s) = RI(s)$$

$$v(t) = L\frac{di(t)}{dt} \qquad\qquad V(s) = sLI(s) - i(0)$$

$$v(t) = v(0) + \frac{1}{C}\int_0^t i(t)\,dt \qquad V(s) = \frac{v(0)}{s} + \frac{1}{sC}I(s)$$

---

* Gardner, Murray F., and Barnes, John L., *Transients in Linear Systems*, Vol. 1 (John Wiley & Sons, Inc., 1958), Chapter 5.

Similarly, Kirchhoff's connection laws can be transformed:

$\Sigma i(t) = 0, \qquad \Sigma I(s) = 0;$

$\Sigma v(t) = 0, \qquad \Sigma V(s) = 0.$

Any linear, time-invariant network problem can be solved by applying the element-defining equations and Kirchhoff's connection laws. For an example, consider the series connection of a voltage source, a resistor, an inductor, and a capacitor. When initial conditions (current through the inductor at $t = 0$ and the voltage across the capacitor at $t = 0$) are zero, Kirchhoff's voltage law applied around the loop yields

$$v(t) = Ri(t) + L\frac{di(t)}{dt} + \frac{1}{C}\int_0^t i(t)dt.$$

By means of Laplace transformation, this integrodifferential equation is converted to an algebraic equation:

$$V(s) = RI(s) + sLI(s) + \frac{1}{sC}I(s),$$

which can be readily solved for the Laplace transform of the current in the loop:

$$I(s) = \frac{V(s)}{R + sL + \dfrac{1}{sC}}.$$

By expanding $I(s)$ in partial fractions (see Appendix C), $I(s)$ is expressed as the sum of simple terms. Using the Laplace transform pairs presented in Appendix A, $i(t)$ is readily obtained.

Appendix C

PARTIAL-FRACTION EXPANSION

Let $F(s)$ be a rational function, i.e., the ratio of two polynomials. Let the degree of the numerator polynomial be m and the degree of the denominator polynomial n. In this text, $0 \leq m \leq n + 1$. Then,

$$F(s) = \frac{N(s)}{D(s)} = \frac{a_m s^m + a_{m-1} s^{m-1} + \cdots + a_1 s + a_0}{s^n + b_{n-1} s^{n-1} + \cdots + b_1 s + b_0}.$$

For the most general case, $m = n + 1$; and $F(s)$ can be written as

$$F(s) = \frac{a_{n+1} s^{n+1} + a_n s^n + a_{n-1} s^{n-1} + \cdots + a_1 s + a_0}{s^n + b_{n-1} s^{n-1} + \cdots + b_1 s + b_0}$$

$$= a_{n+1} s + (a_n - a_{n+1} b_{n-1}) + G(s),$$

where

$$G(s) = F(s) - a_{n+1} s - (a_n - a_{n+1} b_{n-1}) = \frac{c_{n-1} s^{n-1} + c_{n-2} s^{n-2} + \cdots + c_1 s + c_0}{s^n + b_{n-1} s^{n-1} + \cdots + b_1 s + b_0}.$$

The c-coefficients are functions of the a- and b-coefficients.

The function $G(s)$ is in proper form for partial-fraction expansion. Factor the denominator polynomial and obtain

$$G(s) = \frac{c_{n-1} s^{n-1} + c_{n-2} s^{n-2} + \cdots + c_1 s + c_0}{(s - p_1)(s - p_2) + \cdots + (s - p_n)}.$$

When all the poles are simple, i.e., $p_1 \neq p_2 \neq \cdots \neq p_n$, the partial-fraction expansion of $G(s)$ is given by

$$G(s) = \frac{K_1}{s - p_1} + \frac{K_2}{s - p_2} + \cdots + \frac{K_n}{s - p_n},$$

where

$$K_1 = [(s - p_1)G(s)]_{s=p_1},$$
$$K_2 = [(s - p_2)G(s)]_{s=p_2},$$
$$\vdots \qquad \vdots$$
$$K_n = [(s - p_n)G(s)]_{s=p_n}.$$

Note that the partial-fraction expansion contains n terms, one for each pole of $G(s)$.

When some or all the poles of $G(s)$ are identical, the partial-fraction expansion still contains n terms. For example, if pole p_1 is of multiplicity 2 and the remaining poles are simple, then the partial-fraction expansion of $G(s)$ is given by

$$G(s) = \frac{K_{11}}{(s - p_1)^2} + \frac{K_{12}}{(s - p_1)} + \frac{K_2}{s - p_2} + \cdots + \frac{K_n}{s - p_n},$$

where

$$K_{11} = [(s - p_1)^2 G(s)]_{s=p_1},$$
$$K_{12} = \left\{ \frac{d}{ds} [(s - p_1)^2 G(s)] \right\}_{s=p_1},$$
$$K_2 = [(s - p_2)G(s)]_{s=p_2},$$
$$\vdots \qquad \vdots$$
$$K_n = [(s - p_n)G(s)]_{s=p_n}.$$

Poles of multiplicity higher than two do not occur in this text.

For an example, consider the partial-fraction expansion of:

$$G(s) = \frac{1}{(s + 1)(s^2 + 1)}.$$

The expansion has three terms, one due to each pole of $G(s)$:

$$G(s) = \frac{1}{(s + 1)(s - j)(s + j)} = \frac{K_1}{s + 1} + \frac{K_2}{s - j} + \frac{K_3}{s + j},$$

where

$$K_1 = \left. \frac{1}{s^2 + 1} \right|_{s=-1} = \frac{1}{2},$$

$$K_2 = \left. \frac{1}{(s + 1)(s + j)} \right|_{s=j} = \frac{1}{(j + 1)2j} = -\frac{1}{2}\frac{1}{1 - j},$$

$$K_3 = \left. \frac{1}{(s + 1)(s - j)} \right|_{s=-j} = \frac{1}{(-j + 1)(-2j)} = -\frac{1}{2}\frac{1}{1 + j}.$$

Hence,

$$G(s) = \frac{1}{2} \left[\frac{1}{s + 1} \right] - \frac{1}{2} \left[\frac{1}{1 - j} \right] \left[\frac{1}{s - j} \right] - \frac{1}{2} \left[\frac{1}{1 + j} \right] \left[\frac{1}{s + j} \right].$$

Appendix D

CONTINUED-FRACTION EXPANSION

If the numerator and denominator polynomials of a rational function, $F(s)$, are arranged in descending powers of s, then the continued-fraction expansion of $F(s)$ is given by

$$F(s) = \frac{a_m s^m + a_{m-1} s^{m-1} + a_{m-2} s^{m-2} + \cdots + a_1 s + a_0}{b_n s^n + b_{n-1} s^{n-1} + b_{n-2} s^{n-2} + \cdots + b_1 s + b_0}$$

$$= A s^{m-n} + F_{r1}(s)$$

$$= A s^{m-n} + \cfrac{1}{\cfrac{1}{F_{r1}(s)}} = A s^{m-n} + \cfrac{1}{B s^{n-m+1} + F_{r2}(s)}$$

$$= A s^{m-n} + \cfrac{1}{B s^{n-m+1} + \cfrac{1}{C s^{m-n} + \cfrac{1}{D s^{n-m+1} + \cdots}}}$$

The expansion may terminate or continue indefinitely.

An examination of the resulting A, B, C, D, ... coefficients shows that they can be readily obtained by using the following algorithm:

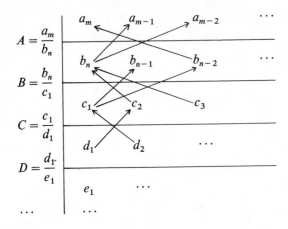

The first two rows of the table are obtained from $F(s)$. The succeeding rows are constructed by using the coefficients of the two preceding rows and the following formulas. The arrows in the table show how the formulas are obtained by using the elements in the first column as pivot elements:

$$c_1 = \frac{b_n a_{m-1} - b_{n-1} a_m}{b_n} \neq 0 \qquad c_2 = \frac{b_n a_{m-2} - b_{n-2} a_m}{b_n}$$

$$d_1 = \frac{c_1 b_{n-1} - c_2 b_n}{c_1} \neq 0 \qquad d_2 = \frac{c_1 b_{n-2} - c_3 b_n}{c_1}$$

$$e_1 = \frac{d_1 c_2 - d_2 c_1}{d_1} \neq 0 \qquad \cdots$$

If any of the a- or b-coefficients is missing, a zero must be placed in its position. However, if $F(s)$ is an odd function divided by an even function, or vice versa, then the alternately missing coefficients can be left out altogether, and the coefficient array constructed as follows:

$$A = \frac{a_m}{b_n} \quad
\begin{array}{cccc}
a_m & a_{m-2} & a_{m-4} & \cdots \\
\end{array}$$

$$B = \frac{b_n}{c_1} \quad
\begin{array}{ccc}
b_n & b_{n-2} & b_{n-4} \quad \cdots \\
\end{array}$$

$$C = \frac{c_1}{d_1} \quad
\begin{array}{ccc}
c_1 & c_2 & c_3 \\
\end{array}$$

$$D = \frac{d_1}{e_1} \quad
\begin{array}{cc}
d_1 & d_2 \\
\end{array}$$

$$e_1$$

The resulting expansion is given by

$$F(s) = As^{m-n} + \cfrac{1}{Bs^{n-m+2} + \cfrac{1}{Cs^{m-n} + \cfrac{1}{Ds^{n-m+2} + \cdots}}}.$$

EXAMPLE 1 Expand

$$F(s) = \frac{s^3 + 6s^2 + 8s}{s^2 + 4s + 3}$$

into continued fractions.

SOLUTION

$$
\begin{array}{r|ccc}
 & 1 & 6 & 8 \\
\hline
1 & 1 & 4 & 3 \\
\cline{2-4}
\frac{1}{2} & 2 & 5 & \\
\cline{2-4}
\frac{4}{3} & \frac{3}{2} & 3 & \\
\cline{2-4}
\frac{3}{2} & 1 & & \\
\cline{2-4}
\frac{1}{3} & 3 & &
\end{array}
$$

Since $m - n = 1$, the continued-fraction expansion is

$$F(s) = s + \cfrac{1}{\dfrac{1}{2} + \cfrac{1}{\dfrac{4}{3}s + \cfrac{1}{\dfrac{3}{2} + \cfrac{1}{\dfrac{1}{3}s}}}}.$$

EXAMPLE 2 Expand

$$F(s) = \frac{s^4 + 10s^2 + 9}{s^3 + 4s}$$

into continued fractions.

SOLUTION   Note that $F(s)$ is an even function divided by an odd function.

$$
\begin{array}{c|ccc}
 & 1 & 10 & 9 \\
\hline
1 & & & \\
 & 1 & 4 & \\
\hline
\frac{1}{6} & & & \\
 & 6 & 9 & \\
\hline
\frac{36}{15} & & & \\
 & \frac{15}{6} & & \\
\hline
\frac{5}{18} & & & \\
 & 9 & &
\end{array}
$$

Since $m - n = 1$, the continued-fraction expansion is

$$
F(s) = s + \cfrac{1}{\dfrac{1}{6}s + \cfrac{1}{\dfrac{36}{15}s + \cfrac{1}{\dfrac{5}{18}s}}} .
$$

If the numerator and denominator polynomials of $F(s)$ are arranged in ascending powers of s, then the continued-fraction expansion of $F(s)$ is given by

$$
F(s) = \frac{a_0 + a_1 s + a_2 s^2 + \cdots + a_{m-1}s^{m-1} + a_m s^m}{b_0 + b_1 s + b_2 s^2 + \cdots + b_{n-1}s^{n-1} + b_n s^n} \qquad (b_0 \ne 0)
$$

$$
= A + \cfrac{1}{B\dfrac{1}{s} + \cfrac{1}{C + \cfrac{1}{D\dfrac{1}{s} + \cdots}}} .
$$

The method of obtaining the A, B, C, D, ... coefficients is the same as in the previous cases. Only the a- and b-coefficients are entered in the array in ascending rather than descending order. (If $b_0 = 0$ and $a_0 \ne 0$, the first term in the expansion is A/s.)

If $F(s)$ is Od/Ev or Ev/Od, then the corresponding expansions take the forms

$$
F(s) = \frac{\text{Od}}{\text{Ev}} = \frac{a_1 s + a_3 s^3 + a_5 s^5 + \cdots}{b_0 + b_2 s^2 + b_4 s^4 + \cdots}
$$

$$
= As + \cfrac{1}{Bs + \cfrac{1}{Cs + \cfrac{1}{Ds + \cdots}}} ,
$$

$$F(s) = \frac{Ev}{Od} = \frac{a_0 + a_2 s^2 + a_4 s^4 + \cdots}{b_1 s + b_3 s^3 + b_5 s^5 + \cdots}$$

$$= A\frac{1}{s} + \cfrac{1}{B\frac{1}{s} + \cfrac{1}{C\frac{1}{s} + \cfrac{1}{D\frac{1}{s} + \cdots}}}.$$

EXAMPLE 3 Expand

$$F(s) = \frac{3 + 4s + s^2}{8s + 6s^2 + s^3}$$

into continued fractions.

SOLUTION

	3	4	1
$\frac{3}{8}$			
	8	6	1
$\frac{32}{7}$			
	$\frac{7}{4}$	$\frac{5}{8}$	
$\frac{49}{88}$			
	$\frac{22}{7}$	1	
$\frac{968}{21}$			
	$\frac{3}{44}$		
$\frac{3}{44}$			
	1		

$$F(s) = \frac{3}{8}\frac{1}{s} + \cfrac{1}{\frac{32}{7} + \cfrac{1}{\frac{49}{88}\frac{1}{s} + \cfrac{1}{\frac{968}{21} + \cfrac{1}{\frac{3}{44}\frac{1}{s}}}}}.$$

EXAMPLE 4 Expand

$$F(s) = \frac{9 + 10s^2 + s^4}{4s + s^3}$$

into continued fractions.

SOLUTION   Note that $F(s)$ is Ev/Od.

	9	10	1
$\frac{9}{4}$			
	4	1	
$\frac{16}{31}$			
	$\frac{31}{4}$	1	
$\frac{961}{60}$			
	$\frac{15}{31}$		
$\frac{15}{31}$			
	1		

$$F(s) = \frac{9}{4}\frac{1}{s} + \cfrac{1}{\cfrac{16}{31}\frac{1}{s} + \cfrac{1}{\cfrac{961}{60}\frac{1}{s} + \cfrac{1}{\cfrac{15}{31}\frac{1}{s}}}}.$$

Appendix E

α or $-\alpha$	Real part of complex root
A	Open-loop gain
A_c	Common-mode gain
A_d	Difference-mode gain
A_0	Open-loop dc gain
AP	All-pass
β	Imaginary part of complex root
BP	Band-pass
BS	Band-stop
BW	Bandwidth in rad/s
CMRR	Common-mode rejection ratio
Ev	Even part of
GB	Gain-bandwidth product in rad/s
GB_n	Normalized GB (with respect to ω_o)
HP	High-pass
i	Time-varying current
I	Laplace transform of i
Im	Imaginary part of
K	Closed-loop gain
K_0	Closed-loop dc gain

665

LG	Loop gain
lhp	Left half-plane
LP	Low-pass
Od	Odd part of
p	Pole
P	Polynomial
Q	The Q of complex poles
Q_a	Actual Q
r	Root
rad/s	Radians per second
Re	Real part of
s	Complex-frequency variable
s_n	Normalized s (with respect to ω_o)
$S, \mathscr{S}$	Sensitivity
T	Transfer function
τ_{gr}	Group delay
τ_{ph}	Phase delay
v	Time-varying voltage
V	Laplace transform of v
ω_a	Open-loop bandwidth of amplifier in rad/s
ω_n	Normalized ω (with respect to ω_o)
ω_o	Magnitude of complex pole, undamped natural frequency
ω_{oa}	Actual ω_o
ω_{3dB}	3-dB frequency in rad/s
$X(\omega)$	Reactance function
Y	Admittance
z	Zero
Z	Impedance
Z_0	Closed-loop output impedance

Answers to Selected Problems

CHAPTER 1

1-1 (a) $\dfrac{V_2}{V_1} = \dfrac{1}{RC} \dfrac{1}{s + \dfrac{1}{RC}}$ (b) $\dfrac{V_2}{I_1} = \dfrac{1}{sC}$

(c) $\dfrac{I_2}{V_1} = \dfrac{1}{L}\dfrac{R_2}{R_1 + R_2} \dfrac{1}{s + \dfrac{1}{L}\dfrac{R_1 R_2}{R_1 + R_2}}$ (d) $\dfrac{I_2}{I_1} = \dfrac{s^2}{s^2 + \dfrac{1}{LC}}$

1-2 $\dfrac{I}{V} = \dfrac{s}{s^2 + 1}$, $\dfrac{V_L}{V} = \dfrac{s^2}{s^2 + 1}$, $\dfrac{V_C}{V} = \dfrac{1}{s^2 + 1}$

1-3 $V_o(s) = \dfrac{\gamma s - \rho/C}{s^2 + \dfrac{1}{LC}}$

1-4 (a) $\dfrac{I}{V} = -sC$ (b) $\dfrac{V}{I} = 0$ (c) $\dfrac{I}{V} = \dfrac{1}{R}\dfrac{s}{s + \dfrac{1 - \alpha}{RC}}$

(d) $V_2/V = 1$ if $R \neq 1$; if $R = 1$, V_2/V is not unique.

1-5 (a) $V_2 = I_1 \dfrac{s}{s^2 + 1} + V_1 \dfrac{s^2}{s^2 + 1}$ (b) $I_2 = I_1 \dfrac{s^2}{s^2 + 1} - \dfrac{\gamma}{s^2 + 1}$

(c) $V_2 = V_1 \dfrac{s^2}{s^2 + 1} - \dfrac{\rho}{s^2 + 1}$ (d) $V_2 = 0$

1-6 (a) $V_{3c} = \dfrac{V_1 + V_2}{2} \dfrac{2R_1}{R + 2R_1}$, $V_{3d} = 0$, $V_3 = V_{3c}$

  (b) $I_{3c} = 0$, $I_{3d} = (I_1 - I_2)\dfrac{1}{2}\dfrac{s}{s + \dfrac{1}{2RC}}$, $I_3 = I_{3d}$

1-7 $I_{3c} = 0$, $V_{3d} = 0$

1-8 (a) $\dfrac{A}{F} = \dfrac{1 - K}{M}$

  (b) $K = 0: \dfrac{A}{F} = \dfrac{1}{M}$ (Newton's law)

$\left.\begin{array}{l}
K = 1: \dfrac{A}{F} = 0 \qquad \text{(Infinite mass)} \\[2em]
K = 2: \dfrac{A}{F} = -\dfrac{1}{M} \qquad \text{(Negative mass)} \\[2em]
K = -1: \dfrac{A}{F} = \dfrac{2}{M} \qquad \text{(Dependent source reduces mass)}
\end{array}\right\}$ Control of mass by means of a dependent source

1-9 (a) $s = -\dfrac{1}{RC}$ (b) $s = -\dfrac{R}{L}$ (c) $s = \pm j\dfrac{1}{\sqrt{LC}}$

  (d) $s = -\dfrac{R}{2L} \pm \sqrt{\left(\dfrac{R}{2L}\right)^2 - \dfrac{1}{LC}}$ (e) $s = -\dfrac{1}{2RC} \pm \sqrt{\left(\dfrac{1}{2RC}\right)^2 - \dfrac{1}{LC}}$

  (f) $s = -1$ (g) No poles, or a pole at $s = 0$, depending on which system function is chosen.

1-10 Poles are at $s = (-3 \pm \sqrt{5})/2$.

1-11 Both functions have a pole at $s = -\tfrac{1}{2}$.

1-12 Pole(s) of $\begin{cases} I_1/V \text{ are at } -1 \text{ and } -\tfrac{1}{2} \\ I_2/V \text{ is at } -1 \\ I_3/V \text{ is at } -\tfrac{1}{2} \end{cases}$ Poles are different because network is not properly excited; i.e., no element appears in series with the voltage source.

CHAPTER 2

2-1 One possible solution is:

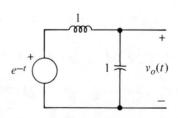

Another solution is:

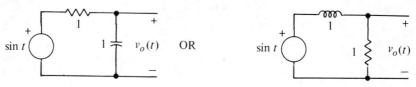

2-2 $r(t) = \alpha + (1 - \alpha)e^{-t}$

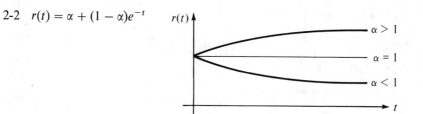

2-3 $i(t) = \dfrac{1}{2}e^{-t} + \dfrac{1}{\sqrt{2}}\sin\left(t - \dfrac{\pi}{4}\right),$ $i_n(t) = \dfrac{1}{2}e^{-t},$

$$i_f(t) = \dfrac{1}{\sqrt{2}}\sin\left(t - \dfrac{\pi}{4}\right)$$

2-4 $\dfrac{V_C}{V} = 0,$ $\dfrac{I_C}{V} = 0,$

where V_C and I_C are the voltage and current variables involved with the capacitor.

2-5 $K = 1$

2-7 (a)

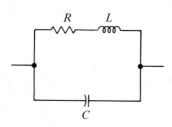

(b)

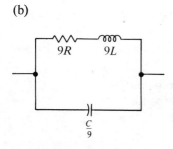

2-8 $v_2(t) = \frac{2}{3}e^{-t/2} - \frac{2}{3}e^{-2t}$ (Natural response only; excitation poles are cancelled.)

2-9

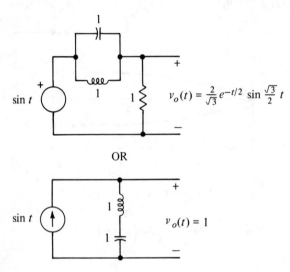

$$v_o(t) = \frac{2}{\sqrt{3}} e^{-t/2} \sin \frac{\sqrt{3}}{2} t$$

OR

$$v_o(t) = 1$$

2-10 $v_2(t) = e^{-t/T_2}$ where $T_2 = RC < T_1$ and $K = \dfrac{RC}{T_1}$.

2-11

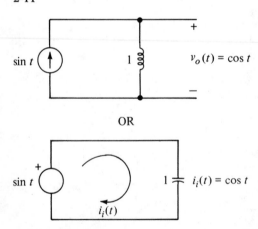

$$v_o(t) = \cos t$$

OR

$$i_i(t) = \cos t$$

2-12 $2v_1(t) + i_1(t)$ will result in zero forced response;
       $v_1(t) + i_1(t)$ will result in zero natural response.

2-13 $\alpha = \frac{1}{2}$.

2-14 (a)

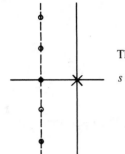

The zeros are located at

$$s = -\frac{\ell na}{T_p} \pm jk\frac{2\pi}{T_p} \qquad (k = 0, 1, 2, \ldots)$$

(b)   When $T_p = \ell na$, the system pole at $s = -1$ is cancelled.

(c)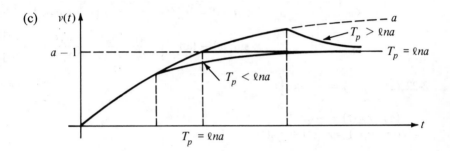

(Note that after $t > T_p$, only the forced response is present.)
$$v(t) = a(1 - e^{-t}) \qquad (0 \le t \le T_p),$$
$$= (a - 1) + (e^{T_p} - a)e^{-t} \qquad (t \ge T_p).$$

2-15 $v(t) = \underbrace{\frac{e^{-t}}{\omega}}_{\text{natural}} + \underbrace{\left(\frac{-e^{-t}}{\omega}\right)\cos \omega t}_{\text{forced}}$

Note large individual
amplitudes as $\omega \to 0$ but
small overall response.

2-16 $i(t) = [\omega/(1 - \omega^2)] (\cos \omega t - \cos t) \qquad (\omega \neq 1); \quad i(t) = (t/2) \sin t$
$(\omega = 1)$.

2-17 $r_f(t) = E_m|G(j\omega)|\sin[\omega t + \theta + \theta_G(\omega)]$ where $G(j\omega) = |G(j\omega)|e^{j\theta_G(\omega)}$.
If $G(s)$ has any poles in the right half-plane or multiple-order poles on the
imaginary axis, the natural response grows with time. To speak of steady-
state under these circumstances is meaningless.

2-18 $r_{ss}(t) = |G(j\omega)|\cos[\omega t + \theta + \theta_G(\omega)].$

2-19 $r_f(t) = |G(\alpha + j\omega)|e^{\alpha t}\sin[\omega t + \theta + \theta_G(\alpha, \omega)].$

2-20 (a) $v_f(t) = 0$ (b) $v_f(t) = -\frac{1}{2}e^{-t}$

  (c) $v_f(t) = -\frac{2}{3}\cos 2t$

  (d) $v_f(t) = -\sqrt{\frac{2}{5}}\,e^{-t}\cos\!\left(t + \tan^{-1}\frac{1}{3}\right)$

2-21 (a) The excitation is $e(t) = B + A\sin \omega t$. The response is $y(t)$. Then

$$G(s) = \frac{Y(s)}{E(s)} = \frac{1}{a_n s^n + a_{n-1}s^{n-1} + \cdots + a_0}.$$

  (b) $y_f(t) = G(0)(B) + A|G(j\omega)|\sin(\omega t + \theta_G)$ (by superposition)

$$= \frac{B}{a_0} + \frac{A}{\sqrt{(a_0 - a_2\omega^2 + a_4\omega^4 - \cdots)^2 + \omega^2(a_1 - a_3\omega^2 + a_5\omega^4 - \cdots)^2}}$$

$$\times \sin\!\left\{\omega t - \tan^{-1}\left[\frac{\omega(a_1 - a_3\omega^2 + a_5\omega^4 - \cdots)}{a_0 - a_2\omega^2 + a_4\omega^4 - \cdots}\right]\right\}.$$

2-22 (a) $r_{f1}(t) = \frac{1}{\sqrt{2}}\cos\!\left(t - \frac{\pi}{4}\right),$

  (b) $r_{f2}(t) = -\frac{1}{2}e^{-t},$

  (c) $r_f(t) = r_{f1}(t) + r_{f2}(t).$

2-23 $v_{ss}(t) = \sum_{n\,odd} V\frac{8}{T}\frac{1}{\sqrt{\left(\frac{1}{RC}\right)^2 + \left(\frac{2\pi n}{T}\right)^2}}\cos\!\left(\frac{2\pi n}{T}t - \tan^{-1}\frac{2\pi nRC}{T}\right)$

2-24 $v_f(t) = \sum_{n\,odd} I\frac{8}{T}\left[\frac{1}{1 - \left(\frac{2\pi n}{T}\right)^2}\right]\cos\frac{2\pi n}{T}t$

2-25

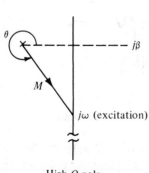

High-Q pole

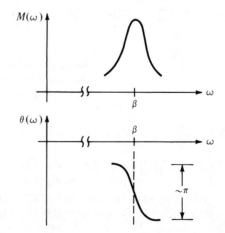

2-26

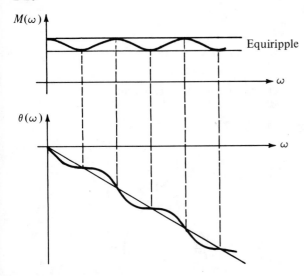

2-27

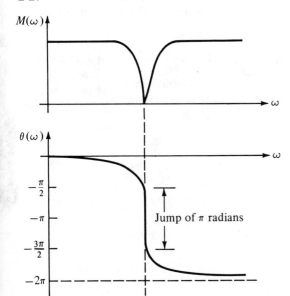

CHAPTER 3

3-1 $Z(s) = H \dfrac{(s - p_{sc1})(s - p_{sc2}) \cdots}{(s - p_{oc1})(s - p_{oc2}) \cdots}$

where *sc* stands for short circuit and *oc* for open circuit.

3-2 (a) $Z = \dfrac{1}{s}$ (b) $Z = 1$ (c) $Z = \infty$

3-3 $Z = 1$

3-4 (a) $Z = -\left(1 + \dfrac{1}{s}\right)$

(b) $V_{oc} = -\dfrac{I_s}{s}$ $I_{sc} = I_s \dfrac{1}{s+1}$ $\dfrac{V_{oc}}{I_{sc}} = -\left(1 + \dfrac{1}{s}\right) = Z$

(c) $V_{oc} \doteq -V_s$ $I_{sc} = V_s \dfrac{s}{s+1}$ $\dfrac{V_{oc}}{I_{sc}} = -\left(1 + \dfrac{1}{s}\right) = Z$

3-5 (a) $Z = \infty$ (b) $Z = \dfrac{1}{s2C}$ (c) $Z = 1 + \dfrac{1}{sC}$

3-6 $Z_1 = s$

3-7 (a) $Z = \dfrac{Z_1}{2} \pm \sqrt{\left(\dfrac{Z_1}{2}\right)^2 + Z_1 Z_2}$ (*Z* is not a rational function.)

(b) $Z = Z_0 \dfrac{1 + \Gamma(s)e^{-s2\sqrt{lc}\,L}}{1 - \Gamma(s)e^{-s2\sqrt{lc}\,L}}$, where $\begin{cases} \Gamma(s) = \dfrac{Z_L - Z_0}{Z_L + Z_0}, \\ l = \text{inductance}/m, \\ c = \text{capacitance}/m, \end{cases}$ $\begin{array}{l} Z \text{ is not a} \\ \text{rational} \\ \text{function.} \end{array}$

3-8 $Y_1 = \dfrac{Y_2 + Y_3}{2}$

3-9 $Z_1 = \dfrac{Z_2 + Z_3}{2}$

3-10 $Z(s) = \dfrac{R}{4} \dfrac{s + \dfrac{3}{RC}}{s + \dfrac{1}{RC}}$

3-11 (a) (b) (c)

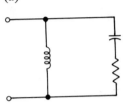

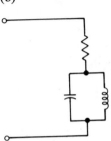

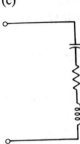

3-12 $v(t) = \dfrac{1 - \dfrac{3+\sqrt{5}}{2}}{\sqrt{\dfrac{3+\sqrt{5}}{2}}(-\sqrt{5})} \sin\left(\sqrt{\dfrac{3+\sqrt{5}}{2}}\,t\right)$

$\qquad + \dfrac{1 - \dfrac{3-\sqrt{5}}{2}}{\sqrt{\dfrac{3-\sqrt{5}}{2}}(\sqrt{5})} \sin\left(\sqrt{\dfrac{3-\sqrt{5}}{2}}\,t\right)$

$\qquad = 0.448(\sin 1.618t + \sin 0.618t)$

3-13 (a) $k_1 + k_2 \sin(t + \theta_2) + k_3 \sin(4t + \theta_3)$
   (b) $k_1 + k_2 \sin(4t + \theta_2)$
   (c) $k_1 + k_2 \sin(3t + \theta_2) + k_3 \sin(4t + \theta_3)$
   (d) $k_1 + k_2 \sin(4t + \theta_2 +) k_3 t \sin(4t + \theta_3)$

3-14 $X(\omega)$

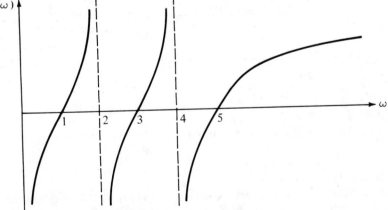

3-15

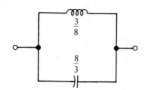

3-16 $X(\omega)$

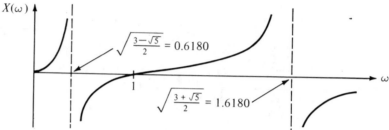

$$\sqrt{\frac{3-\sqrt5}{2}} = 0.6180$$

$$\sqrt{\frac{3+\sqrt5}{2}} = 1.6180$$

3-17 θ_Z

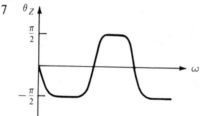

$|Z(j\omega)|$

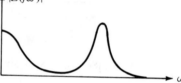

3-18

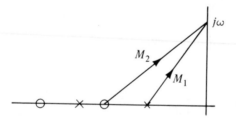

Consider M_2/M_1 and then generalize.

3-19 $Z(s) = \dfrac{2(s+1)(s+3)}{s(s+2)}$,

3-20 $Z_{\text{new}}(s) = Z_{\text{old}}(s)|_{s \to s + 1/RC}$.

3-21 All poles and zeros are shifted to the left by $1/\tau$, where

$$\tau = R_{a1}C_1 = \cdots = \frac{L_1}{R_{a7}} = \cdots$$

3-22 Same as the properties of $Z_{RC}(s)$, except that what applies to poles of $Z_{RC}(s)$ applies for the zeros of $Z_{RL}(s)$, and vice versa.

CHAPTER 4

4-2 (a) $3z, 4p$ (b) $4z, 3p$ (c) $4z, 5p$ (d) $6z, 7p$ (e) $6z, 5p$

4-3 (a) $Z = \dfrac{C_1 + C_2}{C_1 C_2} \dfrac{s\left[s^2 + \dfrac{L_1 + L_2}{L_1 L_2 (C_1 + C_2)}\right]}{\left(s^2 + \dfrac{1}{L_1 C_1}\right)\left(s^2 + \dfrac{1}{L_2 C_2}\right)}$

   (b) $Z = \dfrac{C_1 + C_2}{C_1 C_2} \dfrac{s}{s^2 + \dfrac{1}{L_2 C_2}}$

A pair of poles is cancelled by a pair of zeros; the network reduces to:

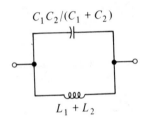

$C_1 C_2 / (C_1 + C_2)$

$L_1 + L_2$

4-4 (a) $X(\omega)$

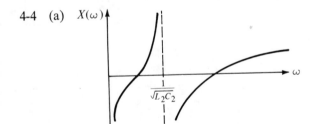

$\dfrac{1}{\sqrt{L_2 C_2}}$

   (b) $X(\omega)$

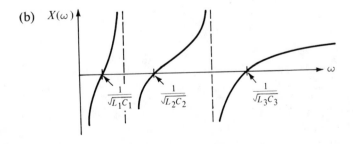

$\dfrac{1}{\sqrt{L_1 C_1}}$ $\dfrac{1}{\sqrt{L_2 C_2}}$ $\dfrac{1}{\sqrt{L_3 C_3}}$

4-5 (a)

(b)

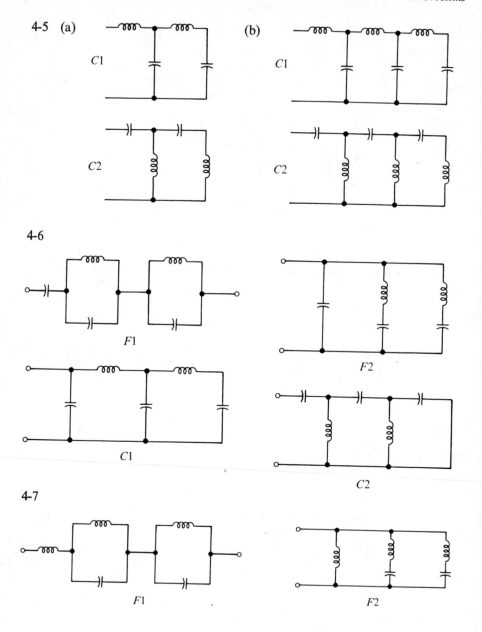

$C1$

$C2$

$C1$

$C2$

4-6

$F1$

$C1$

$F2$

$C2$

4-7

$F1$

$F2$

4-8

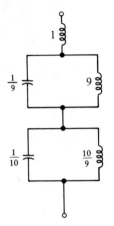

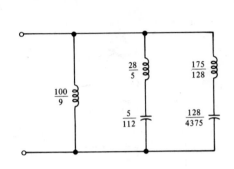

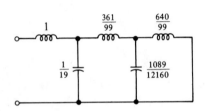

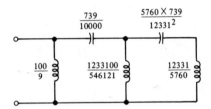

4-9

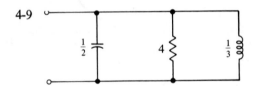

4-10 (a) $Z = \sqrt{1 + s^2}$ (b)

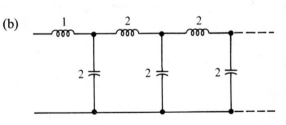

(c) $|Z(j\omega)|$

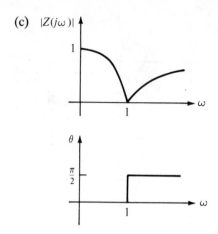

(d) $Z = \dfrac{s}{2} + \sqrt{1 + \left(\dfrac{s}{2}\right)^2}$

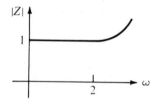   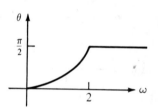

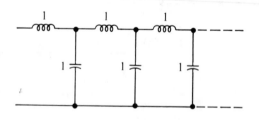

4-11

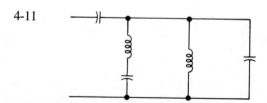

4-12

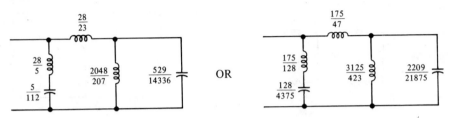

OR

4-13

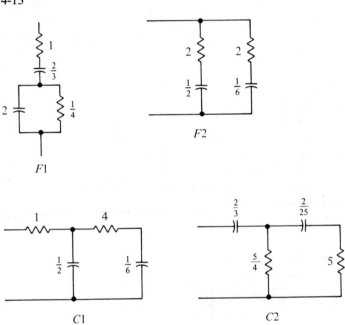

4-14 Yes; by successive steps, convert Z_2 to an all-$F1$ network and compare.

4-15 (a) $\quad Z = \dfrac{1 \pm \sqrt{1 + \dfrac{4}{s}}}{2}$ (b)

4-16

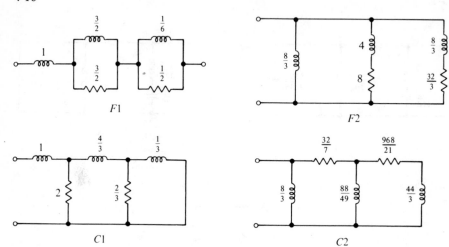

$F1$

$F2$

$C1$

$C2$

CHAPTER 5

5-1 2 zeros at $s = 0$, 2 zeros at $s = \infty$; 4 poles.

5-2 2 zeros at $s = 0$; 2 poles.

5-3 (a) One zero at $s = -1/RC_2$; 1 pole.

     (b) No zeros; no poles (pole-zero cancellation).

     (c) No zeros; no poles (pole-zero cancellation).

     (d) Two zeros at infinity; 2 poles (pole-zero cancellation).

5-4

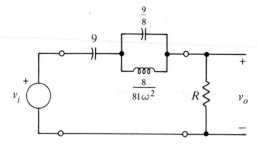

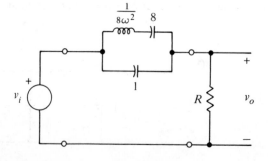

5-5 $N(s) = s^2\left(s^2 + \dfrac{1}{L_2 C_2}\right)$,   6 poles;

$$\left.\frac{V_o}{V_i}\right|_{s\,\text{small}} \cong s^2 L_3 C_1, \qquad \left.\frac{V_o}{V_i}\right|_{s\,\text{large}} \cong \frac{1}{s^2 L_1 C_3}.$$

5-6 $N(s) = s\left(s + \dfrac{1}{R_3 C_3}\right)\left(s + \dfrac{1}{R_4 C_4}\right)$,   4 poles;

$$\left.\frac{V_o}{V_i}\right|_{s\,\text{small}} \cong s C_1 R_2, \qquad \left.\frac{V_o}{V_i}\right|_{s\,\text{large}} \cong \frac{1}{s R_1 C_2}.$$

5-7   The zeros are at

$$s_{1,2} = -\frac{1}{2R_1 C_1} \pm j\sqrt{\frac{1}{L_1 C_1} - \left(\frac{1}{2R_1 C_1}\right)^2},$$

$$s_{3,4} = -\frac{R_2}{2L_2} \pm j\sqrt{\frac{1}{L_2 C_2} - \left(\frac{R_2}{2L_2}\right)^2}.$$

5-8

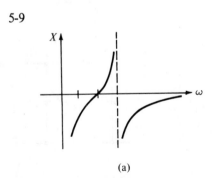

5-9

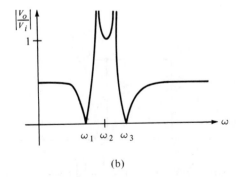

(a)   (b)

5-10 (a) $\dfrac{I_o}{V_i} = T_{LC} \times \begin{cases} Hs \\ \dfrac{H}{s} \end{cases};$ (b) $\dfrac{V_o}{I_i} = T_{LC} \times \begin{cases} Hs \\ \dfrac{H}{s} \end{cases};$ (c) $\dfrac{I_o}{I_i} = T_{LC} \times H.$

5-11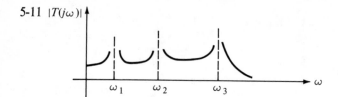

5-12 Solutions are not unique. The following networks result if the poles of Z_i are chosen to be at $s = 0, \pm j\sqrt{2}, \pm j2$:

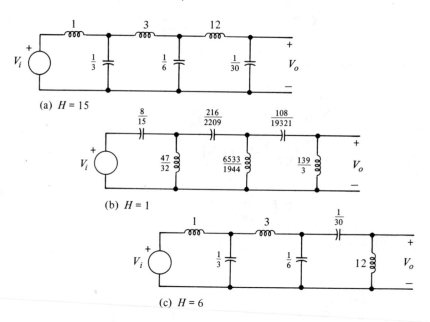

(a) $H = 15$

(b) $H = 1$

(c) $H = 6$

5-13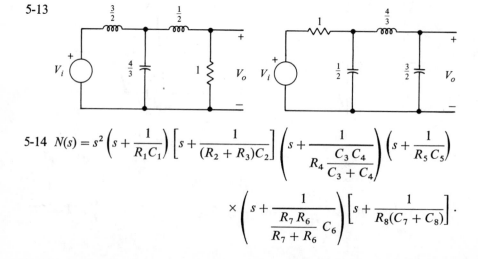

5-14 $N(s) = s^2 \left(s + \dfrac{1}{R_1 C_1} \right) \left[s + \dfrac{1}{(R_2 + R_3)C_2} \right] \left(s + \dfrac{1}{R_4 \dfrac{C_3 C_4}{C_3 + C_4}} \right) \left(s + \dfrac{1}{R_5 C_5} \right)$

$$\times \left(s + \dfrac{1}{R_7 \dfrac{R_6}{R_7 + R_6} C_6} \right) \left[s + \dfrac{1}{R_8(C_7 + C_8)} \right].$$

There is also a zero at ∞ because of $C_7 C_8$. Note that there is only one zero at the origin even though both C_0 and $C_3 C_4$ by themselves would produce zeros at the origin.

5-15 Solutions are not unique. Element values correspond to the choices indicated.

(a) Poles of Z_i are chosen to be at $s = 0, -2$.

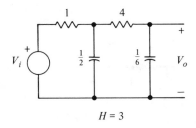

$H = 3$

(b) Poles of Z_i are chosen at $s = 0, -2$

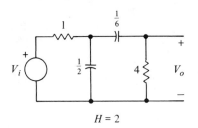

$H = 2$

(c) Zero of Z_i is chosen at $s = -2$

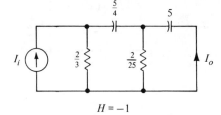

$H = -1$

(d) Poles of Z_i are chosen at $s = 0, -2, -4$.

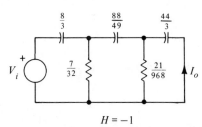

$H = -1$

(e) Zeros of Z_i are chosen at $s = -1, -3$.

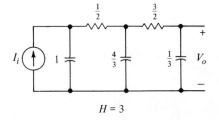

$H = 3$

(f)

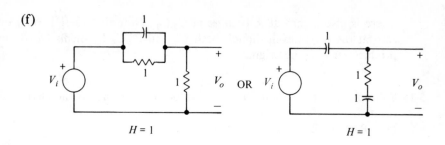

$H = 1$ $H = 1$

5-16 (a)

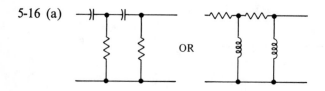

OR

(b)

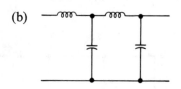

(c)

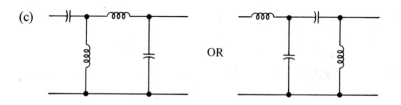

OR

5-18 Poles: $s = -1, -2, -3$; zeros: $s = 0, -3 \pm j\sqrt{2}$.

5-19 $\left|\dfrac{V_o}{V_i}\right|^2_{\max} = \dfrac{19 + 10\sqrt{19}}{19 + 8\sqrt{19}} > 1.$ Maximum occurs at $\omega = 1/RC\sqrt{\dfrac{1 + \sqrt{19}}{2}}$.

5-20 Refer to Fig. 5-18 to see where V_{o1} and V_{o2} are taken.

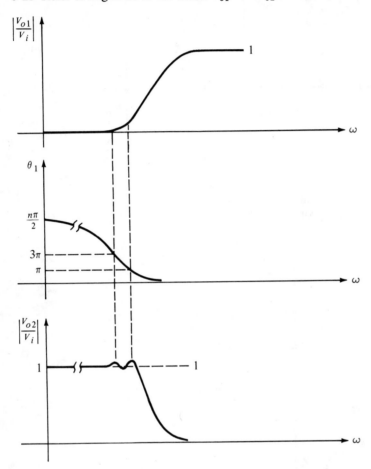

5-21 $R(C_1 + C_2) = \dfrac{L}{R_L}$, $\qquad s = \pm j\sqrt{\dfrac{C_1 + C_2}{C_1 C_2 L}}$

CHAPTER 6

6-3 (a) (b)

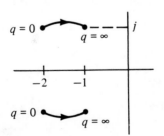

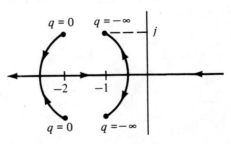

6-4   (a)

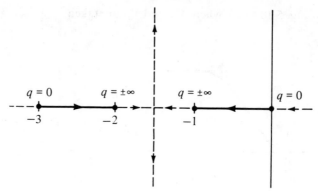

Solid line $q \geq 0$,
Dotted line $q < 0$.

(b)

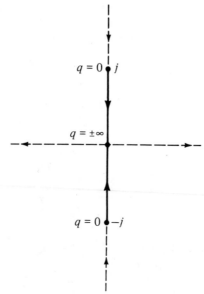

Arrows indicate the direction of increasing q from $-\infty$ to $+\infty$.

6-5 (a) (b) (c)

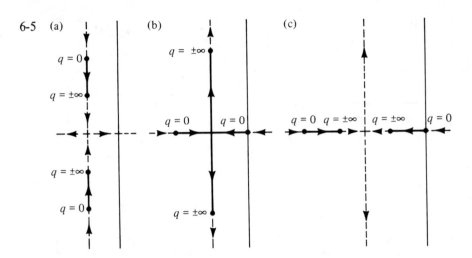

6-6 $P_1 = 0$ or $P_2 = 0$ or $P_1 = kP_2$

6-7 (a)   (b)

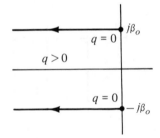

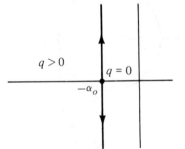

(c)   (d)

6-11 (a) $\dfrac{V_o}{V_c} = \dfrac{s^2 R_1 R_2 C_1 C_2 + s(R_1 C_2 + R_2 C_2) + 1}{s^2 R_1 R_2 C_1 C_2 + s(R_1 C_1 + R_2 C_2 + R_1 C_2) + 1}$

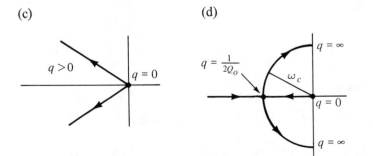

(b)

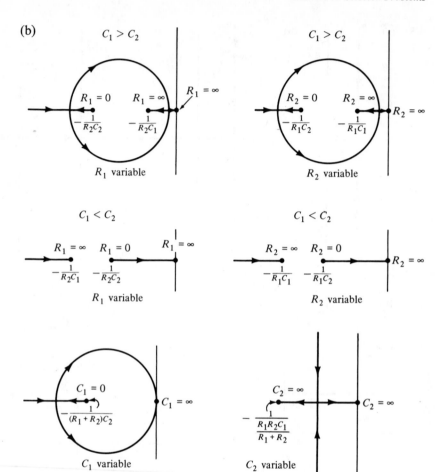

(c) $S_{R_1}^{z_{1,2}} = \dfrac{z_{1,2}R_2C_2 + 1}{D},\qquad S_{R_2}^{z_{1,2}} = \dfrac{z_{1,2}R_1C_2 + 1}{D},$

$S_{C_1}^{z_{1,2}} = \dfrac{z_{1,2}(R_1 + R_2)C_2 + 1}{D},\qquad S_{C_2}^{z_{1,2}} = \dfrac{1}{D},$

where

$D = [2z_{1,2}R_1R_2C_1C_2 + (R_1 + R_2)C_2]|z_{1,2}|,$

$S_{R_1}^{\omega_o} = S_{R_2}^{\omega_o} = S_{C_1}^{\omega_o} = S_{C_2}^{\omega_o} - \dfrac{1}{2},\qquad S_{R_1}^{Q} = -\dfrac{1}{2} + \dfrac{R_2}{R_1 + R_2},$

$S_{R_2}^{Q} = -\dfrac{1}{2} + \dfrac{R_1}{R_1 + R_2},\qquad S_{C_1}^{Q} = -S_{C_2}^{Q} = \dfrac{1}{2}.$

6-12 $S_R^{p_1,2} = \dfrac{1}{20}\left(1 \pm j\dfrac{1}{\sqrt{399}}\right)$, $\qquad S_L^{p_1,2} = \mp j\dfrac{10}{\sqrt{399}}$,

$S_C^{p_1,2} = \dfrac{1}{20}\left(1 \mp j\dfrac{199}{\sqrt{399}}\right)$.

6-16 (a)

where $\begin{cases} LC = \dfrac{1}{\omega_o^2}, \\[2mm] RC = \dfrac{Q}{\omega_o}. \end{cases}$

(b) $S_R^{p_1,2} = \dfrac{1}{2Q}\left(1 \pm j\dfrac{1}{\sqrt{4Q^2 - 1}}\right)$; $\qquad S_L^{\omega_o} = S_C^{\omega_o} = -\dfrac{1}{2}$, $\qquad S_R^{\omega_o} = 0$;

$S_L^{p_1,2} = \mp j\dfrac{Q}{\sqrt{4Q^2 - 1}}$; $\qquad S_C^Q = -S_L^Q = \dfrac{1}{2}$, $\qquad S_R^Q = 1$;

$S_C^{p_1,2} = \dfrac{1}{2Q}\left(1 \mp j\dfrac{2Q^2 - 1}{\sqrt{4Q^2 - 1}}\right)$;

$S_R^T = \dfrac{s}{\omega_o}\dfrac{T}{Q}$, $\qquad S_L^T = T - 1$, $\qquad S_C^T = -\left(\dfrac{s}{\omega_o}\right)^2 T$.

CHAPTER 7

7-1 (a) $V_o = V_1\left(-\dfrac{R_2}{R_1}\right) + V_2\left(1 + \dfrac{R_2}{R_1}\right)$ 　　(b) $V_o = V_1\left(-\dfrac{R_2}{R_1}\right)\left(-\dfrac{R_4}{R_3}\right)$

$\qquad\qquad\qquad\qquad\qquad\qquad\qquad\qquad\qquad\qquad\qquad + V_2\left(1 + \dfrac{R_4}{R_3}\right)$

(c) $V_o = -(V_1 + V_2) + \tfrac{3}{2}(V_3 + V_4)$ 　　(d) $V_o = V_2 - V_1$

(e) $V_o = \dfrac{ab}{1 + b\dfrac{1+a}{1+c}}V_i$ 　　　　(f) $V_o = I_i\dfrac{R_2 R_3}{R_1 + R_3}$

7-5 $V_o = -V\dfrac{\Delta R_2}{R_1 + R_2}$

7-6 $V_o = (V_2 - V_1)\left(1 + \dfrac{2R_2}{R_1}\right)\dfrac{R_4}{R_3}$

7-7 $V_o = \dfrac{\gamma}{s} - V_i \dfrac{1}{sRC}$

7-8 $V_o = -V_1 \dfrac{1}{sRC} + V_2 \dfrac{1}{sRC}$

7-9 (a) $R_a = R_5 + R_3 \left(1 + \dfrac{R_5}{R_4}\right)$, $R_b = R_3 + R_5 + \dfrac{R_5}{R_4}(R_1 + R_3)$

   (b) Choose the R_5/R_4 ratio large.

7-10

$Z_i \longrightarrow$ R_1 $C\left(1 + \dfrac{R_2}{R_1}\right)$

7-14 The pole loci are:

$\beta = \dfrac{R_1}{R_1 + R_2}$; $\beta A_0 = 0$ $\beta A_0 = 0$ $-\alpha_2$ $-\alpha_1$

   Arrow indicates direction of increasing βA_0.

7-15 $a_0 = \dfrac{a_1}{a_3}\left(a_2 - \dfrac{a_1}{a_3}\right)$; $s = \pm j\sqrt{a_1/a_3}$

7-16

$\beta A_0 = 0$ $\beta A_0 = 0$ $\beta A_0 = 0$ $\beta A_0 = 0$ Stable amplifier

$-\alpha_3$ $-\alpha_2$ $-\alpha_1$

7-17 $\alpha < 120\pi$ rad/s.

7-18

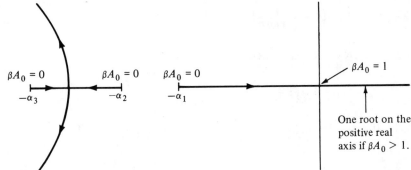

One root on the positive real axis if $\beta A_0 > 1$.

7-19 dc gain $= 5000$, $\omega_{3\,\text{dB}} = 200$ rad/s.

7-20 $v_o(t) = -10^5(1 - e^{-10^{-6}t}) - 10^{-7}e^{-10^6 t} \cong -\dfrac{t}{10} - 10^{-7}e^{-10^6 t}$;

$$v_o(t) = -\frac{t}{10} \qquad \text{(Ideal integrator)}.$$

7-21 $\beta_{eff} A(s) - 1 = 0$, where $\beta_{eff} = \beta_- - \beta_+$.

7-22 $\alpha = \dfrac{R_0 + R_1 + R_2}{R_0(R_1 + R_2)C}$.

7-25

$$\frac{BW_b}{BW_a} = 0.644\sqrt{K_0}, \qquad \frac{BW_c}{BW_a} = 1.272\sqrt{K_0}; \qquad \left(BW_a = \frac{GB}{K_0}\right)$$

7-27 (a) $\dfrac{R_2}{R_1} = \dfrac{R_b}{R_a}$ (b) and (c) (d) $i_{\text{source}} = \dfrac{v_i}{R_a}$ (e) Z in parallel with $\left(\dfrac{R_b}{K_0} + \dfrac{R_b}{s}\dfrac{GB}{K_0^2}\right)$

CHAPTER 8

8-1 $V_o = (I_1 - I_2)R_3\left(1 + \dfrac{R_2}{R_1}\right) + V\left(1 + \dfrac{R_2}{R_1}\right)$

8-3 $v_{oss} = -V\left(1 + \dfrac{R_1}{R_2}\right) \cong -V$; when $+$ terminal is grounded,

$$v_{oss} = V\left(1 + \frac{R_2}{R_1}\right) \cong V\frac{R_2}{R_1} \gg V.$$

8-4 (a) $V_o = (V_2 - V_1)(1 \pm 0.02) \mp 0.02V_2$ (b) $CMRR = 40$ dB

8-5 $V_o = V_2 G\left(1 - \dfrac{1}{CMRR}\right)$

8-6 If $R_1 R_3 = R_2 R_4$, then

$$V_o = (V_1 - V_2)\left(1 + \frac{R_2}{R_1}\right)\left(1 - \frac{1}{CMRR}\right).$$

8-7 If $CMRR = \infty$, $v_o = 0$; if $CMRR$ is 40 dB, $v_o = v_i/100$.

8-8

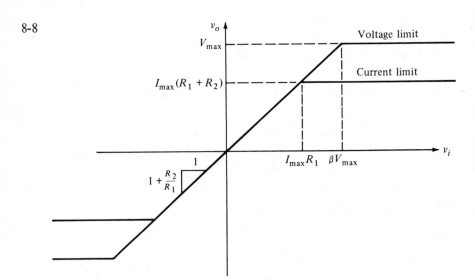

8-9 (a) $V_o = V_2 - \dfrac{V_d}{A}$ (b) $V_a = \dfrac{V_2}{A} - \dfrac{V_d}{A}$

(c) $\dfrac{V_{ad}}{V_{an}} = -\dfrac{V_2}{V_d}$, $\dfrac{V_{od}}{V_{on}} = -\dfrac{V_2}{V_d} A$

The ratio is A times better in the output.

8-10 $\left|\dfrac{dv_o}{dt}\right| = V \dfrac{R_2}{R_1 + R_2} GB$

8-11

8-12 $SR > 0.2\pi$ V/μs

8-13

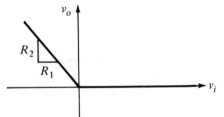

8-14

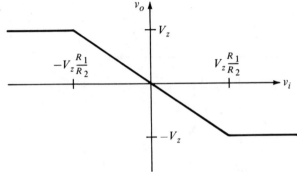

8-15 (a)

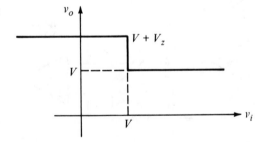

(b)

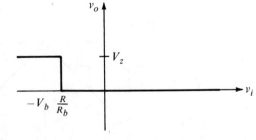

8-17 (a)

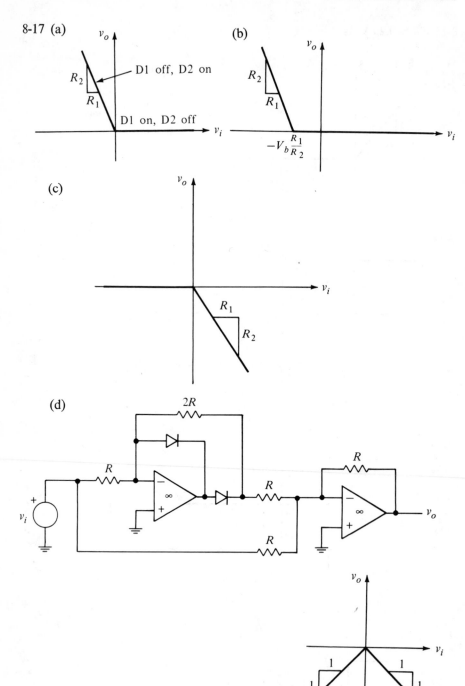

(b)

(c)

(d)

8-18

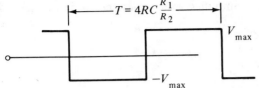

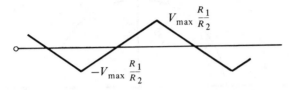

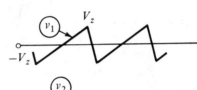

8-19

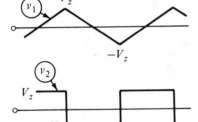

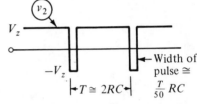

Switch at 1   Switch at 2

CHAPTER 9

9-1 $\dfrac{V_o}{V_i} = \dfrac{KT_{31}}{1 - KT_{32}}$ (*T*'s are transadmittances).

9-4 (b)   Network of Fig. 9-28a

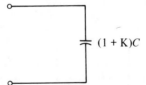

Network of Fig. 9-28b

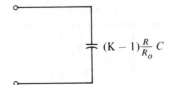

In each case, the amplifier gain varies the value of the equivalent capacitor.

9-5

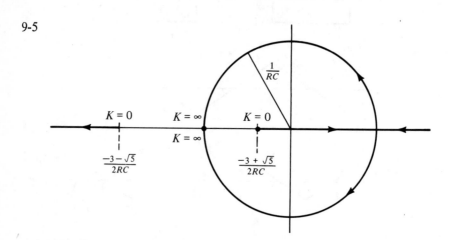

9-7 (a) $\dfrac{V_o}{V_i} = -\dfrac{\dfrac{s}{R_1 C}}{s^2 + s\,\dfrac{2}{R_2 C} + \dfrac{1}{R_1 R_2 C^2}} \; ;$

(b) $\omega_o = \dfrac{1}{C\sqrt{R_1 R_2}} , \qquad Q = \dfrac{1}{2}\sqrt{\dfrac{R_2}{R_1}}$

9-8

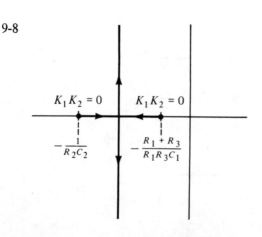

9-9

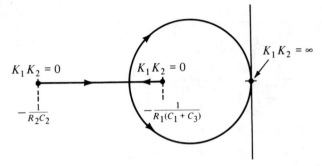

9-11 (b) $\dfrac{V_o}{V_i} = \dfrac{K\left(s + \dfrac{1}{K}\right)}{s^2 + s(3 - K) + 1}$

9-12 (b) $\dfrac{V_o}{V_i} = -\dfrac{T_{31}}{T_{32}}$ (*T*'s are transadmittances).

9-15 $Z_i = \dfrac{R_0 R}{Z}$ (Gyrator).

9-17 (b) $K_1 = 4,$ $K_2 = 6 - \dfrac{2}{Q},$ $RC = \dfrac{1}{\omega_o}$

9-18 $\dfrac{V_o}{V_i} = \dfrac{\left(s + \dfrac{1}{RC}\right)\left(s^2 + \dfrac{K}{R^2 C^2}\right)}{s^3 + \cdots + \dfrac{1}{R^3 C^3}}$

CHAPTER 10

10-3 (a) $\dfrac{V_o}{V_i} = \dfrac{1}{LC} \dfrac{1}{s^2 + s\left(\dfrac{R_1}{L} + \dfrac{1}{R_2 C}\right) + \dfrac{1 + R_1/R_2}{LC}};$

(b) $S_L^{\omega_o} = S_C^{\omega_o} = -\dfrac{1}{2},$ $S_{R_1}^{\omega_o} = -S_{R_2}^{\omega_o} = \dfrac{1}{2}\dfrac{1}{1 + \dfrac{R_2}{R_1}},$

$S_L^Q = -\dfrac{1}{2} + \dfrac{1}{1 + \dfrac{1}{R_1 R_2 C}},$ $S_C^Q = -\dfrac{1}{2} + \dfrac{1}{1 + \dfrac{R_1 R_2 C}{L}},$

$$S_{R_1}^Q = \frac{1}{2}\frac{1}{1+\dfrac{R_2}{R_1}} - \frac{1}{1+\dfrac{L}{R_1 R_2 C}},$$

$$S_{R_2}^Q = -\frac{1}{2}\frac{1}{1+\dfrac{R_2}{R_1}} + \frac{1}{1+\dfrac{R_1 R_2 C}{L}}.$$

10-4 (c) Better than $\frac{1}{20}$ %.

10-6 (a) $\omega_{oa} = 0.673\omega_o$, $\qquad Q_a = 1.18$;

(b) $\omega_o = 440$ Krad/s, $\qquad Q = 0.87$.

10-9 (a) $\dfrac{V_o}{V_i} = -\dfrac{K_0}{R^2 C^2}\dfrac{1}{s^2 + s\dfrac{4}{RC} + \dfrac{2+K_0}{R^2 C^2}}$, $\qquad \omega_o = \dfrac{\sqrt{2+K_0}}{RC}$,

$$Q = \frac{\sqrt{2+K_0}}{4},$$

(b) $\dfrac{\omega_{oa}}{\omega_o} \cong 1 - \dfrac{1}{2}\left(1 - \dfrac{1}{8Q^2}\right)\left(\dfrac{\omega_o}{\omega_a}\right)^2 \cong 1$, $\qquad \dfrac{Q_a}{Q} \cong 1 + Q\left(1 - \dfrac{1}{8Q^2}\right)\dfrac{\omega_o}{\omega_a}$.

10-10 $\dfrac{V_o}{V_i} = -\dfrac{K_1 K_2}{R_1 R_2 C_1 C_2}\dfrac{1}{s^2 + s\left[\left(1+\dfrac{R_1}{R_3}\right)\dfrac{1}{R_1 C_1} + \dfrac{1}{R_2 C_2}\right] + \dfrac{1 + \dfrac{R_1}{R_3}(1+K_1 K_2)}{R_1 R_2 C_1 C_2}}$

$$\omega_o = \frac{\sqrt{2+K_1 K_2}}{RC}, \qquad Q = \frac{\sqrt{2+K_1 K_2}}{3}.$$

10-12 (a) $Y_i = \dfrac{1}{R_1} + \dfrac{1}{R_2} + \dfrac{Z}{R_1 R_2}$.

10-13 (a) $Z_i = (R_1 + R_2) + \dfrac{R_1 R_2}{Z}$.

10-16 Introduce an impulse of value $-\gamma RC$ in series with the input.

10-19 (c) GB $> 4Q\omega_o$ (approximate).

10-21 (a) $\dfrac{V_n}{V_i} = \dfrac{1}{1 + s\tau_n + s^2\tau_n\tau_{n-1} + s^3\tau_n\tau_{n-1}\tau_{n-2} + \cdots + s^n\tau_n\tau_{n-1}\tau_{n-2}\cdots\tau_1}$.

CHAPTER 11

11-1 (a) $\dfrac{V_o}{V_i} = \dfrac{1}{R_1 C}\dfrac{s}{s^2 + s\left(\dfrac{1}{R_1 C} + \dfrac{1}{R_2 C}\right) + \dfrac{1}{LC}}$,

(b) $S_L^{\omega_o} = S_C^{\omega_o} = -\dfrac{1}{2}$, $S_{R_1}^{\omega_o} = S_{R_2}^{\omega_o} = 0$,

$$S_L^Q = -S_C^Q = -\frac{1}{2}, \qquad S_{R_1}^Q = \frac{1}{1+\dfrac{R_1}{R_2}}, \qquad S_{R_2}^Q = \frac{1}{1+\dfrac{R_2}{R_1}}.$$

11-3 (a) $\dfrac{V_o}{V_i} = \dfrac{K_0(sRC)}{(sRC)^2 + (sRC)(5 - K_0) + 2}$, $\omega_o = \dfrac{\sqrt{2}}{RC}$, $Q = \dfrac{\sqrt{2}}{5 - K_o}$,

(b) $s^3 + s^2\left(\dfrac{5\omega_o}{\sqrt{2}} + \omega_2\right) + s\left(\omega_o^2 + \dfrac{\omega_o\,\omega_2}{Q}\right) + \omega_2\,\omega_o^2 = 0.$

11-5 $\dfrac{V_o}{V_i} = -\dfrac{K_0(sRC)}{(sRC)^2 + 4(sRC) + (3 + K_0)}$, $\omega_o = \dfrac{\sqrt{3 + K_0}}{RC}$,

$Q = \dfrac{\sqrt{3 + K_0}}{4}.$

11-8 (a) $\dfrac{V_o}{V_i} = -\dfrac{1}{R_1 C_1} \cdot \dfrac{s}{s^2 + s\left(\dfrac{1}{R_1} + \dfrac{1}{R_2}\right)\dfrac{1}{C_2} + \dfrac{1}{R_1 R_2 C_1 C_2}}.$

(b) C_1 affects the ω_o but not the bandwidth.

11-9 (a) $\dfrac{V_o}{V_i} = -\dfrac{C_2}{C_2 + C_3} \cdot \dfrac{\dfrac{s}{R_1 C_1}}{s^2 + s\left(\dfrac{1}{R_1} + \dfrac{1}{R_2}\right)\dfrac{1}{C_2 + C_3} + \dfrac{1}{R_1 R_2 C_1(C_2 + C_3)}}.$

(b) Peak values are the same. The higher ω_o curve has a wider bandwidth.

11-10 $r = \dfrac{1}{2(Q^2 - 1)}$ $(Q > 1)$.

11-12 $\dfrac{V_o}{V_i} = \dfrac{-\dfrac{1}{R_1 C_1}\left(\dfrac{1 + \dfrac{R_4}{R_3}}{1 + \dfrac{R_5}{R_6}}\right) s}{s^2 + \dfrac{s}{R_1 C_1}\left(\dfrac{1 + \dfrac{R_4}{R_3}}{1 + \dfrac{R_6}{R_5}}\right) + \dfrac{R_4/R_3}{R_1 R_2 C_1 C_2}}.$

11-13 (a) $Z_i = \dfrac{(R_2 + Z)R_1}{R_1 + R_2(1 - K) + Z}.$

(b)

$$R_1$$

$$L = R_1 R_2 C$$

CHAPTER 12

12-3 (a)
$$\frac{V_o}{V_i} = \frac{s^2 \dfrac{R_2}{R_1 + R_2}}{s^2 + s\left[\dfrac{R_1 R_2}{R_1 + R_2}\dfrac{1}{L} + \dfrac{1}{(R_1 + R_2)C}\right] + \dfrac{1}{LC\left(1 + \dfrac{R_1}{R_2}\right)}},$$

(b) $S_L^{\omega_o} = S_C^{\omega_o} = -\dfrac{1}{2}, \qquad S_{R_1}^{\omega_o} = -S_{R_2}^{\omega_o} = -\dfrac{1}{2}\dfrac{R_1}{R_1 + R_2},$

$$S_L^Q = -\dfrac{1}{2} + \dfrac{1}{1 + \dfrac{L}{R_1 R_2 C}}, \qquad S_C^Q = -\dfrac{1}{2} + \dfrac{1}{1 + \dfrac{R_1 R_2 C}{L}},$$

$$S_{R_1}^Q = -\dfrac{R_1}{R_1 + R_2}\left(\dfrac{1}{2} + \dfrac{R_2^2 - \dfrac{L}{C}}{R_1 R_2 + \dfrac{L}{C}}\right),$$

$$S_{R_2}^Q = \dfrac{R_1}{R_1 + R_2}\left(\dfrac{1}{2} - \dfrac{R_2}{R_1}\dfrac{R_1^2 - \dfrac{L}{C}}{R_1 R_2 + \dfrac{L}{C}}\right).$$

12-4

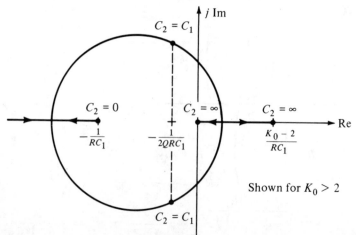

$C_2 = C_1$

$j\,\mathrm{Im}$

$C_2 = 0$ $C_2 = \infty$ $C_2 = \infty$

$-\dfrac{1}{RC_1}$ $-\dfrac{1}{2QRC_1}$ $\dfrac{K_0 - 2}{RC_1}$ Re

Shown for $K_0 > 2$

$C_2 = C_1$

12-5 (a) $\omega_o = \dfrac{1}{C\sqrt{R_1 R_2}}, \qquad Q = \dfrac{1}{2}\sqrt{\dfrac{R_2}{R_1}},$

(b) $\dfrac{\Delta\omega_o}{\omega_o} \cong -Q\,\dfrac{\omega_o}{GB}, \qquad \dfrac{\Delta Q}{Q} \cong Q\,\dfrac{\omega_o}{GB}.$

12-6 $\dfrac{V_B}{V_i} = -\dfrac{R_6}{R_5 + R_6}\left(1 + \dfrac{R_4}{R_3}\right)\dfrac{1}{R_1 C_1}$

$$\times \dfrac{s}{s^2 + s\,\dfrac{1}{R_1 C_1}\left(1 + \dfrac{R_4}{R_3}\right)\dfrac{R_5}{R_5 + R_6} + \dfrac{R_4/R_3}{R_1 R_2 C_1 C_2}} \qquad \text{(Band-pass)},$$

$\dfrac{V_L}{V_i} = \dfrac{V_B}{V_i}\left(-\dfrac{1}{sR_2 C_2}\right) \qquad \text{(Low-pass)},$

$\dfrac{V_H}{V_i} = \dfrac{V_B}{V_i}(-sR_1 C_1) \qquad \text{(High-pass)}.$

CHAPTER 13

13-2 (a) $\dfrac{V_o}{V_i} = \dfrac{R_2}{R_1 + R_2}\,\dfrac{s^2 + \dfrac{1}{LC}}{s^2 + s\,\dfrac{1}{(R_1 + R_2)C} + \dfrac{1}{LC}}\,;$

$$S_L^{\omega_z} = S_C^{\omega_z} = -\dfrac{1}{2}, \qquad S_{R_1}^{\omega_z} = S_{R_2}^{\omega_z} = 0,$$

$$S_L^{\omega_o} = S_C^{\omega_o} = -\dfrac{1}{2}, \qquad S_{R_1}^{\omega_o} = S_{R_2}^{\omega_o} = 0,$$

$$S_L^{Q} = -S_C^{Q} = -\dfrac{1}{2}, \qquad S_{R_1}^{Q} = \dfrac{R_1}{R_1 + R_2}, \qquad S_{R_2}^{Q} = \dfrac{R_2}{R_1 + R_2}.$$

(b) $\dfrac{V_o}{V_i} = \dfrac{R_2}{R_1 + R_2}\,\dfrac{s^2 + \dfrac{1}{LC}}{s^2 + s\,\dfrac{1}{L}\dfrac{R_1 R_2}{R_1 + R_2} + \dfrac{1}{LC}}\,;$

$$S_L^{\omega_z} = S_C^{\omega_z} = -\dfrac{1}{2}, \qquad S_{R_1}^{\omega_z} = S_{R_2}^{\omega_z} = 0,$$

$$S_L^{\omega_o} = S_C^{\omega_o} = -\dfrac{1}{2}, \qquad S_{R_1}^{\omega_o} = S_{R_2}^{\omega_o} = 0,$$

$$S_L^{Q} = -S_C^{Q} = \dfrac{1}{2}, \qquad S_{R_1}^{Q} = -\dfrac{R_2}{R_1 + R_2}, \qquad S_{R_2}^{Q} = -\dfrac{R_1}{R_1 + R_2}.$$

13-3

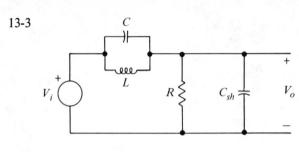

$$\omega_o < \omega_z$$

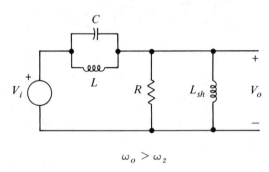

$$\omega_o > \omega_z$$

13-4 (a) $\dfrac{V_o}{V_i} = \dfrac{s^3 RLC_1C_2 + s^2 rRC_1C_2 + sR(C_1 + C_2) + 1}{s^3 RLC_1C_2 + s^2(LC_2 + rRC_1C_2) + s(rC_2 + RC_1 + RC_2) + 1}$

   (b) $R = \dfrac{L}{r(C_1 + C_2)}, \qquad s_{1,2} = \pm j \dfrac{1}{\sqrt{L\dfrac{C_1C_2}{C_1 + C_2}}}$

   (c) $Q < Q_c$

13-7

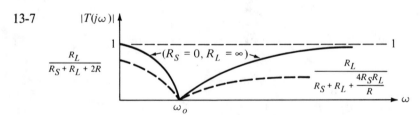

13-8 (b)

$$q = \tfrac{2}{\sqrt{3}} - 1$$

13-10 $k = 12$, $\omega_z = \dfrac{\sqrt{3}}{RC}$

13-11 $\dfrac{V_o}{V_i} = \dfrac{K\left(s^2 + \dfrac{1}{R^2 C^2}\right)}{s^2 + \dfrac{s}{RC}(4 - 2K) + \dfrac{1}{R^2 C^2}}$

13-12 (a) $Z_i = sR_4 R_5 C_2 + R_4 + R_5$

(b) $\dfrac{V_o}{V_i} = \dfrac{s^2 + \dfrac{1}{R_4 R_5 C_1 C_2}}{s^2 + s\,\dfrac{R_3 + R_4 + R_5}{R_4 R_5 C_2} + \dfrac{1}{R_4 R_5 C_1 C_2}}\,;$

$Q = \dfrac{\sqrt{R_4 R_5}}{R_3 + R_4 + R_5}\sqrt{\dfrac{C_2}{C_1}}$ can be made greater than $\dfrac{1}{2}$.

13-13 (a) $\dfrac{V_o}{V_i} = -k_1 \dfrac{s^2 + s\left(1 - 2\dfrac{k_2}{k_1}Q^2\right)\mathrm{BW} + \omega_o^2}{s^2 + s\mathrm{BW} + \omega_o^2}$

(b) $\dfrac{k_1}{k_2} = 2Q^2$, where

$\omega_o = \dfrac{1}{C\sqrt{R_1 R_2}}$, $\mathrm{BW} = \dfrac{2}{R_2 C}$, $Q = \dfrac{1}{2}\sqrt{\dfrac{R_2}{R_1}}$

13-14 (a) $\dfrac{V_o}{V_i} = -\dfrac{R_8}{R_7} \dfrac{s^2 + s\,\dfrac{R_7}{R_4 R_6 C_2}\left(\dfrac{R_6}{R_7} - \dfrac{R_4}{R_5}\right) + \dfrac{R_7/R_1}{R_2 R_6 C_1 C_2}}{s^2 + s\,\dfrac{1}{R_4 C_2} + \dfrac{R_8/R_3}{R_2 R_6 C_1 C_2}}$

(b) $R_5 = \dfrac{R_4 R_7}{R_6}$

CHAPTER 14

14-1 (a) $\dfrac{R_1}{R_2} = 1$ (b) $\dfrac{R_1}{R_2} = 1 + 2\dfrac{R_3}{R}$

14-3 (b)

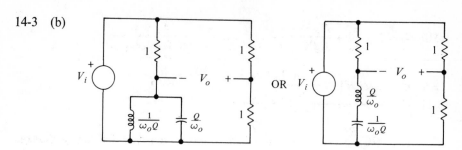

14-4 (a) $\dfrac{V_o}{V_i} = \dfrac{1 - Z_{LC}}{1 + Z_{LC}}$ which is AP (see Problem 14-2),

(b) $Z_i = 1$,

(c)

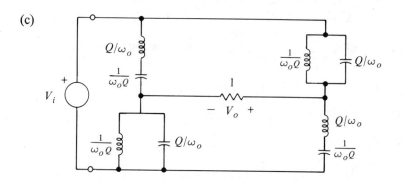

14-5 In either case, let $k_2/k_1 = \dfrac{1}{Q^2}$.

14-6 (a) $k = 1$ (c) $k = 1$

14-7 $\dfrac{R_2}{R_1} = 1 + 2\dfrac{(R_4 + R_5)}{R_3}$

14-8 $R_5 = \dfrac{R_7}{2}\dfrac{R_4}{R_6}$, $\dfrac{R_7}{R_1} = \dfrac{R_8}{R_3}$

CHAPTER 15

15-1 (a) $\dfrac{sRC}{(sRC)^2 + 3(sRC) + 1}$ (c) 1.34

15-2 (a) $K = 1$, $\omega_o = \dfrac{1}{\sqrt{LC}}$ (b) $K = 1$, $\omega_o = \dfrac{1}{\sqrt{LC}}$

15-3 $K = 1 + \dfrac{R_2}{R_1} + \dfrac{R_2}{R_3}\left(1 + \dfrac{C_1}{C_2}\right), \qquad \omega_o = \sqrt{\dfrac{R_1 + R_2}{R_1 R_2 R_3 C_1 C_2}}$

15-4   (a) $v_o(t) = -V\dfrac{(1 + R_4/R_3)}{\sqrt{\dfrac{R_4\,R_2\,C_2}{R_3\,R_1\,C_1}}}\,\sin\sqrt{\dfrac{R_4/R_3}{R_1 R_2 C_1 C_2}}\,t$

      (b)   No, unless the circuit is started anew after each frequency change.

15-5   (a) $K = 2\dfrac{R_c}{R_2}$ (b) $\omega_o = \dfrac{1}{C}\sqrt{\dfrac{R_c + R_d}{R_2 R_c R_d}}$

        Since R_d appears in ω_o and not in K, it can be used to vary ω_o.

      (c)   The bandwidth of the BP network is independent of R_d. Hence the Q of the BP network increases with ω_o.

15-6   (a) $K = 2\dfrac{C_1}{C_2}, \qquad \omega_o = \dfrac{1}{R\sqrt{C_1(C_2 + C_3)}}$

      (b) C_3 appears in ω_o but not in K. Hence C_3 can be used to vary the frequency of oscillation.

15-9   (a) $R_2/R_1 = 2, \qquad \omega_o = \dfrac{\sqrt{3}}{R_2 C}$

      (b)   In the steady state, assume that $v_1 = V_m \sin \omega_o t$. Then

$$v_2 = V_m \sin\left(\omega_o t + \dfrac{2\pi}{3}\right), \qquad v_3 = V_m \sin\left(\omega_o t - \dfrac{2\pi}{3}\right).$$

CHAPTER 16

16-1 $|G(j\omega)| = H$

16-2   (a) $G_{NM} = \underbrace{G_M[(s + \alpha)^2 + \beta^2](s + \gamma)}_{\text{Min. phase}} \times \underbrace{\left\{\dfrac{(s - \gamma)[(s - \alpha)^2 + \beta^2]}{(s + \gamma)[(s + \alpha)^2 + \beta^2]}\right\}}_{\text{All-pass}}$

      (b) $G(s) = \underbrace{\dfrac{s[(s + 1)^2 + 1]}{[(s + 2)^2 + 1](s + 3)}}_{\text{Min. phase}} \times \underbrace{\left[\dfrac{(s - 1)^2 + 1}{(s + 1)^2 + 1}\right]}_{\text{All-pass}}$

16-3 $G(s) = \dfrac{s^2 + 1}{s^4 + s^3 2.613 + s^2 3.414 + s 2.613 + 1}$

16-4 $N(s) = s(s^2 + 1)(s + 1)[(s + 1)^2 + 1]$ (Min. phase)

or

$\qquad = s(s^2 + 1)(s + 1)[(s - 1)^2 + 1]$

or

$\qquad = s(s^2 + 1)(s - 1)[(s + 1)^2 + 1]$

or

$\qquad = s(s^2 + 1)(s - 1)[(s - 1)^2 + 1].$

16-5 $\theta = \cos^{-1}\left[\dfrac{V_{m3}^2 - (V_{m1}^2 + V_{m2}^2)}{2V_{m1}V_{m2}}\right]$

16-7   (a) $\dfrac{(s + 1)^2 + 1}{(s + 1)[(s + 2)^2 + 1]}$ (b) $\dfrac{(s - 2)^2 + 1}{(s + 1)[(s + 1)^2 + 1]}$

16-8   (b) $K\left\{A_c + 2A_m \cos\left[\omega_m\left(t + \overbrace{\dfrac{d\theta(\omega)}{d\omega}\Big|_{\omega = \omega_c}}^{-\tau_{gr}}\right)\right]\right\} \sin\left[\omega_c\left(t + \overbrace{\dfrac{\theta(\omega_c)}{\omega_c}}^{-\tau_{ph}}\right)\right]$

16-9   (a) $\tau_{ph} = \dfrac{1}{\omega}\tan^{-1}\dfrac{\omega}{\alpha},$ $\tau_{gr} = \dfrac{1/\alpha}{1 + (\omega/\alpha)^2}$

        (b) $\tau_{ph} \cong \dfrac{1}{\alpha}\left[1 - \dfrac{1}{3}\left(\dfrac{\omega}{\alpha}\right)^2\right],$ $\tau_{gr} \cong \dfrac{1}{\alpha}\left[1 - \left(\dfrac{\omega}{\alpha}\right)^2\right]$

16-11 $\tau_{ph} = \dfrac{1}{\omega}\left(\tan^{-1}\omega + \tan^{-1}\dfrac{\omega}{1 - \omega^2}\right),$ $\tau_{gr} = \dfrac{2 + \omega^2 + 2\omega^4}{1 + \omega^6}$

CHAPTER 17

17-1   Expand $|G(j\omega)|$ in the Maclaurin's series and show that the first $(n - 1)$ derivatives with respect to ω^2 are zero.

17-2   Choose the n values of k such that the roots are in the left half-plane. [*Hint.* Start with angle $\pi/2$ and add onto it.]

17-3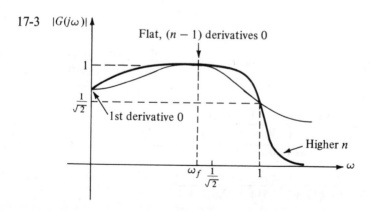

17-4 (a) $|G(j\omega)|^2 = 1 + (a_2 - b_2)\omega^2 + [(a_4 - b_4) - b_2(a_2 - b_2)]\omega^4$
$+ \{(a_6 - b_6) - b_4(a_2 - b_2) - b_2[(a_4 - b_4) - b_2(a_2 - b_2)]\}\omega^6 + \cdots$

(b) Let the coefficients of ω^2, ω^4, ω^6, etc., be equal to 0 to assure as many derivatives as possible to become zero at $\omega = 0$.

(c) Match numerator and denominator coefficients and select b_{2n} to make $|G(j1)|^2 = \frac{1}{2}$.

17-6 $|G(j\omega)|^2 = \dfrac{(\omega_{o1}^2 - \omega^2)^2(\omega_{o2}^2 - \omega^2)^2}{(\omega_{o1}^2 - \omega^2)^2(\omega_{o2}^2 - \omega^2)^2 + (\omega_{o1}^2 - 1)^2(\omega_{o2}^2 - 1)^2\omega^{2n}}$

$(\omega_{o2} \geq \omega_{o1} > 1)$.

17-7 (a) Make use of $\cos\theta = \dfrac{e^{j\theta} + e^{-j\theta}}{2}$, $\cosh\theta = \dfrac{e^{\theta} + e^{-\theta}}{2}$.

(b) Use the expansion form of $\cosh\theta$.

(c) Use $\cos(A \pm B)$ trigonometric expansions.

(d) Use trigonometric identities.

17-10 (a)

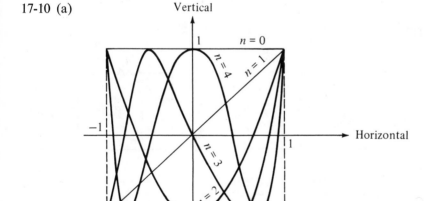

(b) Vertical deflection $= C_n$(Horizontal deflection)

17-11

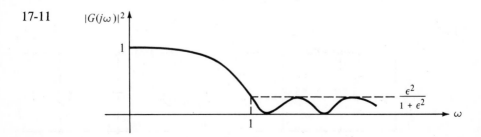

17-12 (a) $|G(j\omega)|^2 = \dfrac{1}{1 + \varepsilon^2 \left(\dfrac{\omega}{\omega_c}\right)^{2n}}$ (b) $|G(j\omega)|^2 = \dfrac{1}{1 + \varepsilon^2 C_n^2 \left(\dfrac{\omega}{\omega_c}\right)}$

(c) $n_B = \dfrac{\log\left(\dfrac{\sqrt{A^2 - 1}}{\varepsilon}\right)}{\log\left(\dfrac{\omega_a}{\omega_c}\right)}$, $n_C = \dfrac{\cosh^{-1}\left(\dfrac{\sqrt{A^2 - 1}}{\varepsilon}\right)}{\cosh^{-1}\left(\dfrac{\omega_a}{\omega_c}\right)}$ (Round n_B and n_C to next higher integer.)

(d) Butterworth $n = 14$;   (e) Butterworth $n = 13$;
    Chebyshev $n = 7$.   Chebyshev $n = 6$.

17-13 $|G(j\omega)|^2$ represents the transitional Butterworth–Chebyshev approximation with equal amounts of Butterworth and Chebyshev functions.

17-14 $\pm 0.7071 \pm j0.7071$.

17-17 $\dfrac{s}{\left(\dfrac{s}{10^3}\right)^4 + 2.6131\left(\dfrac{s}{10^3}\right)^3 + 3.4142\left(\dfrac{s}{10^3}\right)^2 + 2.6131\left(\dfrac{s}{10^3}\right) + 1}$

CHAPTER 18

18-1

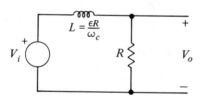
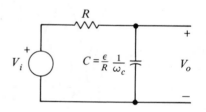

18-2

18-3

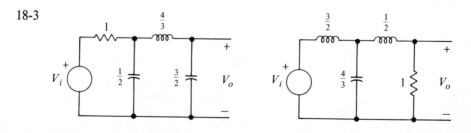

18-4 For resistive-source termination and *n* even, the networks are:

(a)

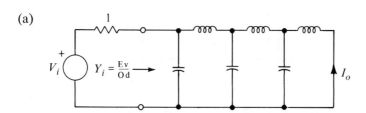

(b)

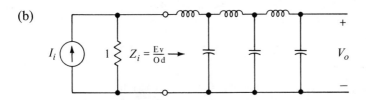

(c)

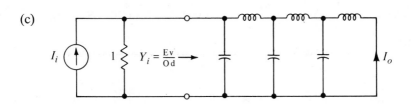

18-5

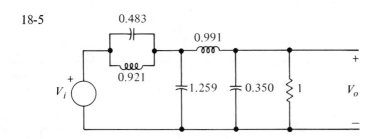

18-6

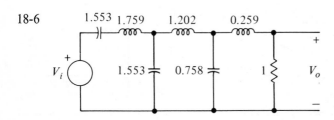

18-9

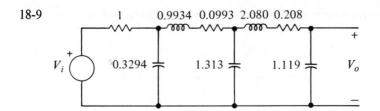

18-10 (a)

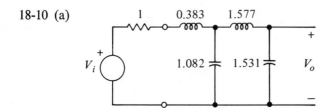

(b)

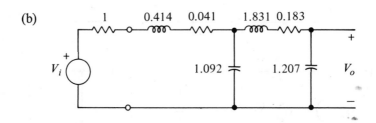

(c)

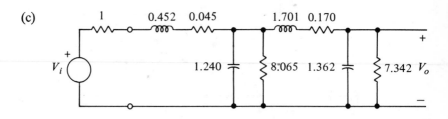

18-11 (a)

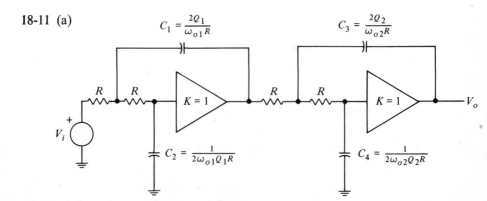

(b)   Let $R = 10$ KΩ; then,
     $C_1 = 17230$ pF, $C_2 = 14700$ pF,
     $C_3 = 41590$ pF, $C_4 = 6091$ pF.

18-12

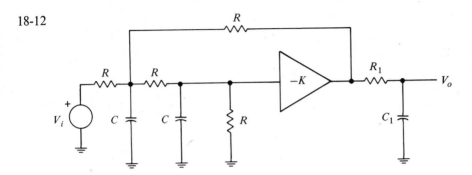

18-13

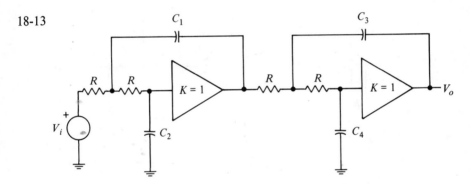

$R = 10$ KΩ, $C_1 = 0.0774$ μF, $C_2 = 0.01669$ μF, $C_3 = 0.1868$ μF,
$C_4 = 0.001502$ μF.

18-14 The desired characteristics are met by the fourth-order, 3-dB Chebyshev
low-pass network shown below.

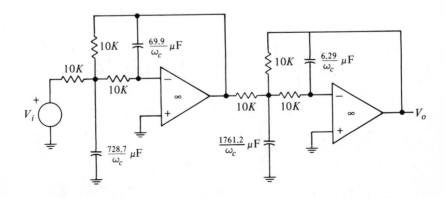

18-15

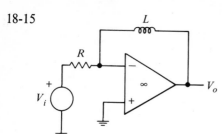
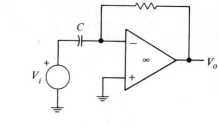

$$\frac{V_o}{V_i} = (-\frac{L}{R})s$$

$$\frac{V_o}{V_i} = (-RC)s$$

18-16

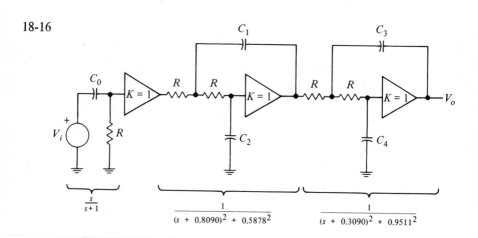

$$\frac{s}{s+1}$$

$$\frac{1}{(s + 0.8090)^2 + 0.5878^2}$$

$$\frac{1}{(s + 0.3090)^2 + 0.9511^2}$$

or

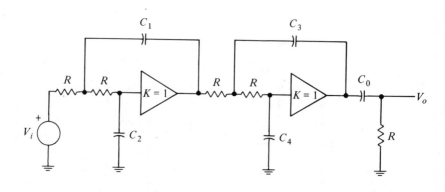

CHAPTER 19

19-2

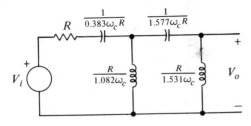

19-3   (a) $G_{HP} = \dfrac{0.2457s^4}{0.2756s^4 + 0.7426s^3 + 1.4539s^2 + 0.9528s + 1}$

(b)

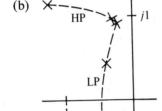

LP poles $\begin{cases} -0.1395 \pm j0.9834, \\ -0.3369 \pm j0.4073. \end{cases}$

HP poles $\begin{cases} -0.1414 \pm j0.9968, \\ -1.2057 \pm j1.4579. \end{cases}$

(c)

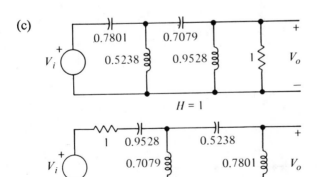

19-4   (a)

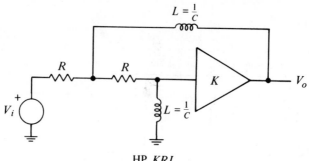

HP *KRL*

(b)

HP *KRC*

(c) $\dfrac{V_o}{V_i} = \dfrac{Ks^2}{s^2 + s(3 - K)RC + R^2C^2}$

19-5

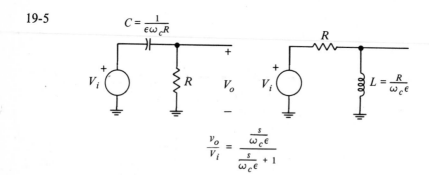

$$\dfrac{v_o}{v_i} = \dfrac{\dfrac{s}{\omega_c\epsilon}}{\dfrac{s}{\omega_c\epsilon} + 1}$$

19-6   (a)

(b)

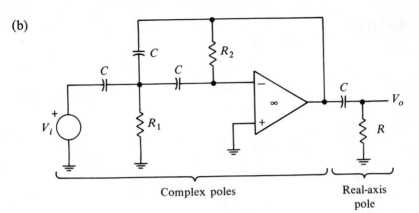

Complex poles   Real-axis
pole

19-7 (a)

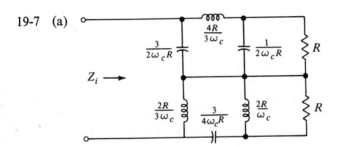

$Z_i \longrightarrow$

(b) $Z_i = R$

19-8

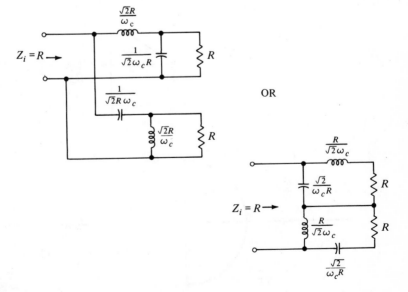

OR

CHAPTER 20

20-2 (a)

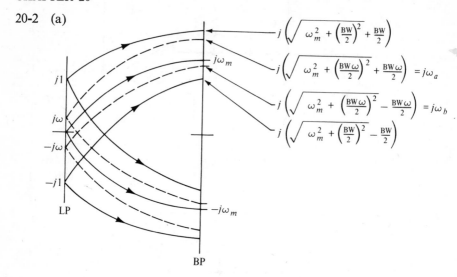

$$j\left(\sqrt{\omega_m^2 + \left(\frac{BW}{2}\right)^2} + \frac{BW}{2}\right)$$

$$j\left(\sqrt{\omega_m^2 + \left(\frac{BW\,\omega}{2}\right)^2} + \frac{BW\,\omega}{2}\right) = j\omega_a$$

$$j\left(\sqrt{\omega_m^2 + \left(\frac{BW\,\omega}{2}\right)^2} - \frac{BW\,\omega}{2}\right) = j\omega_b$$

$$j\left(\sqrt{\omega_m^2 + \left(\frac{BW}{2}\right)^2} - \frac{BW}{2}\right)$$

(b) Refer to diagram in (a) and note the following:

LP magnitude at $j\omega$ = LP magnitude at $(-j\omega)$
= BP magnitude at $j\omega_a$ = BP magnitude at $j\omega_b$.

LP phase at $j\omega$ = $-$[LP phase at $(-j\omega)$]
= BP phase at $j\omega_a$ = $-$[BP phase at $j\omega_b$].

20-3

$$\frac{\alpha}{\omega_m} \cong \frac{1}{2\sqrt{2}} \frac{BW}{\omega_m},$$

$$\frac{\beta_{2,1}}{\omega_m} \cong 1 \pm \frac{1}{2\sqrt{2}} \frac{BW}{\omega_m}.$$

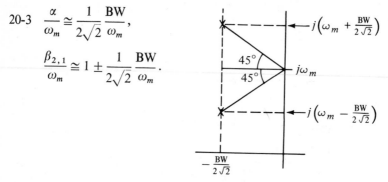

20-4

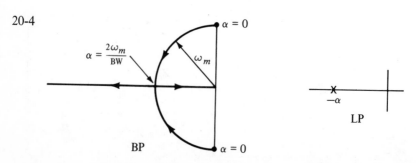

20-7 (b)

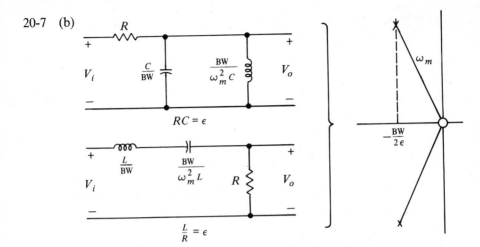

$$RC = \epsilon$$

$$\frac{L}{R} = \epsilon$$

Choose $\epsilon = 1$ for 3-dB pass-band ripple.

20-8 $BW(n) = BW(1)\sqrt{2^{1/n} - 1}$, where $BW(n)$ is the bandwidth of n stages and $BW(1)$ is the bandwidth of one stage.

20-11

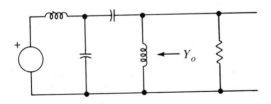

Obtained directly from LP *network* through the LP to BP transformation.

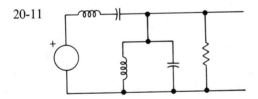

Obtained from BP *function*. Expand Y_o in Cauer II and insert source in series with last inductor.

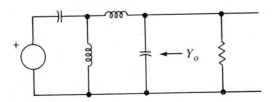

Obtained from **BP** *function*. Expand Y_o in Cauer I and insert source in series with last capacitor.

20-12

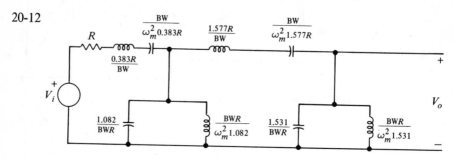

20-13 The envelope frequency is $\frac{1}{2}(\omega_{o2} - \omega_{o1})$.

20-14 (a) $K = 2.96$ **(b)** $\Delta Q = \begin{cases} -10.62 \ (\Delta K < 0) \\ +71.2 \ (\Delta K > 0) \end{cases}$; $\Delta \omega_o = 0$

(c) 0.123% $(\Delta Q > 0)$
    -0.152% $(\Delta Q < 0)$

20-15

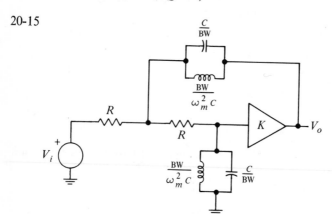

20-16

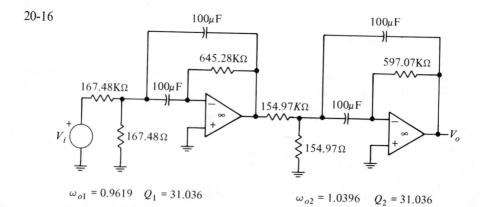

$\omega_{o1} = 0.9619$ $Q_1 = 31.036$ $\omega_{o2} = 1.0396$ $Q_2 = 31.036$

All impedances may be multiplied by the same constant without changing the transfer function. To change the center frequency from 1 to ω_m, divide all C's by ω_m.

20-18 (a) $\dfrac{V_o}{V_i} =$

$$\frac{1}{RC\left(2+\dfrac{C}{C_c}\right)\left[\left(s^2+s\dfrac{\omega_o}{Q}+\omega_o^2\right)\right]}\cdot\frac{s^3}{\left[s^2+s\dfrac{\omega_o}{Q}\dfrac{1}{1+\dfrac{2C_c}{C}}+\dfrac{\omega_o^2}{1+\dfrac{2C_c}{C}}\right]}$$

(b) Fixed pole
Variable pole

20-20 (a) $\dfrac{V_o}{V_i} =$

$$\left[-\left(\dfrac{1}{R_0C}\right)\frac{s}{s^2+s\dfrac{\omega_o}{Q}+\dfrac{\omega_o^2}{1+k\dfrac{R}{R_1}}}\right]\left[-\left(\dfrac{1}{R_0C}\right)\frac{s}{s^2+s\dfrac{\omega_o}{Q}+\dfrac{\omega_o^2}{1-k\dfrac{R}{R_1}}}\right]$$

(b)

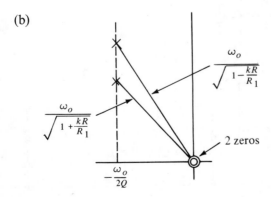

20-21

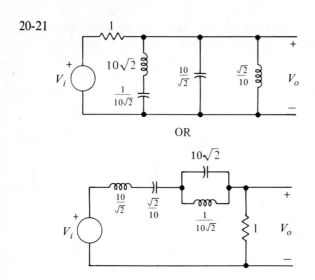

OR

CHAPTER 21

21-2 A circle and the entire real axis.

21-3 (a)

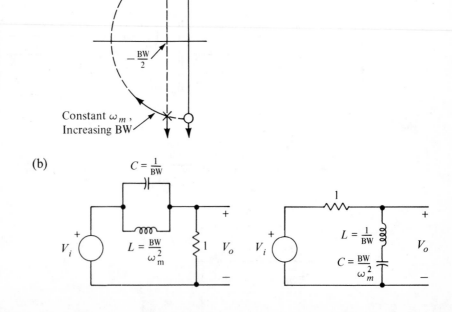

21-4 $|T_{BS}(j\omega)|$

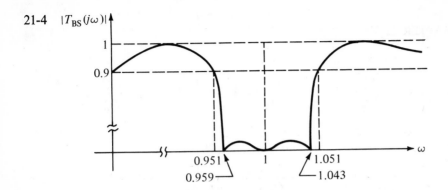

21-5 $|T_{BS}(j\omega)|$

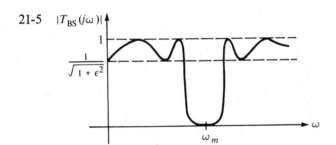

21-6 $|T(j\omega)|$

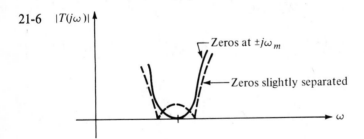

Zeros at $\pm j\omega_m$

Zeros slightly separated

21-7

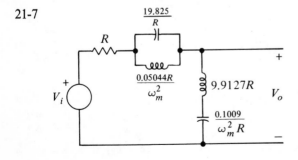

CHAPTER 22

22-1 $Q = 1/\sqrt{3}$

22-3 $\tau(\omega) = \dfrac{1}{Q\omega_o} + \dfrac{3}{Q\omega_o}\left(1 - \dfrac{1}{3Q^2}\right)\left(\dfrac{\omega}{\omega_o}\right)^2 + \dfrac{5}{Q\omega_o}\left(1 - \dfrac{1}{Q^2} + \dfrac{1}{5Q^5}\right)\left(\dfrac{\omega}{\omega_o}\right)^4 + \cdots$

22-4 $g_1 = \dfrac{1}{\alpha_{z1}} + \dfrac{1}{\alpha_{z2}} + \cdots + \dfrac{1}{Q_{z1}\,\omega_{oz1}} + \dfrac{1}{Q_{z2}\,\omega_{oz2}} + \cdots$

$$-\left(\dfrac{1}{\alpha_{p1}} + \dfrac{1}{\alpha_{p2}} + \cdots + \dfrac{1}{Q_{p1}\,\omega_{op1}} + \dfrac{1}{Q_{p2}\,\omega_{op2}} + \cdots\right),$$

$$g_3 = -\left[\dfrac{1}{3\alpha_{z1}^3} + \dfrac{1}{3\alpha_{z2}^3} + \cdots + \dfrac{1}{Q_{z1}\,\omega_{oz1}^3}\left(\dfrac{1}{3Q_{z1}^2} - 1\right)\right.$$

$$\left.+ \dfrac{1}{Q_{z2}\,\omega_{oz2}^3}\left(\dfrac{1}{3Q_{z2}^2} - 1\right) + \cdots\right] + \left[\dfrac{1}{3\alpha_{p1}^3} + \dfrac{1}{3\alpha_{p2}^3} + \cdots\right.$$

$$\left.+ \dfrac{1}{Q_{p1}\,\omega_{op1}^3}\left(\dfrac{1}{3Q_{p1}^2} - 1\right) + \dfrac{1}{Q_{p2}\,\omega_{op2}^3}\left(\dfrac{1}{3Q_{p2}^2} - 1\right) + \cdots\right],$$

$$g_5 = \dfrac{1}{5\alpha_{z1}^5} + \dfrac{1}{5\alpha_{z2}^5} + \cdots + \dfrac{1}{Q_{z1}\,\omega_{oz1}^5}\left(1 - \dfrac{1}{Q_{z1}^2} + \dfrac{1}{5Q_{z1}^4}\right)$$

$$+ \dfrac{1}{Q_{z2}\,\omega_{oz2}^5}\left(1 - \dfrac{1}{Q_{z2}^2} + \dfrac{1}{5Q_{z2}^4}\right) + \cdots - \left[\dfrac{1}{5\alpha_{p1}^5} + \dfrac{1}{5\alpha_{p2}^5} + \cdots\right.$$

$$\left.+ \dfrac{1}{Q_{p1}\,\omega_{op1}^5}\left(1 - \dfrac{1}{Q_{p1}^2} + \dfrac{1}{5Q_{p1}^4}\right) + \dfrac{1}{Q_{p2}\,\omega_{op2}^5}\left(1 - \dfrac{1}{Q_{p2}^2} + \dfrac{1}{5Q_{p2}^4}\right) + \cdots\right].$$

Index